T0271042

FATIGUE DESIGN OF MARINE STRUCTURES

Fatigue Design of Marine Structures provides students and professionals with a theoretical and practical background for fatigue design of marine structures including sailing ships, offshore structures for oil and gas production, and other welded structures subject to dynamic loading such as wind turbine structures. Industry expert Inge Lotsberg brings more than 40 years of experience in design and standards-setting to this comprehensive guide to the basics of fatigue design of welded structures. Topics covered include laboratory testing, S-N data, different materials, different environments, stress concentrations, residual stresses, acceptance criteria, non-destructive testing, improvement methods, probability of failure, bolted connections, grouted connections, and fracture mechanics.

Featuring 18 chapters, 300 diagrams, 47 example calculations, and resources for further study, *Fatigue Design of Marine Structures* is intended as the complete reference work for study and practice.

Inge Lotsberg is a Specialist Engineer and Senior Vice President at DNV GL (merger of Det Norske Veritas and Germanischer Lloyd) in Norway. He has more than 40 years of practical experience with design and verification of steel structures, linear and nonlinear finite element analysis, rule development, fatigue and fracture mechanics analyses, reliability analysis, and laboratory testing. The author or co-author of 120 refereed papers, he has also served as member and chairman in committees for developments of fatigue design standards within NORSOK and ISO.

Fatigue Design of Marine Structures

Inge Lotsberg
DNV GL, Norway

CAMBRIDGE
UNIVERSITY PRESS

CAMBRIDGE
UNIVERSITY PRESS

University Printing House, Cambridge CB2 8BS, United Kingdom

One Liberty Plaza, 20th Floor, New York, NY 10006, USA

477 Williamstown Road, Port Melbourne, VIC 3207, Australia

314-321, 3rd Floor, Plot 3, Splendor Forum, Jasola District Centre, New Delhi - 110025, India

79 Anson Road, #06-04/06, Singapore 079906

Cambridge University Press is part of the University of Cambridge.

It furthers the University's mission by disseminating knowledge in the pursuit of education, learning and research at the highest international levels of excellence.

www.cambridge.org
Information on this title: www.cambridge.org/9781107121331

First published 2016

A catalogue record for this publication is available from the British Library

Library of Congress Cataloging in Publication data
Lotsberg, Inge, 1948–
Fatigue design of marine structures / Inge Lotsberg,Det Norske
Veritas-Germanischer Lloyd, Norway.
New York NY : Cambridge University Press, [2016]
Includes bibliographical references and index.
LCCN 2015046144   ISBN 9781107121331 (hardback : alk. paper)
LCSH:Offshore structures – Design and construction.   Marine steel – Fatigue.
Steel, Structural – Fatigue.   Metals – Fatigue.
LCC TC1665.L68 2016   DDC 627–dc23
LC record available at http://lccn.loc.gov/2015046144

ISBN 978-1-107-12133-1 Hardback
ISBN 978-1-107-54730-8 Paperback

Contents

Preface

This book is intended to act as a guide for students and practicing engineers for fatigue design of dynamically loaded marine structures. Fatigue of structures is a broad and complex area that requires more background than can be included in design standards. Many papers on fatigue of structures are published each year, and different design approaches have also been issued. However, due to the nature of the fatigue phenomena and scatter in test results, it may be difficult for engineers to obtain a good overview of what is found to be a good recommended fatigue assessment methodology.

The purpose of this book is not to give a complete overview of different design approaches, but rather to provide the reader with a sound background for the most common recommendations in design standards for fatigue assessment of marine structures. The content of this book is colored by the experiences by the author, and it may be relevant to consider this textbook in relation to the Standards with which the author has been most heavily involved, including the Recommended Practice DNVGL-RP-C203 Fatigue Design of Offshore Structures and DNV-RP-C206 Fatigue Methodology of Offshore Ships. However, similar content can also be found in a number of other design standards, such as: ISO 19902 (2007), API RP2A (2014), AWS (2010), BS 7608 (2014), Eurocode 3 (EN 1993–1–9, 2009), and IIW (Hobbacher, 2009). Thus, this book might best be considered as providing background for fatigue assessment of welded structures on a broad basis.

Based on the author's main background experience, a number of DNVGL standards are referenced. As these documents can be downloaded for free from the Internet, they are also useful reference documents for students studying fatigue of marine structures.

Much of this book is related to fatigue capacity of steel structures. The book may be seen as complementary to the Naess and Moan's book, *Stochastic Dynamics of Marine Structures*. Thus mainly the fatigue capacity of marine structures is considered in this book. The dynamic loading may be due to different sources such as waves, wind, rotors on wind turbines, dynamic response, vortex-induced vibrations, pile driving, and loading and unloading of content. Although this book will be easier to read and understand if the reader has a sound background in structural mechanics, the derivation of equations and the examples are presented in sufficient detail that it should be also possible to understand for engineers whose background is not rooted in structural engineering. A number of relevant examples are also included for the purpose of education of students.

A number of other test books on fatigue are recommended for a more basic learning about fatigue. Rather than repeating content that has already been well presented elsewhere, the author concentrates on engineering practice based on his own experiences in this book. Other textbooks on fatigue include (listed in alphabetical order by the author): Almar-Naess (1985), Collins (1993), Dowling (1998), Fischer (1984), Forrest (1962), Gurney (1979), Gurney (2006), Haibach (2006), Lassen and Recho (2006), Macdonald (2011), Maddox (1991), Marshall (1992), Nussbaumer et al. (2011), Pilkey (1997), Radaj et al. (2006), Radaj and Vormwald (2013), Schijve (2009), Sines and Waisman (1959), Sors (1971), Stephens et al. (2001), and Wardenier (1982). Books related to fatigue based on fracture mechanics include: Anderson (2005), Broek (1986), Carlsson (1976), Hellan (1984), Knott (1973), Liebowitz (1968), and Taylor (2007).

Although our understanding of the fatigue phenomena has improved over time, the assessment procedures are still strongly related to laboratory fatigue test data. Therefore, some of the author's experiences related to laboratory testing are included in the first section of this book. Careful review of these sections will enable the reader to obtain a better understanding of the remaining part of the book.

Most of the terminology used in this book is defined at first use, and the index may be useful in this respect. Some expressions are used more frequently than others; one example is the term "fatigue strength," which can be defined as magnitude of stress range leading to a particular fatigue life. Fatigue life or the number of cycles to a failure under the action of a constant amplitude stress history may also be denoted "fatigue endurance." A "fatigue strength curve" or "S-N curve" is defined as the quantitative relationship between the stress range (S) and the number of stress cycles to fatigue failure (N), used for fatigue assessment of a particular category of structural detail. Thus, the expression "fatigue strength" needs to be associated with some number of cycles to be fully meaningful. The same comment may be made with respect to expressions as "fatigue resistance" used in some design standards and "fatigue capacity" used by designers to characterize the resistance against fatigue failure in structures. Thus also these expressions may be interpreted as resistance or capacity in relation to an S-N curve. Both the term "fatigue strength" and "fatigue capacity" are used in this book to characterize resistance against fatigue. Normally the word "capacity" may be considered to be more general than "strength" and include more influencing parameters when comparing this also with other failure modes than fatigue for structures. For example, the wording "fatigue strength" is used to describe the resistance to fatigue failure in a single fatigue test or of, for example, a single bolt, and "fatigue capacity" is used to describe the fatigue resistance of a bolted connection where the fatigue capacity is dependent on more parameters such as surface conditions of plates, friction coefficient, and pretension of the bolts. In some literature the S-N curves are also denoted as Wöhler curves.

See Sections I.4 and 4.11.1 for definition of characteristic and design S-N curves. When the accumulated number of cycles is divided by a reference value, such as the characteristic number of cycles to failure as derived from an S-N curve, the wording "fatigue damage" is used. Fatigue damage accumulates with time when a structure is subjected to dynamic loading. Fatigue endurance is similar to fatigue life, which may be measured often in terms of years. Fatigue endurance can be observed during laboratory fatigue tests or can be calculated based on a defined design procedure. The calculated values normally differ from fatigue test data or observed values; therefore, the term "calculated" is often inserted in front of fatigue life in order to make this difference more clear.

In design standards for offshore structures the notation SCF is used for a linear elastic stress concentration factor (see Section 3.2.1 and Chapter 8). In design standards for sailing ships K is used as notation for the same stress concentration factor; see, for example, the IACS common rules from 2013. In this book SCF is used as notation for stress concentration factor, and it should not be mixed with the stress intensity factor used in fracture mechanics analysis that is denoted by K – see Section 16.2.

Some items are presented in more than one section of the book. However, where this occurs, cross-referencing has been used to improve readability.

Acknowledgments

I have worked at the head office of Det Norske Veritas (DNV) in Oslo during the 1979–2013 period and have been working at DNV GL since the merger between DNV and the former German classification company Germanische Lloyd (GL) in 2013. A significant proportion of the work within DNV GL is related to assessment of marine structures. This has given me the possibility of meeting several experienced engineers in different industries. I have had the pleasure of working together with engineers within material technology, fracture mechanics, laboratory testing, reliability analysis, and structural analysis and I would like to acknowledge this support from my colleagues.

During the work with this book I have received valuable input and comments by a number of persons, which is acknowledged. Hege Bang and Arne Fjeldstad have reviewed the proposed text and provided useful comments. Knut Olav Ronold has provided valuable input and comments to the section on derivation of design S-N curves from fatigue test data. Andrzej Serednicki, Arne Eikebrokk, and Vebjørn Andresen have provided useful input to the section on bolted connections. I also acknowledge comments by Professor Torgeir Moan, Professor Stig Berge, and Professor Sigmund Kyrre Aas at the Norwegian University of Science & Technology NTNU Trondheim regarding the technical content of this book. István Szarka helped me establish a template for writing. Harald Rove has prepared the most complex sketches, and Lucy Robertson has polished my language in most parts of this book. I would also like to mention that I have received great help from our library to provide me with most of the literature I have asked for; special thanks to Ingunn Lindvik in this respect. I would also like to acknowledge DNV GL for allowing me to publish background material for our design standards that have been developed in a number of different research and development projects. Furthermore, I acknowledge the financial support from DNV GL for writing this book.

I received my technical education at the Technical University of Trondheim, from which I earned a PhD in 1977. The research was related to the use of finite element analysis within the field of fracture mechanics. I am grateful for the support from the university during my study and the encouragement to publish my research from my supervisor, Professor Pål G. Bergan at the Department of

Structural Mechanics. My later work proved to me that a good education in basic structural mechanics is important for designing reliable and optimal structures. I hope that this book can be a contribution and an inspiration back to the university and to the education of the next generation of students studying design of marine structures.

Introduction

I.1 History of Fatigue

The history of fatigue of metals, components, and structures goes back to the 1830s when failures of chains in mines were reported due to dynamic loading, and fatigue testing of these chains was performed for mitigation (Schütz 1996). In association with this, the first wires were invented to avoid the problems with fatigue of the chains. Since then, up until year 2000, around 100,000 papers related to fatigue were published (Schijve 2003, 2009). With so much published literature available, providing a broad and objective historical overview would be highly challenging. Thus, the historical presentation provided here is limited to those aspects that are of most relevance as background for this book. Reference is made to Schütz (1996), Stephens et al. (2001), and Anderson (2005) for a more detailed historical presentation related to fatigue.

The term "fatigue" is apparently first mentioned in the literature in 1854, by an Englishman called Braithwaite. In his paper, Braithwaite describes many service fatigue failures of brewery equipment, water pumps, propeller shafts, crankshafts, railway axles, levers, cranes, and so on. At about the same time many disastrous railroad accidents occurred, such as one on 5 October 1842 when an axle broke at Versailles due to fatigue and the lives of 60 people were lost. Failures of railway axles became a serious problem and as late as in 1887, an English newspaper reported "the most serious railway accident of the week." In many cases these accidents were due to fatigue failures of axles, couplings, and rails.

In some publications, the fatigue strength in terms of S-N curves is presented as "Wöhler curves" that are named after the work that Wöhler performed in Germany to determine the fatigue strength of railway axles based on fatigue testing in the period from 1860 to 1870. Already in 1858, Wöhler was measuring the service loads on railway axles using self-developed deflection gauges. He also introduced the concept of safety factors, where two sets were needed: one for maximum stress in service in relation to static strength, and the other for allowable stress amplitude under dynamic loading. The safety factors were provided for ensuring design for infinite life. The factors were valid only for un-notched specimens, and fatigue testing was recommended for other geometries. Wöhler presented his test data in tables. His successor, Spangenberg, started plotting these data into curves, using the form of a linear abscissa and ordinate. In 1910, an American named Basquin plotted the same data into a log-log format, in the same way as S-N curves are presented in fatigue design standards today.

At this time constant fatigue life diagrams were also developed in which the effect of static loading (maximum stress or stress ratio) was included in addition to the stress range, as this parameter was necessary for inclusion in bridge design. The first constant fatigue life diagram was published by Gerber in 1874, based on an assessment of Wöhler's test data. Other diagrams provided in the literature are derived from, for example, Smith in 1880, Goodman in 1899, and Haigh in 1917, in which relationships between stress amplitude and mean stress were presented. A more conservative diagram referred to in the literature for this is from Soderberg in 1930 (Sendeckyj 2001).

Fatigue failures in aircrafts from the late 1920s resulted in research to investigate representative long-term loading and fatigue capacity of aircraft structural components. Component testing became part of the qualification procedure for new elements. After World War II, it had become common practice to perform testing of large components of aircrafts in fatigue as opposed to small-scale test data. However, failures continued to occur, resulting, for example, in the Comet crashes in 1954. These accidents were later found to be due to unsatisfactory detailing of the corners of the windows in combination with an erroneous test procedure that had provided compressive residual stresses at the hot spot areas, such that the actual capacity differed from that of the laboratory test component. This fact does not only demonstrate the importance of a good detailed design in order to achieve sufficient fatigue capacity but also shows the need for performing realistic component testing in the laboratory. It is important that the structural geometry of the component and the fabrication are representative for the actual structure in terms of boundary conditions and loading. It was also realized after the failures of the Comet airplanes that these structures were not fail-safe (Edwards 1988). After these accidents, the requirements for component testing in the aircraft industry were improved. It should also be mentioned that a high-strength aluminum alloy had been used in the Comet airplanes, and it is known that the fatigue strength of a material does not necessarily correspond with the material strength. Fatigue crack initiation in the base material is considered to be improved, see Section 4.1.5, while crack growth parameters are considered to be rather constant with increasing material tensile strength, see also Section 16.11. The crack growth parameter is considered to be more a function of the Young's modulus of the material used than of the material strength. Furthermore aircraft accidents from the late 1960s led to the development of enhanced research programs on fatigue strength based on fracture mechanics. A near-fatal accident with a Boeing 737 in 1988 resulted in the initiation of investigative activities regarding the structural integrity of old and poorly maintained aircrafts.

Professor Bauschinger of the Technical University of Munich had already in 1881 observed that the elastic limit of materials changes under repeated stress cycles. This was also the basis for the hypothesis by Manson and Coffin in the 1950s, which is still used for assessment of low cycle fatigue (Manson 1954; Coffin 1954, 1984). Low cycle fatigue is defined as stress ranges leading to repeated plastic strain during loading and unloading, such that elastic shakedown does not occur during a load cycle. Shakedown is understood to mean that the material has elastic behavior after it has yielded during large load amplitude; see also Section 3.3.2.

Manson and Coffin described the behavior of materials under cyclic inelastic strain amplitudes by a four-parameter formula, creating a new field of activity called "low-cycle-fatigue" (LCF) that is further described in Section 3.1. In 1903, two Englishmen, Ewing and Humfrey, described the so-called slip band on the surface of rotating bending specimens. Later, in the 1930s, Polany and Orowan presented the dislocation theory, and this became the basis for many metallurgical papers on fatigue of metals. Orowan suggested that fatigue initiation is a process of cyclic shear hardening that, after depletion of the local ductility, leads to fatigue cracking. The effect of notches on the fatigue capacity of structural components was known before Wöhler performed his work. He thus recommended special fatigue tests for sharply notched specimens. In 1898, Kirch calculated that the elastic stress concentration factor for a hole in an infinite body was equal to 3.0. Analytical derivation of stress concentration factors at notches in machine components were derived by advanced mathematics combined with linear elastic theory of solids; an overview of stress concentrations around holes was presented by Johnson (1961). Measurements of stress concentrations were achieved by photo-elasticity from around 1930, and around 1970 it became normal practice to analyze stress concentrations in detail by the finite element method that had been developed during the 1960s.

In 1937, Neuber published the first comprehensive book on the theoretical calculation of the stress concentration factor K_t and also the fatigue notch concentration factor K_f. The German Thurm had previously observed in experiments that notched specimens showed a longer fatigue life than was predicted by the theoretical elastic stress concentration factor; the difference between these factors increases with increased notch effect, or with a reduced radius at the region showing the largest stress. This difference was also found to be material-dependent and the effect was described by a notch sensitivity factor that is dependent on both material and notch radius. Neuber's book was written in German but was translated into English in 1946, and some of Neuber's work is included in Peterson's Stress Concentration Factors from 1953, which is a renowned book for everyone who works with fatigue of machine components. This book is still recommended for background on the notch effect and notch sensitivity, and for an explanation of elastic stress concentration factors as compared with fatigue notch concentration factors. In the early 1950s, the National Advisory Committee for Aeronautics (NACA) made an attempt to transform the calculation of the fatigue stress concentration factor, K_t, into engineering practice, based on Neuber's material constant. However, the results were found to be of insufficient accuracy and the experimental effort required was too great. Thus, the concept of notch sensitivity is considered to be of limited use for detailed calculations today. Nevertheless, it is useful for a more qualitative understanding of the physical fatigue behavior of notched specimens.

After the 1980s a notch stress method based on a linear elastic finite element analysis has been developed, Radaj (1996) and Radaj et al. (2006). This approach has been linked to a notch stress S-N curve and is frequently being used by researchers and also by the industry in more special cases. A more detailed background for this methodology is presented in Section 9.6.

Fatigue tests are normally performed under a constant amplitude loading. However, many dynamically loaded structures are subject to variable amplitude loading. Therefore, different test spectra representing long-term dynamic loading were developed for fatigue assessment in the automobile industry, the aircraft industry, and, later, in the offshore industry. Reference is made to Section 10.2 for an example of transformation of a long-term stress range distribution into a constant amplitude loading.

Most marine structures are subject to variable amplitude loading, and this is also the load condition for many other structures, such as components in aircrafts, automobiles, and cranes, among others. In addition to requiring load spectra that are representative for long-term cyclic loading, criteria must be defined regarding how to compute fatigue damage under variable amplitude loading. In 1924, Palmgren, a Swede, published a paper (in German) on the lifetime of ball supports. This paper included not only a fatigue damage hypothesis for variable amplitude loading but also a numerical description of a probability of survival for fatigue-loaded ball bearings. An identical damage-accumulation hypothesis was presented in 1937 by Langer, an American who was probably not aware of Palmgren's work. He separated fatigue into fatigue crack initiation and fatigue crack growth, and suggested a damage sum equaling 1.0 for each of these phases. The linear fatigue damage accumulation hypothesis was presented again in 1945 by Miner, and he also performed some testing to check out his hypothesis. This linear damage accumulation has since been named the Palmgren-Miner rule. This is further assessed in Section 3.2.3.

The accuracy of the fatigue damage hypothesis is also linked to how the design S-N curve is established. Fatigue testing under spectrum loading is needed in order to determine the shape of the S-N curve in the high cycle region at the fatigue limit as the number of cycles beyond the fatigue limit becomes "infinite" or very long if the stress range in constant amplitude tests is less than the stress range that corresponds to the fatigue limit. Such test results are also denoted as "run-outs"; see also Section 3.2.2. Fatigue testing relevant for design of steel structures under spectrum loading has been performed since the 1980s as described in Section 3.2.3. The uncertainty related to variable amplitude loading is also assessed in Chapter 12. This issue is still being debated, and different recommendations on the position of the fatigue limit and change in slope in the S-N curves are found in different design standards (e.g., EN 1993–1–9:2009, ISO 19902 (2007), and BS 7608 (2014)).

It has been a challenge to develop a counting method for cycles and stress ranges that represents the physical behavior of a fatigue damaging process in a reliable manner. Several proposals for counting methods have been presented in the literature; seven different methods for cycle counting were described in a paper by Schijve in 1961 (Schijve 2009). He was closely involved in assessment and testing of components under variable amplitude loading, and considered that the rainflow counting method, published by the Japanese Matsuishi and Endo in 1968, fulfills the requirements of an acceptable counting method. At the same time, a Dutchman, DeJonge, also published the "range-pair-range," which is the same as the rainflow procedure. Statistics were needed for assessment of test data due to scatter. Weibull in Sweden carried out thousands of fatigue tests on bolts and aluminum specimens to prove his distribution and to obtain numerical data on the standard deviation of the number of cycles to failure. His work was published between 1947 and 1955. His two-parameter statistical distribution, with a shape parameter and a scale parameter, has also become important for description of long-term stress range distribution for fatigue analysis of marine structures. Reference is also made to Chapter 10 of this book.

Wöhler had already been introducing the concept of safety factors in fatigue design in around 1860. During the 1930s it became more obvious that fatigue test data were associated with uncertainties, while the safety philosophy also became an issue for discussion in the industry. The concepts of "fail-safe" and "safe life" had been introduced in the 1930s, with "safe life" meaning that an aircraft component had to be scrapped after a specified service life. A better alternative to this is to have a "fail-safe" aircraft, which means that failure of a primary member does not endanger flight safety. After the Comet accidents it was found that it was possible for the aircraft industry to achieve design solutions that could be considered "fail-safe." Using this approach, aircraft can be regularly inspected for fatigue cracks and reliability is significantly improved today as compared with the 1950s.

In 1920, Griffith, of the Royal Aircraft Establishment in the United Kingdom, developed the basis for use of fracture mechanics for assessment of overload rupture with cracks present in the material; at that time it was only applicable to brittle material, such as glass. Irwin developed the energy release concept, including the effect of plasticity, in around 1956. Westergaard (1939), Irwin (1957), and Williams (1957) extended the information about stresses and displacements around crack tips and the concept of stress intensity factors was developed. Reference is also made to Sih and Liebowitz (1968) for their overview chapter on mathematical theories of brittle fracture (Liebowitz 1968). In 1961, Paris developed a fatigue crack growth equation demonstrating that an increment in crack growth during a stress cycle can be related to the range in the stress intensity factor during that cycle. This equation soon became popular and was later extended in a number of ways to account better for mean stress effect and material fracture toughness; see also Section 16.2. In the United Kingdom the crack tip opening displacement (CTOD) parameter developed by Wells became extensively used in fracture mechanics during the 1960s. The CTOD design curve was developed and used in design assessment of the first oil platforms built for the North Sea in the 1970s; for further information, reference is made to Wells (1969), Burdekin and Daves (1971), the first draft proposal for CTOD testing by BSI (1972), and Burdekin (1981). In Norway fracture mechanics was used already in the beginning of the 1970s as a tool to establish rational acceptance criteria in relation to non-destructive testing techniques and for documentation of leak-before-failure in welded connections in spherical tanks in ships for transport of liquefied natural gas (LNG) made of nine percent nickel steel and of aluminum alloy 5083-0; ref. Kvamsdal and Howard (1972), Tenge and Solli (1973), Aamodt et al. (1973), and Tenge et al. (1974). More information on the historic development of fracture mechanics can be found in Hellan (1984), Anderson (2005), and Macdonald (2011). Reference is also made to Section 16.6.

After World War II, welding became the normal method for making connections between plates. S-N data were derived from fatigue testing of different types of welded connections, and nominal stress S-N curves were established for design purposes; see, for example, Gurney (1976). These S-N curves were included in the British Standard (BS 5400) for bridge structures (1980) and in the Department of Energy Guidance Notes for offshore structures (1984). Later these S-N curves were used in the Health and Safety Executive (HSE) Standards and were also copied into the ISO Standard for Design of Offshore Structures, ISO 19902 (2007). These curves are still referred to in the latest revision of BS 7608 (2014) issued as a guide for design and fatigue assessment of steel products. Similar S-N curves for different structural details have also been included in fatigue design standards in other countries.

Most design S-N curves in standards are made for air environment assuming that the structures are sufficiently corroded protected. For fatigue assessment of offshore structures also S-N curves for seawater with cathodic protection were developed in the 1980s; see Sections 4.1.6, 4.1.7 and 4.7.

Many of the railway and highway bridges built in the first half of the 19th century are still in use, despite the fact that their planned technical life span is already completed. A large percentage of these bridges are more than 50 years old, and about 30% of them are more than 100 years old (Haghani et al. 2012). These bridges are currently being subjected to increased traffic intensity and higher traffic loads in order to meet the requirements for more efficient transportation systems. This results in fatigue problems in a number of different details. This information is considered useful as feedback for bridge designers such that details can be improved for new constructions. Many of the reported fatigue cracks are caused by secondary effects, so-called deformation-induced cracking. This type of fatigue damage is often the result of secondary restraining forces generated by some kind of unintentional or overlooked interaction between different members in the bridge. Poor detailing, along with unstiffened gaps and abrupt changes in stiffness at the connections between different members, also contributes to fatigue cracking in many details. More details and learning from this can be found in Dexter and Ocal (2013). Reference is also made to the state-of-the-art review on fatigue life assessment of steel bridges by Ye et al. (2014).

During the 1980s it became more common to analyze structural details by the finite element method. In many cases it was difficult to use the calculated stresses from finite element analysis together with nominal stress S-N curves for calculation of fatigue damage, as the calculated stresses also partly included effects from the structural geometries that were included in the analysis models, which also were accounted for in the nominal stress S-N curves. Thus, part of the stress due to the detail became included twice, leading to a conservative fatigue assessment. This resulted in the development of structural stress methods and corresponding hot spot stress methods that can be used when finite element analyses of structural components are performed; see, for example, Fricke and Petershagen (1992) and Fricke (2003). This has also resulted in the development of alternative analysis methods based on local approaches since the beginning of the 1990s, as explained more in detail in Chapter 9.

Installation of fixed offshore structures (jackets) in more harsh environments such as in the North Sea resulted in a need for improved fatigue analysis procedures, and significant developments were made on fatigue assessment of tubular joints from the end of the 1970s through 1980s; Chapter 8 describes this in more detail. The first jacket structures installed in the North Sea were mainly designed with respect to the Ultimate Limit State. However, in the late 1970s it had become practice to analyze jacket structures with respect to the Fatigue Limit State. The same practice was not introduced for floating platforms until after the *Alexander L. Kielland accident*.

In March 1980, the accommodation platform *Alexander L. Kielland* capsized in the North Sea with 212 men onboard (see NOU 1981: 11). The primary reason for this accident was failure of one of the main braces. Fatigue cracks had initiated at a hydrophone support that was welded into one of the main braces and had propagated further until it was so long that it finally fractured in a storm (see Figure I.1). The platform was designed without significant redundancy. Thus, the fracture of this

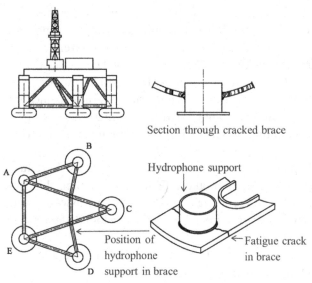

Section through cracked brace

Hydrophone support

Position of hydrophone support in brace

←Fatigue crack in brace

Figure I.1. Sketch of the *Alexander L. Kielland* platform with hydrophone support and fatigue crack into the main brace connecting column D to the structure (based on NOU 1981:11).

brace resulted in failures in other elements, buoyancy was then lost, and the platform turned upside down; 123 men lost their lives. This accident resulted not only in greater research effort being directed toward fatigue of offshore structures but also in revised design standards with stricter requirements regarding documentation of redundancy and survivability of structures in unexpected situations. Please see Sections I.2.1, 5.5.6, and 16.3.2 for more details and assessments related to this incident.

In around 1975, high-strength steel was introduced into the fatigue design of ships. High-strength steel in ship fabrication is understood to refer to steel material with a nominal yield strength of around 315–350 MPa, while normal yield strength is approximately 235 MPa. Thus, the ultimate structural capacity of ships could be documented with use of less steel. The structural details remained similar to those for when normal steel was used, and upon which the design experience was based. This resulted in reduced section modulus of structural elements and increased stress ranges at the hot spot areas, especially in the side longitudinals. Significant fatigue cracking was reported until improved classification standards were developed; see, e.g., Yoneya et al. (1993) and Xu (1997).

Due to residual tensile stresses at welds, it has been normal practice in fatigue design of land and offshore structures to assume that the full stress range is providing the same fatigue damage independent of mean stress level. This means that all stress ranges are included in the design procedure as if they were producing tensile stress at the hot spot during the full load cycle. Until the 1990s, the fatigue design of sailing ships had to a large extent been based on experience, and in some way this had resulted in less required amount of steel than what could be documented based on the same fatigue analysis methodology as used for offshore structures. Thus, to avoid unnecessary increase in scantlings at details with a positive in-service experience without fatigue cracking, it was proposed to include a mean stress effect factor where the compressive part of the stress cycle is less damaging than a similar tensile stress range. Calibration of the fatigue analysis procedure with experience accounts for the mean stress effect and it also accounts for the uncertainty due to variable long-term loading, which has been difficult to fully assess and agree on.

During 2004 and 2005, new design rules for bulk carriers and oil tankers were proposed. These proposals were developed in two joint industry projects supported by different classification societies, and this resulted in two rather different analysis procedures for fatigue assessment (Lotsberg 2006a). Around 2009 it was decided by the International Association of Classification Societies (IACS) to harmonize these rules into one set of recommendations. This work was performed by the classification societies within IACS, and new common recommendations on fatigue assessment of bulk carriers and oil tankers accounting for residual stresses and mean stress effects were issued in 2013.

Long-term use of ship-shaped units permanently installed on the field require improved reliability with respect to corrosion and fatigue than a sailing ship that can be easier repaired. Thus, different recommendations on fatigue assessment of floating production vessels from that of ships have been developed the past 15 years.

As the offshore structures become older, the need for inspection and maintenance increases. Here inspection methods have been improved and probabilistic methods have been developed over the past 30 years to make the inspection planning more optimal, as explained in Chapter 18. Furthermore, it has been interesting to observe a significant development of Remote Operated Vehicles (ROVs) during the past 20 years. This has removed much of the need for using divers for underwater operations related to inspection and repair.

Development of new types of structures has lead to research in new areas. Fatigue strength of girth welds in tethers, risers pipes that also may be reeled, umbilical, and so forth are examples of this. At the end of the 1980s the need for high capacity connectors in tether strings such as shown in Figure 13.13 led to fatigue testing of high strength steel with yield strength larger than 500 MPa. This resulted in S-N curves for such material as shown in Section 4.1.5. Use of high strength steel forgings in riser systems and subsea wellheads has required further fatigue testing; reference is made to Wormsen et al. (2015) and additional S-N curves for this steel were included in DNVGL-RP-C203 (2016). Umbilicals are needed for subsea operations and fatigue testing of these has been performed in a number of projects, see also Section 4.1.11. Fatigue of flexible pipes used in the offshore industry is another item that has required research and development with respect to fatigue; see, for example, Fergestad et al. (2014). Research is also ongoing to improve the understanding between defects and fatigue strength in welded connections, which can be subject to high dynamic loads as explained in Section 4.6.2. Wind turbine structures placed in the sea is another example of new developments after year 2000 where a good fatigue design is required in order to resist long-term dynamic loading from wind, waves, and rotor motions.

Since the 1990s a number of papers have been published on improvement of welded connections using High Frequency Mechanical Impact (HFMI) treatment; see, for example, Statnikov (2004) and Kudryavtsev et al. (2004) for information on historic development. This is based on ultrasonic impact treatment resulting in less noise and vibration than by using standard hammer peening. However, here also some uncertainty on long-term improvement remains due to potential shakedown of beneficial compressive stresses at the hot spots due to variable loading; see also Section 11.7.8.

Research on development of new material, welding methods, and improvement of non-destructive methods is being performed. Also, development of numerical tools for analysis of residual stresses depending on materials and welding methods is an interesting but challenging research topic. Here a further developed analysis methodology may improve the basis for using more advanced analysis methods and taking actual fabrication methods and mean stress effects more into consideration in fatigue design of marine structures.

Thus, the understanding of fatigue has gradually improved over a long period, based on in-service experience, measurements of dynamic loads, laboratory fatigue testing, and theoretical considerations and analyses. However, the area is still significantly influenced by experience, and the fatigue capacity is dependent on a number of parameters that need to be assessed. There is still a need for analysis models that properly link these parameters together. Thus, based on history, it is expected that the need for significant research related to fatigue of structures will remain in the future. The history also shows that it is important that the engineers get a relevant education in fatigue assessment as the basis for design of reliable and optimal structures. It is also important that in-service experience from actual structures and learning from research and developments are transferred into design standards for fatigue assessment such that optimal and reliable structures can be designed and fabricated in the future. Since the 1980s a number of different standards for this purpose have been developed and maintained through revisions such as API RP 2A (2014), NORSOK N-006 (2015), ISO 19902 (2007), DNVGL-RP-C203 (2016), DNV-RP-C206 (2012), DNV CN 30.7 (2014), and IACS (2013); see also reference list for a more detailed description of these standards. During these years the safety philosophy has become better described. It should also be mentioned that the methodology for inspection planning for fatigue cracks during in-service life has been improved significantly since the beginning of the 1980s; see also Section 18.1 for a more detailed description of this development.

I.2 Examples of Fatigue Failures of Marine Structures

I.2.1 The *Alexander L. Kielland* Accident

The hydrophone support in the brace that failed in the *Alexander L. Kielland* accident was designed with a small double-sided fillet weld where the structural strength was significantly less than that of the main plate, as indicated in Figure I.2. Connections with local weak sections may show a brittle structural behavior when subjected to external loading, even if the material in the weld and the base material are ductile; see also Section 16.1. Fabrication of these connections may also be difficult as residual stresses resulting from the fabrication may lead to cracking due to deformation from temperature shrinkage. When the throat thickness is significantly smaller than the thickness of the plates, this deformation will likely be concentrated in the weakest part of the structure, which is the weld. The investigation report after the *Alexander L. Kielland* accident describes coating being observed in a 70 mm length on the inside fillet weld, showing that some cracking had already occurred during fabrication (see NOU 1981: 11). Further cracking of the fillet welds is reported to have been mainly due to overload of the static capacity of the welds until a crack pattern around a

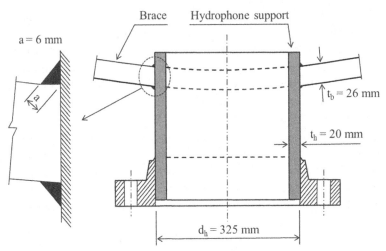

Figure I.2. Hydrophone support inserted in one of the main braces of the *Alexander L. Kielland* platform (based on NOU 1981: 11).

significant part of the circumference of the hydrophone support was reached. From this stage, further cracking into the main brace was by fatigue. At one side of the support a fatigue crack may have continued from the crack tip already present in the fillet weld. At the other side a fatigue crack may have initiated due to the large stresses at the hole when the fillet welds were cracked. The effective crack length was now already equal to the width of the hole, and the crack propagated rather fast. This crack growth is in agreement with calculated crack growth based on fracture mechanics in Section 16.3.2. An improved design methodology for tubular penetrating plates has been developed (see Section 5.5). Based on this methodology, it can be shown that the small fillet welds around the hydrophone support would also not have been adequate due to fatigue, even without the fabrication defects (see Section 5.5.6). A fatigue failure within a time period of less than four years can also be predicted using the analysis methodology available today, even without defects in the weld root. This accident demonstrated:

- that detailed and reliable fatigue analysis of offshore structures is needed;
- that requirements for robustness or damage tolerance in offshore structures in extreme situations are needed, and that the fatigue design criteria for marine structures should be dependent on consequence of a failure;
- that fatigue crack growth may initiate from items that are considered to be of minor importance for the ultimate load capacity;
- that fatigue cracks around a tubular section penetrating a plate result in large stress intensity as soon as the fillet welds are cracked around a significant part of the circumference; thus, the following crack rates correspond to a stress intensity for a crack length equal to the tubular diameter;
- that small welds with less structural strength than the base plate may show a brittle structural behavior, even if the material is ductile (see also Section 16.1);
- that the welding around a stiff and restrained section can be a challenge.

This accident showed that the design standards for offshore structures needed to be revised.

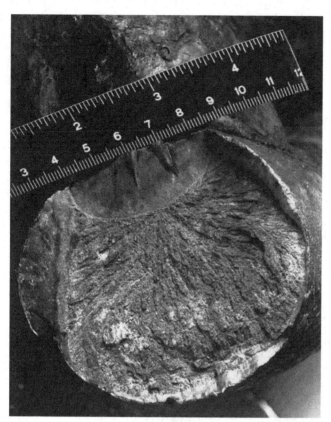

Figure I.3. Photo of fractured K4 chain.

The design of the fillet welds between the hydrophone support and the brace is further described in Section 5.5.6, and further crack growth is assessed in Section 16.3.2.

I.2.2 Fatigue and Fracture of a Mooring Chain

During the 1980s it was normal practice to use chains with studs, as this was considered to simplify handling of the chain onboard the vessels. However, it was reported that the studs became loosened through use and that fatigue cracking initiated at the footprints of the studs. A photo of a fractured 84 mm K4 chain link is shown in Figure I.3. This link was part of a chain that had been used to anchor an FPSO (floating production, storage, and offloading) unit in the North Sea during a period of approximately eight years. This failure is further assessed in Appendix A.

I.2.3 Fatigue Cracking in Ship Side of a Shuttle Tanker

The development of design criteria for fatigue assessment of ships has largely been based on experience. It was not until the 1990s that performing fatigue analysis of full ships using the finite element method became a normal routine. Fatigue design criteria also became included in recommendations and standards for certification of ship structures. The experience providing the basis for the design criteria in the new standards was derived from ships sailing in worldwide trades. These sailing routes are

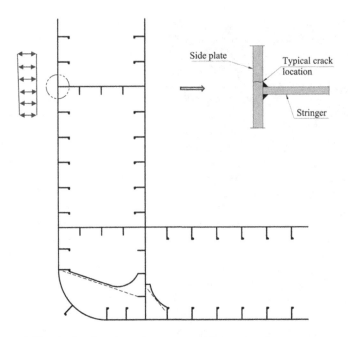

a) Sketch showing dynamic pressure at hot spot region

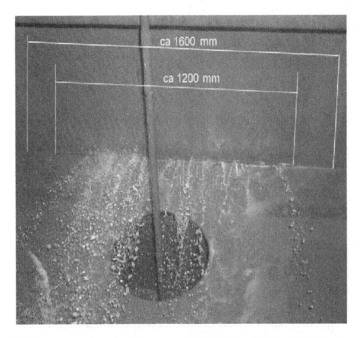

b) Water spray through a fatigue crack in the ship side in a shuttle tanker

Figure I.4. Water spray through a fatigue crack in the ship side in a shuttle tanker.

now considered to result in only half as much fatigue damage as for ships sailing in the North Atlantic. In the 1990s, the use of more high-strength steel in ship fabrication resulted in a number of fatigue cracks, as also explained in Section I.1. Fatigue cracks were also reported in the sides of shuttle tankers in the North Sea, as illustrated in Figure I.4. The ships' sides are subjected to dynamic pressures from passing waves.

Pressure from ballast water inside results in more tensile stresses at the weld toes of the fillet-welded longitudinal sections than the stresses experienced in empty ballast tanks. When the dynamic pressure is known, then the thickness of the plates, t, and the distance between the side longitudinals, s, govern the fatigue life of the ship sides. This is a simple calculation in principle, as is also illustrated in a calculation example in Appendix A. However, the results depend on knowledge of the pressure loading, and therefore experience from in-service has also been used in order to determine recommendations for the maximum distance between the longitudinal stiffeners to plate thickness (s/t–ratio) to be used in ships sailing in harsh environmental areas, such as the North Sea. Vessels have been designed with various s/t-ratios in the ships' sides. Based on this information and on in-service experience from these ships, it has been possible to obtain improved recommendations for maximum s/t-ratios. This has resulted in a recommended maximum s/t-ratio of 46 for FPSOs operating in the North Sea (DNV-RP-C206, 2012). This value may also be used for shuttle tankers.

The section modulus of the plates is proportional to the thickness squared. Thus, an increase in thickness of 1–2 mm can significantly increase reliability with respect to fatigue of the side plates in ships.

I.3 Types of Marine Structures

The term "marine structures" embraces a wide group of structures, including sailing ships, floating production vessels (FPSOs), semi-submersible platforms, tension leg platforms, spar platforms, jack-ups, and the traditional jacket structures. Since around 2003 a number of wind turbine support structures have been installed in the marine environment. The design of these structures is mainly governed by the Fatigue Limit State. Furthermore, pipelines and risers are also marine structures that are subjected to significant dynamic wave loading during installation, and following installation for risers and in pipelines if there are long free spans that are subjected to vortex-induced vibrations. Flare towers and bridges between platforms that are subjected to wind loading and relative displacements are further examples of marine structures where fatigue design is important. In recent years, fatigue assessment of subsea wellheads has also become important due to large consequences of a fatigue failure and the large costs associated with waiting for drilling operations during inclement weather. Various marine structures have been selected for illustration in Figure I.5.

I.4 Design Methodology for Marine Structures

Marine structures are designed according to design standards such as ISO 19900 (2007), API RP 2A (2014), or NORSOK N-001 (2012) for offshore structures. These standards define requirements for fulfillment of different limit states:

- The Ultimate Limit State (ULS)
- The Fatigue Limit State (FLS)
- The Serviceability Limit State (SLS)
- The Accidental Limit State (ALS)

In the ULS design the calculated design load effect is required to be less than the design capacity. In standards using a load and resistance format, the design load effect is derived from the characteristic load multiplied by a load coefficient, and the design

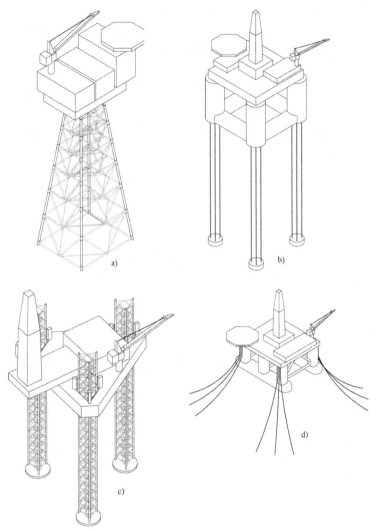

Figure I.5. Example of some marine structures. a) Jacket; b) Tension Leg Platform; c) Jack-up; d) Semi-submersible

capacity is derived from a characteristic capacity divided by a material coefficient. The characteristic values of the load effect and the capacity are specified together with values for the load and material coefficients such that the probability that the limit state is exceeded is acceptably low.

The number of load cycles during planned in-service life is multiplied with a Design Fatigue Factor (DFF) before the characteristic S-N curve is entered for calculation of the number of cycles to failure. Thus, the S-N curve used in design becomes equal the characteristic S-N curve and is therefore often also denoted design S-N curve. The DFF depends on the consequence of a failure and the possibility for inspection and repair, such that potential fatigue cracks can be detected before they become a significant threat to the integrity of the structure. This implies that the structures can be considered to be fail-safe, such that inspection can be relied upon as a safety barrier. Thus, in-service inspection is important for the control of the integrity of marine structures. However, for some structural parts, such as piles in the soil that are supporting jacket structures, in-service inspection is not feasible. Thus, these structural parts have to be designed with sufficiently large Design Fatigue Factors that the probability of a failure occurring becomes acceptably low (or as low as intended for other failure modes in the other limit states). Reference is also made to Section 12.7 with respect to recommended Design Fatigue Factors.

The SLS are associated with:

a) deflections that may prevent the intended operation of equipment,
b) deflections that may be detrimental to connections or non-structural elements,
c) vibrations that may cause discomfort to personnel,
d) deformations and deflections that may spoil the aesthetic appearance of the structure.

Serviceability requirements are normally defined by the operator for a specific project. Limitations with regard to deflections, displacements, settlements, water tightness, vibrations, and operation of the facility are normally defined early during the design process and stipulated in the design premises for the structure.

The ALS is included to ensure that the risk of failure due to accidental damage is kept acceptably low. Similar principles are followed for design with respect to ALS as for design in the ULS. It is important to check that the structure has sufficient ductility to develop the relevant failure mechanism. Furthermore, commonly used design methods are based on the assumption that design values for load effect (actions) and capacity (resistance) can be calculated separately. In cases where integrated nonlinear analyses are used, care should be taken to ensure that the intended levels of safety are achieved. For example, the ultimate capacity of offshore structures may be reduced due to a growing fatigue crack and robustness by using an increased DFF may be needed. The presented design methodology is developed with the purpose of limiting the risks of failure for these structures.

In considering risk of failure of structures, it is practical to group the risks as follows:

- risk of failure of the structure resulting from statistical variations in loads and structural load bearing capacities;
- risk of failure due to accidents;
- risk of failure due to a gross error during design, fabrication, installation, and/or operation of the structure.

The first risk listed is controlled by the ULS, with specified load and material coefficients. When design standards are developed, the aim is to have a low probability of failure. For welded offshore structures, an annual failure probability of below 10^{-4}–10^{-5} is aimed for when load and material coefficients are calibrated for design standards for welded structures (Fjeld 1978). In the FLS the aim is also for a similar low probability of failure due to fatigue. Thus, this probability of failure normally becomes small in comparison with the other risks for structural failure. The risk for failure due to accidents is accounted for by appropriate risk assessment studies, denoted Quantified Risk Analysis (QRA). If the risk from an accidental event is high (typically above 10^{-4} on an annual basis), then this event will be included in the design. The risk of failure due to a gross error is the most difficult to manage, as it cannot normally be removed by additional safety coefficients. Detailed plans for verification and quality assurance of important items in the design, fabrication, and installation process are usually required in order to reduce the probability of gross errors in a project to an acceptable level. However, due to their nature, it is intrinsically difficult to assess the risks for gross error. Gross errors are understood to include:

- lack of human understanding of the methodology used for design;
- negligence regarding information;
- mistakes such as calculation errors; these can include input errors to the analysis programs used and also errors in computer software that are used for design;
- lack of self-check and verification;
- lack of follow-up of material data testing, welding procedures, inspection during fabrication, and so on;
- mistakes resulting from lack of communication or misunderstanding in communication;
- lack of training of personnel onboard the platform that may lead to improper operation of ballasting systems;
- errors in systems used for operation of the platform.

Thus, gross errors are generally understood to be human mistakes. Different people are involved in every project, and it becomes very difficult to predict the probability of a gross error occurring in any given project. However, past history demonstrates that gross errors have been significant contributors to the failures of structures, and focusing on these issues is important for ensuring project success. It should also be kept in mind that many errors are difficult to detect by self-checks. Therefore, many errors will only be discovered through objective review by another person or through an independent analysis. Discipline checking and verification are important for removal of human errors, but the risk is only reduced by a limited extent. Therefore, the possible use of other barriers should also be considered where the consequences of a human error are significant and where such an additional barrier is feasible.

Reliability analyses are used for calibration of load and material factors to be used for ULS design required in standards taking into account normal uncertainties in the parameters that are significant for the calculated probabilities of failure; see, for example, Moan (1999). As pointed out by Kvitrud et al. (2001), reliability analyses underestimate the risks of accidents for offshore structures, as gross errors are not included in the analysis. In general the experiences from the largest accidents in Norway show that human error is the main contributor to the risk. Most accidents may appear to have arisen from a unique set of circumstances, which are unlikely to ever be duplicated, even if similar single errors contributing to the accident may occur again. Furthermore, it is often more than one error that leads to failure. Each error may be linked to details in design and fabrication. Thus, it is important to use educated and experienced engineers in the development of new structures that should meet long-term requirements for safe operation. For several reasons it may be difficult to write precise target safety levels into a design standard. In some design standards this is solved by requiring that a safety level with a failure probability "As Low As Reasonably Practicable" (known as the ALARP principle) is aimed for during design. In some design standards, such as in the DNV Rules for Classification of Compressed Natural Gas (CNG) Carriers from 2012, this principle is included in addition

Table I.1. *Calculation examples in this book related to stress concentration factors and S-N data*

Topic	Description	Section
Stress concentration factors	Butt welds in plated strutures	5.3.2
	Stress Concentration Factor for girth weld at thickness transition in jacket leg	6.6
	Single-sided weld at thickness transition in pile	16.3.4
	Simple tubular joints	8.2.1
	Stress Concentration Factor (SCF) in grout filled tubular joint	8.6.2
	Tubular with external ring stiffener subjected to axial force	6.7.2
	Tubular with external ring stiffener subjected to internal pressure	7.3
	Conical connections	6.8.2
	Riser string	6.9.1
	Fatigue assessment of anode attachment close to a circumferential weld in a jacket leg	6.6
	Skew loading due to lateral pressure	8.12.3
Fatigue design based on S-N data	Selection of S-N curve for detail using nominal stresses	A.1
	Fatigue assessment of a bridge between platforms with dynamic loading due to friction at supports	A.1
	Assessment of S-N detail of a tubular against a plate	6.7.1
	Fatigue assessment of a drum	6.7.2
	Fatigue assessment of a ship's side	A.2
	Effect of thickness on fatigue strength	4.7.2
	Effect of weld length on fatigue strength	4.7.6
	Effect of temperature on S-N strength	10.7.2
	Design using the maximum stress range less than the fatigue limit	6.7.2
	Assessment of the fillet welds around the hydrophone support in the *Alexander L. Kielland* platform	5.5.6
	Derivation of S-N curves with few test data	4.13
	Low cycle fatigue	3.1.4
	Example of use of design chart	10.6.4
Fatigue testing	Example of stress range and the number of cycles for fatigue testing of connector	4.13.4

to a specified high target safety level. This principle is adopted as a failure philosophy to ensure focus on continuous safety efforts and that the overall (and absolute) targets are not misused to argue that sound safety measures are not implemented.

I.5 Overview of Fatigue Analysis Examples in This Book

It is difficult, probably impossible, to become a fatigue expert only from reading the literature. Practice in solving actual problems is important, as the field of fatigue is based on gaining experience of what seems to be a good design solution and what has been shown to be less successful. A number of different examples of fatigue assessment have therefore been included in this book and are recommended for training. The examples are simplified for illustration purposes. The industry uses often more

Table I.2. *Calculation examples in this book related to effect of residual stresses, long-term loading, finite element analysis, probability of failure, fabrication, and fracture mechanics*

Topic	Description	Section
Effect of residual stresses and shakedown	Example of shakedown of residual stresses	3.3.2
	Example of residual stress in tubular after cold forming	3.3.4
	Measurement of residual stresses after pile driving	2.2.2
Long-term loading	Transformation of the 100-year long-term loading to 20-year long-term loading	10.5
	Closed-form solutions for fatigue based on Weibull long-term stress range distribution	10.4
	Fatigue assessment of a hole in deck structure in an FPSO	10.4
	Fatigue design of topside stool on deck of FPSO	10.4
	Combined loading in piping on deck of an FPSO due to operation and waves, including effect of temperature.	10.7.2
	Transformation of load from one probability of exceedance to another	A.2
Finite element analysis	Hot spot stress analysis of gusset connection with partial penetration weld	9.3 9.6.4
	Fatigue analysis using the notch stress concept	9.6.4
	Tubular joints	8.2
Probability of a fatigue failure	Calculation example	12.8
	Fatigue design of pipes for storage of CNG	12.7.3
Fabrication	Example of weld improvement by grinding	11.7.5
Crack growth using fracture mechanics	Fatigue crack growth from maximum defects in different S-N curves that meet required design life	4.6.2
	Example of crack growth from an internal defect in a cruciform joint	16.3.1
	Crack growth calculation from the root of single-sided weld	16.3.4
	Crack growth in brace of *Alexander L. Kielland*	16.3.2
	Crack growth calculation in cruciform joint from weld toe and weld root	16.3.3
	Crack growth and unstable fracture of a large diameter chain	A.3
	Example of use of failure assessment diagram	A.3
Grouted connections	Loads transferred from transition piece to monopile	17.2.4

advanced numerical programs for calculation of fatigue damage. However, when using such programs, the physical behavior of the problem may be difficult to follow. Therefore, it is often recommended that simplified methods, which are based on engineering understanding and assessments, are used before more refined analysis is initiated. The results from simplified analyses may also be of significant value when the results from advanced or refined analyses are assessed in terms of correctness for the problem being considered, and the risk for a gross error becomes reduced.

An overview of the examples, with reference to the relevant sections, is provided in Tables I.1 and I.2.

1 Fatigue Degradation Mechanism and Failure Modes

1.1 General

It is generally accepted that in welded structures subjected to repeated external loads, microcracks may be initiated rather early in the fatigue life. Based on observations on the development of fatigue cracking, it has become common practice to consider fatigue life as consisting of three phases. These are initiation of a crack, propagation of the crack, and final fracture. Final fracture is simply the fracture under the last up-loading tensile load cycle and may be treated by assessment of unstable fractures, as presented in Chapter 16. However, the distinction between the first two phases is not very clear. The question arises of how to determine that the crack is so large that its growth can be properly defined in laboratory testing and in crack growth analysis. Fatigue crack propagation is understood here to mean the growth of cracks that are so large that the continuum mechanics approach can be applied.

The proportion of fatigue life of a structure that is spent in each of the two first phases depends on the material, the geometry of the detail being considered, and the loading. The initiation period for a fatigue crack in the base material without significant notches is relatively long in comparison with the propagation period. In contrast, at weld toes in structures with more severe stress concentrations, the formation of a dominant crack occurs relatively early in the fatigue life, and the propagation phase constitutes the major portion of the total life. Radencovic (1981) and test data on tubular joints reported by Pozzolini (1981) showed that 70–90% of the fatigue life of welded connections is related to fatigue crack growth. However, as indicated in the preceding paragraph, such numbers also depend on the definition of initiation and the type of connection. At start of fatigue crack growth after initiation it is assumed that a sharp crack tip has developed. The size of the crack at this stage is not so easy to define and in literature it ranges from below 0.1 mm up to 1 mm; see also Sections 3.1.3 and 4.7.4.

The basic mechanism of crack nucleation in the base material is cyclic slip and the extrusions and intrusions at the surface of the base material. At weld toes, however, initiation is more usually from defects at undercuts or from other imperfections in welded connections. Thus, fatigue crack growth may start from macroscopic defects, but fatigue cracking may also occur in originally uncracked base material when subject to large cycles of stress or strain. The initiation of a crack in the base material can be observed as slip planes in zones adjacent to the outer surface or to internal voids or inclusions, and thus, microplastic properties may be decisive. Crack initiation is always located at surfaces or phase transitions where a degree of freedom for accumulation of slip exists; see, for example, Hellan (1984). Typical extrusions and intrusions are formed by repeated slip on other planes due to, for example, strain hardening and oxidation of the generated free surface. Thus, some intrusions may act as crack starters, both by concentrating stresses at the resulting surface irregularities and by enabling aggressive environments to take effect. It might be expected that one of the primary planes, slipping at a 45° angle with the load direction, would be the first growing crack. However, it is almost impossible to identify a clear transition between the initiation phase and further crack growth. During further dynamic loading, crack growth normally occurs in the direction normal to the largest principal stress. Fatigue crack growth in ductile material is typically a transcrystalline type of cracking that is identified by striation as the crack front moves forward through each new load cycle. The striations can be explained by local plastic deformation as the crack opens up during a tensile part of the load cycle. This deformed part becomes too large when the crack closes during the next part of the load cycle and will be forced to penetrate into the material during crack closure, leaving permanent ripples at the crack front. Fatigue is a discrete process with the crack "jumping" from one atomic position to another from one load cycle to another. Thus, the lowest possible crack rate is propagation of one atomic spacing per load cycle. Even smaller values may be explained as average values along a crack front.

1.2 Low Cycle and High Cycle Fatigue

When there are few load cycles, a test specimen can be subject to a significant strain range before it fails. This range is so large that it exceeds the yield strain, both in tension and compression. This implies change of yield stress and plastic dissipation during a load cycle, as was observed by Bauschinger during the nineteenth century. Thus, the low cycle fatigue strength is a function of material yield strength. The total strain is the sum of elastic and plastic strain: as the plastic strain amplitude is reduced, the specimens can be subjected to a larger number of load cycles before failure occurs. Thus, there is a gradual transition from fatigue under significant plastic strain to that of elastic behavior. Fatigue associated with significant plastic strain is often referred to as Low Cycle Fatigue (LCF) and is associated with relatively few cycles until failure, as indicated in Figure 1.1. A more detailed description of LCF using a strain-based approach is presented in Section 3.1. The region with mainly elastic strain behavior is denoted High Cycle Fatigue (HCF). Here stress range is used on a similar basis as strain for LCF, as there is only the Young's modulus between the stress and the elastic strain. The transition between LCF and HCF can be considered to be in the region 10^4 to 10^5 load cycles. It may be practical to define this transition at 10^4 cycles, as this is frequently the left starting point in the stress range-based S-N diagrams for calculation of HCF. HCF S-N curves are further described in Section 3.2 and Chapter 4.

Examples of LCF can be found in design standards for ship structures and floating production vessels where large load cycles may occur due to loading and unloading. Some minimum numbers of design load cycles for this type of LCF are provided in Table 1.1 (Urm et al. 2004). Due to uncertainty regarding actual design cycles at the design stage, the number of actual design cycles may be higher than the

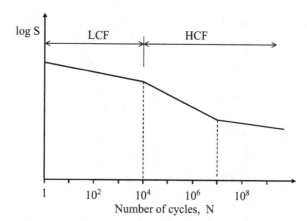

Figure 1.1. Schematic S-N diagram, including the LCF and HCF regions.

minimum design. A design fatigue factor of 2 is usually employed for actual design with respect to LCF of ship structures. The need for design criteria with respect to LCF has increased with subsidence of fixed platforms, such that waves may hit the deck structures during major storms (e.g., NORSOK N-006 2015). Thus, LCF has become an important part of design for the Ultimate Limit State for these structures. A failure is unlikely to be associated with only one single large wave, but more probably occurs following a number of waves during the same storm that results in the accumulation of fatigue damage until failure. A procedure for this has been included in NORSOK N-006 (2015) "Assessment of structural integrity for existing offshore load-bearing structures." Reference should also be made to Section 3.1.3 of this book, where some background for the LCF part of the S-N curve is presented.

The long-term loading from waves and wind entails so many stress cycles that HCF becomes relevant. The relative distribution of fatigue damage along the log N axis in a typical S-N diagram for the air environment of a floating platform is shown in Figure 1.2. This figure is based on a Weibull distribution of long-term stress ranges with the shape parameter h = 1.0. This represents a typical wave loading on a floating structure during a service life of 20 years and with a Palmgren-Miner damage equal to 1.0. The vertical line in the graph is due to the change in slope in the S-N curve at 10^7 cycles, where the negative inverse slope of the S-N curve is m = 3.0 for the left-hand side and m = 5.0 for the right-hand side. Half of the calculated fatigue damage is to the left of $5.79 \cdot 10^6$ cycles in Figure 1.2 and the other 50% is to the right of this number of cycles. The relative fatigue damage is moved to the right in Figure 1.2 for longer calculated fatigue lives. This illustrates the most significant region of the S-N curve for calculation of fatigue for this type of structure. The mean up-crossing response period for offshore structures is typically 6–8 sec, and is somewhat higher

Table 1.1. *Minimum number of load cycles for Low Cycle Fatigue*

Ship type	Minimum number of load cycles
Oil tankers	500
Chemical tankers	750
Panamax bulk carriers	1,000
Shuttle tankers	1,500

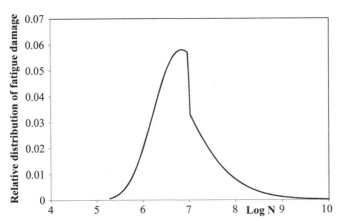

Figure 1.2. Typical contribution to fatigue damage along S-N curve for a floating offshore structure subjected to wave loading with Miner sum equal 1.0 in 20 years service life.

for floating structures than for structures that are fixed to the sea bottom. The number of load cycles can be derived from the total service life divided by the mean up-crossing response period. With an average period equal to 6.3 sec, the number of cycles during 20 years can be calculated from $60 \cdot 60 \cdot 24 \cdot 365 \cdot 20/6.3 = 10^8$ cycles. Wind turbine support structures are subject to approximately twice as many load cycles (from wind, waves and rotor actions) as offshore structures subject to wave loading during the same period of service life.

1.3 Failure Modes due to Fatigue

There are four main fatigue cracking failure modes that should be considered in fatigue design of structures.

1.3.1 Fatigue Crack Growth from the Weld Toe into the Base Material

In welded structures the most frequent failure mode is fatigue cracking from weld toes into the base material. The fatigue crack is initiated at small defects or undercuts at the weld toe where the stress is highest due to the geometric stress at the considered detail and the weld notch geometry as indicated in Figure 1.3. Examples of fatigue cracks that have initiated at weld toes are shown in Figures 1.4–1.6. Fatigue cracks may initiate at different positions, as shown in Figures 1.4–1.5. The cracks may also grow in different transverse planes to the load direction, as shown in Figure 1.6. Much of this book is aimed at how to achieve a reliable design with respect to this failure mode.

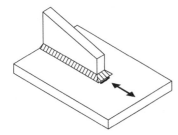

Figure 1.3. Fatigue crack growth from the weld toe into the base material.

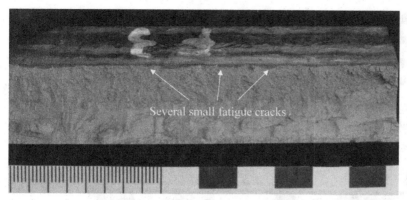

Figure 1.4. Section through crack initiation at the weld toe of a doubling plate (Lotsberg et al. 2014) (Copyright Springer. With kind permission from Springer Science and Business Media).

1.3.2 Fatigue Crack Growth from the Weld Root through the Fillet Weld

Another important failure mode that must be considered is fatigue cracking from the root of fillet welds, with crack growth through the weld, as shown in Figure 1.7. The use of fillet welds should be avoided in connections where the failure consequences are large due to less reliable non-destructive testing (NDT) during fabrication of this type of connection compared with that of a full penetration weld. In addition, it is normally not possible to discover fatigue cracks during service life before they have propagated through the weld. However, the use of fillet welds cannot be avoided for some welded connections. Furthermore, they are considered efficient for fabrication in ship-shaped structures, but are generally used less frequently in fixed offshore structures such as jackets. The design procedure presented in this book should provide reliable connections, including fillet welds.

1.3.3 Fatigue Crack Growth from the Weld Root into the Section under the Weld

Fatigue crack growth from the weld root into the section under the weld, as indicated in Figure 1.8, can be observed both during the service life of structures and also in laboratory fatigue testing. An example of a fatigue crack that initiated from the root of a fillet weld around a bracket toe is shown in Figure 1.9. The fatigue crack had

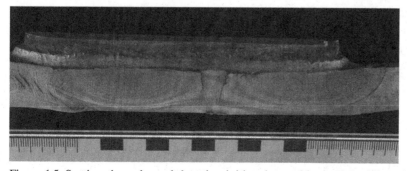

Figure 1.5. Section through crack location initiated at weld toe of doubling plate (Lotsberg et al. 2014) (Copyright Springer. With kind permission from Springer Science and Business Media).

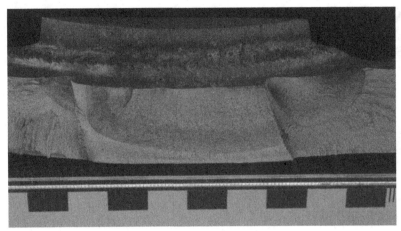

Figure 1.6. Section through crack location initiated at weld toe of doubling plate (Lotsberg et al. 2014) (Copyright Springer. With kind permission from Springer Science and Business Media).

finally grown through the bulb section in a laboratory test. Reference is also made to Section 2.4 of this book.

The number of cycles to failure for this failure mode is of a similar magnitude as that for fatigue cracking from the weld toe in the welded condition. Apart from using alternative types of welds locally (or to reduce the general stress range level), there is no recommended methodology for avoiding this failure mode. This means that if fatigue life improvement of the weld toe is required, the connection will become more highly utilized and it is also necessary to improve the fatigue life of the root. This can be achieved by using a full penetration weld some distance along the stiffener nose. The typical length of a full penetration is 10t, where t is the thickness of the bracket. Use of a full penetration weld at the ends of brackets can improve weld

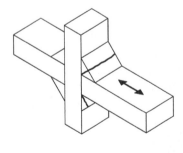

Figure 1.7. Fatigue crack growth from the weld root through the fillet weld.

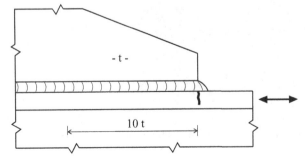

Figure 1.8. Fatigue crack growth from the weld root into the section under the weld.

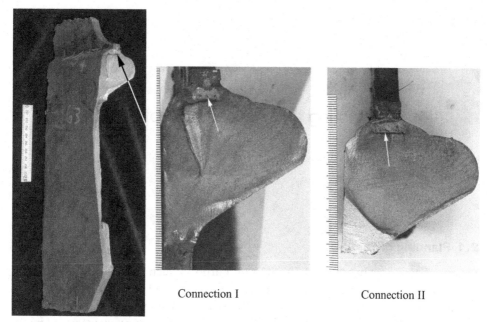

Connection I Connection II

Figure 1.9. Fatigue cracking from the root of the fillet weld at end of the bracket into the bulb section in a laboratory test.

toe fatigue life. Furthermore, with a full penetration weld at the end, the full height of the bracket nose can be ground to a specified radius in order to improve the detail, as shown in Appendix A of DNVGL-RP-C203 (2016) to achieve a higher S-N curve. An alternative to using a nominal stress S-N curve, the maximum stress at a bracket with a radius transition to the plate can be derived more directly from a detailed finite element analysis of the ground geometry, together with an S-N curve for flush ground or machined details such as S-N curve C in DNVGL-RP-C203 (2016).

1.3.4 Fatigue Crack Growth from a Surface Irregularity or Notch into the Base Material

Fatigue cracking in the base material is a failure mode that can occur in components with high stress cycles. In such instances the fatigue cracks often initiate from notches or grooves in the components or from small surface defects or irregularities; see, for example, Figure 1.10. The design procedure presented in this book provides reliable connections with respect to this failure mode also.

Figure 1.10. Fatigue crack growth from a surface irregularity or notch into the base material.

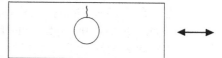

2 Fatigue Testing and Assessment of Test Data

2.1 Planning of Testing

Fatigue testing may be planned for different purposes, such as testing for documentation of a general design S-N curve for a considered detail or for qualification of a detail in a project. More time is usually available for planning of the testing for the first case than for the second case. The first case also normally involves tests of more specimens than time allows for in the second case. When planning the testing, the purpose of the test must be clearly defined as early as possible. When the purpose is defined, the number of tests and the testing time required can be planned. During this initial planning phase, how the test data will be assessed and transferred into a recommended design methodology for the considered project should also be considered. Some examples of fatigue testing of different details are included in Sections 2.2–2.6 for butt welds in plated structures and piles, small-scale specimens for simulation of fatigue strength in sailing ship structures, large-scale specimens from sailing ships and floating production ships, fillet welded connections, and cover plates or doubling plates. In Section 2.7 approaches to how fatigue test data can be used for assessment of a recommended design procedure, where the principal stress direction during load cycling is not normal or parallel with the weld direction, are provided.

2.1.1 Constant Amplitude versus Variable Amplitude Testing

Constant amplitude testing is normally performed for derivation of test data representative for the left section of the high cycle part of the S-N curve (or the part of the S-N curve between the low cycle region and the fatigue limit). The fatigue limit in air is understood to refer to the position of the transition of slope in the S-N curve at 10^7 cycles in Figure 1.1. However, in order to obtain relevant test data for the high cycle region to the right of the fatigue limit, a variable amplitude loading, also known as a spectrum loading, must be used. The reason for this is that some load cycles with a stress range larger than the fatigue limit are needed to initiate a crack, such that crack growth will also occur for stress cycles that are below the fatigue limit. This is further illustrated in Sections 3.2.2 and 3.2.3. It is a challenge that many cycles are required for testing in order to obtain S-N data for the right part of a two-slope S-N curve, as is typically used in fatigue design of marine structures. When the tests using a spectrum load are completed, presentation of the test data in the S-N diagram is also challenging, as, ideally, one test would provide only one data point for the diagram. One way to address this is to calculate an equivalent stress range for the selected number of cycles. However, this poses the question of the number of cycles that should be selected and the slope of the S-N curve that should be used in order to derive the equivalent stress range. Some investigators suggest using an inverse negative slope equal to 3.0 for all stress ranges, but this approach means that, in principle, it has already been predicted that this is also the slope of the S-N curve to be determined.

2.1.2 Fabrication of Test Specimens

Test specimens should be designed and fabricated in a manner similar to that planned to be used for constructing the considered component or structure where the fatigue test data is to be used. Boundary conditions and load application should provide a similar stress distribution at the hot spot stress region as would be expected in the actual structure, although this may be difficult to achieve in all situations. In such cases, a finite element analysis of the fabricated test specimen for comparison with the actual structure, which should be analyzed in a similar way for assessment of differences, is recommended. Further assessment should be based on engineering judgment and experience. When planning the testing, it is also important to design the test specimen such that it will fail at the considered hot spot region, rather than at another region created for testing purposes, such as at the connection to the testing machine. Furthermore, it should be remembered that there might be a significant difference in residual stresses in a scaled test specimen as compared with a real structure. Therefore, measurements of residual stresses at welded connections in the test specimen(s) may be recommended in order to provide better assessment of the test results.

2.1.3 Residual Stresses and Stress Ratio during Testing

Measurement of Residual Stresses

Residual stresses at welded connections are normally tensile at the surface after fabrication. However, there are exceptions, such as for the lug plate type in test specimen no. 5, representing part of a ship structure, which is described in Section 2.3.4. The size of the residual stresses depends on a number of parameters such as restraint of the connection, weld groove geometry in relation to other geometry, and welding procedure. Thus, the residual stresses are normally significantly lower in small-scale test specimens than in actual structures, and therefore it is difficult to account for residual stresses properly in fatigue design standards. In order to address this, design procedures can simply be prepared that are considered to be on the safe side. That is, it can be addressed by stating that there are tensile residual stresses close to the yield strength of the base material and that fatigue testing should be performed at a high maximum stress value in order to obtain representative S-N curves. Knowledge about residual stresses is useful for assessment of test data and for transfer of test data into relevant design S-N curves for a specific design purpose. In such cases, it is also useful to know typical residual stresses in the actual structure. This knowledge is useful both for assessment of fatigue crack growth and for assessment of unstable fractures. Residual stresses are in self-equilibrium, such that if there are significant tensile stresses at a weld toe, there will also be compressive stresses some distance below the surface. This can also imply that there are often compressive stresses at the root in fillet welds and in single-sided girth welds. However, if weld repairs have been performed, the internal distribution of residual stresses in the weld becomes more uncertain and cannot be predicted without direct measurements. Several different techniques can be used to measure residual stresses. The most relevant of these methods are:

- Hole drilling method. This method requires installation of a strain gauge (three-element rosette) on the surface where the residual stresses are to be measured. A small hole, typically 1–4 mm in diameter, is then drilled in the center of the rosette to a depth that is approximately equal to the diameter of the hole. After this localized removal of material, the relief of stresses outside the hole can be measured and a biaxial stress distribution calculated.
- Ring core method. This method also requires installation of a rosette on the surface where the residual stresses are measured. A ring around the rosette is drilled such that the strains at the rosette are completely relieved; the residual stresses can then be calculated from this measured strain relief.
- Sectioning method. The principle for measurement of residual stresses using the sectioning method is similar to the ring core method. The method requires installation of a strain gauge on the surface. The material around the strain gauge is then removed such that the strains are relieved and the residual stresses can be calculated.
- X-ray diffraction method. This method is based on the principle that the distance between the crystal lattices in the material is a function of stress, and that this distance can be measured using X-rays. Stresses at the material surface are mainly measured using this non-destructive technique.
- Neutron diffraction method. This method supplements the data obtained by the X-ray method, such that residual strain information through the thickness of the material can be derived.
- Ultrasonic method. This method is based on the concept that the speed of an ultrasonic wave is a function of the stress level in the material. An average stress is determined through the region through which the wave propagates. This gives a rather low spatial resolution, and the method may be of greatest value where the stress is constant over a larger volume.
- Magnetic method. This method relies on an interaction between strain and magnetization in a ferromagnetic body. The stresses are measured at the surface of the material.

In general, the destructive methods are considered to be the most accurate. The accuracy of the different methods has been presented by Lu (1996). The sectioning method is the most accurate, and within ± 10 MPa is the usual precision in a normal case. Ultrasonic and magnetic methods are the next most accurate and are in the range ± 10–20 MPa, while hole drilling and X-ray methods are within ± 20 MPa. The neutron diffraction method has an accuracy of ± 30 MPa.

There are advantages and disadvantages with each of these methods. The equipment required for hole drilling, sectioning, X-ray, ultrasonic methods, and magnetic measurements is portable. However, the equipment for neutron diffraction is not portable. It is also the most expensive and time consuming for measurements and therefore is mainly relevant for research purpose.

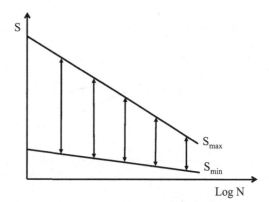

Figure 2.1. Fatigue testing with R = constant.

The first three methods are destructive, in that material is removed. Of these methods, the hole drilling method is the least destructive, and, depending on the situation, the hole may be repaired. The hole drilling method has been the preferred method in the DNV test laboratory for measurements of residual stresses (see also Section 2.2.2). Among the non-destructive methods, the X-ray method has become the most popular measurement method both in industry and for research purposes. Kim and Smith (1994) reported measurements of residual stresses at girth welds in tethers in a tension leg platform using the hole drilling technique and the X-ray diffraction technique. The comparison of these methods showed measured values that were in good agreement. For further, more detailed information about the different measurement techniques, see *Handbook of Measurement of Residual Stress* edited by Lu (1996).

Stress Ratio

The stress ratio during fatigue testing is normally presented in terms of $R = \sigma_{min}/\sigma_{max}$ during a stress cycle. Many small-scale tests described in the literature have been performed with $R = 0.1$. Typically, the testing has been performed under constant R, as indicated in Figure 2.1. Due to differences in residual stresses in small-scale specimens, as compared with actual structures, it has been questioned whether this methodology should be generally recommended for derivation of test data. Based on such considerations, the International Institute of Welding (IIW) has recommended testing where the maximum stress during a load cycle is close to the yield strength of the material, as indicated in Figure 2.2. The difference in derived S-N curves depending on stress ratio is indicated in Figure 2.3.

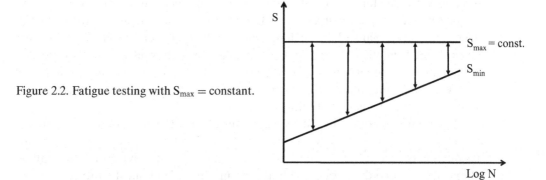

Figure 2.2. Fatigue testing with S_{max} = constant.

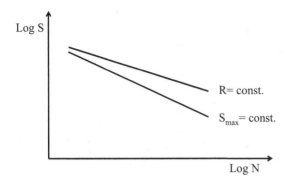

Figure 2.3. Effect of test load on S-N curves.

2.1.4 Number of Tests

The recommended number of fatigue tests depends on the type of specimen to be tested. It is easier and cheaper to test small-scale test specimens than larger prototype specimens. Small-scale specimens can normally be tested in standard test machines, whereas larger prototype testing may require construction of purpose-made test rigs. The number of tests is also dependent on how the assessment is going to be conducted when determining the S-N data to be used for design. For derivation of S-N curves with confidence, see Sections 4.11, 4.12, and 4.13. The recommended number of tests depends on experiences gained from similar tests, and must address whether the scatter in the test data is expected to be similar to that published in literature, or whether the specimen is a new type of fabrication where the scatter in test data is unknown. This may result in different approaches to deriving the design S-N curves, as is presented in Sections 4.11, 4.12, and 4.13.

2.1.5 Instrumentation

Instrumentation of the test specimens is important for several reasons. First, instrumentation is necessary for documentation of the test results, but it is also important as a basis for assessment in cases where the test results are not in line with those that were expected. It is good practice to add strain gauges, such that the main load on the specimen can be controlled, even if the load cells in the test machines have been recently calibrated. Fabrication tolerances cannot be avoided in actual structures, and this is also the case for fabricating test specimens. Therefore, it is recommended to include sufficient strain gauges so that the hot spot stress at the region that fails during fatigue testing can be estimated correctly. In some structures it may be difficult to predict accurately beforehand where the first fatigue crack will initiate. A detailed finite element analysis of the test specimen will improve the basis for placement of strain gauges. In such a situation, a sufficient quantity of strain gauges should be planned for, such that an overall good overview of stress flow can be predicted from the measurements for comparison with results from the finite element analysis. Comparison of these data may enable derivation of reliable hot spot data, even if the measurements have not been derived precisely at the resulting hot spot where fatigue cracks first initiate.

Different types of strain gauges may be installed at the hot spot region, depending on the expected stress condition. If the main stress direction is uncertain, installation of rosettes may be recommended, as these will provide full information about stress in two directions, in addition to information about the principal stress directions. Strains in three directions are measured by a rosette, and these form the basis for derivation of the two stress components and the principal stress direction. If the first principal direction is known, it may be sufficient to use a strain gauge with a cross-type filament. If it is of interest to measure the stress mainly in one direction, then a single filament strain gauge may be sufficient, and the stress is simply derived from the measured strain multiplied by Young's modulus. At hot spots with a significant stress gradient, use of a strip with strain gauges may be recommended for derivation of more strain data.

For a direct derivation of measured hot spot stress, placing the first strain gauge as close as possible to the weld toe, or as close as possible without being affected by the weld notch, is recommended. This is because it is normally assumed that the local stress increase due to the weld toe is accounted for in the hot spot stress S-N curve. Due to such a consideration, IIW recommends placing the first strain gauge at 0.4t from the weld toe, where t = plate thickness. The next strain gauge is positioned 1.0t from the weld toe. The stress at the weld toe can then be derived from a linear extrapolation of the stresses at these two measurement points. For some of the ship classification companies, the practice has been to derive hot spot stress by linear extrapolation of calculated stresses at 0.5t and 1.5t from the weld toe from finite element analysis. In some cases, a similar methodology has also been used in laboratory testing for placing of strain gauges for hot spot stress assessment.

2.1.6 Test Frequency

In air environments, the influence on the fatigue behavior of steel structures has been considered to be small provided that the high-frequency testing do not lead to high temperatures in the test specimens for load frequencies of up to approximately 100–200 Hz. Yokobori and Sato (1976) measured crack propagation rates during fatigue testing at frequencies 0.017, 0.1, and 140 Hz in aluminum alloy and carbon manganese steel. From these data a crack propagation equation including the test frequency was presented. Using this equation for steel, it is found that the crack growth reduces by a factor 1.74–1.90 on fatigue life when increasing the test frequency from 1 Hz to 100 Hz. The effect of frequency on fatigue strength has later been reassessed by Guennec et al. (2013) by fatigue of non-welded bars tested under controlled temperature conditions at different frequencies. A reason for increased fatigue strength with increasing frequency may be explained by increase in yield strength with increased loading rate.

A high test frequency in the order of 100 Hz can be obtained by electromagnetic resonant test machines. This puts limitations to force and size of test specimens, and where a larger force range is required, hydraulic actuators are being used where the test frequency most often is less than 5 Hz. The actual maximum test frequency that can be achieved is dependent on the flexibility of the test specimen in addition to the capacity of the testing machine.

The test frequency in corrosive environment has a significant influence on crack propagation, and for fatigue testing of steel in seawater a frequency of approximately 0.3 Hz is typically used. If the test frequency in seawater condition is increased to 1 Hz, the test results approach that of test data in an air environment. According to Buitrago et al. (2004), different lower frequencies may be recommended for other conditions such as sour environments.

2.1.7 Measurements and Documentation of Test Data

It is important to plot the derived S-N data into graphs to enable comparison with expected data. It is also useful to analyze the test specimen(s) using a finite element analysis, as previously noted in Section 2.1.5. Furthermore, it can be useful to examine the scatter in the derived S-N data. Valuable information may also be derived from opening the cracked test specimen in order to investigate the initiation site. If significant crack growth has occurred during testing, then marks on the surface, together with the number of cycles at times of marking the crack size on the specimen, may help in constructing crack growth curves that can subsequently be compared with mean crack growth curves estimated by fracture mechanics calculations. Measurements and photos may be useful for implementation of the test data into design curves and recommendations. Thus, it is important that all derived test data are stored for later use.

2.1.8 Assessment of Test Data

Examples of assessment of test data are shown in Section 2.2.3 for butt welds in plated structures and piles, in Sections 2.3.3–2.3.6 for small-scale specimens for simulation of fatigue capacity in sailing ship structures, in Sections 2.4.4–2.4.5 for large-scale specimens from sailing ships and floating production ships, in Section 2.5.4 for fillet welded connections, and in Section 2.6.4 for cover plates or doubling plates.

2.2 Butt Welds in Piles

Design S-N curves in design standards are based on fatigue test data, where the stress cycle is due to an external tension load cycle. Due to residual stresses at weld toes, the same S-N curves are also used in design standards for marine structures when part of the stress cycle is compressive – like that of pile driving, where the largest part of the stress cycle is normally compressive. Whether there will be some shakedown of residual stresses in piles during pile driving has been questioned. Fatigue test data from welded connections show that nominal stress cycles in compression are somewhat less detrimental for fatigue than are cycles subjected to tensile stress. The effect of this on fatigue life is considered to be dependent on the maximum tensile stress, which implies yielding at the hot spot area such that shakedown of residual stresses is established; see also Section 3.3.2. During pile driving most of the stress cycle is compressive (dependent to some extent on the soil conditions and reflection of compressive stress cycles from the pile tip), and the design procedure used for fatigue analysis of piles might therefore tend toward the safe side if the design assessment is based on S-N curves derived under cyclic tensile stress ranges. In order to investigate this further, it was proposed that laboratory fatigue testing of specimens that are representative for butt welds in piles could be performed under relevant loading conditions.

During 2006, the *Edda* tripod in block 2/7 in the North Sea was taken ashore. The platform had been in service since 1976, and the piles were considered to be representative for the piles installed in the North Sea jacket structures during the 1970s. It was proposed that a pile should be investigated in detail in order to address the following questions:

- Was there any indication of fatigue cracking at the girth welds?
- How does the actual fabrication of the circumferential weld in a pile from an older jacket appear?
- What is the residual stress at the weld after pile driving and 30 years of in-service life?
- What is the residual fatigue life of the pile?

The purpose of this work was to perform fatigue testing in order to obtain relevant fatigue data for butt welds in piles that can then be used as basis for establishing fatigue design S-N curves for piles.

As residual stresses are considered to be a significant factor governing fatigue life when mean stresses are being considered, a number of residual stress measurements were performed during the preparation of the test specimens. The work also included stress-strain measurements of the material to determine the yield strength, and also hardness measurements and assessment of fatigue crack initiation for the *Edda* pile. The fatigue testing included 30 test specimens of small-scale tests fabricated from welded plates, and also involved testing of specimens from a pile in the *Edda* tripod. The test matrices for the plate specimens have been presented by Lotsberg et al. (2008a). Many of the test specimens were subjected to 100,000 cycles, simulating the loading corresponding to pile driving, before further testing was performed.

2.2.1 Material Data and Fabrication of Test Specimens

Material Data for Test Specimens from Plates

The test program was established before the availability of the *Edda* pile for testing was known. The test specimens were cut from plates that were welded together with a butt weld. Eight plates were welded together to form four plates. As the local bending stress in a plate test specimen may differ from that of a tubular section, the specimens were instrumented with a number of strain gauges for measurements of stress. The fabrication of the test specimens may imply some axial and angular misalignment as indicated in Figure 2.4, where the tensile bending moments due to the resulting eccentricities are shown by the dotted lines. Each test specimen was instrumented with eight single-filament strain gauges for measurement of the bending strains, as shown in Figure 2.5. The same arrangement of strain gauges was made for both sides, with the gauges placed 20 mm and 55 mm from the weld on both sides of the specimen. The strain gauges were positioned such that the bending stress over the thickness due to fabrication tolerances could be measured and extrapolated to the weld toes that cracked. The positions for strain measurements were chosen to enable stress derivation at the weld toes that included the bending stress indicated by the dotted lines; thus, these positions are not directly linked to readout points used for hot spot stress derivation described in Section 9.1.3. The weld was made from both sides, and the welding procedure used in the present fabrication was similar to that used in 1972 (manual welding). The material in the plates was St 52.3 N. The plate thickness was 25 mm. The width of the dog-bone-shaped test specimens used for testing at the welded section was 50 mm. The geometry of each specimen was measured and used for calculation of nominal stress.

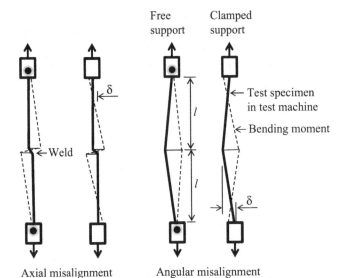

Figure 2.4. Illustration of axial and angular misalignment between welded plates.

Material Data for Test Specimens from the *Edda* Pile

Material testing of the *Edda* pile was performed to determine material characteristics. Material yield strength was 316 MPa and the ultimate strength was 511 MPa; these are material characteristics that might be expected for steel material in piles from the 1970s. In order to assess the welded region in the *Edda* pile, some macro-sections were made with hardness testing at four positions around the circumferential weld at intervals of 90° (see Figure 2.6). Small indentations from the hardness measurements can be seen along the surfaces of the macrograph sections. No indication of fatigue crack initiation was observed. The macro-sections showed a sound weld profile, with some eccentricity at the pile butt weld. In order to improve the assessment, a pile driving analysis of the *Edda* pile was also performed. The damage calculated at the considered section was only 0.012, and therefore fatigue cracking from pile driving would not be expected. Fatigue damage from in-service life has not been calculated.

Strain Measurements

The hot spot strain at the weld toe in Figure 2.5 is obtained by extrapolation from the strains measured at strain gauges 1 and 2 as

$$\varepsilon_b = \varepsilon_1 - \frac{\varepsilon_1 - \varepsilon_2}{a_1 - a_2} a_1 \tag{2.1}$$

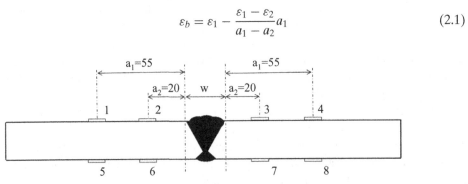

Figure 2.5. Strain gauges on a specimen for measurement of stresses that include effect of misalignment (Lotsberg et al. 2008a) (reprinted with permission from ASME).

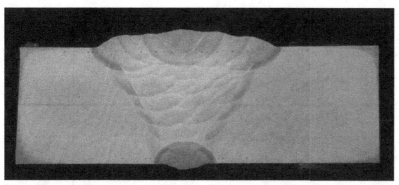

Figure 2.6. Macrograph of section from the *Edda* pile (Lotsberg et al. 2008a) (reprinted with permission from ASME).

where the geometric parameters are presented in Figure 2.5. Similarly the strain values from strains measured at gauges 3 and 4, 5 and 6, and 7 and 8 were extrapolated to the section at the weld toe, as indicated in Figure 2.5. The distances from the weld toe to the strain gauges were measured for all specimens tested, and listed as 20 mm and 55 mm for all the specimens. Based on the strains measured on the two surfaces, the membrane strain and the bending strain over the thickness were calculated. From these results, a stress concentration factor (SCF) at the weld toe was calculated as:

$$SCF = 1 + \frac{\varepsilon_b}{\varepsilon_a} \qquad (2.2)$$

where

ε_b = strain at the weld toe from bending (extrapolated values from the strain gauges shown in Figure 2.5)

ε_a = strain at section of the weld toe due to the axial force.

There was some difference in SCF at the two sides of the weld in Figure 2.5. The SCF derived at the side where the fracture occurred was used for presentation of the test result. The typical testing frequency used was 15 Hz. Reference is made to Lotsberg et al. (2008a) for fatigue test data from the welded plate specimens. The test data without the pile driving simulation tested under R = 0.1 gave stress ranges corresponding to the mean D-curve. The bending stress is included in the stress range. Thus, the stress can be denoted as a hot spot stress or a structural stress. The nominal stress range used for simulation of pile driving was 209 MPa.

Fatigue Testing of Specimens from the *Edda* Pile
The fatigue testing of specimens from the *Edda* pile was made with specimens of similar shape as those of the plate test specimens. The width of the dog-bone-shaped test specimens at the welded section was 50 mm, but the thickness of these specimens was larger than that from the plate specimens (approximately 39 mm). The same arrangement of strain gauges was used for these test specimens as for the test specimens from the plates. Thus, the bending stress over the thickness due to fabrication tolerances could be measured and extrapolated to the weld toes. The fatigue test data from the *Edda* test specimens for R = 0.1 are shown in Figure 2.7. One of the

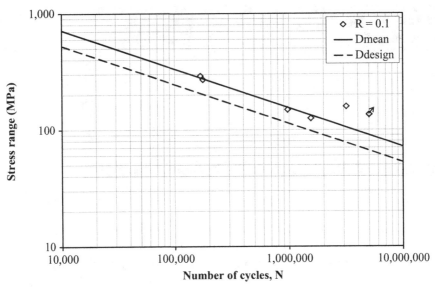

Figure 2.7. Test data from the specimens from the *Edda* pile (Lotsberg et al. 2008a) (reprinted with permission from ASME).

data points was considered as a run-out (where a run-out is understood to be a tested specimen without fatigue failure). Regression analysis of the other five data points gave m = 3.14 for free regression, mean log a = 12.968, and standard deviation = 0.255. Regression analysis with a fixed slope of the S-N curve, m = 3.0, gave mean log a = 12.650 and standard deviation = 0.256. This is not significantly different from that of the D-curve (mean and design) in DNVGL-RP-C203 (2016), where log a for mean D-curve is 12.564, for standard deviation 0.20 based on a log a = 12.164 for the design D-curve. See also Section 4.1.

2.2.2 Measured Residual Stresses

Methodology Used for Measurements of Residual Stresses
Surface residual stresses were measured by the Hole Drilling Strain Gauge Method, according to ASTM E837-01e1 (2001). A hole with diameter 1.8 mm was made in the center of a three-filament rosette strain gauge with a Franklin-Stoller Air-Abrasive Drilling system using aluminum oxide abrasive powder (50 μm). The hole was eroded in steps, down to a total depth of about 2 mm, and the release of strains measured for each step. The residual stresses were calculated from these strain measurements in accordance with ASTM E837-01e1 (2001). The principal stresses and the angle were calculated, as well as stresses transverse and longitudinal to the weld.

Residual Stresses in Welded Plates and Fabricated Test Specimens
Residual stresses were measured after welding from one side when the plates were fixed to strong backs. The plate was then released and the weld on the back side was made. Large residual stresses were measured at the weld after fabrication of the plates, then test specimens were cut from the welded plates. After this the residual stresses in three of the test specimens were measured. The residual stresses were

Table 2.1. *Measured residual stresses in the* Edda *pile*

Location	Position	Outside		Inside	
		σ_T	σ_L	σ_T	σ_L
12 o'clock	1	318	−80	236	−29
	2	195	−153	313	128
3 o'clock	3	240	−166	355	151
	4	304	−110	372	137
6 o'clock	5	265	−46	269	135
	6	239	−65	290	218
9 o'clock	7	199	−85	341	119
	8	214	−108	380	243

T: Transverse refers to the direction normal to the weld.
L: Longitudinal refers to the direction parallel with the weld.

significantly reduced when making the specimens and the residual stresses in all the test specimens were small and differed from those in a welded pile.

Residual Stresses in the *Edda* Pile and Fabricated Test Specimens

The residual stresses at the circumferential weld in the *Edda* pile were measured at four positions around the circumference. The results for measurements on the outside and the inside are shown in Table 2.1. Residual stress measurements were made at positions on both sides of the weld. It is noted that the residual stresses transverse to the weld in the *Edda* pile were close to the material yield stress. Thus, residual stresses at the circumferential weld toes in actual piles welded from both sides are considered significant, both on the inside and the outside of the piles. The tensile stresses from the pile driving are such that shakedown of residual stresses at the hot spot region will not be significant. For further explanation of shakedown reference, see Section 3.3.2. Residual stress measurements in the test specimens made from the *Edda* pile are provided in Table 2.2. As most of the residual stresses at the weld are lost in making the test specimens, it was difficult to use the test specimens for simulation of loading corresponding to pile driving for documentation of fatigue capacity.

2.2.3 Assessment of the Test Data

Early during the project it was realized that deriving fatigue test data from these specimens that were representative for pile driving would be difficult due to the loss

Table 2.2. *Residual stresses measured in a test specimen from the* Edda *pile*

Location	Position	σ_T	σ_L
Cap side	1	109	25
(Outside)	2	22	−4
Root side	3	3	−9
(Inside)	4	−19	−83

of residual stresses at the welds when the small-scale test specimens were fabricated. However, the test data can be used to illustrate the difficulty in deriving reliable design S-N curves from testing of small-scale test specimens. From the definition of design S-N curves, it might seem, from a statistical point of view, straightforward to derive S-N curves for welded structures. However, the difference in residual stresses in small-scale specimens compared with actual structural connections means that this is a complex issue. The derived fatigue test data may also be used as a basis for a fatigue assessment procedure where the residual stresses have been removed by, for example, Post Weld Heat Treatment (PWHT). The fatigue test data were presented as hot spot stress in the S-N diagrams. This means that appropriate stress concentration factors were used, together with a recommended design S-N curve, when welding from outside and inside. The nominal stress S-N curve E is normally used for fatigue design of piles when the circumferential welding is performed from both sides, but this is dependent on the welding position. If the plates are welded in a flat position, the D-curve may be used. The fatigue test data from the tested specimens (plate specimens) may be assessed for equivalent stress range by using the following equation:

$$\Delta\sigma_{eq}(\alpha) = \Delta\sigma_{\text{tension}} + \alpha\Delta\sigma_{\text{compression}} \qquad (2.3)$$

where

$\Delta\sigma_{\text{tension}}$ = part of stress range in tension
$\Delta\sigma_{\text{compression}}$ = part of stress range in compression
α = reduction factor for part of the stress cycles subjected to nominal compression.

By trying different values of α and performing a regression analysis for each value, a best fit was derived for $\alpha = 0.78$. This value of α gave the smallest standard deviation in the regression analyses. Three run-outs in the test data were excluded from the regression analysis. A regression of the test data gives log a = 12.637 for m = 3.00. The standard deviation is equal to 0.161. This gives a design value for log a equaling 12.315. The test data were compared with the D-curve. A log a value for the mean D-curve is equal to 12.564.

When welded details are subjected to tensile loading due to external forces, local yielding at the hot spot area will occur. When the external tensile loading is removed, the residual stress at the hot spot area will be reduced. This is also known as shake-down. The amount of shakedown is a function of material yield strength, local geometry with associated stress concentration factor, and amount of maximum external loading. After shakedown, part of the stress cycle may go into compression more readily, considering the total stress, including that of residual stress. The compressive part of the stress range is less damaging in fatigue than that of the tensile stress. While tensile loading may occur more often in ship structures, piles are not subjected to tensile loading before pile driving, and the stress ranges during pile driving are also mainly compressive. As the tension part of the stress cycles during pile driving is small, the amount of shakedown will be limited for normal pile driving. Hence, the main part of the stress cycle will be on the tensile side, when the residual tensile stresses are also considered and contribute fully to the fatigue damage. Due to the large residual stresses in piles after fabrication, where minor shakedown of residual stresses is expected during normal pile driving conditions, it is recommended that S-N

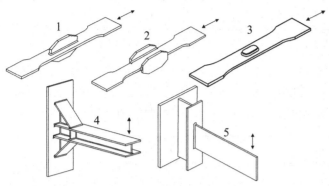

Figure 2.8. Specimens tested by HHI (Kim and Lotsberg 2004) (reprinted with permission from ASME).

curves are used that allow for residual stresses up to yield at the hot spots, including for stress cycles that are mainly compressive, such as during pile driving (see also Buitrago and Wong 2003). However, if the residual stresses in the piles are removed, for example by PWHT, it would be acceptable to use the following equation for calculation of equivalent stress for fatigue design of piles:

$$\Delta\sigma_{eq} = \Delta\sigma_{\text{tension}} + 0.80\Delta\sigma_{\text{compression}} \tag{2.4}$$

where

$\Delta\sigma_{\text{tension}}$ = part of stress range in tension
$\Delta\sigma_{\text{compression}}$ = part of stress range in compression

The stresses in this equation are calculated based on nominal stresses without further consideration of residual stress (here assumed to be close to zero after PWHT).

The test data from the *Edda* specimens for tensile stress ranges are shown in Figure 2.7. It can be seen that the test data correspond well with the mean S-N curve D. This testing also demonstrated that it can be important to instrument the welded connections with strain gauges, as the measured bending stress due to distortion was significant in these specimens.

2.3 Details in Ship Structures

2.3.1 Fatigue Testing

In a Joint Industry Project "FPSO Fatigue Capacity," five different types of welded specimens were fabricated and tested by Hyundai Heavy Industries Co., Ltd. (HHI) in South Korea. These specimens represented typical welded connections in ship-shaped structures (see Figure 2.8). The test specimens were subjected to different load conditions as shown in Table 2.3. The test results for the different load conditions, based on the nominal stress range at the weld toe, were presented in Kim and Lotsberg (2004), and are also shown in Figures 2.9–2.13. Data points with an arrow in the right direction of the diagram show that the fatigue test was stopped at this number of cycles without failure, and the test was denoted as "run-out." Numerical values for the mean S-N curves in Figures 2.9–2.13 are listed in Table 2.4. It was anticipated that it would be possible to correlate the results from the various specimen

Table 2.3. *Load conditions used in testing (Kim and Lotsberg 2004)*

Condition	Models 1, 2, and 3	Models 4 and 5
1	Pre-load = 0 Mean stress = 0 	Pre-load = 0 Mean stress = 0
2	Pre-load = $0.5\,\sigma_0$ Mean stress = 0 	Pre-load = $\pm 0.5\,\sigma_0$ Mean stress = 0
3	Pre-load = $0.85\,\sigma_0$ Mean stress = 0 	Pre-load = $\pm 0.85\,\sigma_0$ Mean stress = 0
4	Pre-load = $0.85\,\sigma_0$ Mean stress = $0.5\,\sigma_0$ 	Pre-load = $\pm 0.85\,\sigma_0$ Mean stress = $0.5\,\sigma_0$
5	Pre-load = $0.85\,\sigma_0$ Mean stress = $0.85\,\sigma_0$ 	Pre-load = $\pm 0.85\,\sigma_0$ Mean stress = $0.85\,\sigma_0$

types using the hot spot stress, and thus justify the use of a single hot spot S-N curve for design. However, the results from two of the models gave high fatigue capacities (Fricke 2001). In order to investigate the reason for this, some additional measurements and tests were performed on identical specimens under the same loading conditions as those used by HHI. The specimens were fabricated by HHI but were tested by DNV in Oslo, Norway, to determine whether the same high fatigue test results would still be obtained from a different laboratory; see Sections 2.3.3 and 2.3.4.

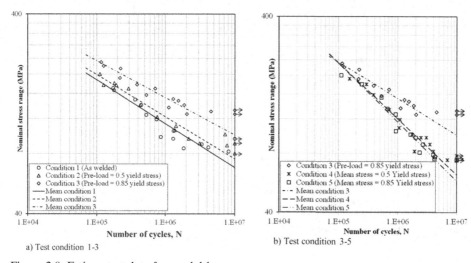

Figure 2.9. Fatigue test data for model 1.

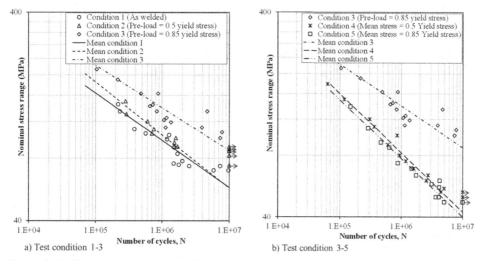

Figure 2.10. Fatigue test data for model 2.

Table 2.4. *Mean S-N data in Figures 2.9–2.13*

Load condition	Mean S-N curves in figures	Specimen type				
		1	2	3	4	5
1	log a	15.365	14.818	18.697	14.567	23.678
	m	4.561	4.433	6.015	3.987	7.202
2	log a	15.895	13.958	18.267	14.411	19.548
	m	4.739	3.950	5.790	3.875	5.555
3	log a	17.752	16.528	21.992	15.290	31.349
	m	5.383	4.910	7.311	4.115	10.194
4	log a	13.299	12.626	14.112	11.969	17.778
	m	3.509	3.460	3.881	2.902	5.103
5	log a	12.869	12.611	12.703	11.344	13.609
	m	3.328	3.502	3.252	2.633	3.563

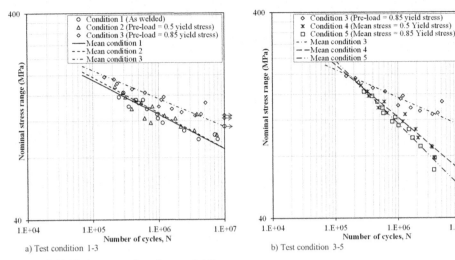

Figure 2.11. Fatigue test data for model 3.

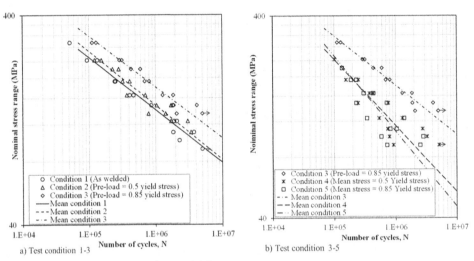

Figure 2.12. Fatigue test data for model 4.

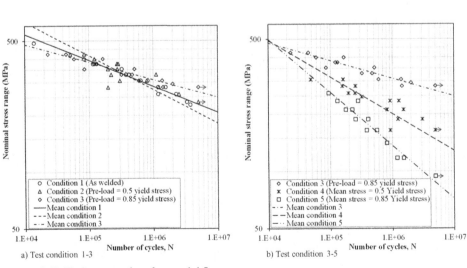

Figure 2.13. Fatigue test data for model 5.

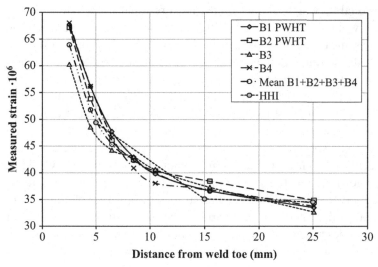

Figure 2.14. Strain data for specimen 5 measured in as-welded and PWHT condition (Kim and Lotsberg 2004) (reprinted with permission from ASME).

From the fatigue test data it is observed that the preloading and the mean stress level in the following cycles may be important for the number of cycles to failure. Thus load conditions 4 and 5 are considered as most relevant to achieve a safe design approach for derivation of a design S-N curve that can be used for different histories and mean stress levels as typically used in fatigue assessment of offshore structures. The other load conditions are relevant for fatigue assessment of ship structures (see, e.g., Lotsberg 2006a). The test results show that the fatigue strength is dependent on residual stresses after fabrication and actual long-term load on the structure during in-service life; see, for example, Ohta et al. (1997) and Section 3.3 on residual stresses where shakedown of these stresses is explained.

2.3.2 Geometry and Fabrication of Specimens

Two specimens of model 4 and four specimens of model 5 as shown in Figure 2.8 were fabricated by HHI in Korea using the same steel and welding conditions as those used to make the original specimens. For reasons that will be discussed later, two of the model 5 specimens were subsequently stress-relieved by PWHT.

2.3.3 Additional Test Results for Model 4

Two of the specimens of model 4 were tested at DNV laboratories in Oslo. The specimens were instrumented with strain gauges prior to fatigue testing and subjected to similar loading conditions as those used by HHI in the original fatigue tests. This involved the application of a static pre-load cycle, followed by constant amplitude cycling loads with a nominal mean stress of about $0.5\sigma_0$, where σ_0 is the characteristic yield strength of the steel (see test condition 4 in Table 2.3). The characteristic yield strength of the AH32 steel was 315 MPa, and the nominal mean stress used for the testing was 154 MPa. This loading sequence was originally chosen to introduce the effect of occasional very high stresses on the local residual stress field, as it was expected that there would be some stress redistribution due to local yielding (Kang and Kim 2003). The same test conditions were used to ensure that the only difference between the new tests and the original ones would be the test laboratory. The original specimens of model 4 failed due to fatigue crack growth from the toe of the weld attaching the inclined tie plate, representing a hopper corner, to the flange of the I-section beam. Therefore, in order to establish the stress distribution in the flange approaching this weld, the specimens were instrumented with strain gauges located on the centerlines of the specimens. A five-element strip gauge was located at the first position, the center of its first element being nominally 2 mm from the weld toe. Single-element gauges were located at the other positions, with their centers nominally 15 mm and 25 mm from the weld toe. Some strange results were obtained from at least one element in the middle in the strip in each of the strip gauges on both of these specimens, as described in Kim and Lotsberg (2004). Similar problems with strips had been observed in the original tests by HHI. Therefore, additional measurements had been taken using single element gauges. The other strain values measured during the tests provided sufficient data to deduce the hot spot stress from these tests. The results obtained from the two fatigue tests of model 4 specimens were plotted in comparison with the original HHI data, both sets of results being expressed in terms of the nominal stress range at the weld toe.

The equation of the mean curve fitted to the original data is: (Stress range)$^{3.211}$ × N = 4.55 × 10^{12} with a standard deviation of log N of 0.250. The two new results from the tests at DNV fell within the natural scatter of the HHI test data.

2.3.4 Additional Test Results for Model 5

Four specimens of model 5 were tested by DNV, two in the as-welded condition and two that had been stress-relieved by PWHT. The latter were designated B1 PWHT and B2 PWHT, while the former were B3 and B4. All the specimens were instrumented with strain gauges. As with the model 4 specimens, the model 5 specimens were subjected to nominally the same loading conditions as those used by HHI in the original fatigue tests. This involved the application of a static pre-load cycle, followed by constant amplitude cycling loads with a nominal mean stress of about $0.85\sigma_0$, where σ_0 is the yield strength of the steel (test condition 5 in Table 2.3). However, one difference that should be noted is that the yield strength assumed by HHI in the original test program was the characteristic value for mild steel (235 MPa), while that assumed for the additional tests was the value measured by HHI for the actual steel (300 MPa). Thus, the mean load levels used in the original and additional tests differed slightly from each other. The corresponding applied stress ratios were R = 0.48 for the additional tests, compared with R = 0.39 in the original tests. According to the HHI test data shown in Figure 2.13b, a higher mean stress reduces fatigue life.

The original model 5 specimens failed as a result of fatigue crack growth from the toe of the weld, around the edge of the weld attaching the bracket to the vertical I-section beam. Therefore, in order to establish the stress distribution in the bracket approaching this weld, the specimens were instrumented with strain gauges located at the weld toe region. All the gauges were fixed onto the edge of the bracket. As with

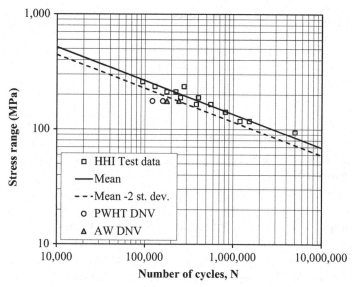

Figure 2.15. Fatigue test results for model 5 specimens (Kim and Lotsberg 2004) (reprinted with permission from ASME).

the model 4 specimens, a five-element strip gauge was located at the first position, the center of its first element being nominally 2.5 mm from the weld toe. Single-element gauges were located at the other positions, with their centers nominally 15 mm and 25 mm from the weld toe.

The strains measured in the four specimens, normalized as before to an applied force of 1 kN, are shown in Figure 2.14. As seen from this figure, there was little scatter in the measured strains. The results also agree well with those obtained in the original tests by HHI, and that are also included in Figure 2.14. The results obtained from the four fatigue tests of model 5 specimens are plotted in comparison with the original HHI data, obtained under test condition 5 in Figure 2.15, with both sets of results expressed in terms of the nominal applied stress range at the weld toe.

The mean curve fitted to the original data, also shown in Figure 2.15, has the equation $(\text{Stress range})^{3.38} \times N = 9.57 \times 10^{12}$ with a standard deviation of log N of 0.095. Due to higher mean stress, the two results for the as-welded specimens from DNV are slightly lower than the HHI test data. The stress-relieved (PWHT) specimens showed somewhat shorter lives than the as-welded specimens. In the first phase of the project, the residual stress near the weld toe was found to be compressive (Kim et al. 2001), and hence beneficial under fatigue loading. However, application of the pre-load cycle, as used in the present tests, relaxed this to a value close to zero. The new test results suggest that any beneficial residual stresses were reduced further by PWHT, resulting in a small reduction in fatigue life.

2.3.5 Effect of Stress Gradient at Weld Toe

For the small-scale test specimens of model 4 that were fatigue tested, the effect of stress gradient on fatigue life may be calculated by crack growth analysis as indicated in Figure 2.16. The reason why fatigue test data from this specimen plotted high in

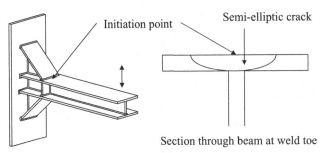

Figure 2.16. Crack growth analysis of hopper corner (Lotsberg and Sigurdsson 2004) (reprinted with permission from ASME).

the diagram of a one hot spot S-N curve is probably due to local plate bending, which involves some redistribution of the stress during crack growth. This demonstrates a limitation in the concept of hot spot stress in relation to one hot spot S-N curve. The reason for this is that the stress to be entered into an S-N curve for fatigue life assessment is that of surface stress, without any information about the stress gradient into the thickness. Thus, details with the same hot spot stress, but with a different stress gradient, will show the same fatigue lives based on S-N data, while the actual lives might be very different. This may be illustrated by crack growth analysis using fracture mechanics. Crack growth of a semi-elliptic crack is considered (see also Figure 16.11).

An example of crack growth for a situation with the same hot spot stress but with different stress gradients through the plate thickness is shown in Figure 2.17. Crack growth using a hot spot stress implies crack growth from a uniform stress field over the plate thickness, as is shown by the crack growth curve ending at a normalized time equal 1.0 in Figure 2.17a. For a situation with a bending stress over the plate thickness that is 80% of the axial stress, the calculated fatigue life changes, as seen from the same figure. This is a similar stress distribution that corresponds to the calculated stress, with a stress concentration factor equal to 1.8. For pure bending, the difference in fatigue life increases further, as shown in this figure. The development of the crack aspect ratio for different stress gradients is shown in Figure 2.17b. The growing crack sizes on the surface for different stress gradients are shown in Figure 2.17c. Fatigue testing for this specimen was performed under load control, and therefore the fracture mechanics analysis was also performed for load-controlled conditions. It is likely that in a real structure, greater redistribution of stresses will occur as the crack grows. Analysis based on displacement control may then be more representative and less conservative, especially for large cracks. Based on fatigue test data under out-of-plane bending loading a reduced effective stress was derived by Kang et al. (2002):

$$\Delta\sigma_{e,\,spot} = \Delta\sigma_{a,\,spot} + 0.592\Delta\sigma_{b,\,spot} \qquad (2.5)$$

where the bending stress at the hot spot is reduced by a factor 0.592. The proposed equation gives similar results as derived by fracture mechanics.

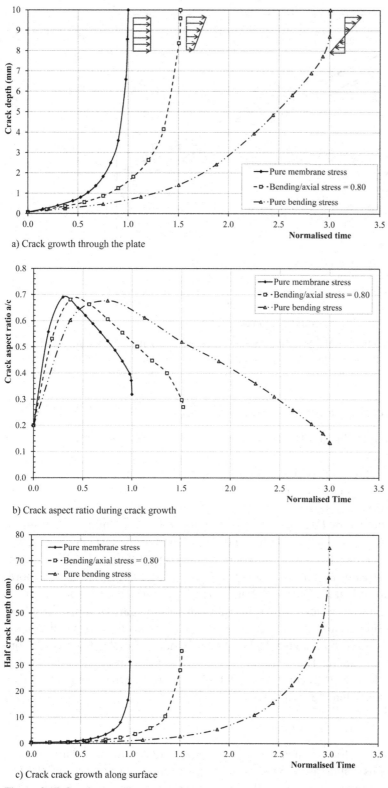

a) Crack growth through the plate

b) Crack aspect ratio during crack growth

c) Crack crack growth along surface

Figure 2.17. Crack growth curves for same hot spot stress with different stress gradients through the plate thickness (Lotsberg and Sigurdsson 2004) (reprinted with permission from ASME).

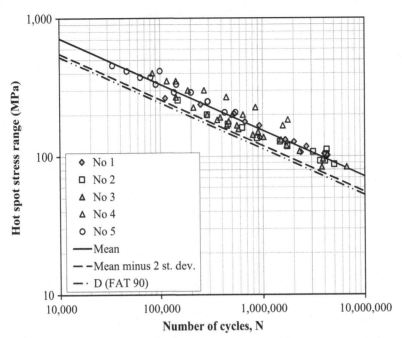

Figure 2.18. Fatigue test results from specimens 1–5 plotted into an S-N diagram, where data for no. 4 are adjusted due to bending over plate thickness and data for no. 5 are adjusted for residual stress (Lotsberg and Sigurdsson 2004) (reprinted with permission from ASME).

2.3.6 Hot Spot Stress for the Tested Specimens

Hot Spot Stress from Finite Element Analysis and Measurements from HHI Specimens

Based on the calculated stress distribution through the plate thickness of the hopper corner detail, the fatigue test data are reduced by a factor of 1.61 in comparison with a hot spot S-N curve. The resulting data points are shown in Figure 2.18. The fatigue test results determined by HHI for specimen number 5 were explained by measured compressive residual stresses at the hot spot area. The fatigue tests performed on stress-relieved specimens showed that stress relief reduced fatigue life by a factor of 2.8. Residual stresses in tension, as may be possible after construction of a more complex structure, might reduce the fatigue capacity further. The hot spot S-N curve should be applicable to welded structures that can be in a state of tensile residual stress at the considered hot spots. Therefore, for specimen no. 5 the original fatigue test data were reduced by a factor of 2.8 on the number of cycles for comparison with a hot spot S-N curve. The resulting S-N curve and the fatigue test data for a high mean stress loading (test condition 5 in Table 2.3) are shown in Figure 2.18. The S-N curve is presented in the format of

$$\log N = \log a - m \log S \tag{2.6}$$

where $\log a = 12.562$ for the mean curve and $\log a = 12.284$ for the characteristic curve (mean minus 2 standard deviations in a logarithmic scale) and $m = 3.0$. This corresponds to a stress range of 98.70 MPa at 2 Mill cycles, which is above the FAT 90 design curve by 9.6%. The tested specimens were fabricated from 10 mm thick plates,

Table 2.5. *S-N curves (Lotsberg and Sigurdsson 2004)*

HHI specimen no	Type of stress data	Loga (mean values)	Standard deviation	Figure
1	Nominal stress	12.200	0.111	
2	Nominal stress	11.686	0.125	
3	Nominal stress	12.186	0.072	
4	Nominal stress as tested	12.123	0.232	
4	Nominal stress – corrected for bending by factor 1.61 on life	11.916	0.232	
5	Nominal stress as tested	12.326	0.104	
5	Nominal stress – corrected for residual stress by factor 2.8 on life	11.879	0.104	
1-5	Hot spot stress S-N curve derived from calculated and measured stress	12.565	0.167	Figure 2.18
1-5	Hot spot stress S-N curve derived from fatigue test data	12.562	0.139	Figure 2.20

which is less thick than normally used as reference thickness in the S-N data (see also Chapter 4). Thus, use of the FAT 90 curve as a hot spot S-N curve was recommended and use of the reference thickness (25 mm) as recommended by IIW (Hobbacher 2009). Increasing the thickness from 10 mm to 25 mm reduces the S-N curve to that of FAT 90 for a rather low thickness exponent, equal to 0.1. It should be noted that this work did not provide sufficient information on the recommended thickness exponent to be used for the hot spot S-N curve. The FAT 90 is considered to be a reasonable lower bound curve for some of the low hopper corner data points in Figure 2.18. Thus, the fatigue test data obtained by HHI supports the FAT 90 curve as a design hot spot S-N curve. The scatter in the test data, including all five data sets, was low. A standard deviation of 0.167 was calculated, indicating a low scatter even when compared with one of the data sets such as the hopper corner where the standard deviation is 0.232. Reference is also made to Table 2.5.

Hot Spot Stress S-N Curve Based on Fatigue Tests

An alternative method of assessing target hot spot stress values is to base the evaluation on the fatigue test data for all five specimens. The mean S-N curves for each of the details were known in terms of nominal stress S-N curves (see Table 2.5). From these, the relative distribution of stress concentration factors (SCFs) for the different details was known. Thus, the main question is related to how to derive an absolute level of a hot spot S-N curve in an S-N diagram.

The different SCFs can be calculated for each data set as

$$SCF = \frac{\sigma_{hot\,spot}}{\sigma_{nominal}} \tag{2.7}$$

Detail 1 is well known and frequently fatigue tested, such that this detail is considered to be one of the most reliably categorized details in design standards. The SCF for specimen no. 1 is found to be 1.32 from finite element analysis. The graph shown in Figure 2.19 is derived by assessing the inherent SCF in S-N curves for this detail for

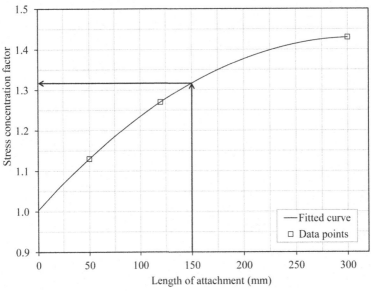

Figure 2.19. SCF as function of attachment length in a fatigue design standard (Lotsberg and Sigurdsson 2004) (reprinted with permission from ASME).

different attachment lengths in DNVGL-RP-C203 (2016). Curve fitting results in a SCF of 1.317 for an attachment length in the test specimen of 150 mm. A similar approach can be followed using IIW (Hobbacher 1996) and Eurocode 3 (EN 1993-1-9 2009). In IIW, FAT 71 is applicable for attachment lengths between 50 mm and 150 mm. In Eurocode 3, FAT 71 applies for attachment lengths between 50 and 100 mm. The results following DNVGL-RP-C203 (2016) lie between these, with FAT 71 between 50 and 120 mm. Thus a SCF for HHI detail no. 1 being equal to 1.32 is assessed to be a good target value. A hot spot stress S-N curve is then derived as:

$$\log a_{hot\,spot} = \log a_{nominal\,1} + 3.0 \log SCF_1$$
$$= 12.200 + 3.00 \log 1.32 = 12.561 \tag{2.8}$$

This is a similar S-N curve to that which was first assessed from the finite element analyses. The resulting SCFs are listed in Table 2.6. For specimen no. 2, a somewhat higher SCF value is obtained from the S-N curve than from finite element analysis. Review of the finite element analysis from the first part of the FPSO Fatigue Capacity

Table 2.6. *Target SCFs (Lotsberg and Sigurdsson 2004)*

Specimen no.	SCF from finite element analyses by Fricke (2001)	SCF from assessment of all finite element analyses supported with measurements	SCF resulting from the HHI fatigue test data
1	1.32	1.32	1.32
2	1.85	1.85	1.96
3	1.22	1.22	1.33
4	1.96	1.82	1.64
5	1.77	1.77	1.69

Joint Industry Project shows that many of the results from the three-dimensional analyses were in the region between 1.92 and 2.03. Therefore, it was reasonable to assume that the target value is around 1.96, and this value was therefore selected as target value for specimen no. 2. For specimen no. 3, a somewhat higher SCF value was also obtained from the S-N curve approach than derived from finite element analysis. It is difficult to argue for such a high value from the three-dimensional analyses. A higher value is obtained from the strain measurements. It was decided to use the target value derived from the fatigue test data. Thus, a target SCF of 1.33 was used in the following. For specimen no. 4, a lower SCF resulted from the S-N data than from the finite element analyses and that was also supported by measurements made at DNV. This may be due to differences in failure criteria for the specimens, where crack growth through the thickness was used as the failure criterion for this specimen. For the HHI specimens 1–3, failure was defined as fracture of the specimens, while the failure criterion for HHI specimen no. 5 was crack growth equaling 20 mm, as a mean value along the two sides. For specimen no. 4 in particular, significant crack growth occurred before the crack went through the plate thickness. An earlier stop in the fatigue testing would result in a lower nominal S-N curve and hence a larger SCF. Thus, it is possible that the target value should be closer to that analyzed and measured, of 1.82. However, to be consistent, a target spot value of 1.64, as derived from the fatigue test data, is used for specimen no. 4. For specimen no. 5, a target value of 1.69 resulted from the S-N analysis data. This is somewhat lower than from that considered to be lower bound from the finite element analysis. Again to be consistent in derivation of target spot values, the value derived from the fatigue test data of 1.69 was used as target hot spot stress value for specimen no. 5.

An alternative approach using the least squares method to assess target SCFs is also used. The quadratic error of the difference between SCFs from finite element analysis and that from S-N data is minimized according to the following expression:

$$\delta = \frac{\partial}{\partial x} \sum_{i=1}^{5} (SCF_{FEi} - xk_i)^2 \tag{2.9}$$

where

x = SCF for detail 1 as $k_1 = 1.0$.
k_i = SCF for considered detail i relative to that for detail no. 1.

SCF_{FEi} = SCF from finite element analysis for detail i as presented in Table 2.6. By putting $\delta = 0$, a SCF for detail no. 1 equal to 1.327 is calculated. This is approximately the same value as derived in the first part of this section based on assessment of design standards. This indicates that the derived SCFs are sound. Based on the target hot spot stresses for the different specimens, a hot spot S-N curve is derived. This curve is shown in Figure 2.20 together with the test data. From Table 2.6 it can be seen that the difference between the target SCFs derived from analyses and measurements correspond well with the SCFs derived from the fatigue test data from the HHI specimens. The difference is in the range of 0–10%, with the largest difference derived for the hopper corner detail as discussed earlier in this section for specimen no. 4.

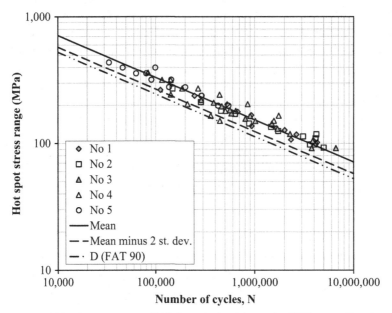

Figure 2.20. Hot spot stress S-N data derived from the HHI tests (Lotsberg and Sigurdsson 2004) (reprinted with permission from ASME).

2.4 Side Longitudinals in Ships

Four different types of full-scale connections between side longitudinals and transverse frames were tested in the FPSO Fatigue Capacity Joint Industry Project; see Figure 2.21 (Lotsberg and Landet 2005). The first two specimens were made with double-sided brackets. The detailed geometry of a connection is shown in Figure 2.22. One specimen was subjected to net pressure loading from the inside, and the other was tested with loading from the outside. Specimen no. 3 was made with similar geometry to specimen no. 1, except for the bracket being on only one side of the transverse frame. Specimen no. 4 was made without any direct welding between the

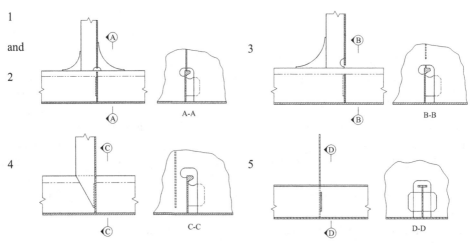

Figure 2.21. Fatigue-tested full-scale connections (Lotsberg and Landet 2005) (reprinted with permission from Elsevier).

Specimens 1 and 2

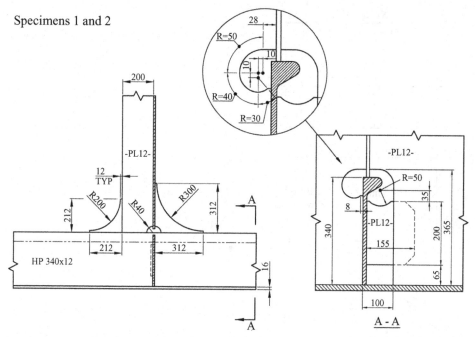

Figure 2.22. Detailed geometry of specimen no. 1 (Lotsberg and Landet 2005) (reprinted with permission from Elsevier).

buckling stiffener in the transverse frame and the longitudinal. The detailed geometry of this specimen is shown in Figure 2.23. Specimen no. 5 was made with symmetrical lug plates (or collar plates). This type of connection is not normally used at areas with large transverse loads.

Specimen 4

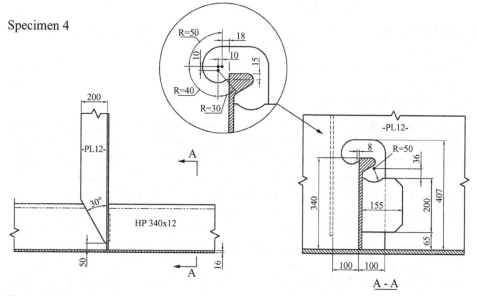

Figure 2.23. Detailed geometry of specimen no. 4 (Lotsberg and Landet 2005) (reprinted with permission from Elsevier).

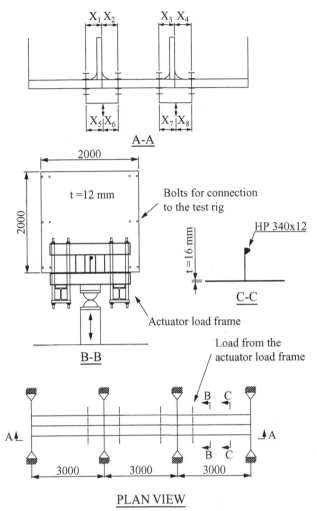

Figure 2.24. Arrangement for fatigue testing of full-scale specimens (Lotsberg and Landet 2005) (reprinted with permission from Elsevier).

2.4.1 Test Arrangement

The loading on the test specimens simulated sideways pressure on the hull in the waterline region. The specimens were subjected to four-point loading. The load application positions were selected to achieve similar shear forces and bending moments as those experienced from a uniform side pressure in an FPSO. This was achieved by calculating appropriate X_i (i = 1-8) values in Figure 2.24, which shows the general test setup. The test frame was constructed for load application from both sides, in order to simulate both external net pressure and internal net pressure. The loading was applied at the hull side plate, thus leaving the HP bulb section free to move sideways due to its asymmetric shape; see also Section 8.12.2. The applied load was recorded from load cells during testing, one at each of the two actuators. Two fatigue load conditions were considered in the project:

a: transverse frame

b: buckling stiffener

c: longitudinal

d: side plate

e: loading frame

Figure 2.25. Test specimen no. 1 installed in test rig.

- "Ballast": ballast condition, as well as net pressure on the ship's side (Specimens 1, 3, 4, 5).
- "Loaded": loaded condition, as well as net pressure on the ship's side (Specimen 2).

The dynamic load amplitude was chosen for both loading conditions in order to obtain an expected fatigue life to crack initiation in the order of 10^5–10^6 load cycles, based on mean S-N curves for similar details from design standards. The maximum load amplitudes from a one-year storm were used for fatigue testing. Fatigue testing was performed under constant amplitude loading. Specimen 1 was fatigue tested at a mean stress level representing the "Ballast" condition (water-filled ballast tanks). Figure 2.25 is a photograph of specimen 1 installed in the test rig; the specimen was oriented at 90° compared with a real FPSO in order to connect the actuators to the strong floor in the laboratory.

2.4.2 Instrumentation

The selection of hot spots for instrumentation with strain gauges was based on experience from several finite element analyses of such details. A number of strain gauges were installed for measurements of strain for two purposes:

- to enable control of the applied load level, distribution, and alignment;
- for measurement of hot spot stresses for fatigue assessment.

In general, each hot spot was instrumented with a minimum of three 6 mm strain gauges positioned 6 mm, 18 mm, and 28 mm from the weld toe. At locations where high stress gradients were expected, strain gauge strips containing five gauges 2 mm apart were installed, with the first gauge 2 mm from the toe. Thus, sufficient information was obtained to plot the stress distribution approaching the weld toe, enabling the hot spot stress to be evaluated.

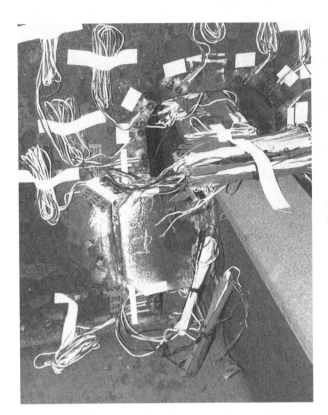

Figure 2.26. Strain gauges in Specimen 5.

Strains were measured in directions normal and parallel to the weld toes (i.e., biaxial strain gauges were used). Strain gauge rosettes were fixed at locations where the principal directions were not well defined from simple assessment of mechanics, and thus a significant number of strain gauges were required on each specimen. Some of the strain gauges installed on Specimen 5 are shown in Figure 2.26.

2.4.3 Testing

Each tested specimen was subjected to a few cycles corresponding to a one-year storm (scatter diagram for the North Atlantic) in order to establish a typical in-service residual stress condition at the hot spots before the main fatigue testing was started. Fatigue testing was performed under the same constant amplitude loading at a frequency of approximately 2 Hz. The number of cycles to crack initiation, defined as a visible crack (visualized with the aid of soap solution and a magnifying glass), was recorded. Thereafter, the crack size was marked on the specimen at regular intervals, and the corresponding number of cycles was noted. Crack growth in the specimens is shown in Figure 2.27. The number of cycles corresponds to the hot spot stress of Specimen 1. This is performed by calculating a modified number of cycles at a given crack size for Specimen i, which is related to the stress range in Specimen 1 as:

$$N_{Spec\,i} = N_{Test\,spec\,i} \left(\frac{\Delta\sigma_{Hot\,spot\,spec\,i}}{\Delta\sigma_{Hot\,spot\,spec\,1}} \right)^{3.0} \qquad (2.10)$$

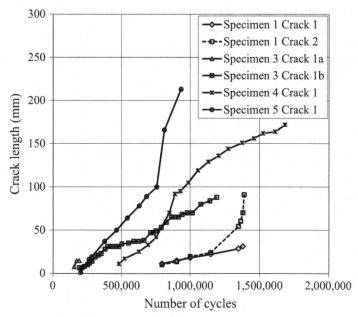

Figure 2.27. Fatigue crack growth of first cracks in the specimens (Lotsberg and Landet 2005) (reprinted with permission from Elsevier).

Crack number 1 in Specimen 1 initiated at the weld toe in the scallop, and stopped after some growth. When the specimen was opened after the test, it could be seen that crack number 1 had merged with crack number 2, which had initiated at the weld toe close to the scallop and had propagated through the bracket; see also Lotsberg and Landet (2005). The first crack in Specimen 3 initiated at the weld toe. After some growth it stopped, and another crack grew from the toe of the bracket from the root of the fillet weld into the bulb section, as shown in Figure 1.9, with the crack length measured over the bulb section on its surface. The first fatigue crack in Specimen 4 occurred at the weld toe at the corner of the lug plate, grew to the edge, and then became rather long as it extended into the plate of the transverse frame. The first crack in Specimen 5 initiated at the weld toe in the lug plate weld to the web of the longitudinal. A fatigue crack also initiated at the weld root in the same region. These cracks merged and propagated in the fillet weld. It can be seen in Figure 2.27 that the steepest gradients with crack growth are, as might be expected, in the connections with the least possibility for redistribution of stress during crack growth.

2.4.4 Assessment of Fatigue Test Data

Hot spot stress refers to the value at the weld toe, usually based on extrapolation from the stress distribution at the weld. It includes the stress concentration factor due to gross geometry, but it does not include that due to weld geometry. Different methods and procedures based on stress measurements can be applied to determine the hot spot stresses. The hot spot stresses in the tests described here were derived from:

- measured stress components normal to the weld toe;

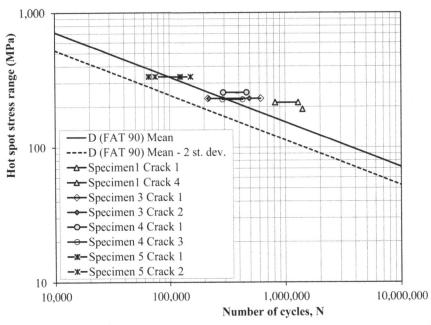

Figure 2.28. Fatigue test data compared with hot spot S-N curves (Lotsberg and Landet 2005) (reprinted with permission from Elsevier).

- linear extrapolation of stresses at distances t/2 and 3t/2 from the weld toe (where t is the plate thickness);
- parabolic curve fitting applied for the three measuring points (i.e., located close to 6 mm, 18 mm, and 28 mm from the toe), to determine the stress levels at t/2 and 3t/2 (when strains were not measured at these locations).

The fatigue cracks in Specimen 5 initiated on the opposite side of the lug plates to the strain gauges. Therefore, the hot spot stress for this case was derived from finite element analysis (Rucho et al. 2001). S-N data from the full-scale fatigue tests are shown in Figure 2.28, together with the mean and design S-N curves for IIW FAT 90. This was proposed as the possible basis for the hot spot design curve (Maddox 2001). This recommendation was also based on the evaluation of other fatigue test data from small-scale specimens obtained in the JIP and relevant published data. This S-N curve is also supported by fatigue test data of small-scale test specimens in this project, with the test data and S-N curves presented in Section 2.3. The first data point in Figure 2.28 corresponds to the number of stress cycles at a crack length of 12 mm, and the last point corresponds to a crack length of 50 mm. Crack no. 4 in Specimen 1 was observed when its length was 26 mm. Shortly afterward, another crack of similar size occurred in the same region and was added to the length of the first one. Thus, only a single point is plotted for this crack, corresponding to a crack length of 50 mm. As seen from Figure 2.28, all the results from the full-scale tests lie above the FAT 90 design curve; the first crack length corresponds to FAT 91 and the final crack length corresponds to FAT 120. The test results for crack lengths of 12 mm are shown together with the recommended hot spot design S-N curve in Figure 2.28. However, as previously explained, the length of crack 4 in Specimen 1 is 50 mm. The FAT number denotes stress to failure at 2 million cycles based on a design S-N curve

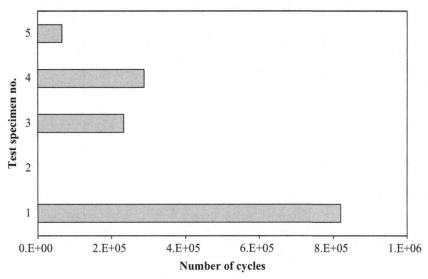

Figure 2.29. Tested number of cycles to fatigue "failure" for different types of specimens (Lotsberg and Landet 2005) (reprinted with permission from Elsevier).

(or characteristic S-N curve). A characteristic S-N curve is established as the mean minus two standard deviations in log N scale. This terminology is used both by IIW (Hobbacher 2009) and Eurocode 3 (EN 1993-1-9 2009).

The full-scale tests are considered to provide verification of the recommended design S-N curve. In this respect it should be noted that the five full-scale test specimens showed fatigue crack growth at very different details, and with crack growth at diverse locations:

- Specimens 1 and 2: crack initiation at weld toes in brackets, with crack growth in the brackets.
- Specimen 3: crack initiation at weld toes of stiffener, with crack growth through the longitudinal bulb section.
- Specimen 4: crack initiation at weld toes of lug plates on the transverse frame, with crack growth into the transverse plate.
- Specimen 5: crack initiation at a weld toe on the edge of a lug plate, with crack growth along the weld toe into the fillet welds.

Despite the different types of connections, the different crack initiation sites, and the different crack growth areas, there was little scatter in the resulting hot spot stress fatigue data, resulting in standard deviations of log N of only 0.15–0.19. The lowest value corresponds to crack size of 50 mm and the largest value to 12 mm. This difference is due to the late detection of one crack, as described earlier in this section for crack no. 4 in Specimen 1. The scatter is not any larger than that which would normally be observed for one specific detail when expressed in terms of nominal stresses. Thus, a single hot spot S-N curve was supported by these full-scale tests. The fatigue endurances to a 12 mm crack are shown in Figure 2.29 for the same ballast loading, and it can be seen that the double bracket design geometry (Specimen 1) is favorable in terms of fatigue life. The information presented in Figure 2.29 may be useful in a variety of ways. For example, should calculated fatigue life not meet the target, then

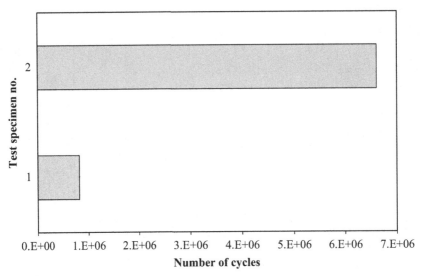

Figure 2.30. Tested number of cycles to fatigue "failure" under compressive versus tensile stresses at the hot spot region for same type of detail (Lotsberg and Landet 2005) (reprinted with permission from Elsevier).

it provides guidance on possible design modifications in order to achieve that target. This comparison is based on sideways pressure; for longitudinal loading, the conclusions are likely to be different. Then the hot spot stresses in Specimens 4 and 5 are likely to be lower than in the other connections with bracket attachments, as shorter attachments attract less hot spot stress. The double-sided bracket geometry (Specimens 1 and 2) was tested in both ballast and loaded conditions. The results from the loaded condition, giving compression at the hot spot, are well above the S-N curve for tensile load cycling. The capacities are shown in Figure 2.30. It should be noted that, in terms of fatigue damage, the ballast condition is significantly more severe than the loaded condition. This may be explained by the different mean stresses at the hot spots for the two loading conditions:

- In Specimen 1, the ballast condition (water-filled ballast tanks) leads to tensile stresses at the hot spots (neglecting the residual stress). ($R = \sigma_{min}/\sigma_{max} = 0.16$.)
- In Specimen 2, the loaded condition (empty ballast tanks) leads to compressive stresses at the hot spots (neglecting the residual stress). ($R = -\infty$.)

The test results from Specimens 1 and 2 show that there is a relatively large effect from mean stress in these tests.

2.4.5 Comparison of Calculated Stress by Finite Element Analysis and Measured Data

In the design of ship-shaped structures it is normal to use shell elements in the finite element analysis. The test specimens in Figure 2.21 were also analyzed by shell finite elements. In general, the agreement between the analyses using shell elements and the measured data was good for all specimens except Specimen 1, in which fatigue cracks initiated at the cope hole in the bracket. The analysis results are shown in Figure 2.31 and are presented along the line normal to the weld toe at the cope hole.

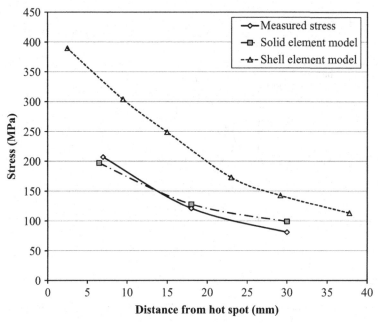

Figure 2.31. Stress from analysis for Specimen 1 compared with measured results (Lotsberg and Landet 2005) (reprinted with permission from Elsevier).

This connection involves eccentric load transfer from the web of the longitudinal to the stiffener in the transverse frame and torsional loading of a bulb section. It can be seen that for this specimen the results predicted by the analysis model with shell elements were relatively dissimilar from the actual measurements. Therefore, an additional analysis was performed using 20-node solid elements. The results from this additional analysis agreed well with the measured stress. This is an example of the difficulty that can occur in providing a proper representation of an actual structure by simplified modeling using shell elements, such as correct moment of inertia of a bulb section together with eccentricity of the connection between brackets and the web of the longitudinal, torsional stiffness of the bulb, and correct local geometry properties. For a more detailed presentation of these analyses, see Storsul et al. (2004b).

Specimen 4 was also analyzed both by shell elements and by 20-node solid elements. The results from these analyses are shown in Figure 2.32 together with measured data. The correspondence between the derived hot spot stresses is generally good.

2.5 Fillet Welded Connections

2.5.1 Fillet Welds Subjected to Axial Load

Test Matrix

A test matrix for fillet welded specimens was developed in the FPSO Fatigue Capacity JIP as shown in Table 2.7. Test numbers 1–8 involved fillet welded cruciform joints of the type illustrated in Figure 2.33. These specimens were used to investigate the effects on fatigue of main plate thickness t (Specimens 1–6) and weld throat

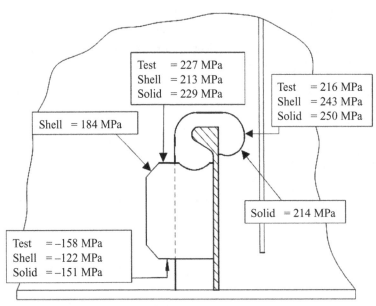

Figure 2.32. Stress from finite element analysis compared with measured stress for Specimen 4.

thickness (Specimens 7 and 8 relative to the base case of Specimens 1 and 2). The remaining tests involved fillet welded tube to plate joints, as shown in Figure 2.34. The purpose of these tests was to investigate the effect of combined normal and shear stresses, with the aim of establishing a design criterion for assessing fillet welds under combined loading. These specimens were subjected to axial loading and/or torsion.

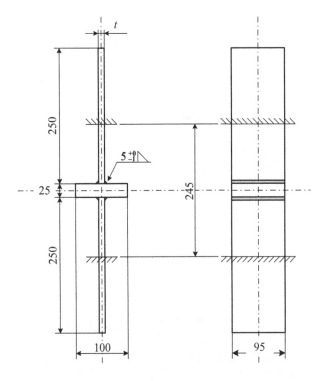

Figure 2.33. Geometry of test Specimens 3 and 4 (Lotsberg 2006b) (reprinted with permission from ASME).

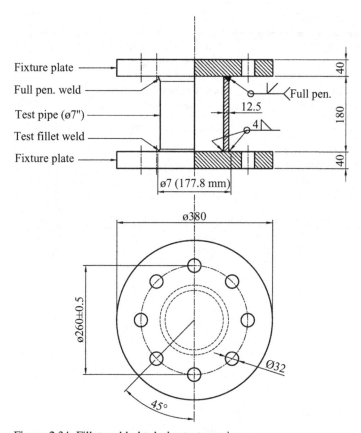

Figure 2.34. Fillet welded tubular test specimen.

Specimens 9 and 10 were included for reference purpose and were subjected to pure axial loading. Specimens 11–20 were used to investigate the effect of mean load level on fatigue strength for fillet welds that were subjected to pure longitudinal shear and for reference purposes for this loading. Specimens 21–33 were included to establish a data basis for assessment of a design criterion for combined stresses. The nominal throat thicknesses used in the fabrication of the specimens are listed in Table 2.7. A fabrication tolerance of +0, −1 was specified on the fabrication drawings for the specimens. The fabrication was specified with this tolerance of +0 as it was necessary to limit the throat thickness in order to ensure crack growth from the weld root. The actual throat thicknesses were measured after testing and used for assessment of the test data.

Test Arrangement
The geometry of test specimens 1–8 is shown in Figure 2.33. The width of the test specimens was 95 mm and the material was St 52.3 N. The specimens were welded by flux cored arc welding (FCAW). The main plate thickness, t, were 12 mm, 25 mm, and 40 mm. The specimens were instrumented with electrical resistance strain gauges in order to determine the extent of any secondary bending due to misalignment of the joints. One strain gauge was attached on the specimen centerline 20 mm from each weld toe such that the bending stress in the specimens could be derived.

Table 2.7. *Test matrix for fillet welded specimens*

Test no.	Axial load in 100 ton actuator (giving axial stress in tube) (kN)		Axial load in 25 ton actuator (giving torsion and hence parallel shear stress in weld) (kN)		Endurance (cycles)
	R	ΔP	R	ΔP	
9	0.5	373.47			2,020,167
10	0.5	450			282,702
11			−1	160	86,856
12			−1	160	90,500
13			−1	120	260,128
14			0.07	140	127,210
15			0.07	140	170,000
16			0.07	140	127,879
17			0.4	120	586,000
18			0.4	120	624,000
19			0.4	120	270,000
20			0.4	120	404,000
21	0.4	360	−1	100	61,730
22	0.4	360	−1	100	91,370
23	0.4	360	−1	100	90,022
24	−1	450	−1	100	126,900
25	0.4	360	0.4	100	95,000
26	0.4	360	0.4	100	64,771
27	0.4	360	0.4	100	60,617
28	0.4	360	0.4	100	165,500
29	0.4	360	0.4	100	164,500
30	0.4	360	0.4	100	120,000
31	0.4	436	0.4	88	47,562
32	0.4	304	0.4	109	71,706
33	0.4	304	0.4	66	754,564

Fatigue Test Data

The failure criterion was a specimen with cracking through two of the fillet welds. Thus, it is not known which of the two fillet welds failed first. All the throat thicknesses were measured. The throat that experienced the highest stress range, including misalignment-induced bending, was used to calculate the stress in the weld that probably failed first. The values for nominal stress in the weld, as listed in Table 2.8, were based on that weld throat thickness. The test results are also shown in Table 2.8.

2.5.2 Fillet Welded Tubular Members Subjected to Combined Axial and Shear Load

Test Arrangement

The test specimen used to investigate the fatigue behavior of fillet welds under combined loading is shown in Figure 2.34. The diameter of the tube was 177.8 mm and the material was St 52.3 N. The fillet welds were made by FCAW in one pass. The test rig used is shown in Figure 2.35; the test frame was braced to the strong floor by an

Table 2.8. *Fatigue test results obtained from fillet welded cruciform joints (Lotsberg 2006b)*

Test no.	Throat thickness in weld of first failure (mm)	t (mm)	Force range (kN)	Nominal "engineering shear stress" in weld (MPa)[1]	Actual stress in weld throat corrected for bending, K_{tw} × nominal stress (MPa)	Endurance (cycles)
1	4.7	12.5	68.93	77.19	77.54	371,684
2	3.8	12.5	109.00	150.97	152.46	90,351
3	4.7	25.0	65.31	73.14	76.63	1,004,392
4	4.1	25.0	64.21	82.43	84.00	524,112
5	4.8	40.0	78.67	86.26	109.50	133,797
6	4.9	40.0	46.79	50.26	53.54	1,673,382
7	8.3	25.0	124.72	79.09	80.72	659,882
8	9.0	25.0	134.70	78.77	79.83	377,758

[1] For definition of engineering shear stress, see Section 3.2.1.

additional member, not shown in the sketch of the rig, to avoid bending moments in the rig from the horizontal actuator. In order to limit friction, a layer of Teflon was used at the support for the torsional loading. Four strain gauge rosettes were installed at 90° intervals around the circumference of the bottom tube for control of loading. Each strain gauge was placed at half a height away from the weld toe in order that the local stress concentration effect from the weld would not be registered. The wall thickness of the permanently installed bottom tubular section was 35 mm. As the torsion load introduced some bending of the test specimen, some additional measurements on two of the test specimens were performed. Additional strain gauges were therefore installed 50 mm above the fillet weld, at the same positions around the circumference as those on the bottom tube.

Test Results
The test loads and the numbers of cycles to failure for specimens 9–33 are presented in Table 2.7. In all cases, fatigue failure occurred in the fillet weld throat, from cracks

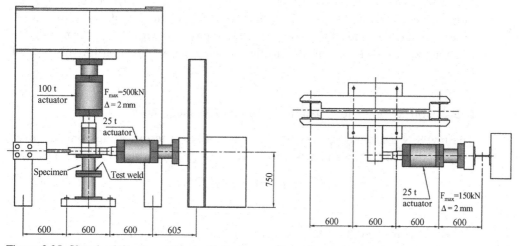

Figure 2.35. Sketch of the rig used to test tubular specimens.

Figure 2.36. Photos of some tested specimens (Lotsberg 2006b) (reprinted with permission from ASME).

that initiated at the weld root. Due to the geometry, with welding to the flange, the fillet weld on the outside of the tubular specimen was the most highly stressed and therefore was the first fillet weld to fail. Some of the cracks were rather long when they were detected, as they were growing fast when they reached the weld surface. Photos of some of the specimens are shown in Figure 2.36. It can be seen that the fatigue crack initiated at the weld root, and in Specimen 17 that was subjected to pure torsional load, the crack grew away from the weld toe notch into the main plate at an angle of 45°. This is normal to the principal stress direction for this type of loading.

2.5.3 Correction of Test Data for Measured Misalignment

The raw S-N test data are presented in Figure 2.37 in terms of the nominal stress range in the weld. The bending strains in the test specimens were calculated from:

$$\varepsilon_b = |(\varepsilon_1 - \varepsilon_3)|/2 \qquad (2.11)$$

where:

ε_1 is the measured strain on one side of the specimen and
ε_3 is the measured strain on other side of the specimen.

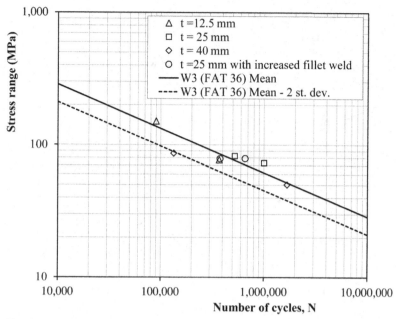

Figure 2.37. Test data before correction for bending stress (Lotsberg 2006b) (reprinted with permission from ASME).

Then the bending stress over the thickness was calculated from:

$$\sigma_b = E\varepsilon_b \tag{2.12}$$

where:

E is the Young's modulus $= 2.06 \cdot 10^5$ MPa.

The total stress was calculated as:

$$\sigma_{tot} = \sigma_a + \sigma_b = \sigma_a \left(1 + \frac{\sigma_b}{\sigma_a}\right) \tag{2.13}$$

where:

σ_a = axial stress
σ_b = bending stress.

Assuming that the bending moment in the plate section at the region of strain gauges resulted from some eccentricity, the total stress can also be calculated as (see also Sections 5.1 and 5.2):

$$\sigma_{tot} = SCF\sigma_a \tag{2.14}$$

where the stress concentration factor SCF is defined as:

$$SCF = 1 + \frac{3\delta}{t} \tag{2.15}$$

where:

t is the plate thickness at strain gauges, and δ is eccentricity.

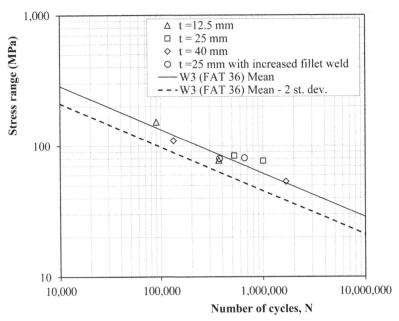

Figure 2.38. Test data after correction for bending stress (Lotsberg 2006b) (reprinted with permission from ASME).

From equations (2.13), (2.14), and (2.15) the equivalent eccentricity is obtained as:

$$\delta = \frac{t}{3} \frac{\sigma_b}{\sigma_a} \tag{2.16}$$

The stress concentration factor for the fillet weld was obtained from Andrews (1996) as:

$$K_{tw} = 1 + \frac{\delta}{t+h} \tag{2.17}$$

where:

    h is the leg length of the fillet weld.

The following stress concentration factor is derived from equations (2.16) and (2.17):

$$K_{tw} = 1 + \frac{t}{3(t+h)} \frac{\sigma_b}{\sigma_a} \tag{2.18}$$

Based on the measured strains, a stress concentration factor for each fillet weld was calculated from equation (2.18). Finally, the actual stress in the weld throat was calculated by multiplying the nominal stress by K_{tw}. The raw fatigue test data are presented in Figure 2.37 and the S-N data corrected for bending stress are shown in Figure 2.38. The data points are distinguished in terms of the main plate thickness, and are compared with the FAT 36 design curve in IIW (Hobbacher 2009), which is the same as W3 curve in DNVGL-RP-C203 (2016).

2.5.4 Assessment of Test Data

As discussed in Maddox (2006), the fatigue lives of fillet welds that fail in the throat depend on the main plate thickness, as well as the weld throat size. Consequently, the design S-N curve for assessing potential weld throat failure applies directly only for plate thicknesses up to some reference value, t_{ref}, above which stresses obtained from the design curve should be reduced by applying a thickness correction. For this, the following has been proposed:

$$S_{mod} = S\left(\frac{t}{t_{ref}}\right)^k \qquad (2.19)$$

The value of the exponent k proposed by Maddox (2006) has been debated. The present results were used to investigate the need for this thickness correction and the corresponding value of k. The procedure was used to calculate the corrected stress range, S_{mod}, for each test specimen, assuming different values of the exponent k, and the fatigue test results were re-plotted in terms of S_{mod}. An S-N curve was then fitted to the results and the value of k that gave the lowest standard deviation of log N was judged to be the most suitable choice. A plate thickness t = 25 mm was used as a reference value. Only the results from Specimens 1–6 were used for this assessment, as these were the only ones considered relevant for this evaluation. Specimens 7 and 8 were fabricated with a different throat thickness, but with t = 25 mm, which is equal to the reference thickness. The standard deviations of log N for the test results were re-analyzed using different thickness exponents. The smallest standard deviation was obtained for the thickness exponent k equal to zero. In other words, these data indicated that for these fillet welds there is no thickness effect at all. However, there are too few results to justify disregarding a possible thickness effect, and therefore, for design, it is recommended that the above correction is retained but used in conjunction with the least severe value proposed for k, namely 0.15 (Maddox 2006). The equation of the S-N curve including the thickness correction is then:

$$\log N = \log A - m \log\left[S\left(\frac{t}{t_{ref}}\right)^{0.15}\right] \qquad (2.20)$$

The test results for Specimens 7 and 8, with increased fillet weld size, in Figure 2.38 do not indicate any size effect related to the fillet weld throat thickness.

The fatigue test results obtained from Specimens 9 and 10 under pure axial loading are presented in Figure 2.39 together with the recommended design procedure from IIW (Hobbacher 2009), in Figure 2.40 together with the recommended design procedure from Eurocode 3 (EN 1993-1-9 2009), and in Figure 2.41 together with the recommended design procedure in DNVGL-RP-C203 (2016). The data points are established from the load range and the mean throat thickness measured after testing. A stress concentration factor of 1.11 is also included in the nominal stress to account for local bending in the tube wall at the transition to the end plate. This factor is derived from equations for stress concentration factors presented for a tubular member welded to a stiff plate using the equations for a stress concentration factor at a bulkhead as presented by equations (6.36) and (6.40). The two data points appear to lie on a higher S-N curve than the design curve. However, they are within the scatter-band enclosing published data for cruciform joints failing in the weld throat,

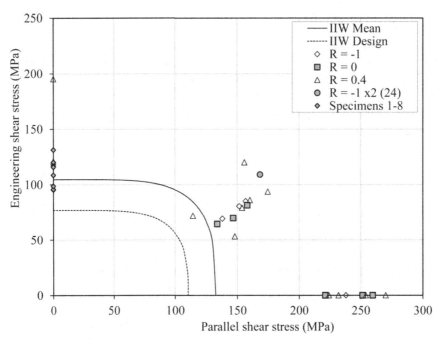

Figure 2.39. Fatigue test data compared with the interaction curve based on IIW at 200,000 cycles (Lotsberg 2006b) (reprinted with permission from ASME).

as presented by Maddox (2006). (The test data are here presented in a linear scale where the scatter becomes more visible than in a logarithmic plot.)

Specimens 11–20 were fatigue tested in pure torsion under three stress ratios (−1, 0, and 0.4), which resulted in parallel shear stress in the fillet weld throat. The

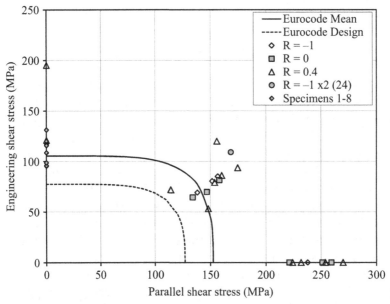

Figure 2.40. Fatigue test data compared with the interaction curve based on Eurocode 3 (EN 1993-1-9 2009) at 200,000 cycles (Lotsberg 2006b) (reprinted with permission from ASME).

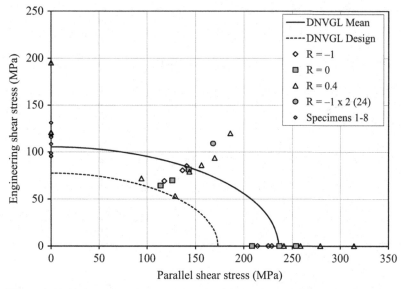

Figure 2.41. Fatigue test data compared with the interaction curve based on DNVGL-RP-C203 at 200,000 cycles (Lotsberg 2006b) (reprinted with permission from ASME).

resulting parallel shear stress was derived from the load range divided by the mean shear area of the fillet welds. The fatigue test results were compared with the Eurocode 3, which is the same as the S-N curve as in IIW for this mode of loading. The test data fell significantly above the mean Eurocode S-N curve according to Lotsberg (2006b). The difference between the mean test data and the mean Eurocode S-N curve is a factor of $10^{17.218}/10^{16.216} = 10.0$ on fatigue life. From the test results it also appeared that the applied stress ratio, with respect to parallel shear stress in the weld, does not significantly affect fatigue life. This was checked by statistical evaluation of the test results. Assuming that for each R value, the results lie on an S-N curve with slope $m = 5$, the corresponding mean S-N curves and standard deviations of log N were calculated. The S-N curve was assumed to have the usual form:

$$\log N = \log A - m \log S \tag{2.21}$$

The combined data was treated in the same way, both for $m = 5$ and by free regression analysis. The resulting S-N curves are detailed in Table 2.9. The standard deviations

Table 2.9. *Statistics of fatigue test data obtained in pure torsion (Lotsberg 2006b)*

Test no.	R	m	log a	Standard deviation of log N
11–13	0	5.0	17.171	0.142
14–16	−1	5.0	17.233	0.182
17–20	0.4	5.0	17.241	0.185
11–20	0, −1, and 0.4	5.0	17.218	0.156
11–20	0, −1, and 0.4	Free 4.115	15.115	0.144

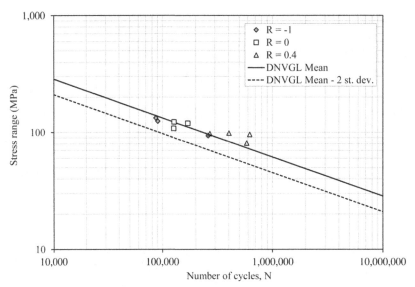

Figure 2.42. Test data for specimens 11–20 compared with DNVGL-RP-C203 (Lotsberg 2006b) (reprinted with permission from ASME).

of log N in Table 2.9 were derived assuming that the test data were log-normally distributed (or normal distributed in a logarithmic diagram). The statistics are based on relatively few test data, and this should be kept in mind when the results are assessed. Nevertheless, it is clear that the statistical evaluation of the test data confirms that R for parallel shear does not have a significant influence on fatigue life. The fatigue test results were also evaluated in terms of the stress used in conjunction with the design curve for shear failure of welds in DNVGL-RP-C203 (2016), as shown in Figure 2.42. As discussed in more detail in the next section, an equivalent stress range derived from equation (2.24) was used for derivation of the test data points in Figure 2.42.

2.5.5 Comparison of Design Equations with Test Data for Combined Loading

The remaining 13 tests, for Specimens 21–33, were performed under combined axial and torsional loading. The test results are compared with the design equations from IIW (Hobbacher 2009), Eurocode 3 (EN 1993-1-9 2009), and DNVGL-RP-C203 (2016). The design equation from IIW reads:

$$\left(\frac{n}{N}\right)_{normal} + \left(\frac{n}{N}\right)_{shear} \leq 0.5 \tag{2.22}$$

where:

n is the number of cycles of applied stress and,
N is the fatigue life at the same stress level obtained from the appropriate design curve (i.e., the design curves for normal and shear stresses).

The S-N curve for parallel shear stress in fillet welds is FAT 80 with slope $m = 5.0$. FAT 45 is used for the engineering shear stress (i.e., the normal stress component) in IIW. This curve has slope $m = 3.0$. The approach for non-proportional loading

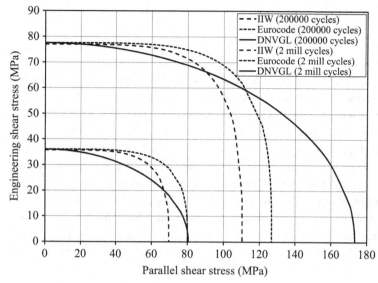

Figure 2.43. Comparison of combined stresses at 200,000 and 2 million cycles using different design standards.

from IIW is used; see Section 4.2 of IIW (Hobbacher 2009). The present test data are derived under proportional loading (or in phase loading). The design equation from Eurocode 3 reads:

$$\left(\frac{n}{N}\right)_{normal} + \left(\frac{n}{N}\right)_{shear} \leq 1.0 \qquad (2.23)$$

The S-N curve for parallel shear stress in fillet welds is FAT 80, with slope $m = 5.0$. FAT 36 is used in Eurocode 3 for engineering shear stress. In DNVGL-RP-C203 an equivalent engineering shear stress is derived as:

$$\Delta\sigma_w = \sqrt{\Delta\sigma_\perp^2 + \Delta\tau_\perp^2 + 0.2\Delta\tau_{//}^2} \qquad (2.24)$$

This stress is then used in conjunction with the FAT 36 curve for calculating the number of cycles to failure. The three approaches are different and lead to different equations to describe the interaction of the engineering shear stress and parallel shear stress. The interaction equations are compared in Figure 2.43 for $N_{ref} = 2 \cdot 10^5$ and $2 \cdot 10^6$ cycles. In order to examine the validity of the three approaches, the test results were evaluated in terms of the corresponding interaction equations. In order to do this, it was necessary to extract the stress range corresponding to a given fatigue life from each set of results. Therefore, the test results were scaled with respect to stress range to achieve the same reference number of cycles; $N_{ref} = 2 \cdot 10^5$ for all test data. This scaling was based on the negative inverse slope of the S-N curve to move the test data points to the same reference number of cycles, N_{ref}. The resulting correction factor on parallel shear stress for comparison with IIW and Eurocode 3 is then:

$$f_{cp} = \left(\frac{N_{test}}{N_{ref}}\right)^{1/5.0} \qquad (2.25)$$

while that on engineering shear is:

$$f_{ce} = \left(\frac{N_{test}}{N_{ref}}\right)^{1/3.0} \qquad (2.26)$$

The latter correction factor is also used to correct the stress to be compared with the interaction equation from DNVGL-RP-C203 (2016). The fatigue test data themselves are compared with the interaction diagrams from IIW and Eurocode 3 in Figures 2.39 and 2.40, respectively. The full line corresponds to that of a mean S-N curve, and the dotted line corresponds to that of the design S-N approach. Data outside the line show a safety margin, while a data point inside the dotted line would not be safe. It should be noted that the fatigue test data are in the safe region using the design approaches of both IIW and Eurocode 3. It should also be noted that the fatigue test data obtained under combined loading should be considered as lower bound values in the way the stress values are derived from the measured data. The torsional loading resulted in some bending of the test specimen that was difficult to account for fully when assessing the stress to be used for plotting the test data. Thus, a rather low stress concentration factor was derived for the fatigue crack initiation point. A higher stress concentration factor from uneven stress distribution around the circumference would lift the data points further away from the design curve, indicating even greater safety.

In Figure 2.41 the test data are compared with the interaction diagram from DNVGL-RP-C203 (2016). It can be seen that the test data are marginally above the design S-N approach of this Recommended Practice. It should be noted that the difference between the standards for relevant values of parallel shear, in combination with engineering shear stress, is reduced with increase in the fatigue endurance considered. For comparison of the different standards at 200,000 cycles and 2 million cycles, see Figure 2.43. As the most significant damage occurs in the low stress range, high cycle fatigue regime, this is relevant to wave loading on structures. There does not seem to be any evidence that the fatigue lives obtained under combined loading depended on the applied stress ratio (R-ratio) with respect to parallel shear in the fillet weld.

There is only one test data for R = −1 for axial loading as well as torsional loading. The test result for this specimen (Specimen 24) is shown as a separate data point in the figures. This result indicates that there is a mean stress effect for the axial loading, as might be expected with compression stresses present in the weld root after fabrication. This may explain why fatigue cracks from fillet welds in ship structures do not occur more frequently. Fillet welds in ships are generally considered to be small in comparison to those normally used in offshore structures.

2.6 Doubling Plates or Cover Plates

2.6.1 Background

Some fatigue testing of fillet welded doubling plates was conducted to investigate possible weld improvement of a doubling plate on a semi-submersible platform. A shorter fatigue life than the target fatigue life had been calculated, and it was questioned whether sufficient documented fatigue life could be achieved by weld toe

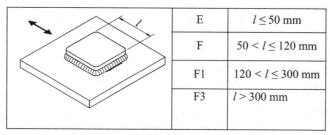

	E	$l \leq 50$ mm
	F	$50 < l \leq 120$ mm
	F1	$120 < l \leq 300$ mm
	F3	$l > 300$ mm

Figure 2.44. Classification of a doubling plate according to DNVGL-RP-C203 (2016).

improvement. "Cover plates" is another notation used in bridge design, and that may also describe the functionality of these plates. In the actual platform these plates were used as supports for ladders to avoid welding the ladders directly to the main load-carrying plates. Different techniques can be used to improve fatigue life at a weld toe, as shown in Section 11.7. However, when fillet welds are used, the efficiency of these improvement methods is debatable. Guidelines have been developed by IIW on details that can be improved, and doubling plates is one example where weld improvement is not recommended (Haagensen and Maddox 2013). However, the fillet welds here are less load-carrying than in other fillet welded cruciform joints, where throat thickness is the main parameter for when fatigue cracks will probably initiate from the weld root. Fatigue cracking at doubling plates may initiate from the weld root through the throat of the fillet weld as indicated in Figure 1.7 and may also initiate from the weld root into the base plate as shown in Figure 1.8. When the fatigue strength of the weld toe is improved, the most likely initiation site may move to the weld root. As the improvement factor on fatigue life given in DNVGL-RP-C203 (2016) has been mainly derived from improvement of weld toes at details with full penetration welds, the relevance of its application for this detail is uncertain. Some guidance on selection of S-N curves for doubling plates in the as-welded condition is provided in DNVGL-RP-C203 (2016), as shown in Figure 2.44. Typical doubling plate geometry results in use of the F1 curve, which is the same as FAT 63 in the IIW recommendations (Hobbacher 2009). When the weld toe is improved, crack growth through the fillet weld or crack growth into the base plate might be possible failure modes. In order to investigate this further, a test program was proposed as described below.

2.6.2 Test Program and Preparation of Test Specimens

Two test specimens were first fatigue tested: one large-scale specimen with three doubling plates was tested in the as-welded condition for reference, and one similar large-scale specimen with ground welds. The fillet welds on these specimens were made with more than one pass. This resulted in a rather large throat thickness, approximately 7 mm, and a leg length slightly less than 10 mm. A third specimen, with welds made in one pass and with ground weld toes, was then tested. The throat thickness for this specimen was approximately 4 mm and the leg length was 5.6 mm. A fourth specimen was tested to study the effect of Ultrasonic Peening (UP) on weld toes with respect to fatigue life improvement. As the purpose was to test efficiency of UP, and a failure from the root should be avoided, this specimen was also made with a large

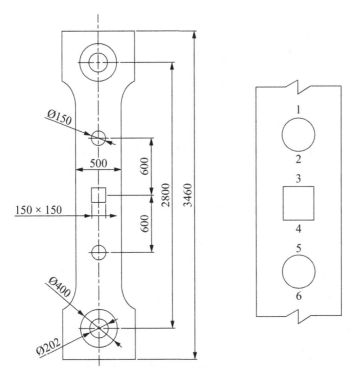

Figure 2.45. Geometry of test specimen and numbering of locations used as reference for presentation of the strain gauge measurements and test results.

throat thickness to enable comparison with the specimen already tested with ground weld toe. During testing of Specimen 4, some unplanned load action occurred that led to strains at the welded regions well above yield. Therefore, it was decided to add another similar test specimen (number 5), which had been improved by UP in a similar way as number 4. Due to the unintended load action incident, the test data from Specimen 4 are not considered further here. The following specimens were tested:

- Specimen 1: As-welded condition with throat thickness of 7 mm
- Specimen 2: Ground weld toes with throat thickness of 7 mm
- Specimen 3: Ground weld toes with throat thickness of 4 mm
- Specimen 5: Ultrasonic peened weld toes with throat thickness of 7 mm

The geometry of the tested full-scale specimens is shown in Figure 2.45. The main plate was 3460 mm long, with a waist width of 500 mm for the two first specimens and of 400 mm for Specimens 3, 4, and 5 and a thickness of 20 mm. The pin holes at each end were reinforced on both sides with 50 mm thick doubling plates. In order to obtain as much information as possible from one fatigue test, three doubling plates with thicknesses of 10 mm were welded to one side of the main plate. Two doubling plates were circular, with diameters of 150 mm, while the doubling plate in the middle was quadratic (150 mm × 150 mm) with rounded corners of radius 25 mm. The reason for including both quadratic and circular doubling plates was that results from finite element analysis showed approximately the same hot spot stress for these geometries; therefore, it was of interest to investigate whether a similar fatigue strength could also be confirmed by relevant fatigue testing. The grinding of the weld toes was

Figure 2.46. Ground weld toes around cover plates (Lotsberg et al. 2014). (Copyright Springer. With kind permission from Springer Science and Business Media.)

based on recommendations in DNVGL-RP-C203 (2016), where it is stated that the grinding should be at least 0.5 mm below any visible undercut. The ground weld toes around plates in Specimen 2 is shown in Figure 2.46. The grinding was done using a 10 mm diameter rotary burr. The person who performed the grinding of Specimen 2 had retired before Specimen 3 had been prepared, and therefore another person performed the grinding of Specimen 3. Initially it was considered that it was important that the same person performed the grinding for both specimens; however, different people conducting this work probably provides a more realistic representation for the tests.

A tensile test of the plate material from Specimen 5 was made, and a yield strength 430 MPa and tensile strength of 492 MPa were measured.

2.6.3 Fatigue Testing

Fatigue testing of the full-scale specimens with doubling plates was performed in a 750 ton servo hydraulic machine (see Figure 2.47). The different hot spot locations were numbered for each specimen, as shown in Figure 2.45. Strain gauges were installed at each of these hot spots. The reason for staggering the strain gauges, as shown in Figure 2.46, was due to their size. It should be added that the hot spot stress was determined by linear extrapolation to the weld toe or grooves from the grinding or peening from the stresses derived at the first and third strain gauge, which were placed on the same line normal to the weld toes. Thus, the strain gauge in the middle was mainly used for control purposes, and there was no indication that the staggering influenced the data measurements. The fatigue test specimens were subjected to a dynamic load with an R-ratio of 0.1 at a constant nominal stress range of 200 MPa ($R = \sigma_{min}/\sigma_{max}$). This stress range represents a maximum stress range at a doubling plate during the lifetime of a typical floating structure with a long-term load distribution, where improvement of the weld toes is required for documentation of sufficient fatigue life. Reference is made to Table 10.3 for maximum allowable

Figure 2.47. Test specimen on the floor and in test machine with numbering of locations.

stress ranges during design life, where the largest allowable nominal stress range is 190 MPa during 20 years of service life in an air environment and 169.6 MPa in seawater (DNVGL-RP-C203, 2016), with cathodic protection for a shape parameter equal to 1.0 in a two-parameter Weibull long-term distribution of stress ranges for a 20-year service life. These values are considered to be representative for a floating platform. A Miner sum of 1.0 is used for this calculation, as this is the design requirement used for floating production platforms where inspection of fatigue cracks are performed during service life. When shorter fatigue lives are calculated, the maximum nominal stress ranges are also higher, and in such cases weld improvement is requested at late fabrication stages. If a fatigue life of 10 years was calculated, the maximum nominal stress ranges during the intended service life increases to 231.4 MPa in an air environment and 199.2 MPa in seawater with cathodic protection. It should be added that in a real structure, this maximum stress range enters the S-N curve only once during the considered lifetime, while the main contribution to fatigue damage is around 10 million cycles, where the stress range is less than a quarter of the maximum stress range. See also Section 1.2 for contributions from wave loading to fatigue damage. The specimens were tested at a frequency of 1 Hz.

Specimen 1

The fatigue testing of Specimen 1 was terminated after 150,800 cycles. At this number of cycles, a fatigue crack that had initiated from the lower weld toe of the quadratic doubling plate (location 4 in Figure 2.45) had grown through the main plate. The welds were regularly inspected for cracks during the fatigue tests. For some

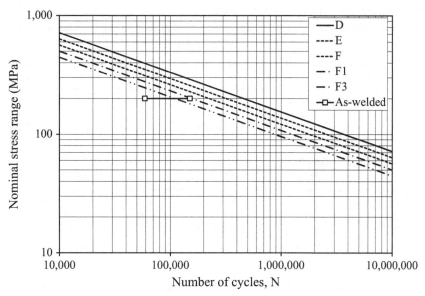

Figure 2.48. Fatigue crack growth in Specimen 1 in as-welded condition compared with mean S-N curves (Lotsberg et al. 2014). (Copyright Springer. With kind permission from Springer Science and Business Media.)

locations, two fatigue cracks were observed that later coalesced into a single crack. Crack depth was measured when the test specimen had been cut up after testing. The crack growth period in specimen 1 is shown in Figure 2.48, with the nominal stress mean S-N curves from DNVGL-RP-C203 (2016) shown in the same figure. These mean S-N curves are derived from the characteristc S-N curves, which are defined as the mean minus two standard deviations for a normal distribution in a logarithmic S-N diagram, by adding two standard deviations. The standard deviations for these S-N curves are presented in Section 4.1.3. The number of cycles is shown for the crack that had grown through the plate thickness at location 4. This number of cycles corresponds to that of the mean F1 curve and is in agreement with the recommendation in DNVGL-RP-C203 for this geometry of doubling plates. At the termination of the test, cracks had also developed at the other five hot spots, with crack depths of between 5.3 mm and 14.5 mm, as shown in Table 2.10. Should the testing have been continued until failure, the fatigue lives at these hot spots would have resulted in data points above the mean S-N curve, with through thickness cracks at these positions. The fatigue lives of the two different geometries of doubling plates (circular and quadratic) were similar, with comparable crack depths at the end of the test.

Specimen 2

The fatigue testing of Specimen 2 was terminated after 553,100 cycles, by which time cracks of between 4 mm and 9 mm deep had been detected at four hot spots at the circular plates, but cracking had not been observed at the hot spots at the squared doubling plate (locations 3 and 4 in Figure 2.45). However, fatigue cracks were observed in the pin holes of the test specimen and the test was terminated. The crack geometries at the end of test are shown in Table 2.10.

Table 2.10. *Measured crack geometries for the specimens*

Specimen no	Location	Hot spot stress (MPa)	Crack at end of test		Number of cycles
			Depth (mm)	Length at surface (mm)	
1	1	302.61	10.0	123	150,800
	2	297.36	14.5	115	
	3	288.96	10.0	155	
	4	298.62	20.0	180	
	5	283.40	5.3	94	
	6	298.52	12.3	108	
2	1	260.40	5	21*	553,100
	2	286.02	6	19*	
	3	276.89	–	–	
	4	263.66	–	–	
	5	290.12	4	48*	
	6	292.64	9**	29*/26**	
3	1	316.89	20	37 + 30 + 20	346,881
	2	324.87	–	–	
	3	299.15	–	–	
	4	302.30	20	83 + 56	
	5	309.96	17.5	18 + 42 + 5	
	6	314.37	5.3***		
5	1	262.61	20	94	140,977
	2	338.21	13.5	68	
	3	288.23	Small cracks	5 + 5 + 5	
	4	332.85		6	
	5	266.70	Small cracks	47	
	6	328.76	Small cracks	10 + 32	

* Crack observed in ground area.
** Crack observed in base metal in front of that ground.
*** Weld repaired after 272,000 cycles when crack length had grown to 22 mm.

Specimen 3

The fatigue testing of Specimen 3 was terminated after 346,881 cycles, when cracks had grown through the thickness at locations 1 and 4 in Figure 2.45, but cracks had not been observed at locations 2 and 3. The crack depth at location 5 was 17.5 mm at the end of the test. The largest crack, which initiated first, was at location 6, and was repaired after 272,000 cycles, by which time it was 22 mm long but did not extend through the plate. As this crack was detected relatively early compared with the tested number of cycles in specimen 2, the reason was investigated by cutting out a "boat sample" from this hot spot location, which included the fatigue crack. A metallographic section through this sample is shown in Figure 2.49, and it can be seen that the crack initiated near to the Heat Affected Zone (HAZ). An enlarged section of the crack initiation site and the crack tip are shown in Figure 2.50. Thus, despite the ground surface appearing smooth in Figure 2.49, there are imperfections that can initiate fatigue cracks as shown in Figure 2.50. The cut was repaired by manual welding in the laboratory before continuing testing. Further information regarding fatigue strength could not be obtained from this location. The crack geometries from

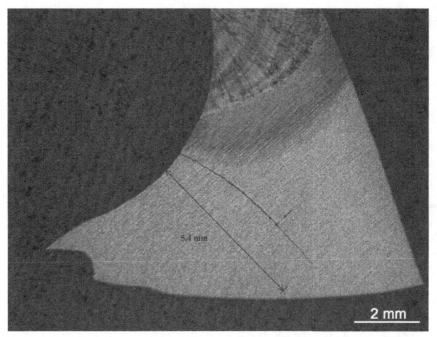

Figure 2.49. Transverse metallographic section from location 6 in Specimen 3 (Lotsberg et al. 2014). (Copyright Springer. With kind permission from Springer Science and Business Media.)

Figure 2.50. Enlarged photo of fatigue crack at location 6 in Specimen 3 (Lotsberg et al. 2014). (Copyright Springer. With kind permission from Springer Science and Business Media.)

Crack initiation adjacent to the HAZ

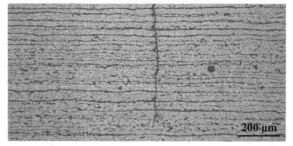

Detail of the crack tip area

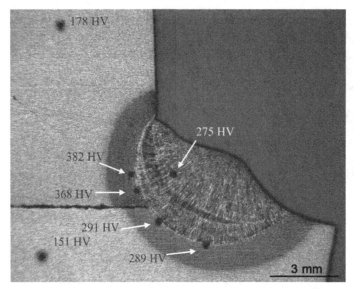

Figure 2.51. Section through weld showing ground profile and measured hardness values.

this test are shown in Table 2.10. The cracked region was opened at a low temperature to make the material brittle; this explains the surface that is seen outside the fatigue-cracked area in Figure 1.4. Two more complete semi-elliptical cracks at location 4 are shown in Figure 1.5, and a rather deep fatigue crack at location 5 is shown in Figure 1.6. A section through the weld showing the ground profile and measured hardness values is shown in Figure 2.51.

Specimen 5

The fatigue testing of Specimen 5 was terminated after 140,977 cycles, when a fatigue crack had grown through the plate at location 1 as shown in Figure 2.45. The crack depth was 13.5 mm at location 2, and only small cracks initiated at locations 3, 5, and 6, as shown in Table 2.10. Cracks were not detected at location 4, and at location 3 only small fatigue cracks along the weld were initiated, as shown in Figure 1.4.

2.6.4 Assessment of Test Data

Different techniques can be used to improve fatigue lives at weld toes in welded structures with full penetration welds. However, full penetration welds cannot be achieved for all details and, in fillet welded structures, the efficiency of weld toe improvement may be reduced by fatigue cracking from the weld root. This was investigated on four large specimens with fillet welded doubling plates. Each of these tests included three doubling plates: two circular plates and one square plate with rounded corners. The crack growth observed in Specimen 1 was compared with the mean S-N curves from DNVGL-RP-C203. The number of cycles for crack growth through the plate thickness corresponds to that of the mean F1 curve, and thus agrees with the recommendations. The ground weld toes (Specimens 2 and 3) showed good improvements as fatigue cracking from the weld roots was not observed during the testing. All locations in the tests were opened after test completion and the weld roots

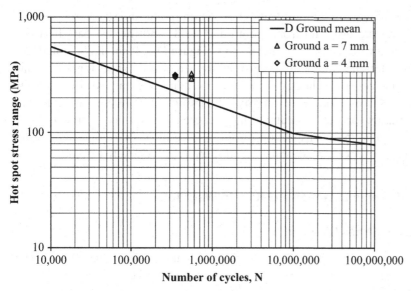

Figure 2.52. Fatigue test data from ground specimens and mean D curve for ground welds (Lotsberg et al. 2014). (Copyright Springer. With kind permission from Springer Science and Business Media.)

examined by microscopy. A small fatigue crack was observed to have initiated from the weld root in one of the doubling plates with throat thickness of 4 mm and also in one of the weld roots in the ground test specimen, with 7 mm throat thickness. A crack indication in one of the root weld samples from the UP tested specimens was also reported. As these crack indications were very small, it is likely that crack growth life is considerable before these cracks had grown large. Thus, the tests showed that weld improvement of fillet welded doubling plates is feasible, as fatigue crack growth from the weld root was not critical in the tests. However, criticality may be affected by fabrication quality and welding procedures, neither of which were considered in this study. The probability of crack growth into the fillet weld is considered to be a function of throat thickness. Thus it is recommended that fatigue from the fillet weld root and add on weld, if necessary, are assessed before improvement of the weld toe is performed by grinding.

The test results for fatigue crack initiation for Specimen 5 are above that for as-welded condition, but not quite as high as for the ground welds. It is well known that the S-N curve for improved welds becomes more horizontal than for as-welded condition. Thus, the effect of improvement for large stress ranges is lower than for smaller stress ranges. Although the test stress range for the test was rather large, it was no greater than might be expected in a newly built floating structure, where weld improvement of a detail is required. However, it is probably larger than that expected in life extension of an existing structure that has been subjected to a large number of load cycles without significant fatigue cracking.

The tests showed that grinding of weld toes is very efficient with respect to improvement by increasing fatigue life as shown in Figure 2.52. These results support the improvement factor of 1.3 on stress at $2 \cdot 10^6$ cycles, as recommended by IIW for ground welds. With a more horizontal S-N curve, further improvement is achieved for smaller stress ranges. This leads to significant fatigue improvement for a typical long-term stress range distribution from wave action on marine structures; see also Figure 11.9. Both the test data from the ground welds and the test data from the UP test are in line with the S-N curves shown in Figure 11.9. It should be noted that the period from fatigue crack initiation to failure is relatively short for the ground welds, and this is a concern related to weld improvements of details with short calculated fatigue lives. This is also a reason why weld improvement is not recommended in design standards for short calculated fatigue lives. Furthermore, this should be reflected in the planning of inspection during service life, as the time interval for detecting fatigue cracks becomes shorter; reference is also made to Chapter 18.

2.7 Effect of Stress Direction Relative to Weld Toe

2.7.1 Constant Stress Direction

Structural details are mostly subjected to proportional loading at the hot spot. This means that the direction of the principal stress within a load cycle is relatively constant. For floating structures in particular, the principal stress direction may depend on the load direction analyzed. In such cases, the calculated fatigue damage can be divided into different damages depending on whether a crack is likely to initiate along a weld toe or more normal to a weld toe. Developing design criteria for non-proportional loading is more complex where the direction of the principal stress may change during a load cycle, as is more typical for machines; see also Section 2.7.5. Assessment of the following design criteria for proportional loading is based on comparison with fatigue test data from Kim and Yamada (2004). The effect of stress direction with respect to fatigue strength is important for situations where the principal stress direction differs from the normal direction to the weld toe. Such stress conditions are found at details in different types of plated structures, such as at connections with soft brackets and at tubular penetrating plates in ship structures. Some fatigue design standards advise using the largest principal stress range within ±45° to the normal to the weld toe, together with an S-N curve derived for stress ranges normal to the weld toe, for fatigue design. One approach by IIW (Hobbacher 2009) uses the largest principal stress range direction within ±60°, together with the same S-N curve as for stress range normal to the weld toe. Different design criteria for proportional loading are assessed in the following sections.

2.7.2 Fatigue Test Data

Fatigue test data from specimens shown in Figure 2.53a and test specimens in Figure 2.53b, where fatigue cracking occurred at the straight part of the weld toe, were selected for assessment. The fatigue test data used in the assessment are listed in Kim and Yamada (2004).

The following notations are used on test specimens: G for gusset specimen and N for non-load-carrying cruciform specimen. The angle α is defined in Figure 2.53, and the specimen number is used as part of the notation. Reference is made to Figure 2.53 for calculation of stresses in the test specimen. Equilibrium in the loading direction of the test specimen gives:

$$(\tau_{//} \sin \alpha + \sigma_\perp \cos \alpha)w / \cos \alpha = \sigma_1 w \qquad (2.27)$$

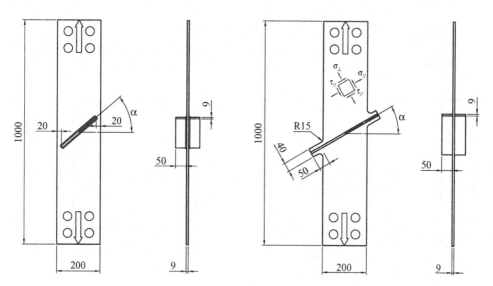

(a) Out-of-plane gusset specimen   (b) Non-load-carrying cruciform specimen

Figure 2.53. Fatigue test specimens (Kim and Yamada 2004) (with permission from Taylor & Francis Ltd., http://www.tandfonline.com).

where w is width of specimen. Equilibrium in the transverse direction to the specimen gives:

$$\tau_{//} \cos\alpha = \sigma_\perp \sin\alpha \qquad (2.28)$$

From these equations, the following stresses are derived:

$$\tau_{//} = \sigma_1 \sin\alpha \cos\alpha$$
$$\sigma_\perp = \sigma_1 \cos^2\alpha \qquad (2.29)$$

The stresses as functions of unit stress for varying values of α are shown in Figure 2.54.

2.7.3 Design Procedures in Different Design Standards

The procedure in Eurocode 3 (EN 1993-1-9 2009) for the combined effect of normal stress and shear stress is a summation of calculated fatigue damages from normal stress range and shear stress range at the weld toe. This can be presented by a design equation as:

$$D_\sigma + D_\tau = 1.0 \qquad (2.30)$$

Two alternative procedures combining the effect of normal stress and shear stress are presented in the recommendations on fatigue design by IIW from 2009:

- Principal stress direction
- Quadratic interaction of allowable normal stress range and shear stress range

The different methodologies are presented in greater detail in the following sections.

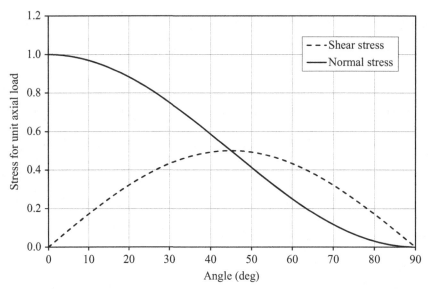

Figure 2.54. Stress as a function of unit stress for varying values of α (Lotsberg 2009a) (with permission from Taylor & Francis Ltd., http://www.tandfonline.com).

Principal Stress Direction in IIW (Hobbacher 2009)

In case of biaxial stress state at the plate surface, use of the principal stress that is approximately in line with the perpendicular to the weld toe – that is, within an angle of $\pm 60°$ – is recommended. The other principal stress may be analyzed, if necessary, using the fatigue class for parallel welds in the nominal stress approach.

Quadratic Interaction of Allowable Normal Stress Range and Shear Stress Range in IIW (Hobbacher 2009)

The effects of combination of normal and shear stress for proportional loading should be verified by:

$$\left(\frac{\Delta\sigma_{S,d}}{\Delta\sigma_{R,d}}\right)^2 + \left(\frac{\Delta\tau_{S,d}}{\Delta\tau_{R,d}}\right)^2 = 1.0 \tag{2.31}$$

where:

   $\Delta\sigma_{R,d}$ or $\Delta\tau_{R,d}$ is the design resistance stress range for the specified number of cycles and the appropriate FAT class for stress normal to the weld and the shear stress at the weld toe.

   $\Delta\sigma_{S,d}$ and $\Delta\tau_{S,d}$ are the corresponding design stress ranges.

A combined stress range (or effective stress range) that takes into account the stress normal to the weld and the shear stress along the weld toe is included in DNVGL-RP-C203. This equivalent stress can be expressed in the following form:

$$\Delta\sigma_{Eff} = \sqrt{\Delta\sigma_\perp^2 + \beta\Delta\tau_{//}^2} \tag{2.32}$$

where the stress components are explained in Figure 2.53.

   The S-N category will depend on the type of detail in relation to the normal stress. This will result in different β values as presented in Table 2.11. The combined

Table 2.11. *Values for β based on S-N curves in DNVGL-RP-C203 (2016) (Lotsberg 2009a)*

Stress direction normal to the weld toe	Stress direction parallel with the weld		
	C	C1	C2
D	0.518	0.646	0.810
E	0.409	0.510	0.640
F	0.322	0.402	0.504
F1	0.254	0.316	0.397
F3	0.201	0.250	0.314
G	0.160	0.199	0.250

stress range should be used together with an S-N curve that is selected as if this stress was acting normal to the weld toe. The detail tested in Figure 2.53 is classified as E according to DNVGL-RP-C203 (2016) for small thicknesses of the attachments and F for larger thicknesses when $\alpha = 0$.

The test results for $\alpha = 0$ are presented in Figure 2.55 together with the E-curve. Reference is made to Table 4.3 for the relation between notations on S-N curves used in DNVGL-RP-C203 (2016), IIW (Hobbacher 2009), and Eurocode 3 (EN 1993-1-9 2009). For presentation of mean S-N curves it is assumed that a standard deviation in logarithmic format is 0.20, and the characteristic curve is defined as the mean minus two standard deviations, assuming that the test data are normally distributed in a logarithmic format. S-N category C2 may be used for continuous shear stress in a full penetration weld, according to table A.8 in DNVGL-RP-C203. Assuming that

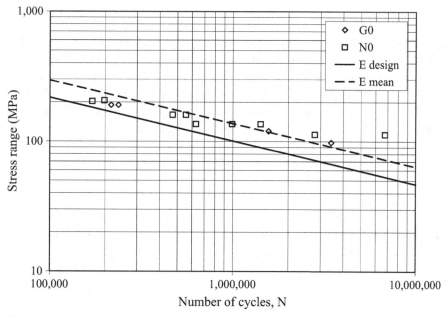

Figure 2.55. Test data for principal stress normal to the weld (Lotsberg 2009a) (with permission from Taylor & Francis Ltd., http://www.tandfonline.com).

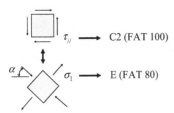

Figure 2.56. Illustration of stresses $\sigma_1 = \tau_{//}$ when $\alpha = 45°$ that are combined with different S-N curves (Lotsberg 2009a) (with permission from Taylor & Francis Ltd., http://www.tandfonline.com).

the shear stress is classified as C2, the following equation for combined or effective stress is derived from Table 2.11 when the effective stress range is combined with S-N curve E:

$$\Delta\sigma_{Eff} = \sqrt{\Delta\sigma_{\perp}^2 + 0.64\Delta\tau_{//}^2} \qquad (2.33)$$

The basis for this equation is also illustrated in Figure 2.56, where $\sigma_1 = \tau_{//}$ at $\alpha = 45°$. The stress components in equation (2.33) are to be combined with different S-N curves, as shown in Figure 2.56. When σ_1 is acting normally to a weld toe it is classified as an E detail or FAT 80 (IIW [Hobbacher 2009]). When the detail is subjected to shear along the weld, then S-N curve C2 or FAT 100 should be applied.

2.7.4 Comparison of Design Procedures with Fatigue Test Data

The detail shown in Figure 2.53 is classified as FAT 80 following Eurocode 3 (2009) and IIW (Hobbacher 2009) for $\alpha = 0$. This is the same as the E-curve in DNVGL-RP-C203 (2016). The test results for $\alpha = 0$ are presented, together with the E-curve, in Figure 2.55. The S-N curves for stress range normal to the weld toe and shear stress can be presented as:

$$\begin{aligned} N_{\perp} &= a_{\perp}\Delta\sigma_{\perp}^{-m} = a_{\perp}\Delta\sigma_{R,d}^{-m} \\ N_{//} &= a_{//}\Delta\tau_{//}^{-m} = a_{//}\Delta\tau_{R,d}^{-m} \end{aligned} \qquad (2.34)$$

For stress normal to the weld, the design S-N curve is FAT 80 with m = 3.0. The design S-N curve for shear stress in Eurocode 3 and IIW is FAT 100 with m = 5.0. The stress ranges in the fatigue test data are scaled such that the test data can be presented for n = 10^6 cycles (similar to the methodology used in Section 2.5.5). The following scaling of stress range is made for comparison with Eurocode 3 and IIW where m = 3.0 for normal stress and m = 5.0 for shear stress:

$$\begin{aligned} \Delta\sigma_{\perp 10^6} &= \Delta\sigma_{\perp\ test}\left(\frac{n_{test}}{10^6}\right)^{1/3.0} \\ \Delta\tau_{//10^6} &= \Delta\tau_{//\ test}\left(\frac{n_{test}}{10^6}\right)^{1/5.0} \end{aligned} \qquad (2.35)$$

From the equation for summation of damage in equation (2.30) and from equation (2.34), the following relation is derived based on Palmgren-Miner rule as expressed in equation (3.40):

$$\begin{aligned} \frac{n}{N_{\perp}} + \frac{n}{N_{//}} &= 1.0 \\ \frac{n}{a_{\perp}}\Delta\sigma_{S,d}^{3.0} + \frac{n}{a_{//}}\Delta\tau_{S,d}^5 &= 1.0 \end{aligned} \qquad (2.36)$$

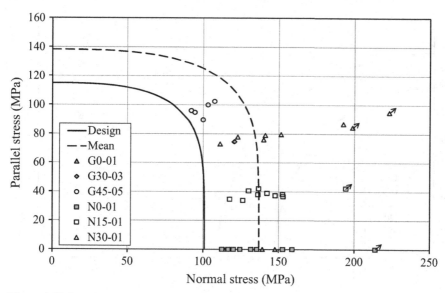

Figure 2.57. Test data presented in the format of the interaction equation in Eurocode 3 (Lotsberg 2009a) (with permission from Taylor & Francis Ltd., http://www.tandfonline.com).

From this equation the expression for shear stress resistance for Eurocode 3 is derived:

$$\Delta\tau_{S,d} = \left[a_{//} \left(\frac{1}{n} - \frac{\Delta\sigma_{S,d}^{3.0}}{a_{\perp}} \right) \right]^{1/5.0} \tag{2.37}$$

This equation, together with fatigue test data, is shown in Figure 2.57. It should be noted that this figure shows a rather small interaction effect between normal stress and parallel stress at $\alpha = 45°$.

IIW (Hobbacher 2009)

The IIW quadratic interaction equation on stress reads:

$$\left(\frac{\Delta\sigma_{S,d}}{\Delta\sigma_{R,d}} \right)^2 + \left(\frac{\Delta\tau_{S,d}}{\Delta\tau_{R,d}} \right)^2 = 1.0 \tag{2.38}$$

From equations (2.34) and (2.38), the following expression for shear stress resistance for IIW is derived:

$$\Delta\tau_{S,d} = \left(\frac{a_{//}}{n} \right)^{1/5} \sqrt{1 - \Delta\sigma_{S,d}^2 \left(\frac{n}{a_{\perp}} \right)^{2/3}} \tag{2.39}$$

This equation, together with test data, is shown in Figure 2.58. The fatigue test data, also at $\alpha = 45°$, agree well with the mean line for quadratic interaction on stress.

DNVGL-RP-C203 (2016)

For comparison of equation (2.33) with fatigue test data, the fatigue test data are scaled with respect to stress range to correspond to 10^6 cycles. The following scaling

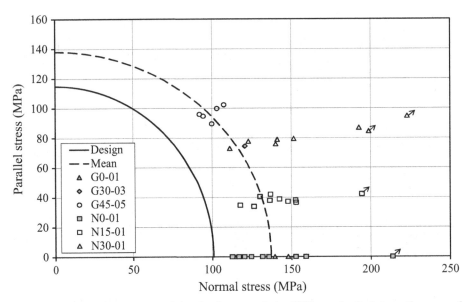

Figure 2.58. Test data presented in the format of the IIW quadratic interaction equation on stress range components (Lotsberg 2009a) (with permission from Taylor & Francis Ltd., http://www.tandfonline.com).

of stress range is made for comparison with DNVGL-RP-C203, with negative inverse slope of S-N curve m = 3.0:

$$\sigma_{\perp 10^6} = \sigma_{\perp \, test} \left(\frac{n_{test}}{10^6} \right)^{1/3.0}$$

$$\tau_{//10^6} = \tau_{// \, test} \left(\frac{n_{test}}{10^6} \right)^{1/3.0} \tag{2.40}$$

The test data for different principal stress range directions are presented in Figure 2.59. There is a good correspondence between the test data and the design equation for effective stress. This analysis approach can be combined with a single S-N curve. Therefore this design approach is also considered to be efficient for using together with stresses read out from finite element analyses (see also Section 9.1.3). The general expression for effective stress is derived from equation (2.32), where β values are derived from Table 2.11. If the hot spot stress is derived by extrapolation of stresses to the weld toe or to the intersection line from readout points t/2 and 3t/2, as explained in Section 9.1.3, this hot spot stress should be combined with S-N curve D. This means that $\beta = 0.81$ in equation (2.32) for calculation of effective stress range. If the hot spot stress is based on readout point at t/2, the hot spot stress should be combined with the E-curve and $\beta = 0.64$ in equation (2.32), or the D-curve can also be used here if the calculated stress is increased by a factor equal to the ratio between the fatigue strength in the D-curve and the E-curve from Table 4.1. This gives a ratio of 1.12. Based on this assessment, a more detailed classification between stress direction and S-N curves is proposed, as shown in Figure 2.60 and Table 2.12 depending on analysis methodology. Figures 2.60a and 2.60b are intended to be used for nominal stress analyses. The selection of E- and F-curves depends

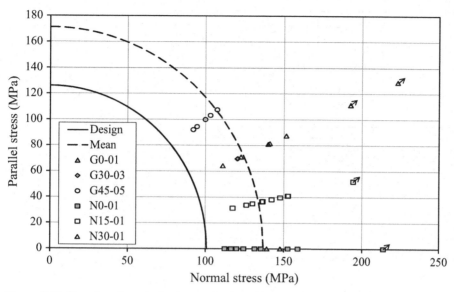

Figure 2.59. Test data for principal stress having different angles with the normal to the weld toe compared with effective stress range in DNVGL-RP-C203 (Lotsberg 2009a) (with permission from Taylor & Francis Ltd., http://www.tandfonline.com).

on the thickness of attachment, as presented in table A.7 of DNVGL-RP-C203 (2016). Figure 2.60c can be used, together with the hot spot stress methodology, or alternatively the more general numerical equations in Section 9.1.3 is used. The stress range in both the two principal directions should be assessed with respect to fatigue. Here a design criterion for α within an angle $\pm 45°$ to the normal to the weld has been assessed against fatigue test data. For the principal stress direction $45° < \alpha \le 90°$, an S-N curve for stress direction parallel with the weld can be used due to the effective stress reduction factor of 0.63 at $\alpha = \pm 45°$ (Lotsberg 2009a). Different design criteria and interaction equations are presented in Figure 2.61 for comparison of design criteria at 10^6 and 10^7 cycles, respectively. The direction of fatigue cracking is dependent on the principal stress directions as indicated in Figure 2.62.

Table 2.12. *Classification of details and selection of S-N curve*

Angle α in Figure 2.60	Detail classified as F for stress direction normal to the weld	Detail classified as E for stress direction normal to the weld	S-N curve when using the hot spot stress methodology
0–30	F	E	D
30–45	E	D	C2
45–60	D	C2	C2
60–75	C2	C2	C2*
75–90	C2*	C2*	C2*

* A higher S-N curve may be used depending on the fabrication. See table A-3 in DNVGL-RP-C203 for further information.

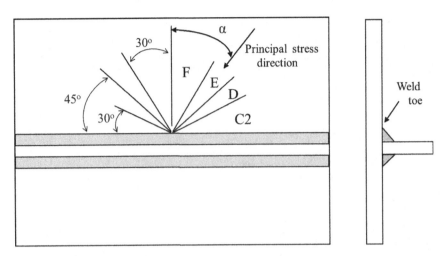

a) Detail classified as F for stress direction normal to the weld

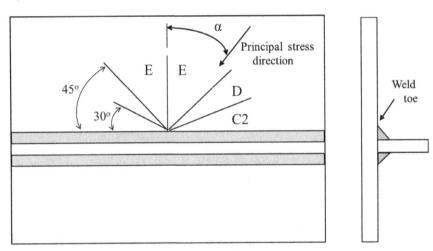

b) Detail classified as E for stress direction normal to the weld

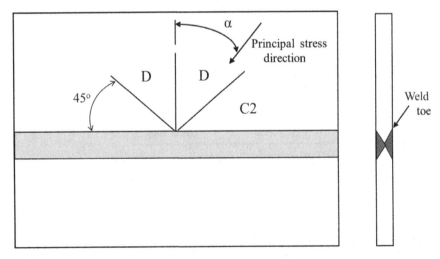

c) S-N curve when using the hot spot stress methodology

Figure 2.60. Classification of details and selection of S-N curve.

Table 2.13. *Recommended β factor for design (Lotsberg 2009a)*

S-N curve for stress normal to the weld	β factor	
	S-N classification C2 (FAT 100) for pure shear stress	S-N classification C1 (FAT 112) for pure shear stress
D (FAT 90) When used as nominal S-N curve or hot spot S-N curve by extrapolation of stress to the weld toe or the intersection line from read-out points 3t/2 and t/2 (t = plate thickness)	0.81	0.64
E (FAT 80) When used as nominal S-N curve or hot spot S-N curve by stress from readout points t/2 (t = plate thickness) from the weld toe	0.64	0.51
F (FAT 71)	0.50	0.40
F1 (FAT 63)	0.40	0.31
F3 (FAT 56)	0.31	0.25
G (FAT 50)	0.25	0.20

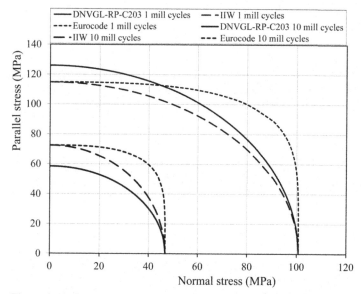

Figure 2.61. Comparison of design equations at 10^6 and 10^7 cycles.

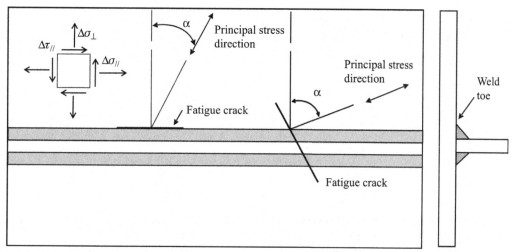

Figure 2.62. Fatigue cracking as function of principal stress range relative to weld toe.

2.7.5 Varying Stress Direction during a Load Cycle

Varying stress direction during a load cycle can be observed in machine type of structures. This is considered to be a topic for further research; see also Hobbacher (2009), where the equation (2.38) is provided also for this type of loading; however, a calculated fatigue damage less than 0.5 is recommended. See Sonsino (1997) and Razmjoo and Tubby (1997) for more background. For aluminum structures, see Wiebesiek and Sonsino (2010).

3 Fatigue Design Approaches

3.1 Methodology for Assessment of Low Cycle Fatigue

3.1.1 Cyclic Strain and Fatigue Strength

The methodology presented in this section is well established for fatigue assessment of machined, notched components typical for machine parts, automotive structures, and so on and has been used for documentation of the fatigue capacity of machined tether connections in a platform in the North Sea (Chen 1989). Furthermore, this methodology is also included in ISO 19902 (2007). Otherwise, it is used infrequently for welded structures, except for assessment of low cycle fatigue.

Fabrication Standards for marine structures require monotonic stress-strain testing of the base material for documentation of the Ultimate Limit State (ULS) (see Section I.4). These standards also give requirements regarding elongation, yield, and ultimate tensile capacity. For monotonic tensile tests, the difference between the engineering stress and strain and the true stress and strain is important for assessment of the ultimate capacity using nonlinear analysis. In a monotonic tensile test, the engineering stress is derived as:

$$S = \frac{P}{A_0} \tag{3.1}$$

where:

P is applied force, and
A_0 is original sectional area of test specimen.

The engineering strain is derived as:

$$e = \frac{\Delta L}{L_0} \tag{3.2}$$

where:

ΔL is increase in elongation at fracture measured over a length L_0.

The true stress is derived as:

$$\sigma = S(1 + e) \tag{3.3}$$

and the true strain is derived as:

$$\varepsilon = \ln(1 + e) \tag{3.4}$$

For stresses less than approximately 2%, the difference between the engineering stress and strain and the true stress and strain is negligible. The strain levels in cyclic loading are normally small compared with strain levels in monotonic loading to failure. Therefore, it is less important to use the concept of true stress and strain for assessment of cyclic loading as it is for assessment of the ultimate capacity. However, some parameters used for fatigue assessment are also based on actual cross-sectional area, A_f, at fracture:

$$\sigma_f = \frac{P_f}{A_f} \tag{3.5}$$

and:

$$\varepsilon_f = \ln(A_0/A_f) \tag{3.6}$$

where:

   σ_f = true fracture strength
   ε_f = true fracture ductility.

The fatigue cracks under low cycle fatigue usually nucleate from localized notch regions, and the plastic straining at such regions can be significant. Therefore, cyclic strain controlled tests can provide a better characterization of the fatigue strength of a material at such notches than can stress-controlled tests. The plastic straining at notches is contained within an elastic region, and therefore the plastic strain is limited by the boundary of the surrounding elastic body. The stress-strain behavior obtained from cyclic loading can differ somewhat from that of a monotonic tension test that is used to characterize the yield strength and the ultimate strength of materials. This behavior had already been observed by Bauschinger (1881) at the end of the nineteenth century and is often called the Bauschinger effect. Thus, cyclic stress-strain curves are needed for low cycle fatigue assessment. Monotonic stress-strain curves should not be used for calculations of low cycle fatigue damage, as the results that are obtained may be too optimistic for strain amplitudes less than about 0.6%, which is within the normal strain range for such cycle loading.

3.1.2 Cyclic Stress-Strain Curve

Three different methods can be used to derive cyclic stress-strain curves according to Stephens et al. (2001). In an industry project on low cycle fatigue that was sponsored by DNV and Daewoo Shipbuilding and Marine Engineering (DSME) in South Korea, eight specimens of different material grade and yield stress were tested to obtain cyclic stress-strain curves. For the results from this project, see also Heo et al. (2004) and Urm et al. (2004). Figure 3.1 shows a cyclic stress-strain curve of AH32 steel obtained in one of the tests. The constant strain was controlled during the cycling test by using an extensometer. A number of stabilized hysteresis loops at different strain amplitudes were used to obtain the cyclic stress-strain curve for the material. The tips of each stabilized loop were used as points in the stress-strain

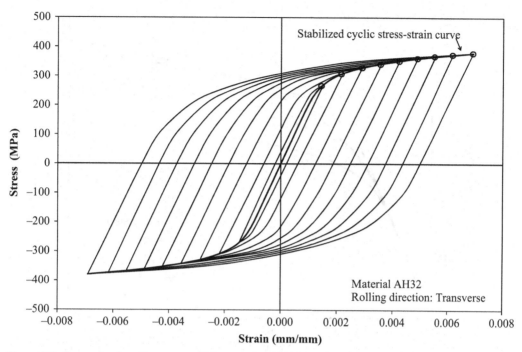

Figure 3.1. Example of cyclic stress-strain curves for AH32 steel.

diagram for derivation of a cyclic stress-strain relationship, as shown in Figure 3.1. Less than half the stress cycles used to derive the stabilized curve are shown in this figure.

A general form of the Ramberg-Osgood relation can be used to represent a cyclic stress-strain curve. This relationship is presented as:

$$\frac{\Delta \varepsilon_T}{2} = \frac{\Delta \sigma}{2E} + \left(\frac{\Delta \sigma}{2K'}\right)^{1/n'} \tag{3.7}$$

where:

K' = cyclic strength coefficient
n' = cyclic strain hardening exponent.

The coefficients in Table 3.1 were obtained for the base metal tested in the transverse rolling direction. Test results of the weld material used for welding of the steel are also included in the same table. These data were derived as the basis for establishing design guidance on low cycle fatigue in ship structures. For the heat-affected zone (HAZ), it was recommended that similar properties as for the weld metal should

Table 3.1. *Coefficients of the cyclic stress-strain curve of base metal*

Material	Mild steel	AH32	AH36	DH36	Welded metal A
n'	0.111	0.105	0.108	0.096	1073.4
K'	582.0	660.1	699.5	731.7	0.159

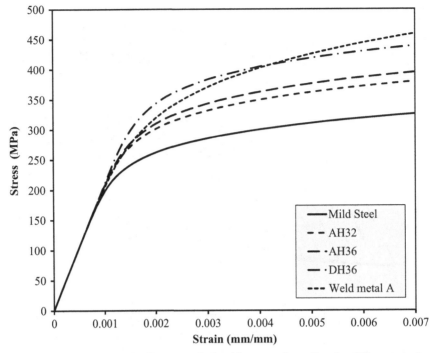

Figure 3.2. Cyclic stress-strain curves derived from strain cycling for different steel grades.

be assumed. The cyclic stress-strain curves for the data in Table 3.1 are presented in Figure 3.2.

3.1.3 Strain-Based Approach for Assessment of Fatigue Life

A stabilized stress-strain hysteresis loop from about half the fatigue life for steel is shown in Figure 3.3.

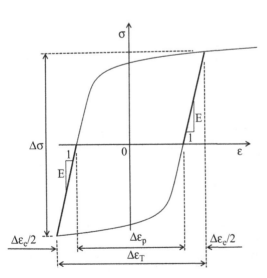

Figure 3.3. Stabilized stress-strain hysteresis loop.

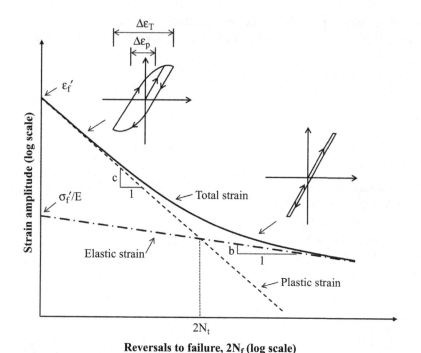

Figure 3.4. Schematic strain-life curves showing elastic, plastic, and total strain.

The total strain range at a notch is a sum of the plastic and elastic strain ranges:

$$\Delta\varepsilon_T = \Delta\varepsilon_e + \Delta\varepsilon_p \tag{3.8}$$

where:

$\Delta\varepsilon_e$ is the elastic strain range, and
$\Delta\varepsilon_p$ is the plastic strain range.

For a complete reversed cycle, the amplitude is half the range and from equation (3.7):

$$\frac{\Delta\varepsilon_T}{2} = \frac{\Delta\varepsilon_e}{2} + \frac{\Delta\varepsilon_p}{2} = \frac{\sigma_a}{E} + \left(\frac{\sigma_a}{K'}\right)^{1/n'} \tag{3.9}$$

And with respect to notation in Figure 3.3:

$$\Delta\varepsilon_T = \frac{\Delta\sigma}{E} + 2\left(\frac{\Delta\sigma}{2K'}\right)^{1/n'} \tag{3.10}$$

Un-notched specimens are cycled to the same strain as the material at the notch root. Thus, the fatigue damage can be related to the local strain, and the methodology is also known as "the local strain approach." Using this approach, the fatigue strength of an un-notched specimen is related to the fatigue strength of a notched specimen. The strain-life curves are often referred to as "low-cycle fatigue data" because much of the data are for fewer than 10^5 cycles. Strain-life curves plotted on a log-log scale are shown schematically in Figure 3.4, where $2N_f$ is the number of load reversals to failure, and N_f is the number of cycles to failure. Both the elastic part and plastic part are represented by a straight line. The elastic part intersects the strain axis at

c_f'/E and at ε_f' for the plastic part. The slopes of the elastic and plastic curves are b and c, respectively. This provides the Coffin-Manson equation for strain-life fatigue assessment:

$$\frac{\Delta \varepsilon_T}{2} = \frac{\Delta \varepsilon_e}{2} + \frac{\Delta \varepsilon_p}{2}$$

$$= \frac{\sigma_f'}{E}(2N_f)^b + \varepsilon_f'(2N_f)^c \qquad (3.11)$$

where:

  N_f = cycles to failure
  $2\,N_f$ = load reversals to failure
  σ_f' = fatigue strength coefficient
  b = fatigue strength exponent
  ε_f' = fatigue ductility coefficient
  c = fatigue ductility exponent

The straight-line behavior between the elastic stress and strain is expressed in terms of a general S-N curve of the Basquin form:

$$\frac{\Delta \varepsilon_e E}{2} = \sigma_f'(2N_f)^b \qquad (3.12)$$

with log N = (1/b)log $(2\sigma_f')$ and m = $-1/b$ when compared with the S-N curve format presented in Section 4.1.2. The fatigue strength coefficient is the intercept of the strain axis at one load reversal ($2N_f = 1$), and hence it is related to σ_f, the true fracture strength in monotonic loading. For most materials it proves to be that $\sigma_f' = \sigma_f$. The plastic strain amplitude, ε_p, is related to cycles to fatigue failure by the following equation:

$$\frac{\Delta \varepsilon_p}{2} = \varepsilon_f'(2N_f)^c \qquad (3.13)$$

The fatigue ductility coefficient is the plastic strain intercept at one load reversal ($2N_f = 1$). The relationship between the ε_f' and the ε_f is not as simple as for the fatigue strength coefficient. Although for most materials investigated $\varepsilon_f' = \varepsilon_f$ is a good approximation (see chapter by Berge in *Fatigue Handbook* in Almar-Naess 1985), for many steels the ratio between these two parameters is less than 1.0. It should be noted that $2N_t$ in Figure 3.4 is at the transition cycle from the plastic strain to the elastic strain. This transition can be calculated by equating the elastic and the plastic parts in equations (3.12) and (3.13), respectively, and results in:

$$2N_t = \left(\frac{\varepsilon_f'}{\sigma_f'}\right)^{\frac{1}{b-c}} \qquad (3.14)$$

For fatigue lives less than $2N_t$, the deformation is mainly plastic, and mainly elastic for fatigue lives longer than $2N_t$. In the high cycle region, strong materials will have a longer fatigue life, but in the low cycle region, ductile materials will be superior. A tough material with an optimum combination of strength and ductility will give the best overall fatigue strength. The transition point appears to be at a strain amplitude of 0.01 and at a fatigue life of about 2,000 reversals. At this strain range, fatigue life is relatively insensitive to material properties (see chapter by Berge in *Fatigue Handbook* in Almar-Naess 1985). The derived fatigue life from equation (3.11) refers to crack initiation in small diameter test specimens. This means that crack size at initiation is not well defined but, according to Stephens et al. (2001), may be approximately 1 mm.

3.1.4 Relationship between Elastic Strain and Nonlinear Elastic Strain

The following two hypotheses were used for finding the relation between the nonlinear elastic strain and the linear elastic stress that is usually obtained from linear finite element analysis.

- Neuber's rule for plane stress
- Glinka's rule for plane strain

The notch stress is obtained from the notch strain, according to the cyclic stress-strain curve:

$$\varepsilon = \frac{\sigma}{E} + \left(\frac{\sigma}{K'}\right)^{1/n'} \tag{3.15}$$

The actual stress and strain concentration factors are difficult to determine without a nonlinear finite element analysis. If a linear elastic finite element analysis is conducted, the nonlinear stress can be obtained by using the calculated linear elastic stress. The hypothesis by Neuber (1961) can be used to predict the nonlinear stress and strain by using the linear elastic strain or stress or vice versa, if the cyclic stress-strain relation is known. Neuber's hypothesis or rule reads:

$$K_t^2 = K_\sigma K_\varepsilon \tag{3.16}$$

where:

$$K_\sigma = \frac{\sigma_{notch}}{\sigma_{nominal}}$$

$$K_\varepsilon = \frac{\varepsilon_{notch}}{\varepsilon_{nominal}} \tag{3.17}$$

Here:

σ_{notch} = the actual stress in the notch root following the cyclic stress-strain relation

ε_{notch} = the actual strain in the notch root following the cyclic stress-strain relation

$\sigma_{nominal}$ = the nominal stress

$\varepsilon_{nominal}$ = the nominal strain

K_σ = local stress concentration factor

K_ε = local strain concentration factor

K_t = elastic stress concentration factor (as frequently used in literature related to this topic; see also Section I.1. Otherwise the notation SCF is used for this factor in this book.)

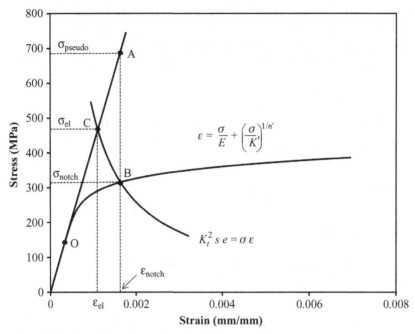

Figure 3.5. Illustration of use of Neuber's rule.

By combining equations (3.16) and (3.17), Neuber's rule can be presented as:

$$K_t^2 \sigma_{nominal}\varepsilon_{nominal} = \sigma_{notch}\varepsilon_{notch} \tag{3.18}$$

This relationship is shown in Figure 3.5. If a nonlinear strain is calculated, where $\varepsilon_{notch} = \varepsilon_T/2$, the pseudo-elastic stress can be derived from the following definition as

$$\sigma_{pseudo} = E \cdot \varepsilon_{notch}$$

The local stress concentration factor decreases sharply when the plasticity effect is significant, that is, when the nominal stress increases. In contrast, the local strain concentration factor increases when the nominal stress increases. It has also been found that Neuber's rule gives larger strain values than those calculated by finite element analysis (Heo et al. 2004). If the cyclic stress-strain relation is combined with Neuber's rule, then Neuber's formula is given as follows:

$$\frac{\sigma_{nominal}^2 K_t^2}{E} = \frac{\sigma_{notch}^2}{E} + \sigma_{notch}\left(\frac{\sigma_{notch}}{K'}\right)^{1/n'} \tag{3.19}$$

The stress concentration factor, K_t, can be derived from K' and n' as determined by experiments or nonlinear finite element analysis. K_t depends on the magnitude of the load and the sharpness of the notch. Neuber's rule is adopted here for analysis of elastic-plastic behavior of steel, as many hot spots in ship structures can be considered to be in the plane stress condition. Neuber's relation is generally considered to provide more conservative results than Glinka's relation (Glinka 1985). If the plane strain behavior is significant, the Glinka relation may be used instead of Neuber's rule:

$$\frac{\sigma_{nominal}^2 K_t^2}{E} = \frac{\sigma_{notch}^2}{E} + \frac{2}{1+n}\sigma_{notch}\left(\frac{\sigma_{notch}}{K'}\right)^{1/n'} \tag{3.20}$$

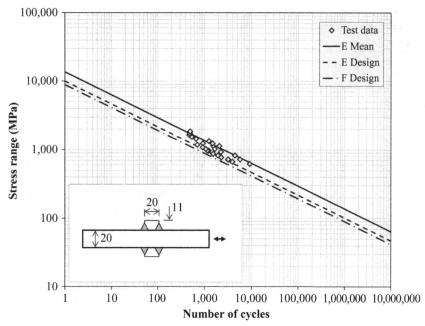

Figure 3.6. Low cycle fatigue test data compared with nominal stress S-N curves from DNVGL-RP-C203.

The fatigue test data for a welded cruciform joint shown in Figure 3.6 are listed in Table 3.2. The elastic amplitude is derived by use of Neuber's rule as:

$$\sigma_{el} = \sqrt{\frac{\sigma_{notch}\varepsilon_{notch}E}{K_t^2}} \qquad (3.21)$$

Here $K_t = 1.0$ as the notch stress is accounted for in the nominal stress S-N curve that is selected for comparison. The test data are presented together with S-N curve E (FAT 80) (Table 4.1) for the considered detail in Figure 3.6. The first data point in Table 3.2 is used as example in the following. The notch strain is derived as $\varepsilon_{notch} = \varepsilon_T/2 = 0.0030$. The stress amplitude is derived as $\sigma_a = \varepsilon_{notch}E = 0.0030 \cdot 2.06 \cdot 10^5$ MPa = 518 MPa. Thus, the stress range as input to Figure 3.6 is $2 \cdot 518$ MPa = 1036 MPa at 937 stress cycles. The test data are compared with the E- and F-curves in DNVGL-RP-C203 (2016). The E-curve applies to transverse plates with thicknesses up to 25 mm, while the F-curve should be used for larger thicknesses of the attachment. Based on the results in Figure 3.6, it is acceptable to use the extrapolated S-N curve into the low cycle region, with the same negative inverse slope m = 3.0, as is used in the high cycle region together with nominal calculated stress ranges from linear elastic analysis.

The notch stress concentration at the weld toe was further calculated by nonlinear finite element analysis based on a fictitious notch radius of 1.0 mm, which is the same as that recommended for linear elastic analysis using the notch method described in Section 9.6. A stress concentration factor (SCF) of 2.2 was calculated for use in the pseudo-stress analysis. The test data shown in Table 3.2, together with this SCF, were then compared with the basic S-N curve for machined welds, the C-curve in DNVGL-RP-C203 (2016). Thus, the fatigue test data were presented in terms of

Table 3.2. *Low cycle fatigue test for cruciform joint – mild steel*

Nominal strain range (ε_T)	Stress range (MPa)	Cycles to failure
0.0060	569	937
0.0055	559	1,541
0.0050	547	1,676
0.0045	534	2,275
0.0040	519	4,468
0.0035	500	5,755
0.0030	475	9,223
0.00325	488	3,962
0.0035	500	3,205
0.00375	510	2,286
0.0040	519	1,954
0.00425	527	1,558
0.00475	541	1,239
0.00475	541	1,172
0.0050	547	1,074
0.00525	553	915
0.00575	564	701
0.0055	559	2,097
0.0060	569	1,500
0.0065	577	1,270
0.0070	585	661
0.0075	593	538
0.0080	599	475
0.0085	606	487
0.0090	611	498

pseudo-notch stress range and are shown in Figure 3.7. It is observed that there is a good correlation between the test data and the basic S-N curve when the S-N curve is extrapolated into the low cycle region, with the same negative inverse slope, m = 3.0, as is used in the high cycle region. Furthermore, it is found acceptable to use Neuber's formula to transform cyclic stress-strain relations into stress ranges to be used for fatigue assessment in the low cycle fatigue region. Alternatively, the rule by Glinka (1985) may be used. Some test specimens with ground weld toes and some with tungsten inert gas (TIG)-dressed weld toes were also fatigue tested in this low cycle range. The improvement was not significant and was lower than for the high cycle range.

The presented methodology was also used to establish a design curve for fatigue assessment of tubular joints subjected to low cycle fatigue in the NORSOK Standard N-006 from 2015. The S-N curve for tubular joints in air is described in this NORSOK standard, with m = 3.0 and log a for design of 12.476 for the region 10^4–10^7 cycles. At a stress amplitude of 1,552 MPa this curve corresponds to 100 cycles. This includes SCF = K_t = 1.0. By using Neuber's rule in equation (3.19), σ_{notch} = 390.59 MPa and $\Delta\sigma$ = 781.18 MPa. It is further assumed that there is a gradual transition from the low cycle region to the high cycle region at around 10^5 cycles. There now being two points in the S-N diagram, a linear curve through these points can be derived, where the negative inverse slope of the curve is m = 7.486 and log a = 23.655. This S-N curve is intended for an additional analysis of the ULS with respect to storm conditions in which the waves are so large that they hit the

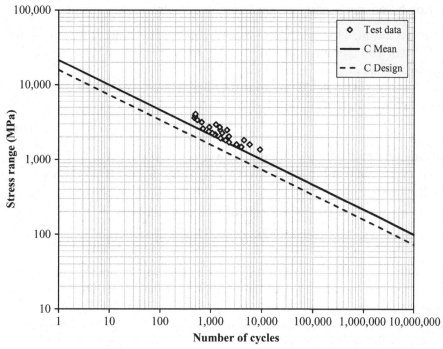

Figure 3.7. Low cycle fatigue test data compared with nominal stress S-N curve for machined weld without notch from DNVGL-RP-C203.

deck structure in fixed platforms, and such that failure in this limit state may occur due to low cycle fatigue.

Where cyclic stress-strain relations are unknown, a plasticity correction factor, k_e, may be introduced into the analysis. This plasticity correction factor can be defined as follows:

$$k_e = \frac{\sigma_{pseudo}}{\sigma_{elastic}} \tag{3.22}$$

where:

σ_{pseudo} = the pseudo-elastic stress, shown in Figure 3.5
$\sigma_{elastic}$ = the elastic stress, shown in Figure 3.5.

The following plasticity correction factor from BS5500 (1997) may be used for correction of the elastic stress range $\Delta\sigma$ when the cyclic stress-strain relation is not known (see Urm et al. 2004):

$$
\begin{aligned}
k_e &= 1 && for \ \frac{\Delta\sigma}{\sigma_y} < 2 \\
k_e &= 0.443\left(\frac{\Delta\sigma}{\sigma_y} - 1\right)^{0.5} + 1 && for \ 2 \leq \frac{\Delta\sigma}{\sigma_y} \leq 3.0 \\
k_e &= 0.823 + 0.164\frac{\Delta\sigma}{\sigma_y} && for \ \frac{\Delta\sigma}{\sigma_y} > 3.0
\end{aligned}
\tag{3.23}
$$

where

σ_y = yield strength.

3.1.5 Notch Sensitivity and Fatigue Strength of Notched Specimens

This section is included to explain the wording used in some of the fatigue literature concerned with notch sensitivity and the fatigue strength of notched specimens. This section should be read after Section 3.1.3.

The local stress at a notch is related to the nominal stress through the theoretical elastic stress concentration factor K_t:

$$\Delta\sigma_{notch} = K_t \Delta\sigma_{nominal} \tag{3.24}$$

This equation states that fatigue strength is inversely proportional to K_t. For a small notch radius this relationship is not directly related to the fatigue notch factor, K_f, derived from such specimens tested in the laboratory. K_f is derived from:

$$K_f = \frac{\text{Fatigue strength of a smooth specimen}}{\text{Fatigue strength of a notched specimen}} \tag{3.25}$$

at the same fatigue life.

Generally, K_f varies with the stress level and the notch acuity. The fatigue strength of notched specimens is related to the size of the fatigue process zone or the volume subjected to large stress or strain ranges at the notch. The size of this zone is dependent on the notch radius. The relationship between the two parameters is presented by Peterson (1974) as:

$$K_f = 1 + \frac{K_t - 1}{1 + \frac{\alpha}{r}} \tag{3.26}$$

where:

α is a material constant, and
r is notch radius.

As indicated by this equation, small notches have less effect on fatigue strength than predicted by the value of K_t. At high stresses, or in specimens with sharp notches, the stress condition at the notch is plastic or nonlinear as shown in Section 3.1.3. The elastic stress concentration factor, K_t, will then not provide an adequate description of the stress at the notch. In general, K_f tends to decrease with increasing stress. As a consequence, a direct measurement of K_f for a given notch will require that several stress and strain levels are tested, and this is not a very attractive solution to practical problems.

3.1.6 Combination of Fatigue Damage from Low Cycle and High Cycle Fatigue

Ship structures experience both static and dynamic loads during their service life. Normally, the fatigue capacity of most joints is checked in view of high cycle fatigue (HCF) due to dynamic loads. Although an HCF design check has been carried out at the design stage, cracks have been reported within a few years after delivery of ships, and these might be due to low cycle fatigue (LCF). Significant yielding has been observed at these areas for static loads, and it is assessed that the linear elastic stress range is more than three times the yield stress of the material. The LCF strength of highly stressed locations that are under repeated cyclic static loads, mainly due to

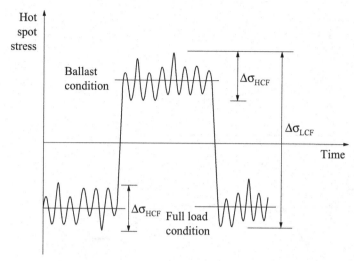

Figure 3.8. Different load conditions resulting in LCF and HCF.

cargo loading and unloading, must be checked as significant yielding can cause cracks and/or paint cracks at hot spots, even if the dynamic stress ranges from wave action are low. Various locations, including web stiffeners on the top of bottom and inner bottom longitudinals due to large frame spacing, heel and end connections of horizontal stringers in transverse bulkheads to longitudinal bulkhead, and lower stool connections to the inner bottom of bulk carriers are the most vulnerable hot spots with respect to LCF.

The fatigue life in low cycle and high stress regions is normally expressed in terms of the total strain range rather than the stress range. Significant plasticity may occur at a notched area due to the high stresses for locations under repeated loading and unloading, thus leading to a consideration that the strain range might be a preferred design parameter. However, an approach based on the pseudo-elastic notch stress range (or hot spot stress range), instead of the total strain range approach, is adopted in the following. In each load condition the structure is subjected to dynamic actions from wave loading, resulting in HCF damage between changes in load conditions. A load cycle causing LCF damage is thus the sum of a stress range representative for the HCF and the stress range resulting from change in loading conditions, as indicated in Figure 3.8. The total fatigue damage is the sum of fatigue damages from HCF and LCF.

3.2 Methodology for Assessment of High Cycle Fatigue

3.2.1 Calculation of Stresses and Relation to Different S-N Curves

General

The maximum principal stress range as defined in Section 9.4.1 is considered to be a significant parameter for fatigue analysis of fatigue crack growth at hot spots in welded structures, and this is the main parameter for description of stress range used in this book. Equivalent stress range similar to von Mises can also be used for calculation of fatigue at notches in the base material where the time for initiation of

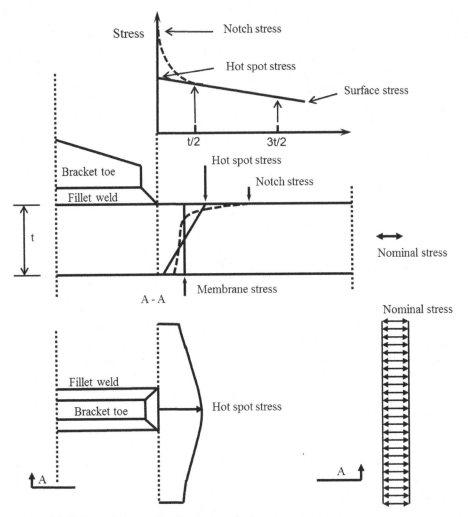

Figure 3.9. Schematic stress distribution at a hot spot.

a crack is a significant part of the fatigue life. Then the calculation of the resulting stress range should be based on the stress range of each component during load cycles:

$$\Delta\sigma_e = \sqrt{\Delta\sigma_x^2 + \Delta\sigma_y^2 + \Delta\sigma_z^2 - \Delta\sigma_x\Delta\sigma_y - \Delta\sigma_y\Delta\sigma_z - \Delta\sigma_x\Delta\sigma_z + 3(\Delta\tau_{xy}^2 + \Delta\tau_{yz}^2 + \Delta\tau_{zx}^2)}$$

(3.27)

where each stress range component can be calculated as the difference between the maximum stress and the minimum stress during a load cycle.

Fatigue analysis may be based on different methodologies, depending on which is found to be most efficient for the considered structural detail. It is important that the stresses are calculated in agreement with the definition of the stresses to be used, together with a particular S-N curve; see Figure 3.9. Three different analysis concepts for fatigue analysis of welded components are defined in the following:

- Nominal stress approach with corresponding S-N curves, which is presented in Section 4.1.
- Hot spot stress approach, which is described in Chapter 9 for plated structures and in Chapter 8 for tubular joints (also referred to as the structural stress method in the literature).
- Notch stress approach, which is presented in Section 9.6.

The nominal stress is understood to refer to a stress in a component that can be derived by classical theory, such as beam or plate theory. In a simple plate specimen with an attachment, as shown in Figure 2.8, the nominal stress is simply the membrane stress that is used for plotting of the S-N data from the fatigue testing. Examples of fatigue design using this procedure are shown in Section 6.7. The hot spot stress is understood to refer to the geometric stress created by the considered detail, and the notch stress due to the local weld geometry is excluded from the stress calculation as it is assumed to be accounted for in the corresponding hot spot S-N curve. The notch stress is defined as the total stress resulting from the geometry of the detail and the nonlinear stress field due to the notch at the weld toe. The determination of the stresses to be used together with the different S-N curves is described in greater detail in the next section. The procedure for the fatigue analysis is based on the assumption that it is only necessary to consider the ranges of cyclic principal stresses in determining the fatigue endurance, that is, mean stresses are normally disregarded for fatigue assessment of welded connections in offshore structures. However, the mean stress is accounted for when assessing fatigue at hot spots in the base material without significant residual stress; see also Section 3.3.3. The effect of mean stress is accounted for when planning inspection for fatigue cracks in vessels as shown in Section 3.3.6. The effect of mean stress is included in the design procedure for both base material and for welded connections in sailing ships (see IACS 2013).

Plated Structures Using Nominal Stress S-N Curves

When a potential fatigue crack is located in the parent material at the weld toe, the relevant hot spot stress is the range of maximum principal stress adjacent to the potential crack location, with stress concentrations being taken into account. The joint classification and corresponding S-N curves allow for the local stress concentrations created by the joints themselves and by the weld profile. The design stress can therefore be regarded as the nominal stress adjacent to the weld under consideration. However, if the joint is situated in a region of stress concentration resulting from the gross shape of the structure, this must also be accommodated. As an example, for the weld shown in Figure 3.10a, the relevant local stress for fatigue design would be the tensile stress, $\sigma_{nominal}$. For the weld shown in Figure 3.10b, the stress concentration factor for the global effect of the hole must, in addition, be accounted for, giving the relevant local stress of $SCF \cdot \sigma_{nominal}$, where SCF is the stress concentration factor due to the hole. Thus the local stress is derived as:

$$\sigma_{local} = SCF\sigma_{nominal} \tag{3.28}$$

σ_{local} shall be used together with the relevant S-N curves, D to G, which are presented in Section 4.1, depending on joint classification. The maximum principal stress range

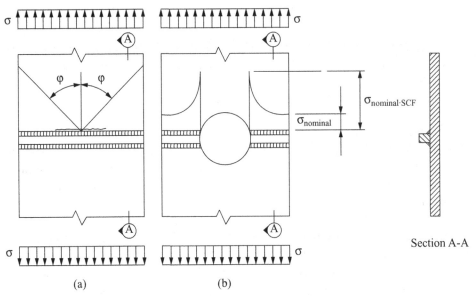

Figure 3.10. Explanation of local stresses at a weld toe.

is a significant parameter for fatigue assessment, together with the shear stress and stress normal to the weld toe, as shown in Section 2.7.4.

Plated Structures Using Hot Spot Stress S-N Curves

For detailed finite element analysis of welded plate connections other than tubular joints, it may be convenient to use the hot spot stress (or structural stress) method for fatigue life assessment; see Section 9.1 for further guidance. The relationship between nominal stress and hot spot stress may be defined as:

$$\sigma_{hot\ spot} = SCF\sigma_{nominal} \tag{3.29}$$

where SCF is the structural stress concentration factor, which is normally denoted as stress concentration factor. For an effect of misalignment during fabrication on hot spot stresses in plated structures, see Section 5.3.3.

Tubular Joints

For a tubular joint – that is, a brace-to-chord connection – the stress to be used for design purposes is the range of idealized hot spot stress, defined as the greatest value of the extrapolation of the maximum principal stress distribution immediately outside the region affected by the geometry of the weld; see also Section 9.5. The hot spot stress to be used in combination with the T-curve for the outside is calculated as:

$$\sigma_{hot\ spot} = SCF\sigma_{nominal} \tag{3.30}$$

where:

SCF = stress concentration factor, as given in fatigue design standards, ISO 19902 (2007), API RP2A (2014), and DNVGL-RP-C203 (2016).

See Section 8.3 for single-side welded tubular joints.

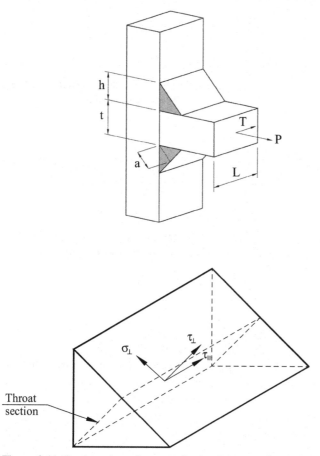

Figure 3.11. Explanation of stresses in the throat section of a fillet weld.

Fillet Welds

The relevant stress range for potential cracks in the weld throat of load-carrying, fillet welded joints and partial penetration, welded joints may be calculated as:

$$\Delta \sigma_w = \sqrt{\Delta \sigma_\perp^2 + \Delta \tau_\perp^2 + 0.2 \Delta \tau_{//}^2} \tag{3.31}$$

where the stress components are explained in Figure 3.11. See Section 2.5.5 for more background for this equation. It is assumed that the cycling of the different stress components is in phase, which is the normal situation for structures subjected to dynamic loading from wind and waves. The problem becomes more complex for out-of-phase loading; see also Section 2.7.5.

The total stress fluctuation (i.e., maximum compression and maximum tension) should be considered as being transmitted through the welds for fatigue assessments. The design criterion includes a combination of different stress components in the fillet weld. The nominal stress in the fillet weld due to the axial force is defined as the combined stress:

$$\sigma_w = \sqrt{\sigma_\perp^2 + \tau_\perp^2} \tag{3.32}$$

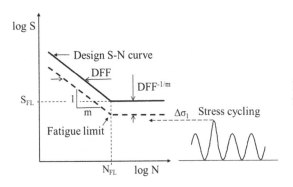

Figure 3.12. Stress cycling where further fatigue assessment can be omitted.

where:

$\sigma_\perp$ = mean normal stress acting on the throat section, as shown in Figure 3.11.

$\tau_\perp$ = mean shear stress acting on the throat section, as shown in Figure 3.11.

In several publications this stress is also referred to as "engineering shear stress" and is derived for the fillet weld in Figure 3.11 as:

$$\sigma_w = \frac{P}{2aL} \tag{3.33}$$

That is, the engineering shear stress is simply the force divided by the weld throat area. By using further calculations based on equilibrium considerations, the stress derived from equation (3.32) can be shown to be equal to that derived from equation (3.33). The parallel shear stress in Figure 3.11 is similarly derived as:

$$\tau_\parallel = \frac{T}{2aL} \tag{3.34}$$

where the notations are illustrated in Figure 3.11.

3.2.2 Guidance Regarding When Detailed Fatigue Analysis Is Required

Detailed fatigue analysis can be omitted if the largest local stress range defined in equation (3.28) for actual details is below the fatigue limit at 10^7 cycles in Table 4.1 for air. A similar approach can be used for seawater with cathodic protection. For Design Fatigue Factors (DFF) larger than 1.0, the allowable fatigue limit should be reduced by a factor $(DFF)^{-0.33}$. For definition of DFF, see Section 12.7; see Section 6.7 for an example of using this methodology for fatigue assessment. The use of the fatigue limit is illustrated in Figure 3.12. Although detailed fatigue assessment can be omitted if the largest stress cycle is below the fatigue limit, nevertheless, if there is one stress cycle $\Delta\sigma_1$ (or in reality a few stress cycles, ref. experience from reeling of pile lines) above the fatigue limit, as in the example in Figure 3.13, then a further fatigue assessment is required in which the fatigue damage from the stress cycles, $\Delta\sigma_2$, must also be included. This is because a fatigue crack may be initiated when some stress ranges exceed the fatigue limit, and when a fatigue crack is initiated, it can grow for stress cycles lower than the fatigue limit, as is also explained in Section 3.2.3. Requirements for detailed fatigue analysis may also be assessed based on the fatigue assessment charts in Figure 10.3 for a Weibull long-term stress range distribution.

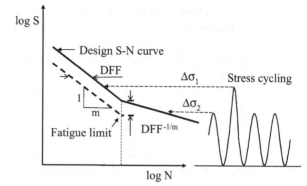

Figure 3.13. Stress cycling where a detailed fatigue assessment is required.

The expression for the reduction factor on the fatigue limit can be outlined as follows. The fatigue limit for the full drawn design S-N curve in Figure 3.12 can be denoted by stress range S_{FL} starting at N_{FL} number of cycles. Then the dotted S-N curve shown in the same figure by DFF cycles to the left of the design curve is considered. This curve is described by cutoff with the log N axis equal log a and negative inverse slope equal m. The load cycles on this curve at stress range S_{FL} can be described by

$$\log\left(\frac{N_{FL}}{DFF}\right) = \log a_d - m\log S_{FL} \tag{3.35}$$

and at stress range $\Delta\sigma_1$ the following relation applies:

$$\log N_{FL} = \log a_d - m\log\Delta\sigma_1 \tag{3.36}$$

By subtracting equation (3.35) from equation (3.36) the following result is obtained:

$$\log DFF = -m\log\Delta\sigma_1 + m\log S_{FL} = \log\left(\frac{\Delta\sigma_1}{S_{FL}}\right)^{-m} \tag{3.37}$$

From this equation it is seen that:

$$DFF = \left(\frac{\Delta\sigma_1}{S_{FL}}\right)^{-m} \tag{3.38}$$

and further from this equation:

$$\Delta\sigma_1 = S_{FL}DFF^{-1/m} \tag{3.39}$$

The existence of a fatigue limit is considered to be dependent on the type of material; see Sonsino (2007). Decreasing fatigue strength beyond the knee point in the S-N curve has been more accepted for materials like aluminum and austenitic steels than for ferritic steels. Under the premise that no environmental or fretting corrosion is present, Sonsino proposed that, provided relevant test data cannot be achieved, the fatigue lives may be extrapolated into the long life range beyond the knee point with the following slope data in the design S-N curves:

- For steels, cast irons, and magnesium alloys, m = 45, corresponding to a decrease of 5% per decade if no tensile residual stresses are present.

- For aluminum alloys, welded magnesium alloys, and welded steels, m = 22, corresponding to a decrease of 10% per decade.

The last recommendation has been included in the IIW recommendations (Hobbacher 2009). This information may be considered more important for design of machinery than for marine structures subject to wind and waves. Marines et al. (2003) indicates the number of cycles for automobile engines n = 10^8 cycles, high speed trains n = 10^9 cycles, and gas turbine disks n = 10^{10} cycles.

3.2.3 Fatigue Damage Accumulation – Palmgren-Miner Rule

The fatigue life may be calculated based on the S-N fatigue approach with the assumption of linear cumulative damage, which is also known as the Palmgren-Miner rule. When the long-term stress range distribution is expressed by a stress histogram, consisting of a convenient number of constant stress range blocks, $\Delta\sigma_i$, each with a number of stress repetitions, n_i, the fatigue criterion reads:

$$D = \sum_{i=1}^{k} \frac{n_i}{N_i} = \frac{1}{a_d} \sum_{i=1}^{k} n_i \cdot (\Delta\sigma_i)^m \leq \eta \tag{3.40}$$

where:

D = accumulated fatigue damage
a_d = intercept of the design S-N curve with the log N axis
m = negative inverse slope of the S-N curve
k = number of stress blocks
n_i = number of stress cycles in stress block i
N_i = number of cycles to failure at constant stress range $\Delta\sigma_i$
η = usage factor = 1/DFF from fatigue design standards.

When a histogram is used to express the stress distribution, the number of stress blocks, k, should be large enough to ensure reasonable numerical accuracy, and should not be fewer than 20. Due consideration should be given to selection of the integration method, as the position of the integration points may significantly affect the calculated fatigue life, depending on the integration method. See Section 10.6.1 for calculation of fatigue damage using design charts, and Section 10.7 for derivation of fatigue damage calculated from different processes.

The Palmgren-Miner rule is not considered perfect, and there have been many suggestions for improvement. However, it is simple to use and has become the main fatigue damage accumulation rule in fatigue design of welded structures that are subject to variable amplitude loading. Miner did not include stress ranges below the fatigue limit, as he believed these stress cycles did not contribute to fatigue damage, and this has been a criticism of his work. However, when a second S-N curve with a slightly different slope is included, it is generally acceptable for design of marine structures. Niemi (1997) assessed the deficiencies in the Palmgren-Miner rule related to fatigue of welded components and concluded that when D = 1.0, this rule works quite well in fatigue analysis of welded joints, provided that the number of cycles to failure, N, is resolved from a lower-bound characteristic S-N curve that fully accommodates the residual stress effects. This is achieved by performing fatigue testing at $\sigma_{max} = \sigma_y$ (yield strength), as described in Section 2.1.3.

This issue may also be linked to the question about the number of cycles in the S-N curve at the fatigue limit as reflected by fatigue test data reported by Dahle (1993), Marquis (1996), and Zhang and Maddox (2009a) where it is indicated that the number of cycles at the fatigue limit in an air environment is beyond 10^7 cycles.

According to Fisher (1984) random-variable testing of details in bridges demonstrated that the variation in the cumulative damage ratio and variable load test data was no greater than the variation in the constant cycle test data. One should not expect a reduction in variability under random loading as compared to the variability experience by constant cycle data.

Zhang and Maddox (2009a) presented the results from an investigation on the effect loading spectra with different mean stress on the validity of Palmgren-Miner's rule on the fatigue performance of two types of welded joints. Significant residual stresses were measured after fabrication of these specimens. Three types of loading spectra were used to investigate the effect of mean stress:

A. Stress cycling down from a constant maximum stress of 280 MPa.
B. Stress cycling at a constant mean stress of 175 MPa.
C. Stress cycling up from a constant minimum stress of 70 MPa.

The spectra included the same stress range distribution with identical number of cycles for each stress range. Based on these tests it was found the Palmgren-Miner's rule was non-conservative for all tests under spectrum A with values down to 0.4. It was modestly non-conservative around 0.8 for spectrum B and conservative around 1.3 for spectrum C.

Spectrum A is considered to be more severe than actual loading conditions in typical marine structures. Spectrum B may be relevant for loading in tethers in tension leg platforms. However, the mean stress is less for most other details in, for example, fixed offshore structures. A value of 0.8 is considered to be well within the range of uncertainty accounted for when establishing Design Fatigue Factors in Sections 12.2 and 12.7.

Use of the Palmgren-Miner rule is relatively straightforward in cases where all the stress cycles are above the fatigue limit. However, in marine structures a significant number of load cycles below the fatigue limit also occur. In the early part of the fatigue life these cycles do not contribute to crack growth, provided that the cracks are small enough that the resulting range of the stress intensity factor is below the threshold value for the stress intensity factor range, based on linear fracture mechanics as presented in Chapter 16. As the cracks grow larger, the stress threshold value in terms of stress range reduces, and more of the stress cycles contribute to crack growth. As previously mentioned, Miner assumed that stress cycles below the fatigue limit do not contribute to fatigue damage, and a modification of the Palmgren-Miner rule was proposed by Haibach (1970). As not all cycles are effective in early fatigue life, the number of cycles can be reduced. However, Haibach found it easier to increase the number of cycles to failure in the denominator by changing the slope of the S-N curve below the fatigue limit to $m_2 = 2m - 1$, where m is the negative inverse slope of the S-N curve to the left of the fatigue limit. (Some design standards also present $m_2 = m + 2$, which gives the same m_2 value for m = 3.0.) This approach has been implemented in a number of later fatigue design standards. Thus, the accuracy of the Palmgren-Miner rule is also linked to the way in which the S-N curve below the fatigue limit is presented in design standards; see studies reported by Dahle (1993), Marquis (1996), and Zhang and Maddox (2009a, 2012). It might be added that the actual slope of the S-N curve below the fatigue limit may be considered to be a function of the long-term stress range distribution as some large stress range are needed to initiate a crack and this crack may grow further at small stress ranges when it has been initiated. Due to uncertainty of the slope of this part of the S-N curve one may keep the m_2 value equal to m when m is as large as 5.0, as then the slope of the second part of the S-N curve becomes less significant for the calculated fatigue life of marine structures subjected to typical environmental loads.

3.3 Residual Stresses

3.3.1 Residual Stresses due to Fabrication

Residual stresses and mean stresses are important factors that may govern fatigue capacity in many structural components. In design standards for land structures and fixed offshore structures it is normally assumed that there are residual stresses at the hot spot at welds that correspond to the yield stress in tension, and therefore it is assumed that the whole stress cycle provides an effective stress cycling at the hot spot in tension, independent of mean stress from external cyclic loading; see, for example, Harrison (1981) and Ohta et al. (1997). In fatigue assessment of ship structures, the nominal stress ranges into compression can be less detrimental than that of tension. The residual stress at the weld root in single-sided girth welds in risers and pipelines is also found to be small as compared with that at the outside weld toe. In fracture mechanics analysis it is also possible to account for residual stress and mean stress through calculation of effective stress intensity ranges (see Chapter 16).

The hot spot stress range in braced structures shows that the mean stress due to the load effect from external loading is not far from zero. Based on experience from fatigue testing of welded structures and long-term load distributions on jacket structures and semi-submersibles, the effect of the mean stress for these structures can be assessed as being small in terms of fatigue life and can be ignored. Mean stress may have a significant effect in butt welds in jacket legs with some topside weights, but these welds normally show long fatigue lives without accounting for the mean stress effect. It might also be of interest for jacket piles, but these cannot be inspected and should be designed with sufficient design fatigue factor such that their reliability is acceptable without in-service inspection. For floating production vessels, the use of a reduction factor depending on nominal mean stress level has been proposed for calculation of an effective stress range for the purpose of inspection planning. However, due to uncertainties about residual stresses after fabrication, the mean stress effect is not recommended for design of welded offshore structures.

3.3.2 Shakedown of Residual Stresses

Measurements of residual stresses at welded connections show that significant tensile stresses may be present after fabrication. The residual stresses at a weld toe may be

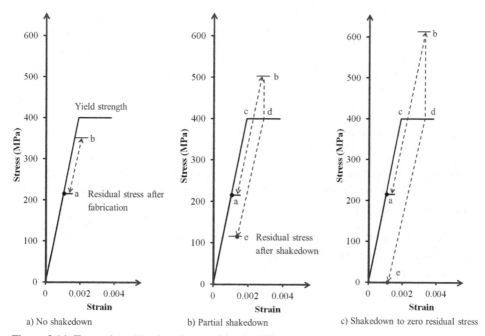

a) No shakedown   b) Partial shakedown   c) Shakedown to zero residual stress

Figure 3.14. Examples of load cycles resulting in different residual stresses at hot spot.

as large as yield stress of the base material. When the structure is subjected to an external loading, some different developments of the stress condition at the hot spot may occur, as illustrated in Figure 3.14. Here, a residual tensile stress normal to a weld toe is assumed to be equal to 215 MPa. An example is shown in Figure 3.14a, in which the increased stress at the hot spot due to external load is not so large that local yielding at the hot spot will occur. The stress at the hot spot is increased from position a to b in the figure, meaning that the residual stresses will not change after unloading. It is assumed here that the stress increase at the considered hot spot is equal to the hot spot stress multiplied by a notch stress concentration factor, K_w, which is due to the local weld geometry. Here a K_w in the order of 1.5 is assumed, but it may actually be larger at weld notches as has been shown in Radaj et al. (2006). Figure 3.14b shows an example where the increased stress at the hot spot due to external load implies local yielding at position c. It is assumed that the load is increased to level b when stresses are calculated elastically. This implies a permanent plastic straining at the hot spot, with a permanent strain increase from c to d before the structure is unloaded, with the same load amplitude from d to e that follows an elastic curve. Due to the permanent strain elongation introduced at the hot spot, the residual stress has now been reduced to position e. This procedure for shakedown can be used, as indicated in Figure 3.14c, to establish a criterion for shakedown of residual stresses to zero, or even below zero if the maximum stress is larger than that at position b. For shakedown to occur it is necessary that the tensile notch stress exceeds the material yield stress. It might be added that Figure 3.14 is made for illustration purpose, and more accurate plastic strains c–d can be derived using Neuber's rule in Section 3.1.4 or a nonlinear analysis program.

Due to differences in load conditions, the effect of loading condition with respect to shakedown of residual stresses may be larger for floating structures than the load

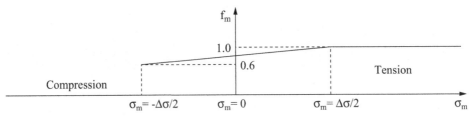

Figure 3.15. Mean stress reduction factor for base material.

from the wave environment only, as illustrated in Figure 3.8. In such a case, the following conditions for a maximum load that is likely to occur during the first year of service life are considered:

Shakedown of residual stress cannot be documented.
Shakedown of residual stress can be documented.
The requirement for documentation of fully shakedown of residual stresses reads:

$$\sigma_{hot\ spot\ \max 1\ year} K_w \geq \sigma_y \qquad (3.41)$$

where:

σ_y = actual yield stress of the base material at the considered hot spot, as derived from material certificates
K_w = 1.5 (local stress increase due to the weld notch)
$\sigma_{hot\ spot\ \max 1\ year}$ = maximum hot spot stress that will probably occur during the first year in service.

3.3.3 Mean Stress Reduction Factor for Base Material

The mean stress effect can be used for fatigue assessment of the base material where residual tensile stresses have not been introduced, such as through fabrication of tubular members by cold forming. The following reduction factor on stress range can be used in design analysis as well as for inspection planning:

$$f_m = \frac{\sigma_t + 0.6\,|\sigma_c|}{\sigma_t + |\sigma_c|} \qquad (3.42)$$

where:

σ_t = maximum tension stress, where tension is defined as positive
σ_c = maximum compression stress, where compression is defined as negative
f_m = 0 if the full stress cycle is compressive, as illustrated in Figure 3.15.

3.3.4 Residual Stress in Shell Plates in Tubular Towers after Cold Forming

The fabrication of shell structures to a circular shape in large-diameter tubular structures is by cold forming of rolled plates. This introduces residual stresses in the circumferential direction and also in the longitudinal direction of the cylinders.

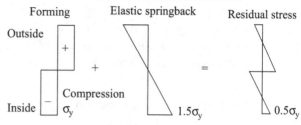

Figure 3.16. Development of residual stress in a monopile cylinder during cold forming.

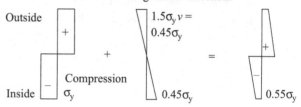

The strain at yielding for a material with yield strength equal 400 MPa is $\varepsilon_y = \sigma_y/E = 400/2.1 \cdot 10^5 = 0.0019$.

The strain at the surface of a cylinder is derived as relative elongation of the outer circumference versus the circumference at the mid-surface that is not strained during cold forming and bending to a circular shape:

$$\varepsilon = \frac{\Delta l}{l} = \frac{2\pi \, (r_0 + t/2) - 2\pi r_0}{2\pi r_0} = \frac{t}{2r_0} \tag{3.43}$$

where:

r_0 = radius of cylinder to the mid thickness, and
t = thickness of the cylinder.

With diameter, D, of 5000 mm and thickness, t, of 80 mm, this gives $\varepsilon = t/D = 80/5000 = 0.016$. Thus, cold forming will imply significant plastic straining in the circumferential direction with compressive stress on the inside and tensile stress on the outside, as indicated in Figure 3.16. This will also result in plastic straining in the longitudinal direction, as indicated in Figure 3.16, by assuming von Mises yield criterion and isotropic hardening of the material. A further development of the residual stresses depends on how the fabrication is performed. If a longitudinal weld is made at this clamped geometry, this will also be the resulting residual stress. However, if following cold rolling the cylinder is released before welding, there will be springback due to elastic deformation, as indicated in the figure. The resulting residual stress in the circumferential direction will be largest close to the mid-thickness of the plates based on a simplified analysis using a perfect plastic material; however, a nonlinear finite element analysis with a more normal stress strain curve shows residual stresses at the mid-thickness close to zero and the main residual stresses similar to those shown here are calculated at the surfaces of the plate. The release of the elastic springback moment in the circumferential direction will also imply a change in the longitudinal stress that is equal to the circumferential springback stress multiplied by

Poisson's ratio. Based on Figure 3.16 and the referred finite element analysis, there are compressive stresses in the longitudinal direction of the tubular sections on the inside of the plate section and tensile stresses at the outer part of the plate. This information may be useful for assessment of fatigue capacity at J-tube holes in monopiles used as support structures for wind turbines.

3.3.5 Mean Stress Reduction Factor for Post-Weld Heat-Treated Welds

The mean stress effect can be used for fatigue assessment of welds that have been subject to Post Weld Heat Treatment (PWHT), such that most of the residual tensile stresses have been removed (see Section 2.2.3), indicating that the following reduction factor on stress range can be used in design as well as for inspection planning:

$$f_m = \frac{\sigma_t + 0.8\,|\sigma_c|}{\sigma_t + |\sigma_c|} \qquad (3.44)$$

where:

σ_t = maximum tension stress, where tension is defined as positive

σ_c = maximum compression stress, where compression is defined as negative.

Regarding whether PWHT is always beneficial, as compared with as-welded condition, it should be noted that not many hot spots have compressive stresses similar to those shown in Section 2.3.4. Niemi (1997) referred to a case where the combined effect of residual stress and the maximum hot spot stress amplitude of 225 MPa probably resulted in a relatively large beneficial compressive stress at the weld toe at a doubling plate. This is in line with the description of shakedown of residual stresses as presented in Section 3.3.2. The stress amplitudes in marine structures are normally not as large as this; see, for example, maximum allowable stress ranges during a typical service life in Section 10.6. Thus, PWHT is a controlled way of reducing residual stresses and may provide the beneficial effect of crack closure for the part of the stress range that is compressive. Other approaches to reducing residual stresses include static loading, which leads to significant tensile stresses at the hot spot, such as by tank testing or by the use of more purpose-made large tensile loads.

3.3.6 Mean Stress Reduction Factor for Inspection Planning for Fatigue Cracks in As-Welded Structures

The following mean stress reduction factor may be used for purpose of inspection planning for fatigue cracks in a floating production vessel (DNVGL-RP-0001 2015):

$$f_m = \begin{cases} 1.0; & \dfrac{\sigma_{mean\ eff}}{\Delta\sigma} \geq 0.5 \\[2ex] \max\left[0.6, 0.9 + 0.2\dfrac{\sigma_{mean\ eff}}{\Delta\sigma}\right]; & \dfrac{\sigma_{mean\ eff}}{\Delta\sigma} \leq 0.5 \end{cases} \qquad (3.45)$$

where:

$\sigma_{mean\ eff} = \sigma_{mean}$ where shakedown to zero residual stress has been documented

$\sigma_{mean\ eff} = \sigma_{mean} + \sigma_{Res}$ otherwise

σ_{Res} = residual stress at the hot spot.

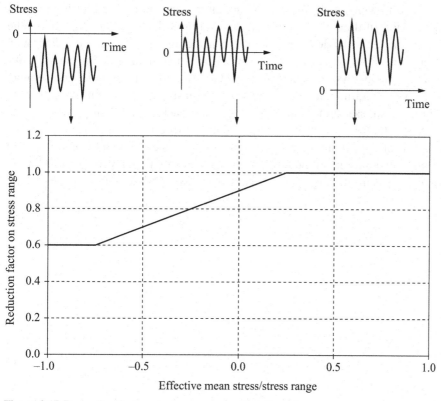

Figure 3.17. Reduction factor on stress range as function of mean stress.

If the amount of residual stress is not known, it may be assumed to be equal to the material yield strength, as derived from material certificates. Tensile stresses are defined as positive, while compressive stresses are negative. This means that if shakedown of residual stress cannot be documented for a relevant load during the first year in service, significant compressive stresses at the hot spot are required before the positive effect of compressive stresses on calculation of fatigue life can be included. The reduction factor on stress range as a function of mean stress (equation 3.45) used for inspection planning for fatigue cracks in DNVGL-RP-0001 (2015) is illustrated in Figure 3.17. Alternatively, crack growth analysis may be performed assuming crack closure for the compressive part of the stress cycle when the crack has grown through 1/3 plate thickness such that the crack tip is grown out of the main tensile residual stress field. Most fatigue tests are performed under pulsating tension loading. Thus, the S-N curves are applicable for this loading condition. Most ship structures are subjected to some permanent loads, which, in some cases, result in significant mean stress. When the mean stress is high in compression, the calculated fatigue damage might thereby be overestimated. Reference is made to fatigue tests performed by Marquis and Mikkola (2001). For information on the mean stress effect in welded connections reference is also made to Shimokawa et al. (1984), Kim et al. (1997), and Sections 2.3 and 2.4.

The Japanese Ship Classification Society (NK) performed a survey of 30 second-generation very large crude oil carriers (VLCC) that had been built between 1987 and 1990 and in which a number of side longitudinals had cracked during the first years of service. In their study of damages, NK found that most cracks were in the cargo tanks near the loaded water line, 2–5 m below the loaded water line, with high stress concentrations, but there was practically no damage in wing ballast tanks (Yoneya et al. 1993). As explained by NK, in a full-load condition, the end connections of the side longitudinals in the empty wing ballast tanks would only be subjected to compressive stresses under the action of sea loads acting on the ships' sides. While the end connections of the side longitudinals in the loaded cargo tanks would be subjected to constant tension under internal pressure, the side pressure from sea would induce stress fluctuation, on the tension side. As a result, the end connections of the side longitudinals in the cargo tanks became more susceptible to fatigue damage.

4 S-N Curves

4.1 Design S-N Curves

4.1.1 General

Fatigue design is based on the use of S-N curves, which are obtained from fatigue tests. They give the number of cycles to failure N at stress range S. The characteristic S-N curves that follow are based on mean-minus-two-standard-deviations for log N curves on log-log scale for relevant experimental data. The S-N curves are thus associated with a 97.7% probability of survival, assuming the test data to be normally distributed on a log N scale. The characteristic S-N curves are also often denoted as design S-N curves as explained in Section 4.11.

4.1.2 S-N Curves and Joint Classification Using Nominal Stresses

For practical fatigue design with the use of nominal stress S-N curves, welded joints are divided into several classes, each with a corresponding design S-N curve. For definition of nominal stress, reference is also made to Sections 3.2.1 and 9.1.1. All tubular joints are assumed to be class T. Other types of joint, including tube to plate, may belong in one of the 14 classes specified in Table 4.1 and Table 4.4, depending on:

- the geometrical arrangement of the detail;
- the direction of the fluctuating stress relative to the detail;
- the method of fabrication and inspection of the detail.

Each construction detail, at which fatigue cracks may potentially develop, should be placed in its relevant joint class in accordance with criteria in the design standard, such as Appendix A of DNVGL-RP-C203 (2016). It should be noted that there are several locations in any welded joint at which fatigue cracks may develop, for example, at the weld toe in each of the parts that are joined, at the weld ends, and in the weld itself. Each location should be classified separately. The basic design S-N curve is given by the following equation:

$$\log N = \log a_d - m \log \left(\Delta\sigma \left(\frac{T}{t_{ref}} \right)^k \right) \qquad (4.1)$$

Table 4.1. *S-N curves in air in DNVGL-RP-C203 (2016)*

S-N curve	N ≤ 10^7 cycles		N > 10^7 cycles log a_{d2} $m_2 = 5.0$	Fatigue limit at 10^7 cycles (MPa)	Thickness exponent, k	Structural stress concentration embedded in the detail (S-N class)
	m_1	log a_{d1}				
B1	4.0	15.117	17.146	106.97	0	
B2	4.0	14.885	16.856	93.59	0	
C	3.0	12.592	16.320	73.10	0.05	
C1	3.0	12.449	16.081	65.50	0.10	
C2	3.0	12.301	15.835	58.48	0.15	
D	3.0	12.164	15.606	52.63	0.20	1.00
E	3.0	12.010	15.350	46.78	0.20	1.13
F	3.0	11.855	15.091	41.52	0.25	1.27
F1	3.0	11.699	14.832	36.84	0.25	1.43
F3	3.0	11.546	14.576	32.75	0.25	1.61
G	3.0	11.398	14.330	29.24	0.25	1.80
W1	3.0	11.261	14.101	26.32	0.25	2.00
W2	3.0	11.107	13.845	23.39	0.25	2.25
W3	3.0	10.970	13.617	21.05	0.25	2.50

See equation (4.1) for notation.

where:

> m = negative inverse slope of the S-N curve (m_1 and m_2 are used to describe the slopes of the left and right parts of two-sloped S-N curves as shown in Table 4.1 and Figure 4.1)
>
> log a_d = intercept of log N axis (log a_{d1} and log a_{d2} are used to describe the intercepts of the log N axis for the left and right part of two-sloped S-N curves as shown in Table 4.1)

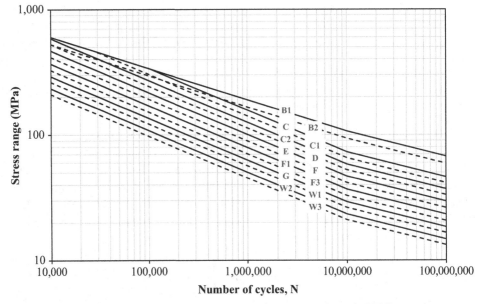

Figure 4.1. S-N curves for steel in air environment (from DNVGL-RP-C203).

Table 4.2. *S-N curves for tubular joints*

Environment	m_1	log a_{d1}	m_2	log a_{d2}	Fatigue limit at 10^7 cycles (MPa)	Thickness exponent, k
	$N \leq 10^7$ cycles		$N > 10^7$ cycles			
Air	3.0	12.48	5.0	16.13	67.09	0.25
	$N \leq 1.8 \cdot 10^6$ cycles		$N > 1.8 \cdot 10^6$ cycles			
Seawater with cathodic protection	3.0	12.18	5.0	16.13	67.09	0.25
Seawater, free corrosion	3.0	12.03	3.0	12.03	0	0.25

See equation (4.1) for notation.

t_{ref} = reference thickness for welded connections (25 mm in many design standards)

T = thickness through which a crack will probably grow

k = thickness exponent.

The reference thickness equals 25 mm for welded connections in plated structures other than tubular joints. For tubular joints, the reference thickness t_{ref} = 16 mm. For bolts, t_{ref} = 25 mm. For larger thickness, the actual thickness is scaled by the reference thickness before it is raised in the power of a thickness exponent k, and the resulting value is used to increase the stress range before the S-N curve is entered for derivation of the number of cycles to failure. For tubular butt welds made from one side, k is 0.10; for girth welds in risers and pipelines, k is 0. These figures are based on test data and fracture mechanics analysis. For threaded bolts subjected to stress variation in the axial direction, k is 0.25. The background for the thickness effect is described more in detail in Section 4.7.

4.1.3 S-N Curves for Steel Details in Air

S-N curves for steel details in air environment are given in Table 4.1 and Figure 4.1. In the low cycle region, the maximum stress range is that of the B1 curve, as shown in Figure 4.1. However, for offshore structures that are subjected to typical wave and wind loading, the main contribution to fatigue damage is in the region $N > 10^6$ cycles, and the bilinear S-N curves defined in Table 4.1 can be used. The last column in Table 4.1 is provided as additional information, as it is not needed for fatigue assessment based on the nominal stress approach. It is included to illustrate the stress concentration factor that is accounted for in the different nominal stress S-N curves. The D-curve is used as a reference curve, as this curve is also the hot spot stress S-N curve (or structural stress S-N curve), and the effect of the weld notch is included in this curve. Thus, if, for example, a G-detail from appendix A in DNVGL-RP-C203 (2016) is analyzed by the hot spot stress method, the target hot spot stress value in that analysis equals 1.80 MPa, with a nominal stress input of 1.00 MPa as shown in Table 4.1. S-N curves for tubular joints from ISO 19902 (2007) are shown in Table 4.2. These are also included in DNVGL-RP-C203 (2016). The S-N curves in Table 4.1 and Table 4.2 are considered to be valid for material with yield strength of up to 960 MPa.

Table 4.3. *Schematic relationship between S-N curves in some design standards*

DNVGL-RP-C203	FAT values in IIW and Eurocode 3[1]	BS 5500:2012 and ISO 19902	BS 7608[2]	API RP2A / AWS
B1	160	B	B	
B2	140			
C	125	C	C	B_{AWS}
C1	112			
C2	100			
D	90	D	D	C_{AWS}
E	80	E	E	
F	71	F	F	D_{AWS}
F1	63			
F3	56			E_{AWS}
G	50	G	G	
W1	45			
W2	40			
W3	36			
T		T		T

[1] The transition in change of slope from m = 3.0 to m = 5.0 in Eurocode 3 is at $5 \cdot 10^6$ cycles, but is at 10^7 cycles in IIW.

[2] The transition in change of slope from m = 3.0 to m = 5.0 in BS 7608 (2014) is at $5 \cdot 10^7$ cycles.

Standard deviations in the S-N curves in Table 4.1 equal 0.20 can be used as representative values unless data indicate otherwise.

4.1.4 Comparison of S-N Curves for Details in Air in Design Standards

Many of the design S-N curves provided in different design standards appear somewhat different from each other. There may be various reasons for this, such as the intended application area of the standard or that the standard is to be used in conjunction with a different fabrication standard, with different details, fabrication tolerances, requirements for non-destructive testing (NDT), and acceptance criteria for defects. The differences in standards for fatigue design also reflect that this is a difficult area to assess. The test data basis in different countries may vary, and committee members drafting the standards may come from a variety of backgrounds and may emphasize different aspects on, for example, how results from small-scale tests should be considered in relation to differences in residual stresses when compared with large-scale structures. S-N curves may be derived from regression analysis of test data. However, further assessment may be recommended if the test data are derived from small-scale test specimens with less residual stresses and tested at a low R-ratio, as described in Section 2.1.3. Other fatigue design standards, such as International Institute of Welding (IIW), are based on a large number of predefined S-N curves, where the classification of each detail is determined to be associated with one of these curves. A schematic relationship between S-N curves in some different fatigue design standards is shown in Table 4.3. However, it should be noted that this table does not provide the complete picture of these relationships in terms of similarity in calculated fatigue lives by the different Standards, as the slopes of the S-N curves may

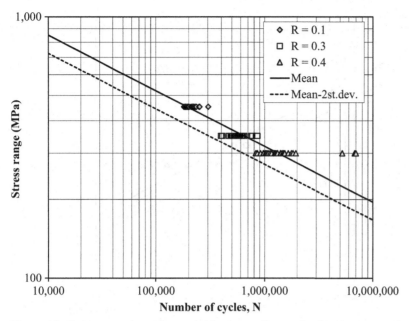

Figure 4.2. Fatigue test data of material from supplier no. 1, with slope m = 4.7.

differ slightly. This also relates to the reference thicknesses and the thickness exponents, and the classification of the detail may belong to different S-N curves. Thus, for a more realistic comparison of a detail, it is necessary to calculate the fatigue lives following the full analytical procedure in each of the standards.

The notations in the different standards may be different. For example, the C-curve in AWS is similar to the D-curve in European standards, and both these S-N curves are used as hot spot stress S-N curves. Fatigue test data presented by Dexter et al. (1994) were considered to be in good agreement with the C-curve in AWS when using this as hot spot stress S-N curve.

4.1.5 S-N Curves for Material with High-Strength Steel

Fatigue test data for high-strength steels, such as machined components in tether connections, are shown in Figures 4.2 and 4.3 (Lotsberg 2005). These test data represent high-strength steel with yield strength above 500 MPa and a surface roughness of $R_a = 3.2$ or better. The resulting S-N curve is based on fatigue test data that were derived as the basis for the design and fabrication of the tethers of a tension leg platform in the North Sea. The design S-N curve for the high-strength steel base material reads:

$$\log N = 17.446 - 4.70 \log S \qquad (4.2)$$

The mean S-N curve is given by $\log N = 17.770 - 4.70 \log S$.

In air, a fatigue limit at $2 \cdot 10^6$ cycles and a stress range of 235 MPa can be used. For variable amplitude loading, with one stress range larger than this fatigue limit, use of a constant slope S-N curve is recommended. It should be noted that these

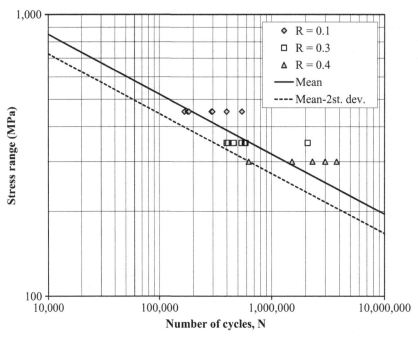

Figure 4.3. Fatigue test data of material from supplier no. 2, with slope m = 4.7.

S-N curves apply only to the base material, and, when welding is performed, the fatigue strength is reduced to the S-N curves for welded structures, as presented in Section 4.1.3.

4.1.6 S-N Curves for Details in Seawater with Cathodic Protection

S-N curves for welded carbon-manganese (C-Mn) steel for the seawater environment with cathodic protection are given in Table 4.4 and Figure 4.4. These S-N curves were earlier considered to be valid for material yield strength up to 550 MPa (Stacey and Sharp 1995). Although fatigue test data for this environment above this yield strength are lacking, these S-N curves have also been used for the design of jack-up legs, where the material yield is around 690 MPa. Concerns have been raised regarding hydrogen cracking when using high-strength steel in seawater with cathodic protection, but to date, no incident due to this failure mode has been reported for these structures. There are also some research reports that now support a higher validity limit for use of high-strength steel in seawater with cathodic protection, see for example, Wormsen et al. (2015); therefore the validity range has now been proposed increased to 690 MPa in DNVGL-RP-C203 (2016).

Slind (1994) reported fatigue testing of T-joints in air and in seawater with cathodic protection. The variable test data gave average Palmgren-Miner fatigue damages less than 1.0 but larger than 0.5 as recommended by Schütz (1981). The crack growth tests for the air environment and seawater with cathodic protection give different crack growth curves. The fracture surfaces showed that variable

Table 4.4. *S-N curves for details in seawater with cathodic protection*

S-N curve	N ≤ 10^6 cycles		N > 10^6 cycles log a_{d2}	Fatigue limit at 10^7 cycles (MPa)	Thickness exponent k	Structural stress concentration embedded in the detail (S-N class)
	m_1	log a_{d1}	$m_2 = 5.0$			
B1	4.0	14.917	17.146	106.97	0	
B2	4.0	14.685	16.856	93.59	0	
C	3.0	12.192	16.320	73.10	0.05	
C1	3.0	12.049	16.081	65.50	0.10	
C2	3.0	11.901	15.835	58.48	0.15	
D	3.0	11.764	15.606	52.63	0.20	1.00
E	3.0	11.610	15.350	46.78	0.20	1.13
F	3.0	11.455	15.091	41.52	0.25	1.27
F1	3.0	11.299	14.832	36.84	0.25	1.43
F3	3.0	11.146	14.576	32.75	0.25	1.61
G	3.0	10.998	14.330	29.24	0.25	1.80
W1	3.0	10.861	14.101	26.32	0.25	2.00
W2	3.0	10.707	13.845	23.39	0.25	2.25
W3	3.0	10.570	13.617	21.05	0.25	2.50

See equation (4.1) for notation.

amplitude tests in air gave a larger number of initiated cracks and a straight crack front, while one or a few elliptical cracks only initiated for the variable tests in seawater at low effective stress ranges. The reduced number of initiated cracks when the environment is changed from air to seawater with cathodic protection may be

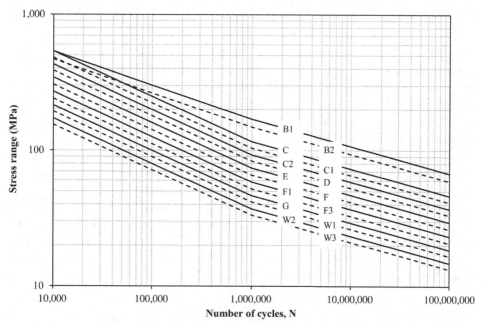

Figure 4.4. S-N curves for steel in seawater environment with cathodic protection (from DNVGL-RP-C203).

Table 4.5. *S-N curves for details in seawater with free corrosion*

S-N curve	log a_d; for all cycles m = 3.0	Thickness exponent, k
B1	12.436	0
B2	12.262	0
C	12.115	0.15
C1	11.972	0.15
C2	11.824	0.15
D	11.687	0.20
E	11.533	0.20
F	11.378	0.25
F1	11.222	0.25
F3	11.068	0.25
G	10.921	0.25
W1	10.784	0.25
W2	10.630	0.25
W3	10.493	0.25

See equation (4.1) for notation.

related to an increase in crack threshold value. The seawater environment gives a slower crack growth due to crack shape but has a higher crack growth rate at medium and high stress intensity levels. The fatigue life in seawater environment depends strongly on the R value and the average value of ΔK. The beneficial effects dominate under long-life conditions (low stress ranges), while the detrimental effect of hydrogen dominates under short-life conditions (Eggen and Bardal 1994). The beneficial effect of cathodic protection at low stress ranges seems to be the result of calcareous deposit formed inside the crack, which is considered to depend on the seawater composition, temperature, crack geometry, loading mode, protection potential and water velocity. The positive effect of calcareous deposit and crack closure is less at higher R-ratios. This means that the crack initiation time tends to be longer in seawater with cathodic protection than in air and the crack growth shorter than in air. This means that the time window for inspection of fatigue cracks may be less than predicted from in-air data (Berge et al. 1994).

Other papers referred to with respect to effect of corrosive environments are presented by Oliver et al. (1981a, 1981b) for constant and variable amplitude loading, as well as Schütz (1987), Scholte et al. (1989), and Wollin et al. (2005).

It may be noted that the fatigue limit in Table 4.4 is at 10^7 cycles and is the same as for air environment in Table 4.1 even if the transition in slopes of the curves are moved to 10^6 cycles in Table 4.4. The reason for this is that there is no available test data to substantiate a different fatigue limit for seawater with cathodic protection from that of air environment.

4.1.7 S-N Curves for Details in Seawater with Free Corrosion

S-N curves in seawater with free corrosion from DNVGL-RP-C203 (2016) are provided in Table 4.5. Fatigue of details in seawater in combination with free corrosion

Table 4.6. *Notch stress S-N curves*

Environment	$\log a_d$	
Air	$N \leq 10^7$ cycles, $m_1 = 3.0$ 13.358	$N > 10^7$ cycles, $m_2 = 5.0$ 17.596
Seawater with cathodic protection	$N \leq 10^6$ cycles, $m_1 = 3.0$ 12.958	$N > 10^6$ cycles, $m_2 = 5.0$ 17.596
Seawater with free corrosion	For all N $\log a_d = 12.880$ and $m_1 = 3.0$	

See equation (4.1) for notation.

should normally be avoided due to the significant reduction in the fatigue strength in the high cycle region, as explained in Section 4.9.3.

4.1.8 S-N Curves for Sour Environment

Sour environments may significantly reduce the fatigue strength of steel structures; see, for example, Buitrago et al. (2004). Due to reduction of fatigue strength in the order of one magnitude on fatigue life, it is difficult to give general recommendations on fatigue strength. The fatigue strength should be considered on a case-by-case basis and should be supplemented by fatigue testing in the relevant environments as necessary.

Clad pipes may be used for transportation of corrosive constituents, with the main pipe made of C-Mn steel and a 3-mm thick clad layer of stainless steel 316L, Incoloy 825, or Inconel 625, typically used on the inside, as presented by, for example, Jones et al. (2011).

4.1.9 S-N Curves for the Notch Stress Method

The calculated notch stress from Section 9.6 should be linked to a notch stress S-N curve for fatigue design, as presented in Table 4.6. The S-N curves are presented as mean minus two standard deviations in a logarithmic S-N format. The thickness effect is assumed to be accounted for in the calculated notch stress, such that a further reduction of fatigue strength for larger thicknesses is not required. The S-N curves are presented in the standard format:

$$\log N = \log a_d - m \log S \tag{4.3}$$

where:

a_d = the intercept of the design S-N curve with the log N axis
m = the negative inverse slope of the S-N curve

4.1.10 S-N Curves for Stainless Steel

For Duplex and Super Duplex steel, the same classification as for C-Mn steels can be used; see also Razmjoo (1995). This classification can also be used for austenitic steel, although it was recommended for some years to downgrade the classification by one S-N class relative to C-Mn steels, mainly due to limited test data.

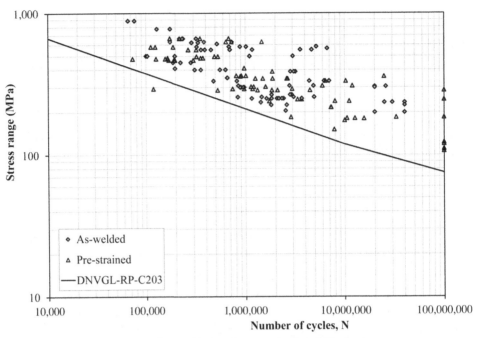

Figure 4.5. Fatigue test data of as-welded and pre-strained umbilical tubes.

4.1.11 S-N Curves for Umbilicals

Fatigue test data from umbilical steel tubes were collected from different participants in a Joint Industry Project, as presented by Kristofferesen and Haagensen (2004). These data also became the basis for a design S-N curve for welded umbilical tubes in DNV-RP-C203 in 2005. Later, more test data were published: see, for example, Buitrago et al. (2006; 2008) and Dobson (2007). A significant quantity of fatigue test data of umbilical tubes has also been derived at the DNV test laboratories in Oslo for Nexans and AkerSolutions. These data were used for a further assessment of a design S-N curve for as-welded and pre-strained umbilical tubes in Lotsberg and Fredheim (2009). A thickness effect, with a thickness exponent of 0.25 for welded umbilical tubes, was proposed by Kristofferesen and Haagensen (2004). DNV also used 115 of these fatigue tests, tested at R = 0.05 to R = 0.1, to assess the thickness effect for welded umbilical tubes. The standard deviation in the test data was calculated for different values of the thickness exponent, k. The standard deviation as a function of thickness exponent showed that k = 0.25 is a reasonable value to be used in a design equation for welded umbilical tubes. Fatigue test data that substantiate the slope of the S-N curve for variable amplitude loading in the region to the right of 10^7 cycles are lacking. However, use of m = 5.0 as the slope of S-N curve in this region is suggested, similar to that used for other S-N curves for Super Duplex steel in DNVGL-RP-C203 (2016).

Additional fatigue test data of welded umbilical tubes from the DNV laboratories in Oslo are shown in Figure 4.5. These include a database of 179 fatigue test data. One data point, belonging to a pre-strained test specimen with a failure from the weld root, falls below the S-N curve. With a design S-N curve defined as mean minus two standard deviations, and with the quantity of test data shown in Figure 4.5, around 4 or 5 test data points would be expected to fall below the curve. The testing was performed at R = 0.1, with a few specimens tested at R = 0.33. The S-N curve for umbilicals should be applicable for a design with a significant mean tension. The effect of mean stress on fatigue strength was considered when the design curve in DNVGL-RP-C203 (2016) was established. From Figure 4.5, no significant differences between the data sets for pre-strained and the as-welded data are seen in terms of result and scatter. Altogether, 109 of the test data have failures outside the weld, or the testing was stopped for other reasons, as explained later in this chapter. These failures were handled as run-outs, and the data were removed in the regression analysis for derivation of the mean line and the standard deviation. However, the run-outs are also valuable contributors to a database, provided that the data are significantly above the design curve. Thus, the resulting S-N curve should be considered more as a lower-bound curve relative to the test data. The large number of run-outs is due to the way in which the tubes are tested by tension load cycles; it is difficult to avoid fatigue failure in the tubular members at the grips in the testing machine, where a stress concentration due to Poisson's ratio is basically unavoidable.

The resulting design S-N curve reads:

For $N \leq 10^7$:

$$\log N = 15.301 - 4.0 \cdot \log \left\{ S \left(\frac{t}{t_{ref}} \right)^{0.25} \right\}$$

and for $N > 10^7$

$$\log N = 17.376 - 5.0 \cdot \log \left\{ S \left(\frac{t}{t_{ref}} \right)^{0.25} \right\}$$

(4.4)

where:

t = actual thickness of the umbilical

t_{ref} = 1.0 mm

A standard deviation of 0.225 was derived from regression analysis of test data representing fatigue failure.

The question arises regarding which S-N curves can be used for base material in umbilical tubes at, for example, bell mouths, where all cycles occur with the same constant stress range. In order to answer this, the industry has invested in fatigue testing. However, here it is even more difficult to avoid run-outs due to failures at the grips of the testing machine, as described earlier in this section.

Some fatigue test data from different test series of base material of umbilical tubes are presented in Figure 4.6. All these test data are run-outs. There are also several run-outs from welded umbilicals that lie above the proposed design S-N curve for the base material. Based on these data, a recommended design S-N curve for base material of umbilical tubes reads:

For $N \leq 10^7$:
$$\log N = 15.301 - 4.0 \cdot \log S$$
and for $N > 10^7$
$$\log N = 17.376 - 5.0 \cdot \log S$$

(4.5)

Standard good fabrication of the umbilicals is assumed as a basis for these design S-N curves. The welds on the inside and outside of the pipes should have a smooth

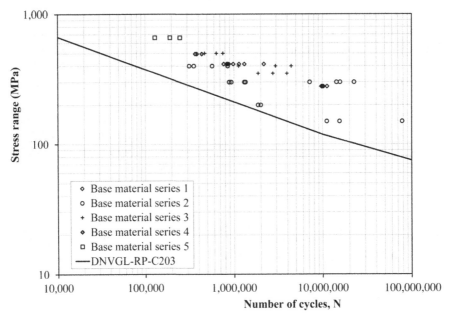

Figure 4.6. Fatigue tests of base material tubes.

transition from the weld to the base material, without notches and/or undercuts. A detailed NDT inspection for each connection is assumed, using NDT methods of visual inspection and X-ray. For single-pass welds, any indication of defects is unacceptable. For multipass welds, the acceptance criteria are according to ASME B31.3–2014, chapter IX. Dye penetrant should be used as a surface test, in addition to visual inspection, when relevant indications (as defined by ASME VIII Division 1, Appendix 4 (2013)) are found by X-ray. The S-N curve is based on fatigue testing of specimens subjected to a mean stress of up to 450 MPa. The S-N curve is established from test specimens that are not pre-strained from reeling. However, based on a few test data with pre-strained specimens, it is also considered acceptable to use the S-N curve for umbilicals that have been reeled. Thus, this S-N curve also applies when the number of cycles under reeling is fewer than 10, and the total strain range (elastic plus plastic) during reeling is less than 2%.

4.1.12 S-N Curves for Copper Wires

Fatigue test data on a 300 mm^2 stranded copper conductor were reported by Nasution et al. (2014). On the basis of nominal stress, the bending tests of full section conductors gave shorter fatigue lives than the fatigue tests of individual wires.

4.1.13 S-N Curves for Aluminum Structures

S-N curves for aluminum structures are presented by IIW (Hobbacher 2009). Design S-N curves for aluminum structures are also presented in Eurocode 9 (EN 1999–1–3:2007). Some background for the recommended S-N curves can be found in Jaccard and Ogle (1997). Another approach to deriving S-N curves for the design of aluminum structures is to follow the guidance in BS 5500:2012. Here, the fatigue

Table 4.7. *Characteristic S-N curves for chains and steel wire ropes*

	log a_d	m
Stud chain	11.079	3.0
Studless chain (open link)	10.778	3.0
Stranded rope	14.531	4.0
Spiral rope	17.230	4.8

strength for aluminum structures is obtained by a scaling of the fatigue strength curves for steel using a scaling factor that is equal to the Young's modulus for aluminum divided by the Young's modulus for steel. A comparative analysis of existing test data on welded aluminum tubular joints is presented by Kosteas and Gietl (1996).

4.1.14 S-N Curves for Titanium Risers

S-N curves for the design of titanium risers are presented in DNV-RP-F201 (2002). The background for these S-N curves is presented in Ronold and Wästberg (2002).

4.1.15 S-N Curves for Chains

Nominal stress S-N curves for chains from DNV-OS-E301(2010) are shown in Table 4.7 and Figure 4.7. These S-N curves are intended for application in seawater. The nominal magnitudes of the stress cycles are computed by dividing the magnitudes of the corresponding tension force cycles by the nominal cross-sectional area of the chain; that is, $2\pi d^2/4 = \pi d^2/2$ where d is the chain diameter.

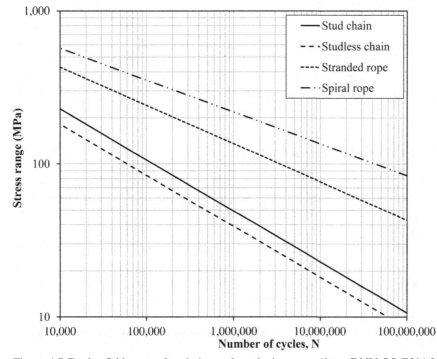

Figure 4.7. Design S-N curves for chains and steel wire ropes (from DNV-OS-E301 2010).

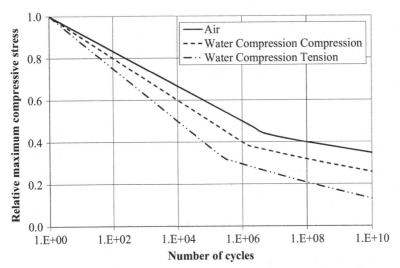

Figure 4.8. Characteristic S-N curves for concrete structures based on DNV-OS-C502 (2012).

4.1.16 S-N Curves for Wires

Nominal stress S-N curves for wires from DNV-OS-E301 (2010) are shown in Table 4.7 and Figure 4.7. It is assumed that the wires are protected from the corrosive effects of seawater. The nominal magnitudes of the stress cycles are computed by dividing the magnitudes of the corresponding tension force cycles by the nominal cross-sectional area of the wire; that is, $\pi d^2/4$ where d is the wire diameter.

4.1.17 S-N Curves for Concrete Structures

Characteristic S-N curves for concrete structures are based on DNV-OS-C502 (2012). The fatigue strength is presented in a diagram that is linear along the vertical axis and logarithmic along the horizontal axis (see Figure 4.8). The fatigue strength is presented in terms of relative maximum compressive strength and is a function of environment and type of loading. The strength is lower in water than in air and is lowest for stresses cycling from compression to tension. The background for this recommendation was presented by Waagaard (1981).

4.2 Failure Criteria Inherent in S-N Curves

Most of the S-N data are derived by fatigue testing of small-scale specimens in test laboratories. For simple test specimens, testing is performed until the specimens fail. In these specimens there is no possibility for redistribution of stresses during crack growth. This means that most of the fatigue life is associated with the growth of a small crack that grows faster as the crack size increases until fracture. The last part of the crack growth in these specimens can be relatively fast. The stress on the net remaining area increases as the crack area grows. The final fracture is a function of maximum stress during a load cycle and material fracture toughness.

Initiation of a fatigue crack takes longer time at a notch in the base material than at a weld toe or weld root. This means that with a higher fatigue resistance of the base material than of welded details, crack growth will be faster in the base material when fatigue cracks are growing due to a higher crack driving stress range. For practical purposes, these failures can be defined as crack growth through the thickness. When this failure criterion is transferred into a crack size in a real structure, where some redistribution of stress is more likely, this failure criterion corresponds to a crack size that is somewhat smaller than the plate thickness. This difference between small-scale specimens and large-scale specimens was also noted during assessment of fatigue test data of the longitudinals in Section 2.4.4. Fracture mechanics, as presented in Chapter 16, may be used for more accurate assessment of crack size at failure in real structures.

Fatigue tests with tubular joints have normally been performed with large sizes. These joints also show a greater likelihood for redistribution of stresses as a crack grows (especially for simple tubular joints). Thus, during testing, a crack can grow through the thickness and also along a part of the joint before a fracture occurs. The number of cycles at a crack size through the thickness was used when the S-N curves for tubular joints were established. As these tests are relatively similar to the actual behavior in a structure, this failure criterion corresponds approximately to the thickness at the hot spot (chord or brace, as appropriate).

4.3 Mean Stress Effect

S-N curves are derived from testing at a high R-ratio or they are adjusted to be representative for a high R-ratio. For design of offshore structures, it is normally recommended that these S-N curves are used directly for fatigue assessment, due to presence of potentially high tensile residual stresses at the hot spots after fabrication. For the base material, it has been standard practice to account for the mean stress in design, as a compressive part of a load cycle is less damaging than a tensile part of same value. Acceptance of the use of mean stress depends on residual stresses at the hot spot during typical long-term loading; for further information, see Sections 3.3.3 and 3.3.4.

4.4 Effect of Material Yield Strength

4.4.1 Base Material

The fatigue strength of base material increases with increasing material strength. For marine structures, it has been practical to distinguish between S-N curves for normalized steels, with yield strength of below 500 MPa, and steel with yield strength larger than 500 MPa.

4.4.2 Welded Structures

The effect of yield strength with respect to fatigue strength in normalized base material is considered to be small when it is welded. Therefore, the same S-N curves are normally used for welded steel structures, independent of material yield strength.

In around 2001, a new steel material was developed that showed improved fatigue initiation life, as well as improved crack growth life in welded structures

(Katsumoto et al. 2005), and a new alternative for fatigue life improvement was introduced. This steel was named Fatigue Crack Arrester (FCA) steel. Two mechanisms are responsible for the improved fatigue resistance and crack arrest in FCA steel. One mechanism is increased fatigue initiation resistance at the weld heat-affected zone (HAZ) due to specific microstructures. The other mechanism is decreased crack growth rate in base material when a fatigue crack passes a grain boundary present in these dual-phase steels, moving from a soft phase (ferrite) to a hard phase (bainite). An S-N curve for this steel was presented in Konda et al. (2010). The slope of the S-N curves for FCA steel and conventional steel differ from each other, and FCA steels show significantly longer fatigue lives than conventional steel for details that are subjected to typical stress ranges from wave loading when looking at the high cycle region of the S-N diagram. According to Konda et al. (2010), the calculated fatigue life is more than three times longer for FCA steels than for conventional steel for a typical long-term stress range distribution from wave actions for a ship structure. FCA steel has now been used for details in a number of new-build ships, where superior fatigue properties are necessary. The differences between FCA steel and conventional steel are insignificant for the left part of the S-N diagram, when approaching the low cycle region. This indicates that FCA steels should mainly be considered for details subject to wave type dynamics, and are of less relevance for areas subject to low cycle fatigue (high stress ranges), such as may result from loading and unloading.

4.5 Effect of Fabrication Tolerances

The effect of fabrication tolerances on fatigue life assessment is considered to be of most significance for butt welds and cruciform joints. There are few measurement data from built structures, and this has made it difficult to decide which tolerances should be accounted for in fatigue analysis. In principle, there is also a link between the recommended S-N curves and the tolerances that should be used together with these curves. Some standards indicate that fabrication tolerances are accounted for in the S-N curves, while other standards indicate that tolerances are not accounted for in the S-N curves. Further considerations on tolerances in the different design standards are presented in Section 5.3.4.

4.6 Initial Defects and Defects Inherent in S-N Data

4.6.1 Types of Defects in Welded Connections

An overview of potential defects in welded connections is shown in Table 4.8. With respect to reduction in fatigue strength, planar defects are considered to be more severe than three-dimensional or volume defects. Therefore, planar defects are generally not accepted in welded connections subjected to dynamic loading. This applies to un-intended lack of penetration, cracks, and lack of fusion. Lack of fusion is understood to mean an absence of proper binding between the weld deposit and the base material, but can also indicate a lack of binding between layers of weld, as indicated in the sketch in Table 4.8.

Table 4.8. *Overview of potential defects in welds*

Type of defect	Description	Schematic illustration of defect	Comments
Three-dimensional defects	Weld undercut		The size of weld undercut depends on the welding method. Some undercuts can be accepted, but any microcracks in the bottom of the undercuts should be avoided.
	Wormhole		
	Clustered pores		Some pores in welded connections may be accepted. Acceptance criteria for these can be found in fabrication standards.
	Irregular slag		Some slag in welded connections may be accepted. Acceptance criteria for this can be found in fabrication standards.
Planar defects	Lack of penetration in double-sided weld		Lack of penetration may be accepted if the connection fulfills the requirements for fatigue life with an S-N curve with lack of penetration. Defects from lack of penetration may be difficult to detect. Therefore, planned lack of penetration should be avoided in connections of importance for structural integrity. Reasons for lack of penetration in planned full-penetration welds may be due to the welding procedure used.
	Lack of penetration in single-sided weld		
	Lack of fusion between weld and base material		Lack of fusion may be due to the welding procedure used.
	Lack of fusion between weld beads		
	Lack of fusion in the area of the weld root		

The presence of defects is more important for welded connections designed according to the higher S-N curves than for the lower curves, due to the large stress range associated with the highest curves if the fatigue capacity is fully utilized. This was observed by Wästberg and Salama (2007) in fatigue testing of butt welds in tether elements, where improvements from grinding the welds were limited due to fatigue crack growth from internal defects. Thus, the requirements for NDT and acceptance criteria are very strict for welded connections designed with use of the highest S-N curves, for which it is required that the welds are machined smooth in order to remove any notches on the outside that may otherwise be where fatigue initiation occurs. If cracks are observed in or at the weld after welding, it should be ascertained whether the welding procedure has been followed correctly or if something is wrong with the material that has been used. Cracks in welds are normally not acceptable, and repair welding must be performed.

In cruciform joints, where forces are transferred normal to the plate, such as in the left connection in Figure 4.10, it is important that lamellar tearing in the transverse plate is avoided. The starting point for this may be defects that are parallel to the rolling direction of the plate. The risk of lamellar tearing can be minimized by using steel with controlled chemistry ("Z" quality steel), such that potential defects are avoided. This steel is also tested for these defects by NDT at the steel mill. Furthermore, the ductility of the material in the transverse direction is documented by tensile testing of specimens prepared for static loading in the transverse direction. Even using steel that fulfills all the requirements for "Z" quality steel, a few cases have been reported in which lamellar tearing has occurred during fabrication of highly restrained connections. It is thus preferable that the main load is transferred in the longitudinal direction of the steel plates if possible.

Weld undercuts are formed due to the surface tension of molten base material that is drawn into the weld and, due to rapid solidification, does not wick back again.

4.6.2 Acceptance Criteria and Link to Design S-N Curves

The acceptance criteria for defects in fabrication standards for welded offshore structures have not changed significantly in the past 40 years for the most typical connections; see, for example, NORSOK M-101 (2011) and DNV (1977). However, some refinement has been made to make material selection and requirements to inspection more efficient and better linked to consequence of failure. A significant part of the basis for the acceptance criteria was developed in the 1960s and 1970s; see, for example, Boulton (1976), Harrison et al. (1978), and Ogle (1991).

Most fatigue cracks are initiated from undercuts at weld toes. There are various reasons for undercuts. Preparation of a good welding procedure is recommended to avoid large undercuts during production welding. According to Hobbacher (2009), no undercut is included in S-N classes better than the D-curve or FAT 90. The maximum allowable size of undercuts of 0.5 mm in depth may be considered inherent in the different design S-N classes D, E, and F. An undercut of 1.0 mm may be accepted for the S-N class F1 (FAT 63) and lower. For a comparison of S-N curves in different design standards, see Section 4.1.4.

Bell et al. (1989) investigated the effect of weld toe undercut depth on the fatigue life of welded plate T-joints. The fatigue life was significantly reduced by undercuts deeper than 0.5 mm. Shallower undercuts do not appear to affect the fatigue life. The deleterious effect of undercuts is governed by stress concentration, determined by depth and root radius rather than depth alone. Deep, sharp undercuts produces short crack initiation lives. A high density of such undercuts along the weld toe generates a high density of initial cracks, which coalesce very early into one straight-fronted crack, and this results in a reduction in crack propagation life. Therefore, larger short undercuts than long continuous undercuts are accepted in fabrication standards (NORSOK M-101 2011).

Moving to the higher S-N classes, the weld details are subjected to larger stress ranges. A double-sided butt weld belongs to D or E, depending on fabrication. Butt welds are more critical with respect to internal defects than many other details. Planar defects, like cracks and lack of fusion, are not admissible. It should be noted that use of S-N classes above that of D implies that there are special considerations with respect to requirements for NDT and acceptance criteria. Thus, reference is made to the requirements in annex C of ISO 5817 (2014) and quality levels from here for S-N details above S-N curve D in DNVGL-RP-C203 (2016). As an example quality level ISO 5817-B125 is used for S-N class C (or FAT 125 following Hobbacher 2009). See NORSOK M-101 (2011) and ISO 5817 (2014) for requirements regarding volumetric defects like porosity and slag inclusions. See also Jonsson et al. (2011) on development of acceptance criteria that are related to fatigue performance.

Girth welds in risers and tethers may be subjected to large dynamic stresses. Therefore, fatigue testing of actual production welded connections are often performed for qualification of welding procedures where the consequence of a failure is large (Buitrago et al. 2003). In these connections planar defects are not accepted. However, it is difficult to fully rely on NDT methods for small defects and it is difficult to document proper fatigue lives based on fracture mechanics analysis when considering initiation and thresholds against crack growth. Therefore, some investigators have tried to build simulated internal defects into the welds before they are fatigue tested; see Buitrago and Zettlemeyer (1999) and Ørjasæeter et al. (2007). It is also considered to be a challenge to simulate this in such a way that the test results are representative for relevant defects that might be present after a fabrication. A number of fatigue tests shows that a high fatigue strength can be achieved for girth welds without defects. At the same time one should also remember that failing to detect some larger defects may significantly reduce the fatigue strength as shown in Wästberg and Salama (2007) for ground welds.

An example of crack growth through a plate thickness of 25 mm from maximum surface defects that fulfill some different S-N curves is shown in Figure 4.9. A semi-elliptic surface defect with half axes $a/c = 0.2$ is assumed as the initial crack (crack length along surface is 2c and a is crack depth). This calculation is based on a Weibull long-term stress range distribution with 10^8 cycles during a lifetime of 20 years. The shape parameter is $h = 1.0$, and the scale parameter is determined such that the accumulated Palmgren-Miner damage is equal to 1.0 during the lifetime of the structure. The same long-term loading for nominal stress is used in the fracture mechanics assessment as in the fatigue assessment using S-N curves. The M_k factors from Bowness and Lee (2002) and the geometry functions by Newman and Raju (1981, 1983) were used for the fracture mechanics analyses; see also Chapter 16. It should be noted that larger surface defects can be accepted for an F detail than for

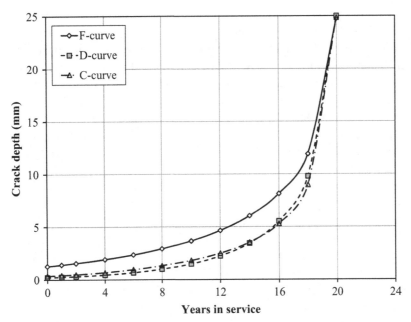

Figure 4.9. Crack growth development from maximum surface defects in some different S-N classes.

D and C. In Figure 4.9 $a_{initial}$ = 1.28 mm for F class, 0.23 mm for D, and 0.37 mm for C. The reason for the low value for the D class as compared with C class is the notch effect at the weld toe, which does not occur in a machined surface (C class). The situation becomes different for internal defects, with very small defects acceptable for C class due to high stresses inside the weld.

4.7 Size and Thickness Effects

4.7.1 Base Material

As long as the maximum calculated surface stress range includes the stress effect due to notches before the S-N curve is entered for calculation of fatigue damage, there is no need for a size effect in fatigue for the base material. This is because the effect of the notch radius is already accounted for in the calculated stress range. However, a size effect has to be accounted for in fatigue assessment of, for example, bolts, based on a nominal stress approach where the radius at the threads is not accounted for in the nominal stress.

4.7.2 Welded Connections

As-Welded

The effects of thickness and size on fatigue strength of welded structures were first published at a conference on offshore structures in 1979 (Gurney 1979; Wylde and McDonald 1979), and, since then, they have been the subjects of several research projects. The research has been largely related to laboratory fatigue testing and

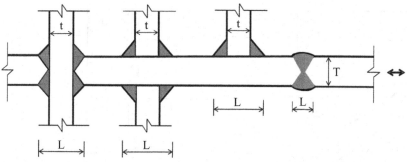

Figure 4.10. Illustration of different types of welded cruciform joints and butt weld (Lotsberg 2014) (reprinted with permission from Elsevier).

numerical simulation by crack growth analysis using fracture mechanics. A number of papers related to this issue have been presented at international conferences on Offshore Mechanics and Arctic Engineering (OMAE) conferences during the 1980s and in journals: see, for example, Lotsberg (2014) for a more detailed reference list. Most of the laboratory tests have been performed under cyclic bending load, as that requires less capacity of the testing machines than for axial testing of thick connections. The effect of plate thickness on fatigue strength is larger in bending than under axial membrane load, due to the stress gradient in bending. In properly designed structures, most of the force transfer in welded joints is by axial force. However, in simple tubular joints, a significant bending stress through the wall thickness also occurs. It is well known that fatigue test data show significant scatter, and this has made provision of reliable design recommendations difficult, with respect to size effect and also with ensuring cost-efficient structures.

The size effect is related to the size of the main plate and to the size of the attachment, such as the thickness of a transverse plate at a cruciform joint, as indicated for different welded connections in Figure 4.10. As the thickness of the attachment increases, more stress moves toward the weld toe region, and fatigue crack initiation occurs earlier. The size effect is also related to the geometry of the weld toe, such as weld toe angle, radius, and undercut. This geometry is thus a function of the acceptance criteria for fabrication; requirements to allowed undercut and weld shape at the transition from the weld to the base material. The effect of the weld toe geometry and the attachment length indicate that a butt weld of limited width will show less size effect than a cruciform joint will. The effective attachment lengths, L, for some welded joints are shown in Figure 4.10. Butt welds are most often subjected to membrane load, which is also less affected by the size effect than bending loads are. The size effect is significantly decreased by grinding the weld toe region, such that the weld notch is reduced. The thickness effect may be entirely removed by machining the weld such that the weld notch is fully removed; this can be demonstrated for a perfect flush ground weld by using fracture mechanics.

In many fatigue design standards (e.g., ISO 19902 2007; API RP2A-WSD 2014) the size effect is accounted for by a modification on stress range, such that the design S-N curve for thicknesses larger than the reference thickness can be presented by equation (4.1). The effect of the plate thickness on calculated fatigue life as a

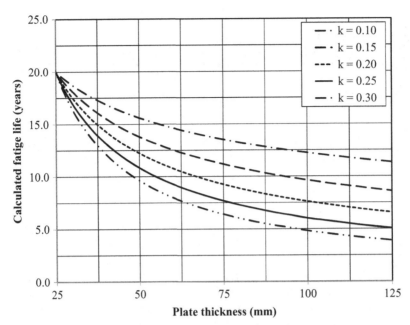

Figure 4.11. Effect of plate thickness and thickness exponent on calculated fatigue life (Lotsberg 2014) (reprinted with permission from Elsevier)

function of thickness and thickness exponent resulting from the design procedure is illustrated in Figure 4.11. The calculations have been performed for a typical floating platform, with a two-parameter Weibull long-term stress range distribution, with a shape parameter of 1.0 (see Chapter 10). A butt weld is analyzed as an example (Design D-curve according to DNVGL-RP-C203 2016). A fatigue life of 20 years without thickness correction is used as a reference, with the total number of cycles $n_0 = 10^8$. The reference thickness is 25 mm. The calculated fatigue life is significantly reduced, with a larger thickness exponent for increasing thickness. It should be noted that there is a thinness effect below the reference thickness, which leads to increased calculated fatigue life for thicknesses lower than the reference thickness (this is not shown in the figure). However, this effect is not normally accounted for in design standards for plated structures.

In general, the size effect is included in a design equation to encompass cases in which the actual size of the considered structural component differs in geometry from typical test specimens from which the S-N data are derived. The mean thickness, or a thickness in the upper part of the test data used to derive S-N data, corresponds to the reference thickness t_{ref} in equation (4.1). The size effect is considered to account for different sizes of plate through which a crack will probably grow. It also accounts for the size of weld and attachment, as previously explained. However, it does not account for weld length in components that differ from those tested, such as seam welds in long pipes as described in Section 4.7.6.

Berge (1984, 1985) described the size effect, with reference to experience from fatigue testing, as that increasing the size of a given type of specimen while maintaining all other parameters will, in general, cause a decrease in fatigue strength. The size effect is explained by two different mechanisms:

1. A statistical effect that is based on the fact that fatigue is a weakest-link process, where crack initiation and growth are initiated by a combination of stress, geometry, material, and defects that provides the shortest fatigue life. For welded joints, this size effect will essentially be a function of weld length, as fatigue cracking is most likely to occur from the weld toe.
2. A thickness effect that is explained by the increase in plate thickness when the local geometry at the weld remains constant.

The size effect might also be explained by residual stresses and, thus, by the fabrication process, as larger residual stresses are expected at thicker plate welds. However, tests of stress-relieved welds show that much the same size effect is observed in test data from stress-relieved specimens as from welded specimens (Palm et al. 1984; Berge and Webster 1987; Yamamoto et al. 2012). Berge (1984) presented a model for describing the size effect in weld toe fatigue failures, based on the following assumptions:

- Welded joints of the same type in various plate thicknesses are geometrically similar (typical for load-carrying welds).
- Initial conditions of fatigue crack growth are independent of plate thickness (crack depth a_i = constant).

Under these assumptions, the effect of plate thickness may be understood from a simple geometrical model of a cruciform joint, as shown in Figure 4.10. The first assumption leads to a steeper stress gradient at the weld toe for the thinner plate, as shown in Figure 4.12. The vertical axis represents the geometry function due to the weld notch, M_{kma}, for membrane loading at axis a in a semi-elliptic crack, as shown in Figure 16.11. The second assumption leads to less stress at the crack tip region in the thinner plate, and hence to a lower crack growth rate here. This can be seen from the different curves in Figure 4.12 by entering the horizontal axes of Figure 4.12a and Figure 4.12b by a small crack (e.g., 0.2 mm deep), to see the difference in geometry function. The difference in initial crack growth rate outmatches the difference in crack growth length to cause a final fracture; hence a thinner joint will have a longer fatigue life. This model for explaining the thickness effect can be substantiated by fracture mechanics analysis for the crack growth part of the fatigue life, based on the geometry functions shown in Figure 4.12. The model can also be used to explain the longer fatigue initiation time for the thinner plate, due to less stress at the weld toe, as indicated in Figure 4.12a. The total fatigue life at welds can be considered to be the sum of fatigue to create an initial crack and to grow this crack further until failure. Initiation of a fatigue crack is typically 10–30% of the total fatigue life for tubular joints; see, for example, de Back (1981) and Radenkovic (1981). This may depend on different parameters, such as weld toe profile and the stress range used in testing. Based on this, the size effect can be explained by the following parameters:

1. The magnitude of the stress concentration at the weld toe, as mainly determined by the local geometry: weld toe angle, weld toe radius, and main plate thickness and transverse plate thickness in cruciform joints, and weld size in butt welds.
2. The stress gradient in the plane of the crack growth, as mainly determined by the plate thickness and type of loading. A bending load leads to a steeper stress

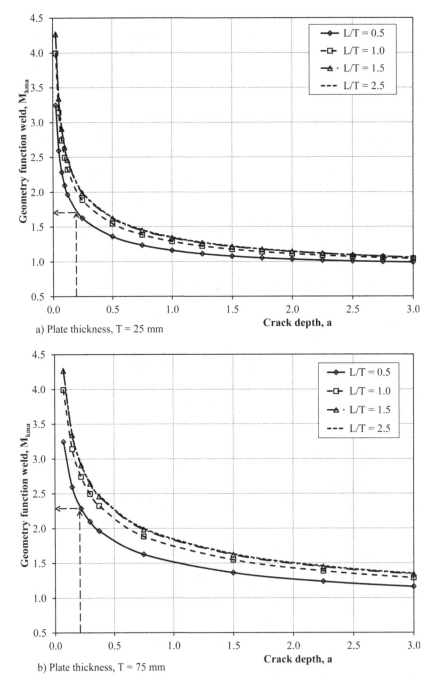

a) Plate thickness, T = 25 mm

b) Plate thickness, T = 75 mm

Figure 4.12. Examples of geometry function for crack growth from the weld toe (Lotsberg 2014) (reprinted with permission from Elsevier).

gradient, and thus to a larger size effect, than does membrane loading. However, fatigue life in bending is longer than in membrane loading. It is standard practice in design to use the same nominal S-N curves for details subjected to bending as for those subjected to membrane loading. Thus, it may be unnecessarily conservative for many types of joints to combine the size effect, established from bending tests, with S-N data, established from dynamic membrane loading (under displacement-driven loading).

3. The number of cycles in crack growth through the region of a steep stress gradient, relative to the total number of cycles to failure.

Post-Weld Treated Joints

Grinding of weld toes reduces the local stress magnification at the weld toes, as shown in HSE report 2000/077 by Bowness and Lee (2002). Grinding also improves fatigue lives as shown by large-scale laboratory tests by Lotsberg et al. (2014), and as accounted for in recommendations by IIW (Haagensen and Maddox 2013). A reduced thickness exponent in joints with a good weld profile has been argued for by Marshall (1993). A reduced thickness exponent for ground welds is included in API RP2A-WSD (2014) and DNVGL-RP-C203 (2016).

4.7.3 Size Effect in Design Standards

The thickness effect was first introduced in the UK Department of Energy's Guidance Notes in 1984 by scaling the calculated stress range by a factor $(T/t_{ref})^k$, where the thickness exponent $k = 0.25$. The reference thickness, $t_{ref} = 22$ mm, was used for plated structures, and $t_{ref} = 32$ mm was used for tubular joints. The reference thickness is not a significant parameter, provided that it is properly related to the log a value in the S-N curve (where log a is the intercept of the S-N curve with the log N axis). The same thickness effect was also introduced in the DNV Classification Note No. 30.2 for welded structures in 1984 and in NS 3472 Steel Structures in 1984, which were used for the design of offshore structures in the Norwegian part of the North Sea until 1998, when the NORSOK standards were developed. However, one difference between these standards was that the thickness effect was not included for the base material in the DNV Classification Note No. 30.2 (DNV 1984) and NS 3472 (1984). This was because a thickness effect had not been observed, based on fracture mechanics crack growth analysis, in connections without a weld notch.

The thickness exponent was increased from 0.25 to 0.30 by the UK Health and Safety Executive (HSE) in 1990 (HSE Offshore Installations, fourth edition). At the same time, the reference thickness was reduced to 16 mm for plated structures and tubular joints. The reason for this reduction is explained by Stacey and Sharp (1995); the size effect is more complex than can be represented by a single equation, as it depends on several parameters. However, information on the various parameters that influence the thickness effect is limited, and therefore a thickness exponent of 0.3 was recommended for all types of joints. In recognition of possible conservatism for particular joints, a mitigating clause was introduced stating that alternative thickness corrections might be included if they were supported by results from experiments or from appropriate fracture mechanics analysis. The same procedure for fatigue assessment as that of HSE (1995) was proposed for inclusion in ISO 19902 (2007) Standard for design of steel structures (Bærheim et al. 1996). However, in response to comments on the Draft Standard, the thickness exponent was reduced to k = 0.25, before the final version was issued in 2007. Here it is stated that this thickness exponent shall be used for all type of details, which also includes fatigue assessment of the base material. The thickness exponent in the fatigue design recommendations by IIW (Hobbacher 2009) for plated structures is dependent on the severity of the weld notch, due to different thickness exponents and calculation of an effective plate thickness similar to that proposed by Gurney (1989). This is represented by equation (4.6). There is a rather severe size effect for tubular joints from IIW (Zhao and Packer 2000), as has also been commented upon by Schumacher et al. (2009). It was based on the UK Department of Energy guidance notes from 1990, according to van Wingerde (1997b). The Eurocode 3 standard originally based their S-N curves on the recommendation from IIW (1996), but a different methodology for assessment of the thickness effect is used. This has resulted in the development of correction factors to be used for some joints with $t_{ref} = 25$ mm and thickness exponent k = 0.25. For other joints, the effect is already included in the S-N classification. The 22nd edition of API RP 2A -WSD (2014) presents an S-N curve for tubular joints with a 16 mm reference thickness and a thickness exponent, k, of 0.25, except for weld-controlled profiles for which the exponent is 0.20. If the weld toe has been ground or peened, the thickness exponent may be reduced to 0.15. In the commentary section to this document, it is explained that the scatter in published data on the thickness exponent is probably due to differences in fabrication, where some of the test specimens may have been of inferior quality as related to the weld toe profile. The NORSOK and DNVGL-RP-C203 (2016) present S-N curves in air that are similar to those of IIW (Hobbacher 2009), with a reference thickness of 25 mm for plated structures. A reference thickness of 16 mm is used for tubular joints in ISO 19902 (2007), API RP2A-WSD (2014) in DNVGL-RP-C203 (2016). The thickness exponent in DNVGL-RP-C203 (2016) starts at zero for details in base material, except for threaded sections. The thickness exponent is also low for weld-improved joints, and higher for lower S-N classes as shown in Table 4.1.

BS 7608 (2014) presents S-N curves with a reference thickness of 16 mm for welded joints and 25 mm for bolts. The thickness exponent is k = 0.25. A thickness correction is not needed for the base material or for butt welds with the weld reinforcement machined flush. In the IACS harmonized Standard from 2013, the following equation is included to account for the size effect in plated structures by calculating an effective thickness together with a thickness exponent k = 0.25:

$$T_{eff} = \min\left(\frac{L}{2},\ T\right), \text{ but } T_{eff} \geq t_{ref} \tag{4.6}$$

where:

T = main plate thickness
L = attachment length, as defined in Figure 4.10.

This size effect is equivalent to the procedure proposed by Gurney (1989) and recommended by IIW (Hobbacher 2009). A somewhat similar size effect has been presented by Gordon et al. (1997) based on assessment of fatigue test data.

4.7.4 Calibration of Analysis Methods to Fatigue Test Data

Initial crack size is a significant parameter as input to a crack growth analysis based on fracture mechanics. The initial defect depth ranges from small values, $a_i = 0.01 - 0.03$ mm from Radenkovic (1981), to larger values, with 0.5 mm as an upper bound for safe assessments in DNVGL-RP-C203 (2014). Bokalrud and Karlsen (1981) presented an exponential distribution with mean 0.11 mm depth, based on measurements of undercuts at weld toes. Ayala-Uraga and Moan (2007) used the same distribution in a fatigue reliability assessment of welded joints using fracture mechanics formulations. Gurney (1991) used an initial depth of 0.15 mm and a surface length of 0.3 mm. Schumacher et al. (2009) used an initial crack depth of 0.20 mm for calculation of the size effect. Similar initial crack sizes have also been used by other researchers. Calibration of initial crack depth may be performed to achieve a good correlation between fatigue test data and fracture mechanics. This is because initiation is an important part of the fatigue life for calculation of the thickness exponent, and therefore large initial cracks may result in non-conservative values being calculated for the thickness exponent. An assessment of the size effect based on crack growth analyses is made below. These analyses were based on available geometry functions obtained from the literature for calculation of stress intensity factors for semi-elliptic cracks, as presented by Newmann and Raju (1981, 1983) and geometry functions for the weld notches presented by Bowness and Lee (2002). The Paris crack growth formula (Paris et al. 1961) was used to calculate crack growth along the two half axes in the semi-ellipse, using the Proban program (1996). Material constants, $m = 3.0$ and $C = 1.83 \cdot 10^{-13}$ (N, mm), were used for calculation of the mean or expected crack growth. An initial semi-elliptic crack at the weld toe with axis ratio $a_i/c_i = 0.2$ was assumed. Fatigue test data or S-N data of welded specimens can be considered to consist of a crack initiation phase and a crack growth phase. The latter can normally be assessed using fracture mechanics. Here, this analysis methodology is calibrated to that of fatigue test data, such that it also accounts for the initiation phase by calculation of crack growth from small fictitious cracks. It is well known that calculated crack growth is sensitive to assumptions made for the analysis, and therefore it is important that the same methodology is used for the following analysis as used for calibration of the initial fictitious fatigue cracks. However, the proposed methodology was not used for calculation of absolute values of the size effect; it was used only to assess the effect of different parameters on the format of the size effect in a relative sense. Based on these input parameters, an initial crack size of $a_i = 0.03$ mm was calculated to correspond with fatigue test data for a cruciform joint, with a reference thickness of 25 mm. This defect size was then used for crack growth analysis of other geometries to enable further assessment of the size effect (Lotsberg 2014). This resulted in the following equation for effective thickness:

$$T_{eff} = \min(14\,mm + 0.66\,L, T) \qquad (4.7)$$

where the geometry is measured in mm for the attachment length, L, that includes the weld leg lengths, and for the main plate thickness T (Figure 4.10). T_{eff} should not be less than t_{ref} and is inserted in equation (4.1) in the place of T. The assessment has mainly been made for cruciform joints, but this procedure can also be used for butt welds and short attachments, like shear keys welded to steel plates to achieve connection with concrete in grouted connections. Butt welds made from both sides can be fabricated with a rather short weld width. Thus, this modification of the size effect is considered to be of significant importance for the fatigue design of piles used in offshore structures and for large diameter monopiles that are used to support offshore wind turbine structures.

4.7.5 Cast Joints

Relatively high S-N curves can be derived from cast specimens without significant defects; see, for example, background document HSE OTH 92390 (1999). However, defects may easily occur in casting larger joints, and the possibility of a weld repair needs to be considered. This implies that the highest S-N curve that can be applied for cast joints is that of a welded joint, where the surface has been made flush by machining or grinding after a weld repair. Cast joints are often used at complex joints in structures showing a significant stress gradient over the thickness. Thus, the fatigue strength of the surface region becomes a governing parameter in design standards such as DNVGL-RP-C203 (2016). Here it is stated that a reference thickness of 38 mm may be used for cast joints, provided that any possible weld repair has been ground to a smooth surface. For cast joints with a stress gradient over the thickness, a reduced effective thickness can be used for assessment of the size effect. This effective thickness can be calculated as:

$$T_{eff} = T_{actual}\left(\frac{S_i}{S_0}\right)^{1/k} \tag{4.8}$$

where:

S_0 = hot spot stress at surface

S_i = stress at the hot spot 38 mm below the surface

T_{actual} = thickness of cast piece at the considered hot spot measured normal to the surface

T_{eff} = effective thickness; T_{eff} shall not be less than 38 mm if the actual thickness is larger than 38 mm

k = thickness exponent = 0.15

By this expression, the stress gradient is accounted for at the same time as the fatigue strength at the surface of the casting is accounted for in the fatigue assessment. It should be noted that as T_{eff} must be larger than 38 mm, equation (4.8) cannot be directly inserted into equation (4.1), and therefore the fatigue calculation needs to be performed in steps.

4.7.6 Weld Length Effect

Design Equation

A longer weld length implies a greater probability of a poorer local geometry, a larger undercut, and a possible defect that reduces fatigue strength when subjected to the same dynamic loading. The effect of this becomes important in long tubular structures, such as for design of girth welds in tendons used for anchoring tension leg platforms and for pipes used for transport and storage of gas, where the dynamic loading along the seam weld is due to filling and emptying of the pipes. This effect may be said to be part of the size effect, in addition to the effect of the thickness. A girth weld refers to the circumferential weld between consecutive pipes, while a weld seam means the longitudinal weld used to fabricate pipes with larger diameters, with diameters being typically larger than 16–18 inches. Seam-welded pipes are made from rolled plates that are shaped into tubular members before a longitudinal

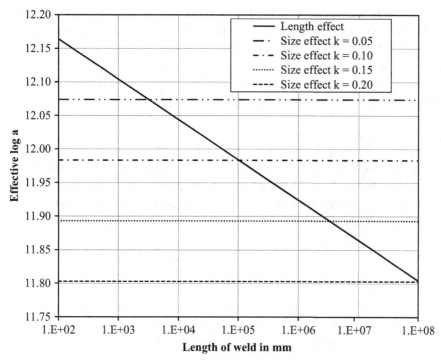

Figure 4.13. Illustration of length effect as compared with the effect of thickness exponent on log a value (Lotsberg 2014) (reprinted with permission from Elsevier).

seam weld is made. Smaller-diameter pipes less than 16–18 inches are made from a single steel piece, and a seam weld is not needed; these are termed seamless pipes. This fabrication process means larger thickness tolerances than in pipes fabricated from rolled plates, where the tolerances are easier to control.

An equation for fatigue strength accounting for the length effect can be presented as:

$$\log N = \log a_d - \frac{s_{\log N}}{2} \log\left(\frac{l_{weld}}{l_{ref}} n_s\right) - m \log\left(\Delta\sigma\left(\frac{T_{eff}}{t_{ref}}\right)^k\right) \qquad (4.9)$$

where:

$s_{\log N}$ = standard deviation of fatigue test data in log N
l_{weld} = weld length
l_{ref} = reference weld leg length = 100 mm
n_s = number of similar welds.

The other parameters are defined in relation to equation (4.1). The effect of length on fatigue strength is illustrated in Figure 4.13, as compared with different values for the size or thickness exponent, k. Here a rather small standard deviation, $s_{\log N} = 0.12$, which may be representative for high-quality butt welds, is assumed in equation (4.9). It should be noted that the reduction in fatigue strength of a 100 m long weld subject to the same dynamic loading along its length is similar to a reduction in fatigue strength due to a size effect with k = 0.10. The fatigue strength is decreased to that corresponding to k = 0.20 for a 100 km long weld. The hot spots in most structures extend only locally, and therefore the effect of the length of welds can normally be neglected for most types of structures other than long pipes subject to varying internal pressures, and for tendons. The diameters of monopiles used for support of wind turbine structures are also quite large, and if the fatigue utilization around the circumference is rather similar, it is recommended that this is considered in design if the thickness effect is otherwise minimized for circumferential butt welds.

Background for Design Equation

Test specimens used for fatigue testing are normally smaller than prototype details or components. The applicability of the test data to the prototype depends on the stress distribution at the hot spot region. For a typical welded joint in a steel structure, there is little difference in the extent of the hot spot region (and the weld length) between specimen and prototype, whereas for a pipe with a longitudinal weld that is subject to the same stress range normal to the weld along its length, the difference is significant.

Fatigue crack growth is normally initiated from small defects at the transition between the weld material and the base material. Thus, the probability of a large defect at the hot spot region rises with increasing weld length, and a specimen with a long weld would be expected to have a shorter fatigue life than a specimen with a short weld. Test results support these observations. The length effect can be included in a design format in a similar way as for the thickness effect by:

$$\log N = \log a_d - m \log \left(\left(\frac{l_{weld}}{l_{ref}} n_s \right)^{k_2} \right) - m \log \Delta\sigma$$

and

$$\log N = \log a_d - m\, k_2 \log \left(\frac{l_{weld}}{l_{ref}} n_s \right) - m \log \Delta\sigma$$

$$(4.10)$$

where:

k_2 is a length exponent factor.

Based on the capacity of a weakest link system, Lotsberg and Larsen (2001) determined k_2 to be about 0.12. The value of k_2 was also assessed based on 120 test data from cruciform joints with weld lengths from 50 mm to 450 mm. By calculating standard deviations for different trial values of k_2, the best fit with the test data was achieved with $k_2 = 0.07 - 0.08$. When considering the statistics from a weakest-link perspective, the results in Figure 4.14 can be derived. It can be seen that the length effect is related to the standard deviation in the test data. As the standard deviation in test data decreases, the length effect distribution will also be narrowed. As $m \cdot k_2$ is approximately equal to a typical standard deviation in cruciform fatigue test data of 0.20, the standard deviation can be used in the test data instead of $m \cdot k_2$. From Figure 4.14 it can be observed that new distributions, including long welds, are narrower than the basic distribution for $n = 1$. Thus, the distance between the mean value and a characteristic value (mean minus two standard deviations) decreases with increasing weld length. The following equation was derived by curve fitting through the characteristic values of the different distributions in Figure 4.14:

$$\log N = \log a_d - 0.10 \log \left(\frac{l_{weld}}{l_{ref}} n_s \right) - m \log \Delta\sigma \qquad (4.11)$$

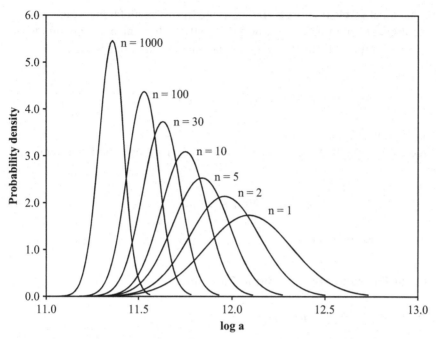

Figure 4.14. Distributions of the lowest fatigue strength as a function of the number of small-scale test specimens.

where:

l_{weld} = length of weld subject to the same stress range
l_{ref} = reference length corresponding to typical length of weld in tested specimen
n_s = number of similar connections subjected to the same stress range
$\log a_d$ = constant in the design S-N curve or cut-off value with the log N axis defined as mean minus 2 standard deviations in test data.

Figure 4.14 was derived by simulation of a new distribution for minimum values drawn from the basic distribution for n = 1 in the figure. For example, the distribution for n = 1000 is derived as a new distribution by repeat plotting of minimum values drawn out of n = 1000 test results. Based on this, it can also be seen that the resulting distribution is a function of the standard deviation in the basic distribution (n = 1). When generalizing equation (4.11) with respect to standard deviations, it reads:

$$\log N = \log a_d - \frac{s_{\log N}}{2} \log\left(\frac{l_{weld}}{l_{ref}} n_s\right) - m \log \Delta\sigma \qquad (4.12)$$

where:

$s_{\log N}$ = standard deviation in the S-N curve.

4.8 Effect of Temperature on Fatigue Strength

The S-N curves presented can be used for temperatures below zero, provided that the steel through which a fatigue crack might propagate has sufficient fracture toughness to ensure that an unstable fracture will not initiate from a small fatigue crack. Above 100°C, the fatigue strength should be lower due to the reduction in Young's modulus. For higher temperatures, the fatigue strength data may be modified using a reduction factor given as:

$$R_T = 1.0376 - 0.239 \cdot 10^{-3}T - 1.372 \cdot 10^{-6}T^2 \qquad (4.13)$$

where T is given in °C.

The reduced strength in the S-N curves can be derived by a modification of the log a_d as

$$\log a_{dRT} = \log a_d + m \log R_T \qquad (4.14)$$

where $\log R_T$ is a negative number.

4.9 Effect of Environment on Fatigue Strength

4.9.1 Condition in Fresh Water

Fatigue strength is lower in a fresh water environment than in an air environment. However, fresh water is not considered to be as severe as seawater and free corrosion; see, for example, Nishida and Urashima (1984).

4.9.2 Effect of Cathodic Protection in Seawater

Cathodic protection of steel in seawater normally uses anodes connected to the steel surface, such that the electric potential is more negative than -900 mV versus an Ag/AgCl reference cell. During the 1980s a considerable amount of research was performed in a number of countries on corrosion fatigue of steel in seawater. A significant part of this work was presented at a "Steel in Marine Structures" conference in Amsterdam in 1987. The proceedings from this conference were edited by Noorhook and deBack and published by Elsevier Science Publishers B.V. Amsterdam, and it is a recommended reading for those who wish to study corrosion fatigue in greater depth; see, for example, Berge et al. (1987a), Berge and Webster (1987), Berge et al. (1987b), Iida (1987), Lachman et al. (1987), Mohaut et al. (1987), Nibbering et al. (1987), and Vosikovski et al. (1987). Some background information can also be found in other literature, such as HSE Background Document OTH 92390 (1999). As seen in Figure 4.15, the S-N curve in seawater with cathodic protection is the same as that in air when the number of cycles exceeds 10^7. This is based on an assessment of fatigue crack growth data as relevant fatigue test data to failure for specimens in seawater with cathodic protection are lacking at this high number of cycles. The reason for the rather high fatigue strength in the high cycle region is the development of calcareous deposit in the crack during crack growth. This is due to the cathodic protection, as the same does not occur in seawater with free corrosion. The difference between air and seawater with cathodic protection is largest for high stress ranges. Thus, there is a difference in endurance of 2.5 for the number of cycles less than 10^6, as indicated in Figure 4.15.

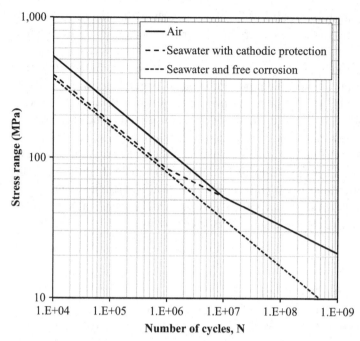

Figure 4.15. S-N curves in air and seawater.

4.9.3 Corrosion Fatigue

Corrosion reduces fatigue strength in various ways. The most important mechanisms are:

1. Development of corrosion pits on smooth surfaces reduces crack initiation time in these areas.
2. Corrosion causes a chemical process at the most highly stressed region at the crack tip, and this enhances the crack propagation rate.

Corrosive action on the surface of metals may cause a general roughness of the surface, with further formation of corrosion pits. These can become initiation sites for fatigue cracks when the material is subjected to fluctuating stress. The combined effects of corrosion and fluctuating stress reduce fatigue strength more than from either of these factors acting separately, and the term "corrosion fatigue" is used to describe this behavior (Forest 1962). According to Kitagawa et al. (1983), the geometric development of corrosion pits is dependent on the level of cyclic stress range and the number of stress cycles. Nakai et al. (2004) compared the geometries of corrosion pits in the hold of a 14-year-old bulk carrier and on the bottom shell of a 22-year-old tanker, and found that they were rather different from each other. This, together with test results being dependent on test frequency, makes it difficult to provide reliable S-N data. Fracture mechanics testing indicates that the threshold value disappears for small stress ranges. Thus, S-N curves for seawater without corrosive protection are linear, without any slope transition at a high number of cycles. The development of corrosion pits for initiation of crack growth is mostly a concern for the highest S-N curves, above the D-curve (or FAT 90). This is supported by fatigue test data from artificial pits, as reported by Grøvlen et al. (1989). Corrosion pits are considered to reduce the fatigue strength of ground welds. Thus, to achieve a long-term positive effect of grinding on fatigue strength, preparation for appropriate corrosion protection is recommended. Grinding reduces the thickness effect. This is reflected in the thickness exponent for the highest S-N curves for welded structures in DNVGL-RP-C203 (2016). However, the thickness exponent is made larger in seawater without corrosive protection than for S-N curves for air and seawater with cathodic protection. The effect of corrosion at large stress cycles is considered to be more severe in seawater with free corrosion than with cathodic protection. Thus, in Figure 4.15 there is a factor of 3.0 on the number of cycles between the free corrosion S-N curve and the S-N curve for the air environment for large stress cycles. It should also be noted that based on the results of fatigue crack growth testing, there is no significant threshold for crack growth in seawater with free corrosion. This is indicated in Figure 4.15 by a straight S-N curve for seawater with free corrosion.

4.9.4 Effect of Coating

S-N Curve for Joints in the Splash Zone
It is standard practice in Norway to use the seawater S-N curve with cathodic protection for joints in the splash zone, and a larger Design Fatigue Factor (DFF) in the splash zone than in other areas in jacket structures. It is also assumed that these joints have a good coating. A high DFF is used because it is difficult to conduct inspection and repair in this area. Increasing the DFF implies that there is a reduction in the probability of fatigue cracking. For example, for DFF = 10, the probability of a fatigue crack during the lifetime becomes very small (the accumulated probability being below 10^{-4} and the annual probability less than 10^{-5} for the last year in service, as shown in Section 12.2). With a DFF as large as 10, the stress ranges are limited and crack growth life is considerable while the cracks are small. Provided that the coating has some ductility, the cracks must grow relatively large before the coating is broken. As long as the coating remains unbroken, the condition is considered to correspond to that of air. The probability of a fatigue crack that is so large that the coating breaks is considered to be low during most of the design life, using a DFF = 10. Based on this, it is considered acceptable to calculate the fatigue damage with an S-N curve that is somewhat reduced compared with that for air, and, in the absence of documented S-N curves in seawater for coated joints, the seawater curve with cathodic protection has been used for joints in the splash zone. This approach is based on the use of a high DFF and a good coating, which reduces the probability of an open fatigue crack subjected to free corrosion during most of the service life of the platform.

S-N Curve for Details in Tanks in FPSOs
Tanks in floating production, storage, and offloading units (FPSOs) are not designed with the same high DFF as used for splash zone joints in jackets. Furthermore, the coating used in tanks in ship structures is not as durable as that used in offshore platforms, becoming more brittle with time and more likely to crack at hot spot regions with large strain cycles, mainly due to loading and unloading; see, for example, Ringsberg and Ulfvarson (1998) and Melchers and Jiang (2006). In tanks without anodes, the efficiency of the coating should be of particular concern. In DNV Classification Note 30.7 (2014), it is assumed that the coating is efficient for several years, and then

that the condition becomes that of free corrosion for the remaining lifetime. A similar procedure may be used for the design of tanks in FPSOs. The length of time with efficient coating depends on the type and quality of the coating used; for further details, see DNV CN 30.7 (2014).

4.10 Selection of S-N Curves for Piles

4.10.1 S-N Curves for Pile Driving

The transition of the weld to base material on the outside of tubular girth welds can normally be classified as S-N curve E. If welding is performed in a flat position, it can be classified as D. If welding is performed from the outside only, it should be classified as F3 for the weld root. S-N curve E applies to weld beads. S-N data corresponding to the air environment condition is used for the pile driving phase due to the resulting short stress cycles.

4.10.2 S-N Curves for Installed Condition

S-N data corresponding to the seawater environment with cathodic protection is used for the operational life of the piles.

4.11 Derivation of Characteristic and Design S-N Curves

4.11.1 General

Qualification of new characteristic S-N curves for fatigue life assessment of structures is an engineering challenge. First, representative fatigue test data for the actual structural connections have to be derived, and these test data then have to be transferred into characteristic S-N curves that represent a predefined probability of survival. Characteristic S-N curves are also often referred to as design S-N curves, as they are often used directly for fatigue life assessment of structures. A typical load and resistance factor design format defines a characteristic capacity as a percentile value, such as 5% (or 2.3%), not to be exceeded. This value is divided by a safety factor in order to achieve a design value for the structural capacity. In fatigue standards for marine structures, it is normal practice to use a safety factor on capacity of 1.0, as safety is introduced into the design format through DFFs on the accumulated damage that is function of the number of cycles, as presented in Section 12.7. Thus, the design S-N curves become equivalent to the characteristic S-N curves. Most fatigue test data are derived from testing of small-scale test specimens. The residual stresses at the weld toes of small-scale test specimens are normally small in comparison with those in full-size structures, due to the different restraints during fabrication. Furthermore, if small-scale test specimens are produced by cutting pieces from larger structures, then the magnitude of the residual stresses will be significantly reduced; see, for example, the experiences described in Section 2.2. Low residual stresses in test specimens may be compensated for by performing fatigue testing at a high R-ratio, as recommended by the IIW (Hobbacher 2009). See also Section 2.1.3 for additional discussion. A high R-ratio is understood to mean a ratio $R = \sigma_{min}/\sigma_{max}$ of around 0.5, while a low

R-ratio is in the region 0 to 0.1, where $\sigma_{\min}$ = minimum stress and $\sigma_{\max}$ = maximum stress during a load cycle. As the maximum nominal stress is limited by the yield strength of the material, it is easier to achieve a high R-ratio for longer fatigue lives at low stress ranges than for high stress ranges at shorter fatigue lives. Another approach that has been used to derive design S-N curves for welded structures, and is based on fatigue test data from small-scale test specimens, has been to force the S-N curves through the center of gravity of the test data with a predefined slope that is considered representative for crack growth in large-scale welded structures with large residual stresses from fabrication. Furthermore, if the fatigue testing is performed at a low R-ratio, the IIW recommends reducing the fatigue strength at $2 \cdot 10^6$ cycles by 20%.

Fatigue testing is normally performed under constant amplitude loading, but actual structures are most often subjected to variable amplitude loading. This introduces additional uncertainty, as shakedown of residual stresses can occur after high maximum loads in details, with a significant stress concentration leading to local yielding at the hot spot area. Most fatigue test data are obtained for fewer than 10^7 cycles to failure. A fatigue limit is often observed under constant amplitude testing around this number of cycles, leading to test data being denoted as "run-outs," as fatigue testing is often stopped at this number of cycles due to time and cost considerations. A variable amplitude loading is required to achieve fatigue test data that are representative for a larger number of cycles in this part of the S-N curve, because some load cycles above the fatigue limit are required to initiate the fatigue crack growth process, as explained in Section 3.2.2. A typical contribution to fatigue damage occurs at around 10^6 to 10^8 cycles for ships and offshore structures subject to wave and wind loading, as shown in Section 1.2. It is thus important that test data are obtained in a region of the S-N curve that is representative for the actual structural behavior, such that additional uncertainties from extrapolation of the test data into design values are minimized.

Small-scale test specimens are generally superior to real structures with respect to tolerances and defects. It might also be questioned whether the planned S-N curve is intended for use for one specific fabrication in one particular yard, or whether it should be acceptable for general use. A larger standard deviation may be expected when production is performed at different yards than at one specific yard. The production quality is also likely to be a function of the welding method and welder qualifications and experience. Thus, an engineering assessment is also required for derivation of recommended design S-N curves in a design standard for reliable evaluation of fatigue strength. It is therefore important that the test program is planned and how the test data will be used is defined before fatigue testing is started.

4.11.2 Requirements for Confidence for Fatigue Assessment in the Literature and in Design Standards

According to DNVGL-RP-C203 (2016), which is a design standard that is frequently used for fatigue assessment of offshore structures, the design S-N curve shall provide a 97.7% probability of survival. On the logarithmic scale the characteristic curves can be derived as mean minus two standard deviations, as obtained from a plot of experimental data, assuming that the data follow a Gaussian distribution on the logarithmic

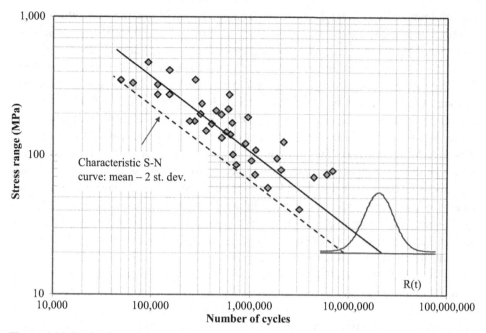

Figure 4.16. Derivation of characteristic S-N curve from fatigue test data.

scale, as shown in Figure 4.16. This representation of a characteristic curve, or a design S-N curve, may be acceptable when a regression analysis is based on many fatigue test data. However, when only limited test data are available, it is questionable whether a characteristic S-N curve can be achieved with sufficient confidence to maintain the same safety level in design. In order to investigate this further, a search was made of design standards on fatigue and other literature (Ronold and Lotsberg 2012). This revealed that recommendations are often based on different procedures and different assumptions, as demonstrated in the following examples. The IIW Recommendations for Fatigue Design of Welded Joints and Components defines the characteristic value of an S-N curve at the 5% quantile at a confidence level of the mean of 87.5% (12.5% probability of being above the extreme values of the confidence interval and 12.5% probability of being below). ASTM E739–91 (1998) presents a methodology to estimate confidence and provides an example with 97.5% confidence. Eurocode EN 1993–1–9 (2009) defines the characteristic S-N curves as follows: "When test data were used to determine the appropriate detail category for a particular construction detail, the value of the stress range $\Delta\sigma_c$ corresponding to a value of $N_c = 2$ million cycles was calculated for a 75% confidence level of 95% probability of survival for log N, taking into account the standard deviation and the sample size and residual stress effects." It should be noted that this approach also extends to considering the complexity of the effect of residual stresses. This may be explained by the S-N curves in Eurocode being based on S-N curves from IIW, and IIW has reduced their S-N curves due to low residual stresses in the test data, as explained above. ISO 19902 (2007) states that "the S-N curves used in design are representative curves" and gives 95% confidence of 97.5% survival as an example. According to ISO 12107 (2003), so many test specimens are required for the derivation of S-N curves that the minimum value of test data can be expected to fall below the true value for the population at a given probability of failure at various confidence levels, 50%, 90%, and 95%. The number of specimens corresponding to a confidence level of 95% is stated to be used for reliability design purposes, those at 50% confidence level are for exploratory tests, while those at 90% are for general engineering applications. For various specified quantiles used to define characteristic values, ISO 2394 (1998) provides tables of the number of standard deviations to be subtracted from the mean value, as a function of the number of tests to obtain characteristic value estimates with specified confidence. This number is dependent on whether the standard deviation is "known" or "unknown," as is discussed further in this section and in Sections 4.12 and 4.13. Schneider and Maddox (2006) presented a guide on how to derive a characteristic curve based on tolerance limits. These limits are determined such that they estimate the characteristic curve with a specified confidence. A confidence of 90% is likely to be of most interest in connection with fatigue analysis, and reference is made to estimation of characteristic curves defined by 95% and 97.5% survival probability. In BS EN ISO 21329 (2004), a different approach from those mentioned earlier in this section is used, as it states that the purchaser should define the required confidence level. Two confidence levels are specified with increasing consequence of failure: Normal and High. Normal involves a total of six fatigue tests, with two tests at each of the three different stress levels, and High means a total of nine fatigue tests, with three tests at each of the three different stress levels. These test data are then used to obtain the mean line and standard deviation on a log-log scale, and a characteristic design curve, as mean minus two standard deviations, can subsequently be derived. With many test data, the final results may be less dependent on confidence level. However, a well-defined procedure becomes more important when only a limited number of tests are performed to establish the characteristic S-N curve. It is then necessary to estimate the characteristic S-N curve with some confidence.

The characteristic curves are derived as mean minus two standard deviations, as obtained from a log S-log N plot of experimental data that are assumed to follow a Gaussian distribution in a logarithmic format. This corresponds to 97.7% probability of survival. The uncertainty in fatigue test data should be accounted for when only a limited number of tests have been performed to establish the characteristic S-N curves. According to DNVGL-RP-C203 (2016), the characteristic (or design) curve should be estimated with at least 75% confidence. When a total of n observations of the number of cycles to failure, N, are available from n fatigue tests carried out at the same representative stress range, S, then the characteristic S-N curve can be derived as:

$$\log a_d = \log a - c s_{\log N} \tag{4.15}$$

where:

$\log a$ = intercept of the mean S-N curve with the log N axis as estimated from the n test data. The term "mean S-N curve" refers here to the mean curve, as derived from a regression analysis of test data presented on a $\log N - \log S$ scale.

$s_{\log N}$ = the sample standard deviation of the n test data of log N

Table 4.9. *Number of standard deviations to be subtracted from the mean to derive a characteristic S-N curve with 97.7% probability of survival*

Number of tests	Standard deviation known		Standard deviation unknown	
	75% confidence	95% confidence	75% confidence	95% confidence
3	2.39	2.95	3.78	9.24
5	2.30	2.74	2.95	5.01
10	2.21	2.52	2.53	3.45
15	2.17	2.42	2.40	3.07
20	2.15	2.37	2.33	2.88
30	2.12	2.30	2.25	2.65
50	2.10	2.23	2.19	2.48
100	2.07	2.16	2.12	2.32
∞	2.00	2.00	2.00	2.00

Note: This table is derived for independent variables. See Ronold and Echtermeyer (1996) for expressions for independent and dependent variables for estimation of characteristic S-N curves. Example of independent variable: The slope of the S-N curve is kept constant when regression analysis is performed, such that the regression is performed with only one variable. This is often the assumption for assessment of S-N curves for welded structures, such as that performed by IIW. Example of dependent variables is when the regression is performed with two variables: the cutoff with the log N axis and the slope of the S-N curve.

> c is a factor whose value depends on the required confidence, on the number of fatigue test data, and on whether the true standard deviation $\sigma_{\log N}$ is "known" or "unknown."

c is shown in Table 4.9 and Figure 4.17 for cases where the standard deviation is either "known" or "unknown." See also EN 1990 (2002) and Ronold and Echtermeyer

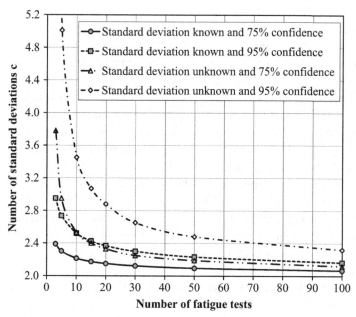

Figure 4.17 Number of standard deviations for derivation of characteristic design S-N curves corresponding to 97.7% probability of survival.

(1996). When there are 15–20 test specimens, the number of standard deviations for c in equation (4.15) is in the range 2.1–3.0. Whether the standard deviation is "known" or "unknown" can be debated. If it is considered as "known," then the testing is mainly performed for derivation of a mean value from which a characteristic value can be derived, based on further assessment of the test results. Provided that the fatigue test data do not include any "strange results," it is likely that the answer is close to that of "known" condition, taking earlier experience from fatigue testing of similar connections into consideration. However, it is obviously safer to treat the test data as "unknown." It should be noted that $s_{\log N}$ is an estimate of the true standard deviation $\sigma_{\log N}$.

A characteristic S-N curve may be derived by various different approaches, and therefore it could be useful to understand the difference between a standard deviation that is "unknown" and a standard deviation that is "known," as described in EN 1990 (2002) and ISO 2394 (1998) (see also DNVGL-RP-C203 2016). A mean S-N curve and a standard deviation of test data can be estimated by following a purely statistical approach. This corresponds to the standard deviation being "unknown" when determining c values in equation (4.15) and will normally require a significant number of test specimens. Although this recommendation is sometimes used in small-scale testing, it is generally prohibitively expensive for full-scale testing. When a lot of testing is performed that results in much the same values for standard deviations, then the standard deviation can be assumed to be "known." Assessment of whether the standard deviation should be assumed to be "known" or "unknown" depends on the expected physical behavior of the problem. For example, more typical test data can be expected for fatigue cracks growing from weld toes into the base material than from cracks growing from potential defects in a weld that is dependent both on fabrication, including the welding method and performance, and the NDT. Assuming a "known" standard deviation is more acceptable in the former case than in the latter. For a "known" standard deviation, the main challenge is derivation of mean values. This reduces the requirements for c values in equation (4.15). This may also be perceived as an engineering approach that may be used in well-defined and controlled tests (instrumented, full-scale test specimens with strain gauges). For complex connections, a larger standard deviation can be expected than for simple small-scale test specimens. The homogeneity of the fatigue test data should be checked, and the scatter should not be any greater than that normally observed in fatigue testing. The number of standard deviations to be used for c in equation (4.15) is shown in Figure 4.17 and Table 4.9, dependent on whether the standard deviation is "known" or "unknown." It should be noted that a few large-scale tests can add significant confidence to a characteristc S-N curve, depending on the type of structural detail to be designed. A prototype test specimen fabricated in a similar way to the actual connection is considered to provide a representation of the physical behavior, as it is similar in geometry, material characteristics, residual stress, and fabrication tolerances. In addition, it can probably be subjected to more relevant loading and boundary conditions than on small-scale test specimens. Thus, even if only a single large-scale test is performed, it may be reasonable to place significant weight on that test result, based on a total engineering judgment, even if this cannot be supported from a statistical perspective.

The quantity of fatigue test data necessary in order to establish new characteristic S-N curves for the purpose of fatigue assessment has been questioned. Due to lack of specific recommendations for confidence levels to be used when only a few test data are available, further investigations were performed in order to substantiate recommendations for confidence levels and the required number of test data for derivation of characteristic S-N curves. This work is reported in Ronold and Lotsberg (2012), and some of this work is presented in the following section.

4.12 Requirements for Confidence Levels, as Calculated by Probabilistic Methods

4.12.1 Probabilistic Analysis

A study was carried out to establish which confidence level is required in the estimation of the characteristic S-N curve from limited data, while maintaining the safety level equal to that achieved under an assumption of perfect knowledge (infinitely many fatigue tests); see Ronold and Lotsberg (2012). The study was conducted for the case that the characteristic value of the number of cycles to failure is defined as the 2.3% quantile; the safety level is represented by a specified target failure probability. Probability calculations for the situation with perfect knowledge, as well as for the situation with limited fatigue test data, were based on a first-order reliability method (FORM); see, for example, Madsen et al. (1986). In the former case (perfect knowledge), the characteristic number of cycles to failure was set according to its definition. In the latter case (limited test data), it was set to a reduced value that was chosen such that the probability of failure remains unchanged from the situation with perfect knowledge. Tolerance-bound theory was then used to calibrate the required confidence level from the results obtained from the FORM analyses. This was achieved by interpreting the reduced characteristic value as a characteristic value estimated with confidence, and then determining the corresponding confidence by applying tolerance-bound theory. Calculations were first carried out for assumptions of no uncertainty in the loading and an unknown standard deviation of log N. Variations were then explored to study the influence of uncertainty in the loading and to investigate the effect of assuming that the standard deviation in log N is known. For further explanation of theory and limit state functions used for the probabilistic analyses, see Ronold and Lotsberg (2012).

4.12.2 Analysis Results for a Design-Life Approach to Safety

Calculations were carried out for assumptions of no uncertainty in the loading and an unknown standard deviation of log N. The results are shown in Figure 4.18. A range of standard deviations of log N, from 0.07 to 0.24, and a range of reliability indexes, β, between 2.00 and 4.71, are covered. These indexes correspond to the probability of failure during lifetime from 10^{-2} to 10^{-6}. The calculated probability of failure is related to the reliability index through the equation $P_f = \Phi(-\beta)$; see, for example, Naess and Moan (2013). If the reliability target is to be met when test data are limited, then it appears from Figure 4.18 that the greater the requirement for the

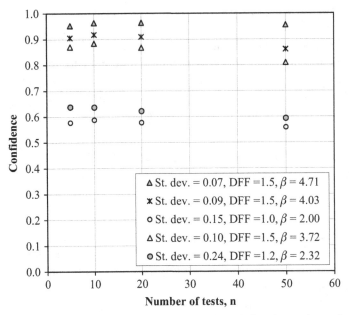

Figure 4.18. Required confidence for characteristic value estimate for the number of cycles to failure in order to maintain reliability over design life when data are limited and the standard deviation is unknown based on Ronold and Lotsberg (2012) (reprinted with permission from Elsevier).

reliability index, β, the larger the requirement for the confidence when estimating the characteristic number of cycles to failure. It is also observed that the requirement for the confidence is not particularly sensitive to the number of observations used for estimating the characteristic S-N curve. Likewise, for a given value of the reliability index, β, the requirement for the confidence is not sensitive to either the standard deviation of log N or to the value of the DFF. The DFF is used as a parameter in the probabilistic analysis to achieve the required calculated safety level. A low DFF is needed for assessment when only uncertainty in S-N test data is present, while a larger DFF is required when there are more uncertain parameters in the fatigue analysis. The requirement for the confidence for a failure probability, $P_F = 10^{-2}$, is in the 59–64% range. For $P_F = 10^{-4}$, the requirement for the confidence is in the 81–88% range. The quoted failure probabilities refer to the probability of failure during a reference period that is equal to the design life of the structure under consideration.

4.12.3 Analysis Results for a Per Annum Approach to Safety

For a fatigue problem in which failure is associated with the accumulation of damage over time, the annual probability of failure increases from one year to the next during the design life. The requirement for the failure probability shall be met in every year of the design life. This implies that if the requirement is fulfilled for the last year of the design life, then it will be fulfilled for all years. In the following realistic example, the design life is taken as 20 years. Two structural reliability analyses are necessary in order to calculate the failure probability in the last year of the design life, one for a reference period of 20 years and one for a reference period of 19 years. The calculated

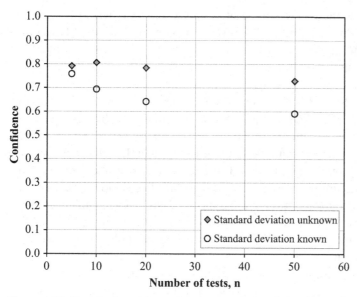

Figure 4.19. Required confidence for characteristic value estimate for the number of cycles to failure in order to maintain an annual failure probability of 10^{-4} in the last year of a 20-year design life based on Ronold and Lotsberg (2012) (reprinted with permission from Elsevier).

failure probability in the last year of the design life is the difference between the calculated failure probability for 20 years and the calculated failure probability for 19 years.

In an example case, the requirement for the annual probability of failure is set at 10^{-4}. The standard deviation in log N is set to 0.15. The DFF is adjusted to 1.68, to produce an annual probability of failure of 10^{-4} in the last year of a 20-year design life, under the assumption of perfect knowledge. The results from this exercise are shown in Figure 4.19 for standard deviation unknown. From Figure 4.19 it can be seen that when the design life is 20 years, and the requirement for the annual failure probability is 10^{-4}, the requirement for the confidence is in the 73–81% range when the available number of test results varies between 5 and 50. The sensitivity to the number of tests is thus not particularly pronounced, and the same holds for changes in the standard deviation and in DFF (Figure 4.18). Different results will be obtained for failure probabilities other than the preselected 10^{-4} per year.

4.12.4 Effect of Uncertainty in Loading Included

Uncertainty in loading is considered at a level corresponding to a 50/50 split in uncertainty importance between load and fatigue strength; this is representative of fatigue in marine structures governed by wave loading. All the results presented in Sections 4.12.2 and 4.12.3 refer to the situation where uncertainty in the loading is not considered and all modeled uncertainty is associated with capacity against fatigue failure: natural variability in capacity, expressed in terms of the number of cycles to failure in the S-N curve; and possible statistical uncertainty in the capacity parameters when these are estimated from limited data. This would be the typical situation for fatigue in a tank for transport of compressed natural gas (CNG), as described in

Section 12.7.3. However, there may well be situations where there will also be uncertainty in the loading, due, for example, to uncertainty in the load calculation and uncertainty in the determination of the appropriate stress concentration factor in the structure. This would be the typical case for fatigue in offshore structures with fatigue loads governed by wave loading.

The example from Section 4.12.3 is considered again. The standard deviation in log N is 0.15, and the DFF = 1.68. Without uncertainty in the loading, this parameter set results in a calculated failure probability of 10^{-4} in the last year of a 20-year design life. Uncertainty in the loading can then be included in the reliability analysis, as represented by a lognormal distribution. This analysis is performed for prediction of the probability of failure over the 20-year design life. The standard deviation in the load distribution is adjusted until the reliability analysis produces an uncertainty importance factor of 50% that is associated with the uncertainty in the loading and an uncertainty importance factor of 50% that is associated with the fatigue strength, when no statistical uncertainty in capacity parameters is included due to limited data. This is assumed to be representative for the situation for fatigue in marine structures with fatigue loads governed by wave loading. Uncertainty importance and uncertainty importance factors are defined in Ditlevsen and Madsen (1996). Then the failure probability in the last year of a 20-year design life increases to $2.02 \cdot 10^{-3}$ from 10^{-4}, when no uncertainty in the loading is included. The failure probability of $2.02 \cdot 10^{-3}$ refers to the situation with perfect knowledge about the number of cycles to failure. With standard deviation $\sigma_{\log N} = 0.15$ and DFF = 1.68, and with the effect of statistical uncertainty owing to limited data included in the probabilistic analysis, the result is a confidence requirement of 0.51 for an exercise for a case of n = 10 fatigue tests. Compared with the requirement of 0.81, when all uncertainty importance is on capacity, this is a rather moderate confidence requirement. This reflects a reduced need for estimation with confidence when the importance uncertainty factor associated with capacity is halved, and also reflects a reduced need for estimation with confidence when the probability of failure is increased. If the DFF is increased from 1.68 to 2.73 and all other input remains unchanged, the failure probability in the last year of the 20-year design life will reduce from $2.02 \cdot 10^{-3}$ to the intended 10^{-4}. When the exercise for determination of the confidence requirement is repeated for this case, a confidence requirement of 0.66 is obtained. This is lower than in the case where all the uncertainty importance is on capacity, but higher than in the case with 50% importance uncertainty on fatigue strength and 20 times higher probability of failure.

4.12.5 Case with Known Standard Deviation

The calculations presented in Section 4.12.4 were all performed under the assumption that the standard deviation of the logarithm of the number of cycles to failure is unknown. However, it is often assumed that the standard deviation is known; see Eurocode EN 1990 (2002). For situations in which the standard deviation is known, the requirement for the confidence for estimation of the characteristic number of cycles to failure is smaller than when the standard deviation is unknown, as presented in Figure 4.18. The requirement for confidence also decreases somewhat with increasing number of test data. The resulting confidences are between 0.69 for n = 5 and 0.56 for n = 50 for a failure probability of 10^{-2}. Furthermore, the larger the requirement for the reliability index, β, the larger the requirement for the confidence when estimating the characteristic number of cycles to failure, if the reliability level in the situation with perfect knowledge shall also be achieved when test data are limited. Results for standard deviation known are given in Figure 4.19 for the case that the annual probability of failure in the last year of a 20-year design life is maintained at 10^{-4}. In this case it also appears that when the standard deviation is known, the requirement for the confidence for estimation of the characteristic number of cycles to failure is somewhat smaller than when the standard deviation is unknown, as also presented in Figure 4.19. Likewise, with an increasing number of test data, the requirement for confidence decreases somewhat. The results for standard deviation known in Figure 4.19 for an annual failure probability of 10^{-4} in the last year of a 20-year design life and standard deviation known vary from a confidence of 0.76 for n = 5 to a confidence of 0.59 for n = 50.

4.12.6 Combination of Cases

With uncertainty in the loading included and an assumed 50/50 split of uncertainty importance between loading and capacity, the study for n = 10 fatigue tests, when the standard deviation of the logarithm of the number of cycles to failure is unknown, produced a requirement for confidence for estimation of the characteristic number of cycles of 0.66, when the failure probability in the last year of a 20-year design life is $P_F = 10^{-4}$. When the standard deviation is known, the requirement for the confidence is 0.64, when the probability of failure in the last year of a 20-year design life is $P_F = 10^{-4}$. Based on this, the use of 75% confidence for derivation of S-N curves for marine structures with 97.7% survival probability is recommended.

4.13 Justifying the Use of a Given Design S-N Curve from a New Data Set

4.13.1 Methodology

This section presents a methodology for validating the use of a particular design S-N class on the basis of a limited number, n, of new fatigue tests. The basis for this was presented in Lotsberg and Ronold (2011) and is included in DNVGL-RP-C203 (2016) for information. It is assumed that the design S-N curve is described in a format as presented in Section 4.1.2 and that the standard deviation in the S-N curve is known. The fatigue testing is performed until a well-defined failure is achieved, such as a fatigue crack through the connection, and not a run-out due to other failures such as leakage or loss of pretension in connectors. However, this procedure may also be used when the fatigue testing is stopped prior to failure, provided that the test load range gives stress range cycles above the fatigue limit, such that a run-out due to stress levels below the fatigue limit is not expected. Due to the limited number of new test data, it is assumed that the slope of a mean S-N curve for the new data is the same as for the considered design S-N curve. Based on this assumption of a fixed m value for the test data, a mean value of log N and $\Delta\sigma$ can be established. A mean S-N curve based on the new test data can be established as:

$$\log N = \log a - m \log(\Delta\sigma \; SMF) + \frac{x_c}{\sqrt{n}} s_{\log N} \qquad (4.16)$$

where:

> N = number of cycles (mean value of new test data at stress range $\Delta\sigma$)
> log a = intercept of considered mean standard S-N curve with the log N axis
> m = slope of standard S-N curve
> $\Delta\sigma$ = stress range
> SMF = Stress Modification Factor
> n = number of test samples
> x_c = factor for confidence with respect to mean S-N data, as derived from a normal distribution; $x_c = 0.674$ for 75% confidence level and 1.645 for 95% confidence level. A 75% confidence level is used in the following equations.
> $s_{\log N}$ = standard deviation in the standard S-N curve

From equation (4.16), the SMF can for 75% confidence be derived as:

$$SMF = 10^{\left(\log a - \log N - m \log \Delta\sigma + \frac{0.674}{\sqrt{n}} s_{\log N}\right)/m} \tag{4.17}$$

And, with n fatigue test data to derive the mean S-N curve from test data:

$$SMF = 10^{\left(\log a - \frac{1}{n}\sum_{i=1}^{n}\log N_i - \frac{m}{n}\sum_{i=1}^{n}\log \Delta\sigma_i + \frac{0.674}{\sqrt{n}} s_{\log N}\right)/m} \tag{4.18}$$

A revised design S-N curve can then be derived as:

$$\log N = \log a - 2 s_{\log N} - m \log(\Delta\sigma\ SMF)$$
or $\tag{4.19}$
$$\log N = \log a - 2 s_{\log N} - m \log\ SMF - m \log(\Delta\sigma)$$

A Stress Modification Factor (SMF) can be considered to account for a nominal stress concentration factor, in addition to items that are specific for the details tested, including fabrication tolerances and workmanship of tested specimens, such as related to connectors (Figure 13.13).

One reason for performing fatigue testing of, for example, connectors is that the high-strength S-N curve is derived from testing of smooth specimens; thus in actual specimens with threads there is a notch effect that implies a less effective stress concentration factor than the theoretical stress concentration factor (see also Section 3.1.4). Thus, the test result may come out positive even if there is some uncertainty in fatigue capacity related to fabrication tolerances.

4.13.2 Example of Analysis of Testing of Connectors, Case A

It is assumed that testing of connectors as shown in Figure 13.13 with the test data shown in Table 4.10 has been performed. It is assumed that the test data follow a shape of the S-N curve similar to the high-strength curve given in Section 4.1.5, with m = 4.7, as connectors are normally machined from high-strength steel. These test data are presented in Figure 4.20, together with the mean curve derived from the six available test results and the mean-minus-two-standard-deviations S-N curve derived from the same test data. The testing is assumed to have been performed in the order listed in Table 4.10, from S1 to S6. In principle, the testing could be stopped after any number of tests to perform an assessment of the resulting SMF, as illustrated in the following example. The SMFs derived from statistics, using Student's

Table 4.10. *Examples of test data and derivations of SMF (standard deviation of six test data = 0.161)*

	Stress Range (MPa)	Number of test cycles	SMF derived from Student's t-distribution for the estimate of log N (standard deviation unknown)	SMF derived from assumed known and constant standard deviation (75% confidence)
S1	80	490,000		4.90
S2	36	9,400,000	19.88	5.25
S3	54	2,700,000	6.34	5.12
S4	47	3,100,000	5.61	5.19
S5	60	900,000	5.47	5.26
S6	40	12,000,000	5.29	5.17

t-distribution for variability in the estimate of log N and considering the standard deviation as "unknown," are shown in Table 4.10 and Figure 4.21. SMFs derived from equation (4.18), by considering the standard deviation as "known," are also shown in Table 4.10 and Figure 4.21. The complete data set (from all six tests) shows a resulting standard deviation of 0.161 in log N. The standard deviation in the high-strength S-N curve in Section 4.1.5 is 0.162. This standard deviation is derived from testing the base material only. Thus, if uncertainties in actual fabrication, fabrication tolerances, and makeup torque of connections are also considered, it is likely that a larger standard deviation should be used. This may be assessed based on general fatigue test data of such connections, if available; see also Example Case B in the next section. Based on this example, it can be seen that a few tests may provide useful information about

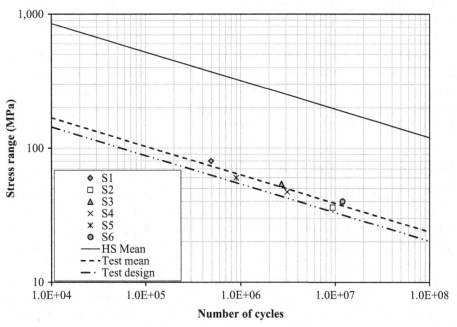

Figure 4.20. Test data used for derivation of SMF in example (standard deviation of six test data = 0.161).

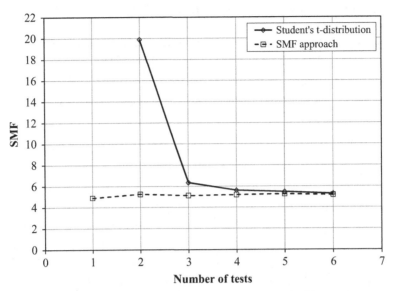

Figure 4.21. SMF factor as function of the number of test specimens (standard deviation of six test data = 0.161) (one test refers to test S1 in Table 4.10, two tests refers to S1 and S2, etc.).

the fatigue design procedure. Even results from a single test may provide useful information and an estimate of SMF to be used for design. The recommended number of tests will, in principle, depend somewhat on the results from the first tests that are performed. If the test results are plotted in a graph, like that shown in Figure 4.21, the convergence in the derived SMF can be evaluated before a decision is reached on further testing. In this example, a significant confidence in the estimated SMF can be seen to have been achieved after the first three tests. From Table 4.10 it can be seen that if the testing is stopped after the first two tests (S1 + S2), then an SMF of 5.25 is derived; after three tests (S1 + S2 + S3), the SMF is 5.12, which is similar to the SMF that is derived after six tests in this example. If only two tests are planned, then stress ranges should be selected, such that failure of the first specimen is aimed at 100,000–500,000 cycles, and the second at 1–10 million cycles. Tables for Student's t-distributions can be found in statistical handbooks, Ang and Tang (1975), DNV-RP-C207 (2012), and also from spreadsheets.

4.13.3 Example of Analysis, Case B

This example is used for checking the robustness of the proposed analysis procedure. The test data from Table 4.10 are modified in a fictitious manner, by approximately doubling the standard deviation to 0.318 for the six test data, such that the scatter is significantly increased. The mean S-N curve from the test data is kept the same as in example Case A. In addition, the ordering of the test data is changed into a worst-case situation, with the first tests furthest away from the basic high strength S-N curve; see also Figure 4.22. The SMF from Student's t-distribution includes the actual calculated standard deviation after two, three, and more tests. Test number 6 falls significantly outside the mean S-N curve, which increases the standard deviation significantly. This explains the SMF based on Student's t-distribution after six tests, as compared with five tests; see Table 4.11 and Figure 4.23. First, it is assumed

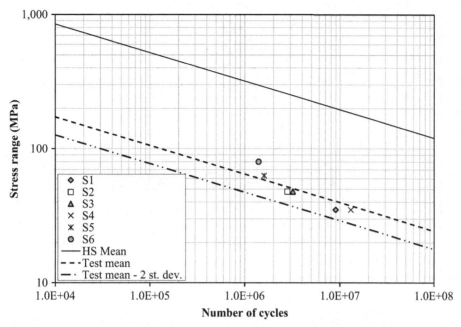

Figure 4.22. Test data used for derivation of SMF in example (standard deviation of six test data = 0.318).

that the standard deviation in the new test data follows that of the high-strength steel curve (with the standard deviation equal to 0.162, while the calculated standard deviation for the six test data is 0.318). The effect of different confidence levels for this assumption is illustrated in Figure 4.23; the difference in calculated SMF is not that large for different confidence levels. The resulting SMFs are also shown in Table 4.11 and Figure 4.23 when the standard deviation in equation (4.18) is increased to 0.318 (for 75% confidence level). This results in a higher SMF, illustrating that a relevant value of an expected (or conservative) standard deviation should be used in equation (4.18). From Figure 4.23 it can be seen that the calculated SMF decreases with number of test data. This effect is likely to be augmented due to the sequence of the test data used in this fictitious example. From the example in Figure 4.23, it might be concluded that it is conservative to use one test only; however, the decrease in SMF

Table 4.11. *Example of test data and derivation of SMFs (standard deviation of six test data = 0.318)*

	Stress Range (MPa)	Number of test cycles	Student's t-distribution for log N	SMF		
				75% conf. and $S_{\log N} = 0.162$	95% conf. and $S_{\log N} = 0.162$	75% conf. and $S_{\log N} = 0.318$
S1	35	9,000,000		6.03	6.51	7.39
S2	48	2,800,000	40.66	5.74	6.06	6.63
S3	48	3,200,000	7.93	5.58	5.84	6.28
S4	35	13,000,000	6.52	5.53	5.74	6.12
S5	63	1,600,000	5.90	5.34	5.53	5.85
S6	80	1,400,000	6.27	5.04	5.20	5.47

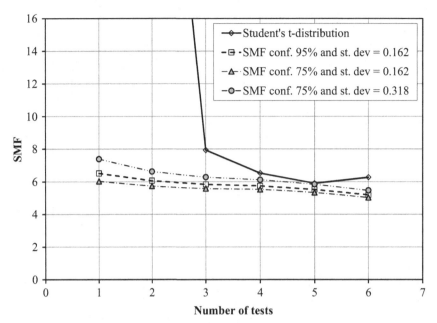

Figure 4.23. SMF factor as function of confidence, standard deviation, and the number of test specimens for test data shown in Figure 4.22.

with the number of tests is due to the sequence of the test data used. Another approach to demonstrate this would be to present the test data in reverse order, which would result in the lowest calculated SMF being obtained for one test and the SMF increasing as the additional test results were included, as shown in Figure 4.24.

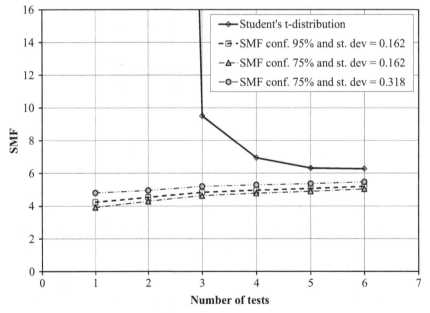

Figure 4.24. SMF factor as function of confidence, standard deviation, and the number of test specimens with test data in reverse order compared with those in Figure 4.23.

4.13.4 Example of Fatigue Proof Testing of Connector in Tethers of a Tension Leg Platform

It is assumed that a connector to be used in the tethers of a tension leg platform has been designed and one connector is planned fatigue tested for documentation of the design; see, for example, Figure 13.13. The design has been performed to account for relevant fabrication tolerances of the threads and a stress concentration factor has been calculated by finite element analysis as SCF = 1.5 relative to the nominal stress in the outside tubular section. It is considered difficult to perform fatigue testing for dynamic axial load due to limitation in capacity of available dynamic actuators. Thus the testing may be performed under bending moment loads. This makes it possible to perform testing at a relevant R-ratio using actuators with less capacity. However, axial testing would be preferred as that would be more representative for the actual physical behavior. The girth welds between the connector and the tubulars may be the most critical regions with respect to fatigue capacity; thus the testing arrangement needs to be considered with this in mind. Here the main focus is on establishing load level and the number of cycles to be used in the fatigue test of the connector until a fatigue failure or the fatigue test can be stopped with a conclusion that the test has demonstrated that the connector has sufficient fatigue capacity. With only one test under constant amplitude loading, one should use so large a stress range that the load is larger than the fatigue limit. When assessing this, one should also consider what is an expected SCF in a connector as that derived from the analysis may be to the large side when considering maximum fabrication tolerances. It is assumed that the connectors are made with high-strength steel and that the S-N curve in Section 4.1.5 applies. Due to fabrication tolerances one may want to use a larger standard deviation than that due to the S-N data only, and in this case it is suggested to use $s_{logN} = 0.25$ in equation (4.16). Then with 75% confidence level the following relation between the number of test cycles and stress range is obtained:

$$\log N = 17.770 - 4.7 \log(\Delta\sigma \cdot 1.50) + \frac{0.674}{\sqrt{1}} 0.25$$
$$\log N = 17.110 - 4.7 \log(\Delta\sigma) \tag{4.20}$$

If one aims to stop the testing after $5 \cdot 10^5$ cycles, the testing should be performed with a stress range $\Delta\sigma = 267.95$ MPa based on this equation. If one selects to use 95% confidence level in equation (4.16), the stress range should be increased to 301.76 MPa.

One may wonder if it is likely that one will achieve test results that are above the mean S-N curve here. One reason for this being likely is the conservatism built into the design using an elastic stress concentration factor for the threads; reference is made to the difference between notched and smooth specimens presented in Section 3.1.4. It may also be due to better tolerances in the fabricated test specimen as compared with those used as basis for calculation of the elastic stress concentration factor.

5 Stresses in Plated Structures

5.1 Butt Welds in Unstiffened Plates

Butt welds are important connections in most types of welded structures. Examples are plated structures used in ships, floating offshore platforms, frame structures fabricated from tubular sections, tethers connecting tension leg platforms to the sea bottom, shell structures used in semi-submersible platforms for oil and gas production, towers for support structures for wind turbines, foundation piles, pipelines, risers, umbilicals, and so on. Many of these structures are subject to dynamic loading, and a reliable assessment of the hot spot stress at these connections is required for calculation of fatigue life.

A stress concentration factor (SCF) can be defined as a stress magnification at a detail due either to the detail itself or to a fabrication tolerance, with the nominal stress as a reference value. The maximum stress is often referred to as the hot spot stress and is used in conjunction with S-N data for fatigue life calculation. It is also termed geometric stress or structural stress. This hot spot stress is derived as the SCF multiplied by the nominal stress. The effects of misalignment due to eccentricities at plate thickness transitions and fabrication tolerances on the hot spot stress also need to be considered. In scaled plate test specimens, additional stress due to angular misalignment resulting from welding distortion may also be of importance for the resulting hot spot stress, as explained in Section 2.2.1. However, such an effect is normally considered to be of less significance for actual structures than laboratory test specimens due to differences in boundary conditions during fabrication and loading. The angular deviation may be more significant for structures with thinner plates.

SCFs due to misalignment at butt welds in plates were presented by Maddox (1985) and have been included in fatigue design rules for plated structures for many years. A simple butt weld between two plates, as shown in Figure 5.1, is considered as an introduction to the derivation of SCFs for butt welds. It is assumed that the plates are welded together from plates of the same size, with an eccentricity, δ, and without angular misalignment. The plates are subjected to a membrane loading per unit width $N = \sigma_{nominal} \cdot t$, where $\sigma_{nominal}$ is nominal stress and t = thickness of the plates.

Due to the asymmetry of the static system shown in Figure 5.1a, the moment will be zero at the middle of the weld, which is also denoted as the inflection point. Moment equilibrium then results in the maximum moment becoming $M = N \cdot \delta/2$,

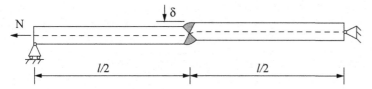

(a) Static system: Plates with axial misalignment at weld

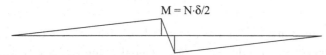

$$M = N \cdot \delta / 2$$

(b) Moment distribution

Figure 5.1. Moment distribution in plates with an eccentric butt weld subject to membrane load.

as shown in Figure 5.1b. The elastic section modulus for a unit plate width is $W = t^2/6$. The maximum bending stress over the plate thickness is then obtained as:

$$\sigma_b = \frac{M}{W} = \frac{N\frac{\delta}{2}}{W} = \frac{\sigma_{\text{nominal}}\, t\frac{\delta}{2}}{\frac{t^2}{6}} = 3\frac{\delta}{t}\sigma_{\text{nominal}} \tag{5.1}$$

and the stress concentration, frequently referred to at an unstiffened plate weld joint, is obtained from the definition given above as:

$$SCF = \frac{\sigma_{tot}}{\sigma_{\text{nominal}}} = \frac{\sigma_{\text{nominal}} + \sigma_b}{\sigma_{\text{nominal}}} = 1 + 3\frac{\delta}{t} \tag{5.2}$$

The resulting moment distribution along the surface of the plate is shown in Figure 5.1b. If the plates are of different lengths, then one of the moments at the weld will be larger than that shown above. If the largest plate length is L_1 and the total length is L, the maximum moment at the weld is $M_1 = N \cdot \delta/2 \cdot L_1/L$, and the stress concentration factor reads:

$$SCF = 1 + 3\frac{\delta}{t}\frac{L_1}{L} \tag{5.3}$$

The angular misalignment may be of importance for fatigue testing of simple plates, in addition to eccentricity, as shown in Figure 2.4. The SCF is a function of axial force (Hobbacher 2009).

For pinned ends:

$$SCF = 1 + \frac{6\,\delta_0}{t}\frac{\tanh(\beta)}{\beta} \tag{5.4}$$

where:

$$\beta = \frac{2l}{t}\sqrt{\frac{3\sigma_m}{E}} \tag{5.5}$$

δ and l are described in Figure 2.4
t = thickness
σ_m = membrane stress
E = Young's modulus.

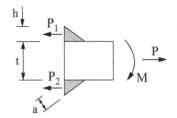

Figure 5.2. Forces acting on a fillet weld connection.

This equation can be outlined in a similar way as for SCF at girth welds in tethers, as shown in Section 6.9, and for seam welds in pipes subjected to internal pressure, as described in Section 7.2.

For fixed ends:

$$SCF = 1 + \frac{3\,\delta_0}{t}\,\frac{\tanh(\beta/2)}{\beta/2} \tag{5.6}$$

The tanh correction term allows for reduction of angular misalignment due to straightening of the connection under tensile loading. For tensile loading it is always less than 1.0, and thus can be ignored. Similarly, the misalignment is amplified for compressive loading, but another expression should be used to account for this. The SCF due to a combination of eccentricity and angle deviation is derived as:

$$SCF = 1 + (SCF_{eccentricity} - 1) + (SCF_{angle} - 1) \tag{5.7}$$

The reason for adding the two contributions together is that these are two separate effects and an SCF due to one factor does not have any influence on the other (these factors would have to be multiplied together if the stress increase due to one factor also increases the other stress concentration factor).

5.2 Fillet Welds

The nominal stresses in a fillet weld over leg length h due to an acting moment that are shown in Figure 5.2 can be calculated in a similar way as for the axial force acting on a full penetration weld. The nominal stress over a unit width of the weld due to axial force (where $P_1 = P_2 = P/2$) is calculated as:

$$\sigma_a = \frac{P}{2h} \tag{5.8}$$

The nominal stress due to bending moment can be calculated from requirement to moment equilibrium as (assuming also here that $M = P \cdot \delta/2$):

$$\sigma_b = \frac{M}{h\,(h+t)} = \frac{P\delta}{2h\,(h+t)} \tag{5.9}$$

And thus the total stress is derived as:

$$\sigma_{tot} = \sigma_a + \sigma_b = \frac{P}{2h} + \frac{P\delta}{2h\,(h+t)} = \frac{P}{2h}\left(1 + \frac{\delta}{h+t}\right) \tag{5.10}$$

From this equation, the SCF is derived, based on its definition in equation (5.2) as:

$$SCF = 1 + \frac{\delta}{h+t} \tag{5.11}$$

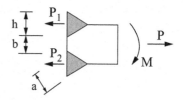

Figure 5.3. Forces acting on a partial weld penetration.

Similarly, for design assessment of a partial penetration weld, the SCF (or K_{tw} in Section 2.5.3) for the detail shown in Figure 5.3 is derived as:

$$SCF = 1 + \frac{\delta}{h+b} \tag{5.12}$$

Equation (5.11) was used by Andrews (1996) to reduce the scatter in fatigue test data derived from specimens with different eccentricities to achieve an acceptable comparison with a design S-N curve for fillet welded connections. See also Section 3.2.1.

5.3 Butt Welds in Stiffened Plates

5.3.1 Background

The effect of stiffeners on SCFs for butt welds in plated structures has not been a parameter in existing design standards. Instead, it has been indicated that fabrication tolerance has little effect on stresses in stiffened plates. This was, therefore, investigated further by some detailed finite element analyses, in order to provide further evidence-based guidance on this subject. The purpose of the finite element analyses of plated structures was defined as:

   To calculate the geometric SCF for typical butt welds in longitudinally stiffened plates in floating production, storage, and offloading vessels (FPSOs) and ships.
   To assess the effect of size of longitudinal stiffeners on plate butt weld SCF.
   To assess the effect of distance between transverse frames on plate butt weld SCF.
   To assess the effect of distance between longitudinals on plate butt weld SCF.
   To assess the effect of the position of the weld between transverse frames for plates with different thicknesses on SCF.
   To assess the effect of cope holes in longitudinals on SCFs for butt welds in the plate.
   To assess how to account for differences in thicknesses in expressions for butt weld SCFs.
   To assess how to account for differences in thicknesses in a general expression for butt weld SCF, where the combined effect of eccentricity (due to the transition in thickness and fabrication tolerance) is included.

The typical geometries of stiffened plates in ships and FPSOs were investigated. Similar geometries are also of relevance for semi-submersibles. Typical distances between longitudinal stiffeners (simply denoted as longitudinals) in an FPSO are

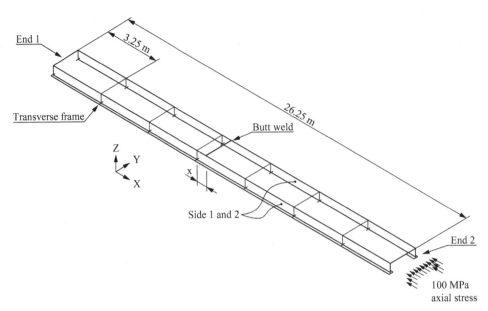

Figure 5.4. Finite element model for analysis of a plated structure with cope hole.

in the 700–800 mm range, and plate thicknesses are in the 12–30 mm range. The distance between transverse frames is typically 3.0–4.0 m. In ship tanker structures, the distance between transverse frames is often larger.

5.3.2 Finite Element Analysis of Stiffened Plates

Geometry and Analysis Models
Fifteen different finite element analyses were performed. Twelve analyses were performed with a typical cope hole (radius = 50 mm) in the longitudinals at the butt weld in the plate. The analysis geometry is shown in Figure 5.4 (Lotsberg and Rove 2014). Three analyses, numbered 13–15, were performed without cope holes in the longitudinals. The distance between longitudinals was 800 mm for analyses 1–14, and 400 mm for analysis 15. Finite element analyses for three different positions of the butt weld relative to the transverse frames were performed with different x values; see Figure 5.4: x = 400 mm, x = 1875 mm, and x = 3650 mm. The geometries are presented in Table 5.1. The effect of boundary conditions in the transverse direction (y direction of Figure 5.4) was investigated by using different boundary conditions in analyses 1–4. The thicknesses and eccentricities of the joined plates are shown in Table 5.1. The transition in slope was 1 to 4 between the thick and the thinner plates; see Figure 5.5. In all analysis models, there is a longitudinal stiffener at each side of the plates. The stiffeners were modeled as a T-bar profile (T320 × 13/112 × 16). They also represent section properties of typical bulb sections used in deck areas. The length of the models was 26.25 m (i.e., spanning over seven frames). The distance between frame/bulkhead was 3.75 m. The plate with the butt weld was placed in the middle of the analysis model; see also Figure 5.4. The frames/bulkheads were not modeled but simulated by use of boundary conditions with x = free, y = fixed,

Table 5.1. *Geometries analyzed and calculated stress concentration factors at butt welds*

Analysis No.	T (mm)	t (mm)	Eccentricity (mm)	Weld position x in Figure 5.4	SCF in the middle of the plate field	SCF in way of cope hole
1	30	19	5.5 mm due to difference in thickness	400	1.56	1.63
2	30	19	5.5 mm due to difference in thickness	400	1.58	1.67
3	30	19	5.5 mm due to difference in thickness	1,875	1.55	1.64
4	30	19	5.5 mm due to difference in thickness	1,875	1.53	1.73
5	24	19	2.5 mm due to difference in thickness	400	1.34	1.37
6	19	19	2.85 mm due to shift of plates	400	1.45	1.48
7	19	19	2.85 mm due to shift of plates	1,875	1.46	1.49
8	30	19	5.5 mm due to difference in thickness	3,650	1.55	1.64
9	24	19	2.5 mm due to difference in thickness	3,650	1.37	1.36
10	24	19	2.5 mm due to difference in thickness	1,875	1.34	1.38
10	24	19	2.5 mm due to difference in thickness	1,875	1.34	1.38
11	30	19	5.5 mm due to difference in thickness + 2.85 mm due to shift of plates	1,875	1.83	2.04
12	24	19	2.5 mm due to difference in thickness + 2.85 mm due to shift of plates	1,875	1.68	1.78
13[2]	30	19	5.5 mm due to difference in thickness	400	1.55	
14[2,3]	30	19	5.5 mm due to difference in thickness	400	1.52	
15[1,2]	30	19	5.5 mm due to difference in thickness	400	1.57	

[1] Distance between longitudinals 400 mm.
[2] Without cope holes.
[3] Geometry as analysis 13, but bottom flange of longitudinals fixed in the z direction to simulate stiff longitudinals or longitudinal bulkheads.

and z = fixed in the analysis model shown in Figure 5.4. All degrees of freedom were fixed at end 1, and at end 2, x = free, y = fixed, and z = fixed. Sides 1 and 2 were free in y direction in analyses 2 and 4 and fixed in all the other analyses. The structures were modeled by using solid, 20-node elements; see Figure 5.6. The finite element mesh was refined at the plate thickness transition and at the butt weld in the middle of the plate field. Outside this region, one element is used over the plate thickness, as recommended for standard hot spot stress analysis, in order to achieve a

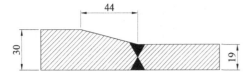

Transition in thickness from 30 mm to 19 mm

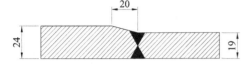

Transition in thickness from 24 mm to 19 mm

Figure 5.5. Thickness transition and eccentricity used in analyses.

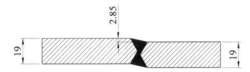

Axial misalignment of 2.85 mm

linear stress distribution over the plate thickness. The load in all analyses was put as "Normal Pressure" of 100 MPa at the end section in the longitudinal direction. Young's modulus for steel of $2.1 \cdot 10^5$ MPa and Poisson's ratio of 0.3 were used in the analyses.

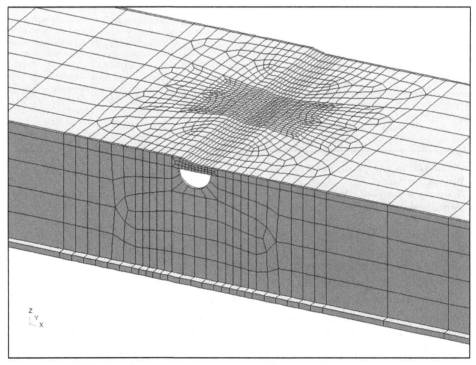

Figure 5.6. Finite element model of butt weld connection in plated structure (Lotsberg 2009b) (reprinted with permission from Elsevier).

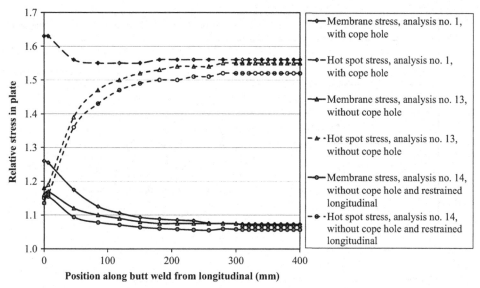

Figure 5.7. Stress at top of plate along butt weld and membrane stress (position 0 is at longitudinal and position 400 is midway between longitudinals).

Analysis Results

The geometric SCFs were derived by dividing the hot spot stress at the weld by the nominal stress of 100 MPa, according to the definition of an SCF. The stress distribution (stress in x direction) over the thickness of the plate was evaluated. It was observed that the stress distribution over the plate thickness at the nodes that join the first and second elements away from the weld toe is approximately linear. Therefore, the stress at the surface at the node one element from the weld was selected as hot spot stress; the stress outside this region is not significantly different. Thus, extrapolation of stress to determine a hot spot stress was not required (as recommended in general in Section 9.1.3). The results of the analyses are listed in Table 5.1. The results are presented for the butt weld in the middle of the plate, between the longitudinals where the largest stress occurs for structures without cope holes, and at the butt weld in way of cope holes, where these were included in the analysis models. (The expression "in way of" is used when the welding is going through the hole. Cope hole is also denoted as scallop; see Section 5.4.4). It should be noted that the calculated SCFs are related to the stress at the thinnest plate section, t.

Effect of Boundary Conditions

From analyses 1–4 it can be noted that the difference in the boundary conditions along sides 1 and 2 in the y direction (Figure 5.4) has little effect on the results.

SCF for Butt Welds without Cope Holes in the Longitudinals

The results from the finite element analyses without cope hole are summarized in Table 5.1. Comparison of analyses 1 and 13 demonstrate that the cope hole has negligible effect on the SCF in the middle of the plate. This is also shown in Figure 5.7. The largest SCF is at the butt weld in the middle of the plate in stiffened plates with cope holes. This means that analyses of SCF for geometries with cope holes are also applicable for SCF calculation for plates without cope holes, when the

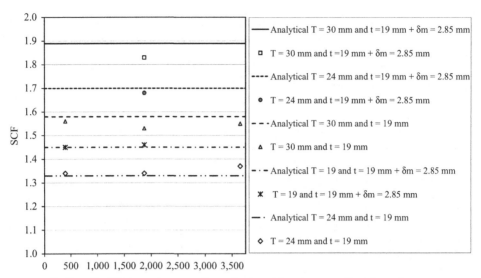

Figure 5.8. Comparison of SCFs from finite element analyses and analytical expressions.

SCF is evaluated in the middle of the plate field. It should also be noted that the cope hole itself results in an increased SCF at the butt weld in way of the cope hole, as compared with plates without cope holes. Figure 5.7 shows the stress distribution along the butt weld for analysis 1 (with cope hole) and analysis 13 (without cope hole); in these two analyses the highest stresses are located at different positions.

The question arises of whether the SCF is decreased by reducing the distance between the longitudinals when cope holes are not present, due to a possible stiffening effect from longitudinals on the plate rotation at the butt welds; in order to investigate this, an analysis with a small distance between longitudinals (400 mm) was included as analysis 15. In Table 5.1 it can be seen that the SCF from this analysis is slightly greater than that of analysis no. 13.

The results from analyses 6 and 7, with equal plate thicknesses and fabrication eccentricities, are shown in Figure 5.8 around the line for the analytical SCF = 1.45. The results from analyses 1, 3, and 8, with thickness transition from 30 mm to 19 mm, are shown in the same figure slightly below the analytical SCF = 1.58; the weld position is the only parameter that is varied in these analyses. The results from analyses 5, 9, and 10 with thickness transition from 24 mm to 19 mm are also shown in Figure 5.8 slightly above the analytical SCF = 1.33. It should be noted that the calculated SCFs are fairly constant, indicating that the position of the weld between the frames is not a significant parameter. The results for combined eccentricity and thickness transition are shown in Figure 5.8 below the analytical values with SCF = 1.70 and SCF = 1.89.

SCF at Butt Welds in Way of Cope Holes

From the finite element analyses in Figure 5.7 it can be seen that the SCFs in way of cope holes are larger than at butt welds without cope holes. In order to investigate this further the membrane stress in the plate is also plotted for the same conditions in Figure 5.7. It should be noted that the membrane stress is larger than the loading, partly due to some global bending of the plate and the longitudinals, as one beam system between transverse frames. In fact, a transition from a plate thickness of 30 mm to one of 19 mm implies a shift in neutral axis for this beam system of 17 mm. This results in some global bending moment of the plated structure. Due to this, a further analysis (14) was included to investigate the effect of longitudinal size. In this analysis, the longitudinals are supported in the vertical direction (z direction) to simulate a support by a longitudinal bulkhead. From Figure 5.7 it can be seen that this did not change the force flow significantly from that of analysis 13. Thus, rotation of the global beam system due to the introduced eccentricity adds only 2–3 MPa to the plate membrane stress, as can be seen from Figure 5.7. This stress is a relatively small fraction of the maximum hot spot stress at the cope hole, as shown in Figure 5.7. This indicates that the calculated SCFs are not very sensitive to the size of the longitudinals or to the distance between transverse frames.

5.3.3 Analytical Equations for Stress Concentrations at Butt Welds in Plated Structures

The finite element analysis results were compared with the following equation for eccentricity, with a shift in neutral axis equal to δ for plates of equal thickness:

$$SCF = 1 + 6\,\delta\,\frac{t^{0.5}}{T^{1.5} + t^{1.5}} \tag{5.13}$$

It should be noted from Figure 5.8 that there is a very good correlation between the finite element analyses and this equation. The finite element analysis results were compared with the following equation for eccentricity, with a shift in neutral axis due to the difference in plate thicknesses:

$$SCF = 1 + 3\,(T - t)\frac{t^{0.5}}{T^{1.5} + t^{1.5}} \tag{5.14}$$

This equation was proposed by Maddox (1985) and is also used by the International Institute of Welding (IIW) (Hobbacher 2009) and in BS 7910 (2013). From Figure 5.8 it can be seen that there is a good correspondence between the finite element analyses and this equation. The finite element analysis results were compared with the following equation for eccentricity, with a shift in neutral axis of δ due to eccentricity of plates, in addition to a shift in neutral axis due to the joining of plates of different thicknesses:

$$SCF = 1 + 6\,\delta\,\frac{t^{0.5}}{T^{1.5} + t^{1.5}} + 3\,(T - t)\frac{t^{0.5}}{T^{1.5} + t^{1.5}} \tag{5.15}$$

This equation can be seen as a combination of equations (5.13) and (5.14), taking into account some dissimilarity in stiffness resulting from the difference in thickness between the plates on the bending effect due to eccentricity. It should be noted from Figure 5.8 that there is a good correspondence between the finite element analyses and this equation.

From the analyses it can be seen that the cope holes imply an increased stress in that region. A reduced area in way of the cope hole will imply an increase in SCF, which can be expressed as:

$$SCF = 1 + \frac{A_{Section}}{A_{Section} - A_{Cope\,hole}} \tag{5.16}$$

where:

$A_{Section}$ = sectional area of the plate and longitudinals without cope hole

$A_{Cope hole}$ = area of cope hole in a section normal to the force

$A_{Section}$ is here = 20944 mm^2 at the plate thickness, t = 19 mm, to which the SCF is related

$A_{Cope hole}$ = 650 mm^2

SCF = 1.032.

By combining equation (5.16) with the other equations to derive SCF for the butt weld in way of cope hole, a good correspondence between SCFs derived from analytical expressions and from finite element analysis was obtained, as explained in Lotsberg and Rove (2014). The results from these analyses can be summarized as:

For geometry with cope holes in the longitudinals, the highest SCF is at the butt weld in way of the cope hole of the longitudinal stiffener.

Without a cope hole in the longitudinal, the highest SCF is in the middle of the plate, between the longitudinals.

There is little stiffening effect from the longitudinals on plate bending in the middle of the plate.

The hot spot stress is not very sensitive to the distance between the longitudinals.

The calculated SCFs are apparently not very sensitive to the size of the longitudinals.

The SCF is relatively insensitive to the position of the butt weld between the transverse frames (position measured in the longitudinal direction).

A simple equation for SCF can be used for the design of butt weld connections in stiffened plates with eccentricity, or fabrication tolerance between plates with equal thicknesses.

The equation for SCF proposed by Maddox (1985) and used in the British Standard BS 7910 (2013) fits the data fairly well for butt weld connections in plates of different thicknesses. This applies for plates with a slope in transition in thickness of 1:4.

For butt welds in way of cope holes, an additional SCF accounting for the reduced area in way of cope holes (Equation 5.16) should be included, in addition to the factors discussed above. This is to enable a better fit of the calculated values if the hole area is significant in comparison with the total area of longitudinal stiffener and plate.

5.3.4 Effect of Fabrication Tolerances in Plated Structures in Fatigue Design Standards

The link between fabrication tolerances for plated structures and design standards is acknowledged to be problematic, and therefore different design standards may vary considerably on this subject. Going back to 1977, the effect of fabrication tolerances was not mentioned in relation to fatigue design of plate butt welds in regulations from the Norwegian Petroleum Directorate (NPD 1977) or from the UK Department of Energy's guidance notes (1977). In 1980, a sentence was included in the British Standard for steel bridges stating that classification of the detail was considered to include the effect of any misalignment, up to the maximum allowable value (BS 5400 1980). However, in a book by Gurney (1979), it was mentioned that an additional stress concentration factor, similar to equation (5.2), should be multiplied by the nominal stress, before entering the S-N curve for butt welds to calculate the number of cycles to failure. At this time, the effect of fabrication tolerances at butt welds was of less concern because it was considered that there would always be some attachment welded on the actual member with a lower S-N classification that would be governing for the calculated fatigue life.

Misalignments in cruciform joints have been measured and significant misalignments were found (Bokalrud and Karlsen 1982). However, to the best of the author's knowledge, measurements of actual tolerances in fabricated butt welds have not been published in the literature.

Tolerances are considered most important for butt welds in plates and cruciform joints, but less important for other types of welded connections. In the DNV Classification Note (DNV CN 30.2 1984) on fatigue strength of offshore units, it was stated that fabrication tolerances were accounted for in the S-N classification. This was also recommended in BS 7608 from 1993. In 1990, the Health and Safety Executive (HSE) in London issued new design guidance on fatigue design of offshore structures in which it was stated that the calculation of hot spot stress should include SCFs to account for geometric effects, including mismatch and changes in thickness. This document was later used as the basis for ISO 19902 on design of offshore structures that was published in 2007; see also Bærheim et al. (1996). In the ISO standard the word "mismatch" is omitted, such that it can be assumed that the allowable fabrication tolerances are accounted for in the S-N classification.

One reason for this confusion was, perhaps, the lack of instrumented test data; such data have been published by Maddox (2001, 2003). In the latter of these (Maddox 2003), it is indicated that an SCF of 1.05 can be considered to be accounted for when using the S-N curve D for butt welds. This is also in line with the IIW recommendations (Hobbacher 2009). However, according to BS 7608 (2014), tolerances are not included in the D curve for butt welds in plated structures.

The new IACS standard for ship structures (2013) recommends accounting for fabrication tolerances only if standard tolerances are being exceeded. In the first version of the NORSOK Standard on design of steel structures (1998), it was proposed that the nominal stress should be multiplied by an SCF that corresponds to the maximum allowable fabrication tolerance; see, for example, NORSOK–M101 (2011) with a minimum tolerance requirement (0.15t, 4.0 mm) that is also shown in Figure 5.9. It was observed that this would be a governing fatigue criterion at thickness transitions in plated structures. As these areas had not been observed to be a problem with respect to fatigue cracking, this design recommendation was reevaluated. At this time IIW (Hobbacher 1996) had stated that a stress concentration of 1.1 was included in the butt weld S-N classification, so this factor was also included in the NORSOK standard that was subsequently developed. It was later determined that a smaller SCF is included in the S-N curve for butt welds; see Maddox (2003) and IIW (Hobbacher 2009). Nevertheless, the expression has been kept in DNVGL-RP-C203 (2016) for plated structures, partly due to the effect from the combination of S-N data and tolerances, based on a probabilistic assessment of the effect of poorer S-N data combined with inferior tolerances.

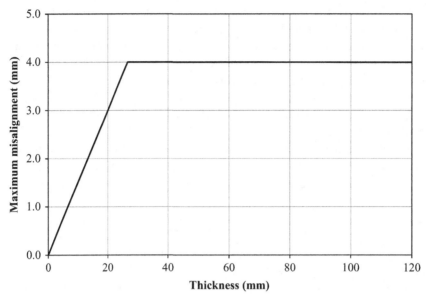

Figure 5.9. Maximum allowable misalignment in butt welds, according to NORSOK M-101.

The eccentricity in equation (5.2) is assumed to be normally distributed, with a 95% probability that the fabrication tolerance is within the prescribed fabrication tolerance of 0.15t. This information, together with the resulting S-N curve D with mean log a = 12.564 and standard deviation of 0.20, provides the basis for the derivation of Figure 5.10 and Figure 5.11; see Lotsberg et al. (2008b) and Section 12.7.3. In Figure 5.10, the horizontal axis represents the mean fatigue test data, including uncertainties due both to tolerances and to S-N data. Mean S-N data are shown up

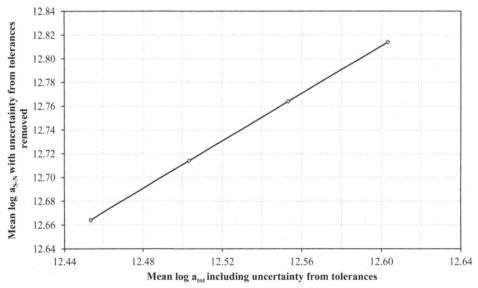

Figure 5.10. Relationship between fatigue test data in terms of mean log a values and fatigue test data that also include the effect of misalignment on fatigue strength in addition to the fatigue data (Lotsberg et al. 2008b) (reprinted with permission from ASME).

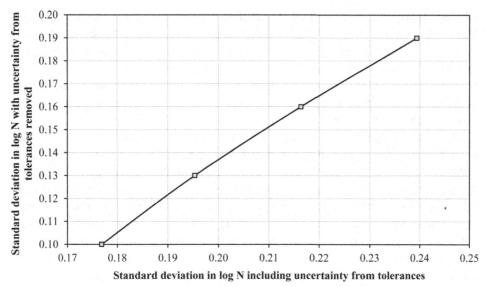

Figure 5.11. Relationship between standard deviations in fatigue test data and fatigue test data that also include the effect of misalignment on fatigue strength in addition to the fatigue data (Lotsberg et al. 2008b) (reprinted with permission from ASME).

the vertical axis, where uncertainties due to tolerances are removed, such that the vertical axis represents S-N data that are similar to those derived from instrumented tests. This reduces the uncertainty in S-N data and increases the mean values from the tests if only S-N data are considered. Similarly, it can be seen from Figure 5.11 that the standard deviation in the S-N data is reduced when uncertainties due to tolerances are removed.

The recommended format of the equation for SCF to be used in design depends on the maximum tolerance included in the fabrication standard versus the quality of the actual fabrication and as-built tolerances. This is illustrated schematically in Figure 5.12, showing one distribution representative for that inherent in the

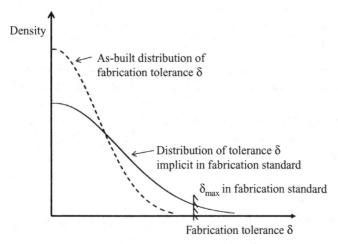

Figure 5.12. Schematic illustration of probability density distributions for as-built tolerances and tolerances in a fabrication standard.

fabrication standard and another distribution representing the as-built tolerances that can be achieved in a good fabrication. If a narrower tolerance is specified as the maximum fabrication tolerance, then the relationship between the two distributions may differ. This may also lead to different requirements for how the SCFs should be calculated in order to meet a predefined safety level. For the situation indicated in Figure 5.12 it may be acceptable not to use a full stress concentration that corresponds to that of δ_{max}. Instead, a reduced factor could be used by subtraction of a δ_0, like that used in DNVGL-RP-C203 (2016) for plated structures, while a full value is used where narrow tolerances are specified for fabrication of risers, pipelines, and tethers in tension leg platforms. This shows that it is important to assess the required fabrication tolerances for butt welds in relation to the tolerances to be used in design calculations. This substantiates that $\delta_0 = 0.1t$ can be subtracted from the axial misalignment in equations (5.13)–(5.15) in order to calculate an effective SCF to be used in fatigue design of butt welds in plated structures. However, if the actual tolerance is known at a butt weld, the SCF for this tolerance should be calculated and an SCF factor of 1.05 (or 1.0 to be safer) should only be assumed to be accounted for in the S-N curve D for butt welds. This is also the procedure for connections with many welds, such as tethers, risers, and pipelines, where there is a greater possibility that the actual fabrication tolerance in one connection is in the outer region of that accounted for in the design procedure.

5.4 Openings with and without Reinforcements

5.4.1 Circular Hole in a Plate

Stresses at a hole from the solution in an infinite plate, first published by Kirsch (1898) and based on Timoshenko and Goodier (1970), are presented as:

$$\sigma_r = \frac{\sigma_n}{2}\left(1 - \frac{R^2}{r^2}\right) + \frac{\sigma_n}{2}\left(1 + \frac{3R^4}{r^4} - \frac{4R^2}{r^2}\right)\cos 2\theta$$

$$\sigma_\theta = \frac{\sigma_n}{2}\left(1 + \frac{R^2}{r^2}\right) - \frac{\sigma_n}{2}\left(1 + \frac{3R^4}{r^4}\right)\cos 2\theta \qquad (5.17)$$

$$\sigma_r = -\frac{\sigma_n}{2}\left(1 - \frac{3R^4}{r^4} + \frac{2R^2}{r^2}\right)\sin 2\theta$$

The notations in these equations are shown in Figure 5.13. Stress contours around a hole from using these equations are also shown in Figure 5.13. The largest stresses at a section for y = 0 for a far-field stress $\sigma_n = 1.0$ are shown in Figure 5.14. It can be seen that the hole has influenced the stress by a factor of 1.2 at a distance of a hole-radius away from the edge of the hole. This should be kept in mind when welded connections are placed close to such holes or penetrations.

5.4.2 Elliptical Hole in a Plate

The following SCF is applicable for an elliptical hole, where the fluctuating stress is parallel to the longest half-axis (semi-major axis), c, in the ellipse:

$$SCF = 1 + 2\sqrt{\frac{a}{c}} \qquad (5.18)$$

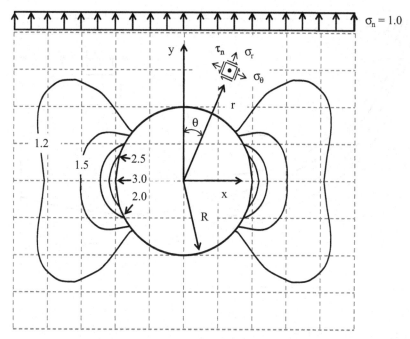

Figure 5.13. Stress contours around a hole in a plate subjected to far-field stress.

The following SCF applies for an elliptical hole, where the fluctuating stress is parallel to the shortest half-axis (semi-minor axis), a, in the ellipse

$$SCF = 1 + 2\sqrt{\frac{c}{a}} \qquad (5.19)$$

From these equations, it can be seen that the smallest SCF is obtained if the elliptical hole can be placed in a plate with its longest half-axis parallel to the main

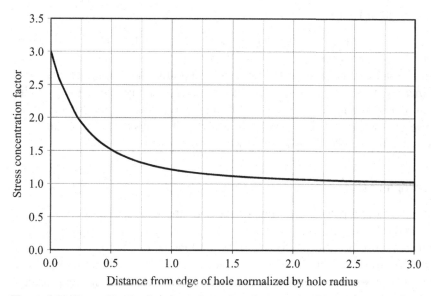

Figure 5.14. Stress distribution through section along x axis of hole in Figure 5.13 for far-field stress $\sigma_n = 1.0$.

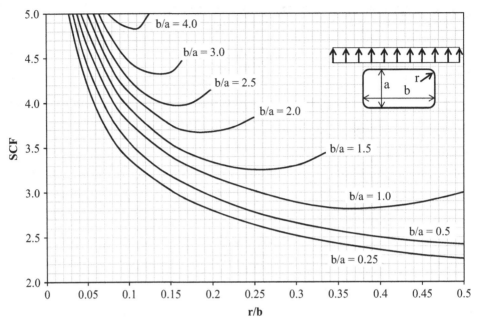

Figure 5.15. SCFs for rectangular holes.

fluctuating stress direction. The same principle may also be applied at thickness transitions in connections with high dynamic loading for reduction of the surface stress. Here also the longest half-axis in the ellipse is placed in the main fluctuating stress direction. This methodology has been used to design transitions from riser pipes to compact flanges, using a profile of a quarter ellipses with half axes in the ratio 1:4. This methodology is also used in the fatigue design of tether connections, as shown in Section 13.9.

5.4.3 Rectangular Holes

SCFs for rounded rectangular holes subjected to a dynamic load normal to one of the sides in the hole are presented in Figure 5.15.

When the same hole is subjected to a shear loading, the stress concentration factor is increased as a shear load also give horizontal compressive load on the rectangular hole shown in Figure 5.15 (in addition to the vertical tensile load), similar to the loading shown in the lowest sketch in Figure 2.56. SCFs for shear loading can be found in Peterson (1974) or Pilkey (1997).

5.4.4 Scallops or Cope Holes

SCFs for scallops (or cope holes) at transverse welds are shown in Figure 5.16. These SCFs are derived from finite element analyses of the scallop geometries shown. SCFs for scallops have also been presented by Fricke and Paetzold (1995).

Cope holes are also being used in bridges and reference is made to Miki and Tateishi (1997), in which an analytical approach is used to show the relationship between the local stress at cope hole details as function of the influential parameters.

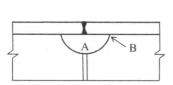

SCF = 2.4 at point A (misalignment not included)
SCF = 1.27 at point B

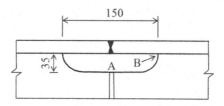

SCF = 1.17 at point A (misalignment not included)
SCF = 1.27 at point B

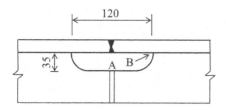

SCF = 1.27 at point A (misalignment not included)
SCF = 1.27 at point B

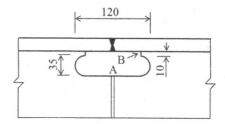

SCF = 1.17 at point A (misalignment not included)
SCF = 1.27 at point B

For scallops without transverse welds, the SCF at point B will be governing for the design.

Figure 5.16. SCFs for scallops from DNVGL-RP-C203 (2016).

5.5 Fatigue Assessment Procedure for Welded Penetrations

5.5.1 Critical Hot Spot Areas

Fatigue cracking around a circumferential weld may occur at various locations in reinforced rings and tubular penetrations in plates, depending on the geometry of the ring or tubular member and on the weld size; see Figure 5.17:

> Fatigue cracking transverse to the weld toe, in a region with a large stress concentration giving considerable stress parallel to the weld. This may be the situation at a flexible reinforcement; see Figure 5.17a.
>
> Fatigue cracking parallel to the weld toe, at a stiff reinforcement with a full penetration weld or a large fillet weld; see Figure 5.17b.
>
> Fatigue cracking from the fillet weld root, at a stiff reinforcement with a small fillet weld; see Figure 5.17c, Figure 5.18, and Figure 5.19.

All these potential regions for fatigue cracking should be assessed in a design with a welded penetration, as shown in the following sections. Appropriate SCFs for holes with reinforcement should be used.

5.5.2 Stress Direction Relative to Weld Toe

The procedure for fatigue analysis is based on the assumption that is necessary to consider the ranges of the stress components at the weld toe. For the stresses shown in Figure 5.17a, it is the stress range along the weld toe that governs fatigue life. For stresses along the weld toe, as shown in Figure 5.17a, the C curve in DNVGL-RP-C203 or the FAT 125 curve (IIW) may be used. However, if there are starts and

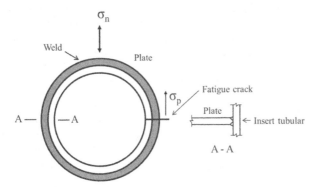

a) Fatigue crack growing normal to the weld toe at thin insert tubular

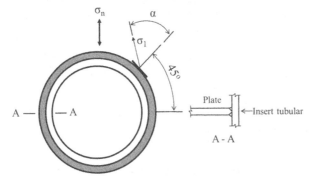

b) Fatigue crack initiating from the weld toe for thicker inserted
tubular member

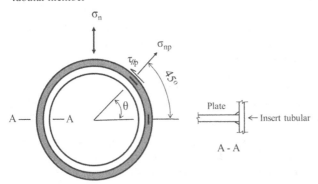

c) Fatigue crack initiating from the weld root in fillet welds.
   Hot spots at $\theta = 0°$ and $\theta = 45°$.

Figure 5.17. Potential fatigue crack locations at welded penetrations.

stops in the weld, then a lower S-N curve should be used. For a fatigue crack at
the position in Figure 5.17b, the fatigue design may be based on the range of maximum principal stress adjacent to the potential crack location, with stress concentrations taken into account. The hot spot stresses shown in Figure 5.17b are linked to
the D-curve following DNVGL-RP-C203 (2016), or the FAT 90 curve following the
IIW notation for fatigue assessment of the weld toe location (Hobbacher 2009). For
fatigue cracking at the 45° position in Figure 5.17c, it is the stress range normal to the
weld, in combination with the shear stress along the weld toe, which determines the
fatigue endurance; see Lotsberg (2004a). For fatigue cracking from the weld root in

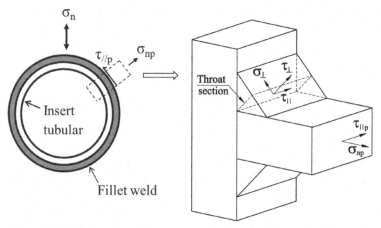

Figure 5.18. Explanation of stresses on the throat section of a fillet weld.

Figure 5.17c, the W3 curve in DNVGL-RP-C203 (2016) or the FAT 36 curve following IIW (Hobbacher 2009) should be used. The basis for these recommendations is comparison of analysis with fatigue test data presented by Skjeggestad et al. (1969) and Hannus (1985). These tests were performed under dynamic tensile loading. From analysis it is observed that the stress condition in Figure 5.22 may be compressive at $\theta = 0$ and this may explain why failure at this position is not observed during testing. However, actual structures may also be subjected to compressive loads, and it is also recommended to perform design for the stress condition in Figure 5.22.

5.5.3 Stress Concentration Factors for Holes with Reinforcement

SCFs for holes with reinforcement are presented in a number of graphs in DNVGL-RP-C203 (2016). Finite element analyses of selected geometries were performed to establish SCFs in this design standard and in CN 30.7 (2014). For analysis of tubular members penetrating through the plates, eight-node shell elements were used.

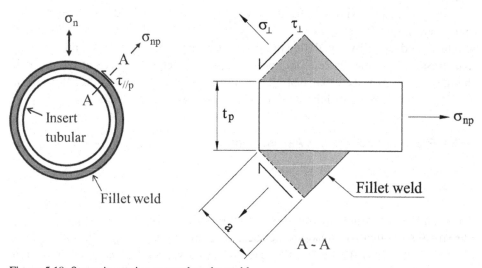

Figure 5.19. Stress in section normal to the weld.

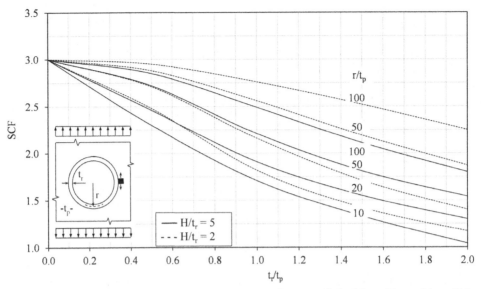

Figure 5.20. SCF at welded penetration; stresses in plate, parallel with weld toe. *Note*: H is defined as length of pipe on each side of plate (total pipe length: 2H).

For analysis of holes stiffened by flat rings, 20-node, three-dimensional isoparametric elements were used. The welds were not included in the shell models. Typical weld sizes were included in the three-dimensional models. A large width for the main plate was included in the finite element model for simulation of an "infinite" width. The size of the element mesh at the hot spot region was t × t, where t = plate thickness. One element was used over the main plate thickness for a one-sided flat ring stiffener. For double-sided flat ring stiffeners, the condition of symmetry was used, which means two elements over the main plate thickness. The hot spot stress in the shell model was derived from extrapolation of the surface stress at point $t/2$ and $3t/2$ to the intersection line. The hot spot stress in the three-dimensional model was derived from stresses at the elements' nodal points, which were derived in the computer program from the stresses calculated at the Gaussian points; see also Section 9.1.3. The stresses at points $t/2$ and $3t/2$ from the weld toe were then used for a linear stress extrapolation for derivation of the hot spot stress at the weld toe. The SCFs in the graphs were derived by calculating the ratio between the hot spot stress and the nominal stress. Finally, the discrete data points were connected by curve fitting through the data points at $t_r/t_p = 0.0, 0.5, 1.0, 1.5,$ and 2.0. Examples of graphs for SCFs at tubular members penetrating plates are shown in Figures 5.20–5.27.

5.5.4 Procedure for Fatigue Assessment

The fatigue assessment procedures for welded penetrations are described as follows.

Potential Fatigue Cracking Transverse to the Weld Toe in Figure 5.17a

For stresses parallel to the weld, the local stress to be used, together with the C–C2 (FAT 125 – FAT 100, ref. Hobbacher 2009) curve, is obtained with SCF from Figure 5.20.

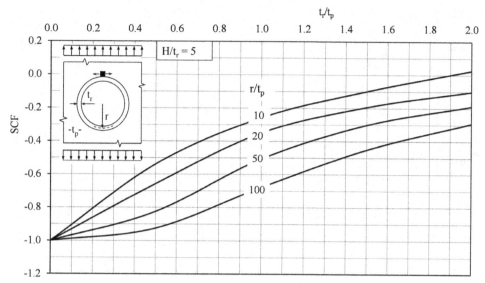

Figure 5.21. SCF at welded penetration; stresses in plate, parallel with weld toe.

Potential Fatigue Cracking Parallel to the Weld Toe in Figure 5.17b

For stresses normal to the weld, the resulting hot spot stress to be used, together with the D curve (FAT 90), is obtained with SCF from Figure 5.27, which gives the principal stress in the plate. The hot spot stress is then derived as:

$$\sigma_{\text{hot spot}} = SCF \, \sigma_{\text{nominal}} \tag{5.20}$$

A similar procedure using stresses derived from finite element analysis of plated structures without including the penetration in the analysis model is shown in Section 9.4. Similar results would be achieved from Section 9.4 as from equation (5.20) with α factor from Section 9.1.3 equal 1.0. Thus, the procedure using the principal stress from

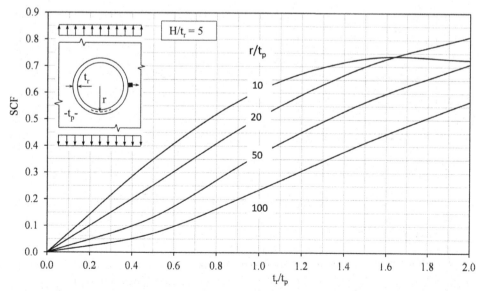

Figure 5.22. SCF at welded penetration; stresses in plate, normal to weld toe.

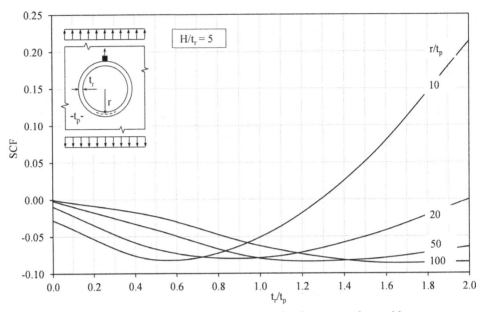

Figure 5.23. SCF at inserted tubular member; stresses in plate, normal to weld toe.

equation (5.20) together with the D-curve is considered to be slightly conservative. Reference is also made to Section 2.7 for further reading related to this.

Potential Fatigue Cracking from the Weld Root in Figure 5.17c

The relevant stress range for potential cracks in the weld throat of load-carrying fillet-welded joints and partial-penetration welded joints may be calculated as:

$$\Delta\sigma_w = \sqrt{\Delta\sigma_\perp^2 + \Delta\tau_\perp^2 + 0.2\,\Delta\tau_{//}^2} \tag{5.21}$$

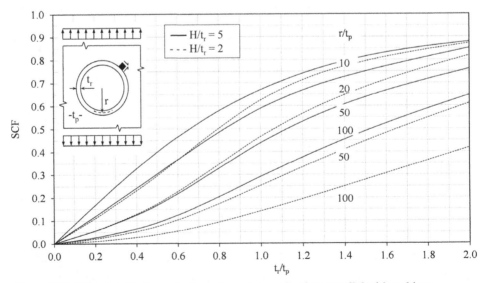

Figure 5.24. SCF at welded penetration; shear stresses in plate parallel with weld toe.

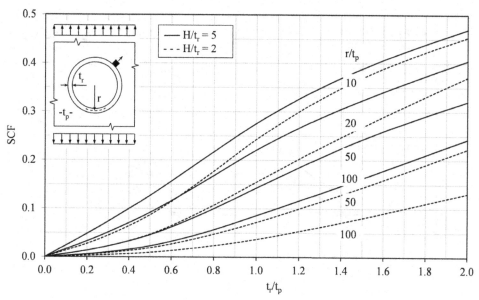

Figure 5.25. SCF at welded penetration; stresses in plate normal to the weld toe.

The stress components used in this equation are shown in Figure 5.18. Equation (5.21) is a general equation for fatigue design of fillet welds subject to complex loading; for comparison with experimental data for components subject to a complex dynamic loading, see Section 2.5.5. At some weld locations stress occurs in the plate normal to the fillet weld, σ_{np} (see Figures 5.18 and 5.19), and a shear stress occurs in the plate parallel with the weld, $\tau_{//p}$. The section at the 45° position in Figure 5.17c is selected for assessment. Force equilibrium of a plate section in the direction parallel with the weld gives:

$$\tau_{//} \, 2a = \tau_{//p} \, t_p \qquad (5.22)$$

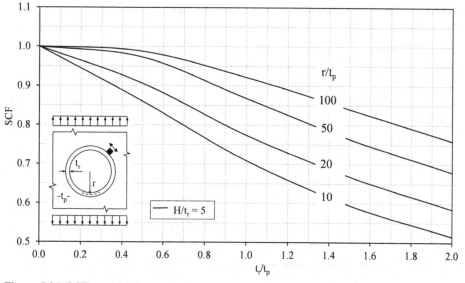

Figure 5.26. SCF at welded penetration; stresses in plate parallel with the weld toe.

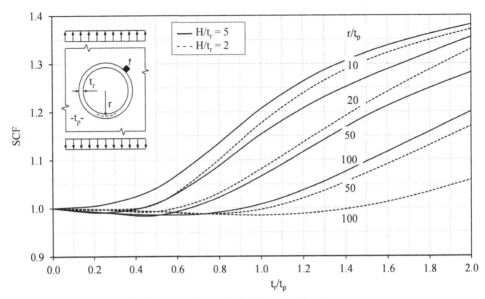

Figure 5.27. SCF at welded penetration; principal stresses in plate.

where:

$\tau_{//}$ = mean nominal shear stress in the weld, as shown in Figure 5.18
$\tau_{//p}$ = shear stress in the plate, as shown in Figure 5.18
a = throat thickness of weld
t_p = plate thickness

The shear stress in the weld is then obtained from equation (5.22) as:

$$\tau_{//} = \frac{\tau_{//p}\, t_p}{2a} \qquad\qquad (5.23)$$

Equilibrium of the plate in a section normal to the weld at the 45° position in Figure 5.17c, in the direction of σ_{np}, as shown in Figure 5.19 gives:

$$(\tau_{\perp} + \sigma_{\perp})\frac{1}{2}\sqrt{2}\, 2\, a = \sigma_{np}\, t_p \qquad\qquad (5.24)$$

Due to the symmetry of the connection in Figure 5.19, it may be assumed that $\tau_{\perp} = \sigma_{\perp}$.
    Then from equation (5.24):

$$\tau_{\perp} = \sigma_{\perp} = \frac{\sigma_{np}\, t_p}{2\sqrt{2}\, a} \qquad\qquad (5.25)$$

From equation (5.21) a combined stress is obtained by using equations (5.23) and (5.25). The resulting stress range should be used together with the W3 curve (FAT 36 in IIW, 2009) for the air environment, as it is assumed that the fatigue crack initiates at the weld root. Then the fatigue analysis for fatigue cracking from the fillet weld

Table 5.2. *Geometries of tested specimens*

	Specimen numbers (Hannus, 1985)		
	3	6–10	11–12
Throat of fillet weld a (mm)	4.0	4.0	4.0
Plate thickness t_p (mm)	8.0	8.0	8.0
Height of tubular 2H (mm)	100	100	100
Thickness of tubular t_r (mm)	4.0	11.2	8.0
Radius r (mm)	48.0	44.4	46.0
r/t_p	6.0	5.55	5.75
t_r/t_p	0.50	1.40	1.0
H/t_r	12.5	4.46	6.25

root at $\theta = 0°$ in Figure 5.17c can be performed in the same way as shown here, but with the shear stress in the plate equal to zero due to symmetry.

5.5.5 Comparison of Analysis Procedure with Fatigue Test Data

The analysis procedure has been compared with fatigue test data from the literature. The fatigue test data were derived for dynamic loading in one direction. The fatigue test data from Skjeggestad et al. (1969) were compared with the design procedure in Lotsberg (2004a). Typical test geometries were reinforcements around holes in plates and tubular penetrations in plates. Fatigue test data from Hannus (1985) were also compared with the design procedure in Figure 5.17.

It should be noted that the numerical design approach presented in Section 9.1.3 gives a lower calculated stress, depending on the value selected for α in equations (9.4) and (9.6). Thus, in order to obtain correlation with fatigue test data for welds around circular penetrations in plated structures, the principal stress range, as shown in Figure 5.17b, should be used, or a large β in equation (2.32). This β factor is said to be dependent on fabrication, such as stops and starts during the welding process. Good fabrication can probably be best achieved in real production welds, rather than by the manual welding used to fabricate the test specimens that are used for comparison of the analysis procedure. Twelve specimens with tubular penetrations were fatigue-tested at Kungliga Tekniska Høgskolan (KTH) in Stockholm, Sweden (Hannus 1985). In the following, the experimental fatigue test data from KTH are reassessed in view of the design procedures described in Section 5.5.4.

Geometry of Test Specimens
The dimensions of a typical test specimen are shown in Figure 5.28. The geometry parameters are listed in Table 5.2.

Fatigue Test Results
The fatigue test data from Hannus (1985) are presented in Table 5.3. The nominal stress in Table 5.3 refers to the gross section of the specimen. These data are further assessed here.

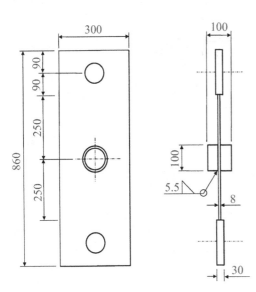

Figure 5.28. Dimension of tested detail from Hannus (1985).

The number of cycles to failure following the design procedure for each type of specimen is calculated. The graphs available for SCFs for $H/t_r = 5.0$ are used. The hot spot stress for each test is calculated for comparison with the recommended S-N curve. The calculated stresses for weld toe assessment are shown in Table 5.4 and for weld root in Table 5.5. The resulting S-N diagrams are shown in Figure 5.29 for weld toe failure and in Figure 5.30 for failure from the weld root. The results derived are discussed in the following sections.

Specimen 3

As there is little stiffness in the inserted tubular member for Specimen 3, fatigue initiation is probably most likely to occur at the point with maximum tangential stress (Figure 5.17a). Here, S-N curves C to C2 apply, depending on the start-stop positions of the welds, according to DNVGL-RP-C203 (2016). A SCF = 2.15 is obtained from Figure 5.20 for this position by extrapolation in this figure. The test result is compared with the mean S-N curves in Figure 5.29. This hot spot will be the most likely initiation point for fatigue cracks when also considering other potential initiation points. This

Table 5.3. *Test results from Hannus (1985)*

Specimen number (according to Hannus 1985)	$\sigma_{nominal}$ (MPa)	Number of cycles to failure
3	187.5	227,000
6	187.5	232,000
7	187.5	267,000
8	187.5	268,000
9	108.3	1,574,000
10	108.3	1,324,000
11	187.5	333,000
12	187.5	209,000

Table 5.4. *Data from failure from the weld toe*

Specimen number	Figures	SCF	$\sigma_{hot\,spot}$ (MPa)	S-N curve
3	Figure 5.27 (Failure at the 45° position)	1.08	202.50	D-curve
	Figure 5.20 (Failure as in Figure 5.17a)	2.15	403.13	C-C2 curve
6–10	Figure 5.27 (Failure at the 45° position)	1.35	253.13	D-curve
11–12	Figure 5.27 (Failure at the 45° position)	1.26	236.25	D-curve

Table 5.5. *Data from failure from the weld root (Lotsberg 2004a)*

Specimen number	Figures	SCF for σ_{np}	SCF for $\tau_{//p}$	Equivalent stress range in weld σ_w (MPa) from equation (5.21)
3	Figure 5.25	0.16		50.22
	Figure 5.24		0.48	
6–10	Figure 5.25	0.42		106.22
	Figure 5.24		0.85	
11–12	Figure 5.25	0.32		86.92
	Figure 5.24		0.75	

Note: SCFs are derived by extrapolation to line for $r/t_p = 5.55$ in the associated figures.

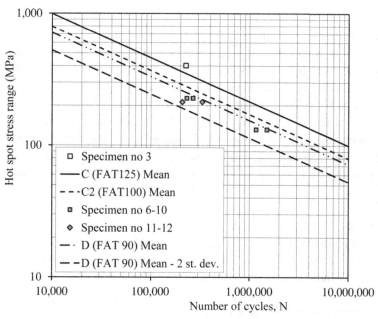

Figure 5.29. Fatigue test data from Hannus (1985) compared with the assessment procedure for hot spot stress parallel and normal to the weld toe. *Note*: There are two data points in the diagram at $2.7 \cdot 10^5$ cycles from Specimens 7 and 8.

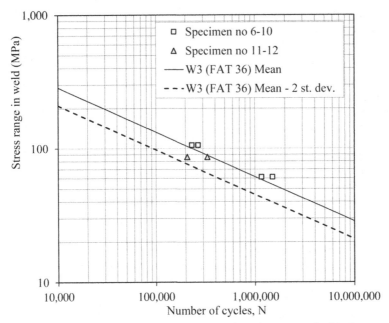

Figure 5.30. Fatigue test data from Hannus (1985) compared with the assessment procedure for root failure (Lotsberg 2004a) (reprinted with permission from Elsevier). *Note*: There are two data points in the diagram at $2.7 \cdot 10^5$ cycles from Specimens 7 and 8.

is also where the fatigue cracking actually occurred, based on a photograph of the cracked specimen (Hannus 1985).

Specimens 6–10

From Figure 5.29 and Figure 5.30, the scatter in the test data for Specimens 6–10 can be observed to be small. From Figure 5.29, it can be seen that a fatigue crack starting from the weld toe is likely for these specimens, with testing of the thickest insert tubular member. From Figure 5.30, fatigue crack growth in the fillet weld can be seen to be also likely for these specimens. According to Hannus (1985), fatigue cracks initiated at the weld toe for the specimen with the large insert thickness tubular member. Based on the photograph of the specimen being tested (Hannus 1985), the question might also arise of whether there has been some fatigue cracking from the root, as the fatigue crack did not grow into the tubular member.

Specimens 11 and 12

From Figure 5.29 and Figure 5.30, the scatter in the test data for Specimens 11 and 12 can be seen to be small. From the same figures, fatigue crack growth starting from the weld toe, as well as from the weld root, can be observed as being likely for these specimens.

From the test report, it should be noted that one of the specimens failed from a fatigue crack starting at the weld toe that grew into the inserted tubular member. The other crack also started at the weld toe but followed the fillet weld without growing into the inserted tubular member.

5.5.6 Example Calculation of the Fillet Welds in the *Alexander L. Kielland* Platform

Based on the procedure presented here, the fatigue capacity of the fillet welds around the hydrophone support in the *Alexander L. Kielland* platform has been calculated. The following data are derived from the NOU 1981:11 report:

> Radius of hydrophone support: 162.5 mm
> Thickness of hydrophone support: 20 mm
> Thickness of brace: 26 mm
> Throat thickness a $= 6$ mm
> Equivalent constant stress range during each load cycle: $\sigma_{eff} = 30$ MPa
> The following values can then be calculated:
> Ratio $r/t_p = 162.5/26 = 6.25$
> Ratio $t_r/t_p = 20/26 = 0.77$
> The graphs for $H/t_r = 5$ for derivation of SCFs are used.

The geometry is somewhat outside the graphs, and therefore some extrapolation has been used for estimation of SCFs. From Figure 5.24, the SCF for the shear stress in the brace is 0.65. From Figure 5.25, the SCF for the normal stress in the brace to the weld toe is 0.25. The normal shear stress and the normal stress in the fillet weld are derived from equation (5.25):

$$\tau_\perp = \sigma_\perp = \frac{\sigma_{np}\, t_p}{2\sqrt{2}\, a} = \frac{30 \cdot 0.25 \cdot 26}{2\sqrt{2} \cdot 6} = 11.49 MPa \tag{5.26}$$

The parallel shear stress in the fillet weld is derived from equation (5.23):

$$\tau_{//} = \frac{\tau_{//p}\, t_p}{2a} = \frac{30 \cdot 0.65 \cdot 26}{2 \cdot 6} = 42.25 MPa \tag{5.27}$$

The resulting stress range in the fillet weld is then derived from equation (5.21):

$$\Delta\sigma_w = \sqrt{11.49^2 + 11.49^2 + 0.2 \cdot 42.25^2} = 24.92 MPa \tag{5.28}$$

This is now compared with the mean W3 curve, see Table 4.1. The standard deviation from Section 4.1.3 is used for calculation of mean S-N curve from the characteristic values. The expression for the S-N curve from equation (4.1) is used for calculation of the number of cycles to failure. Then the number of cycles to expected failure is derived as:

$$N = \frac{10^{\log a}}{\Delta\sigma_w^m} = \frac{10^{(10.97+2 \cdot 0.20)}}{24.92^{3.0}} = 15.14 \cdot 10^6 cycles \tag{5.29}$$

This is equivalent to an expected fatigue life of approximately three years, assuming $5 \cdot 10^6$ cycles every year, which corresponds with a mean up-crossing cycle rate of 6.3 sec.

Thus, the throat thickness around the hydrophone holder in the *Alexander L. Kielland* platform was too small, even if the fabrication of the welds had been free of defects. A mean S-N curve is used for this assessment of an actual fatigue failure; however, the characteristic or design S-N curves as defined in Section 4.1 are normally used for design analysis. When using the design (or characteristic) S-N curve

W3, the calculated fatigue life is reduced to 1.2 years with a Design Fatigue Factor of 1.0.

In this calculation, fatigue strength in the air environment is assumed, as the crack is expected to grow from the weld root. In the actual case, however, there was already a crack in the fillet weld after fabrication, such that an S-N curve for the seawater environment would be a more accurate representation of the real situation. This would have resulted in even fewer calculated cycles to failure.

6 Stress Concentration Factors for Tubular and Shell Structures Subjected to Axial Loads

6.1 Classical Shell Theory

The differential equation for the radial displacement of the cylindrical shell shown in Figure 6.1 can be found in the literature, such as Timoshenko and Woinowsky-Krieger (1959):

$$D_k \frac{\partial^4 w}{\partial x^4} + \frac{Et}{r^2} w = p(x) - \frac{v}{r} N \tag{6.1}$$

where:

D_k = stiffness of the shell defined as:

$$D_k = \frac{Et^3}{12(1 - v^2)} \tag{6.2}$$

E = Young's modulus
v = Poisson's ratio
t = shell thickness
r = radius of the cylindrical shell, measured from the axis of the cylinder to the mid-surface
$p(x)$ = internal radial loading
N = axial force per unit circumferential length.

The form of the solutions given in the following is based on notes from a graduate course at the Technical University of Trondheim in 1973. The solution of the homogenous part of the differential equation (6.1) can be expressed as:

$$w_h = \frac{l_e^2}{2D_k} \left(M_0 g_4(\xi) + Q_0 l_e g_1(\xi) \right) \tag{6.3}$$

where an elastic length is defined as:

$$l_e = \frac{\sqrt{rt}}{\sqrt[4]{3(1 - v^2)}} \tag{6.4}$$

For steel with $v = 0.3$, the elastic length becomes:

$$l_e \cong 0.78 \sqrt{rt} \tag{6.5}$$

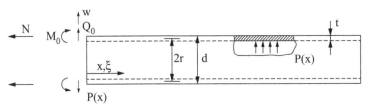

Figure 6.1. Circular cylindrical shell loaded symmetrically with respect to its axis (Lotsberg 1998) (reprinted with permission from Elsevier).

And, with reference to Figure 6.1, the following definitions are made:

$$\xi = \frac{x}{l_e}, \quad g_1(\xi) = e^{-\xi}\cos\xi, \quad g_2(\xi) = e^{-\xi}\sin\xi, \quad g_3(\xi) = g_1(\xi) + g_2(\xi),$$

$$g_4(\xi) = g_1(\xi) - g_2(\xi) \tag{6.6}$$

The particular part of the differential equation can be expressed for p(x) as a polynomial in less or equal to third degree as:

$$w_{part} = \frac{r^2}{Et}p(x) - \frac{vr}{Et}N \tag{6.7}$$

The total radial displacement, w, is the sum of the homogenous and the particular part. For constant, p, the derivative of equation (6.3) is then derived as:

$$\frac{\partial w}{\partial x} = -\frac{l_e}{2D_k}\left(2M_0g_1(\xi) + Q_0l_eg_3(\xi)\right) \tag{6.8}$$

The moment per unit circumferential length at a section, x, as shown in Figure 6.1, is related to the second derivative of the displacement as:

$$M_x = -D_k\frac{\partial^2 w}{\partial x^2} = M_0g_3(\xi) + Q_0l_eg_2(\xi) \tag{6.9}$$

and the shear force per unit circumferential length is derived as the derivative of the moment as:

$$Q_x = \frac{\partial M_x}{\partial x} = -\frac{2M_0g_2(\xi)}{l_e} + Q_0g_4(\xi) \tag{6.10}$$

6.2 Girth Welds

6.2.1 Circumferential Welds in Tubular Members

In the following, a stress concentration in a cylinder is derived taking into account the effect of the stiffness in the circumferential direction and the transition length, L, where the neutral axis is shifted a tolerance, δ. This also becomes a significant parameter when taking the circumferential stiffness into consideration, in addition to that of the butt welds in plates, as presented in Section 5.1; see Figures 6.2a and 6.2b. Figure 6.2b shows the geometry being analyzed. A static model for analysis is shown in Figure 6.2c. The radial springs in Figure 6.2c indicate the radial stiffness due to the stiffness of the tubular member in the circumferential direction. An equivalent model for simplification of the analysis from Figure 6.2c is shown in Figure 6.2d.

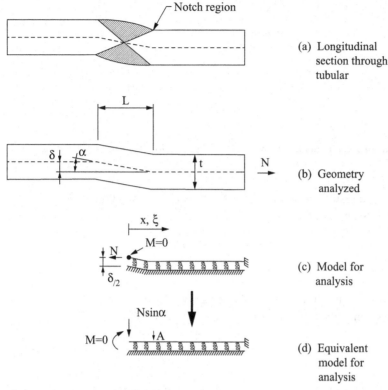

Figure 6.2. Model of fabrication tolerance over a section length, L (Lotsberg 1998) (reprinted with permission from Elsevier).

The load condition in Figure 6.2d gives the following moment and shear force at $x = \xi = 0$: $M_0 = 0$ and $Q_0 = -N \sin \alpha \cong -N \cdot \delta / L$.

Fatigue cracks usually initiate from the notch region at the transition from the base material to the weld, as indicated by point A in the analysis model of Figure 6.2d. Therefore, the bending moment at point A is considered. Here, the reduced coordinate (equation 6.6) is obtained as:

$$\xi_A = \xi(x = L/2) = \frac{L}{2l_e} = \frac{L\sqrt[4]{3(1 - v^2)}}{2\sqrt{rt}} \tag{6.11}$$

For steel with Poisson's ratio $v = 0.3$, the reduced coordinate becomes:

$$\xi_A = \xi(x = L/2) \cong \frac{0.64L}{\sqrt{rt}} \tag{6.12}$$

Then from equation (6.9):

$$M_x(x = L/2) = -Nl_e \frac{\delta}{L} e^{-\xi_A} \sin \xi_A \tag{6.13}$$

It is assumed that the notch region being considered in Figure 6.2a is on the outside of the tubular member. With the definition of positive moment from Figure 6.1, a

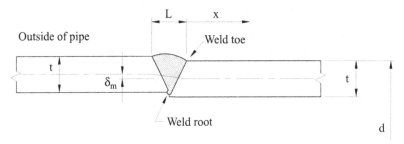

Figure 6.3. Section through a single-side weld (Lotsberg 2009b) (reprinted with permission from Elsevier).

positive moment will result in compression at the notch region. Hence, the resulting tension stress at A is obtained as:

$$\sigma = \frac{N}{t} - \frac{M_x}{W} = \frac{N}{t} SCF \tag{6.14}$$

where:

W = elastic section modulus = $t^2/6$ over a unit width
and the stress concentration factor (SCF) can be deduced as:

$$SCF = 1 + 4.68 \frac{\delta}{L} \sqrt{\frac{r}{t}} e^{-\xi_A} \sin \xi_A \tag{6.15}$$

Or, expressed in terms of the outer diameter, d, the SCF becomes:

$$SCF = 1 + 3.30 \frac{\delta}{L} \sqrt{\frac{d-t}{t}} e^{-\xi_A} \sin \xi_A \tag{6.16}$$

where ξ_A can be obtained from equation (6.12) or from equation (6.17) expressed in terms of the diameter:

$$\xi_A = \frac{0.91L}{\sqrt{(d-t)t}} \tag{6.17}$$

The following approximation can be made for practical use: $\xi \cong \sin \xi$, and equation (6.16) becomes:

$$SCF = 1 + 3 \frac{\delta}{t} e^{-\xi_A} \tag{6.18}$$

where the factor 3 is now an exact number. This equation expresses increased stress at the hot spot from bending of the shell wall with thickness, t, due to eccentricity, δ.

The following example is shown with results from these equations.

A tolerance, $\delta = 0.15t$, as frequently used in fabrication standards for girth welds in offshore structures is assumed (see Figure 6.3), and L is put as equal to t. The SCF for different tubular geometries are then derived as shown in Table 6.1. These SCFs apply to the notch between the weld and the base material on the outside tubular. The SCF due to the weld notch itself is normally accounted for in the S-N curve. The SCF at the root on the inside applies to a region that is close to the intersection point, and hence becomes close to 1.0 for the weld geometry shown in Figure 6.3. It should be noted that the calculated SCF approaches that of the plate equation (5.2) as the diameter/thickness ratio increases (for plates, SCF = 1.45).

Table 6.1. *SCFs for tubular butt welds (Lotsberg 1998)*

Diameter /thickness geometry: (d–t)/t	SCF
15	1.35
30	1.38
60	1.40
300	1.42

6.2.2 Closure Welds at Stubs

Fatigue cracks sometimes occur at circumferential closure welds in jacket bracing members, demonstrating the importance of a sound design and fabrication for these connections (Stacey et al. 1997). Due to the S-N curve being less severe for the outside than the inside of single-side circumferential welds, tubular butt weld connections should be designed such that any thickness transitions are placed on the outside, as shown in Figure 6.4. For this geometry, the SCF for the outside is greater than 1.0. For the inside, the SCF is thus less than 1.0. Thickness transitions to be subjected to a dynamic loading are normally fabricated with a slope of 1:4. It is reasonable to assume that the inflection point (zero moment) of Figure 6.4 can be determined from the distribution of an eccentricity moment to each of the tubular sections. From equation (6.8), the rotational stiffness is proportional to the thickness raised to the power of 2.5. Thus, the distance, x_0, in Figure 6.4 is obtained as:

$$x_0 = L\frac{t^{2.5}}{T^{2.5} + t^{2.5}} = L\frac{1}{1 + (t/T)^{2.5}} \tag{6.19}$$

Following a similar development as for equation (6.12), the reduced coordinate length corresponds to:

$$\xi_0 = \xi(x_0) = \frac{1.82L}{\sqrt{(d-t)t}}\frac{1}{1 + (T/t)^{2.5}} \tag{6.20}$$

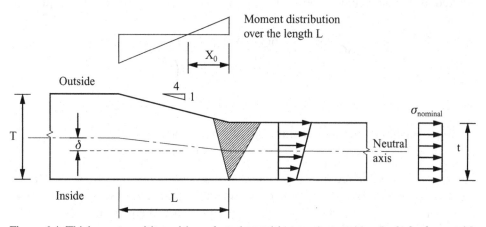

Figure 6.4. Thickness transition with preferred transition on the outside of tubular butt weld (Lotsberg 1998) (reprinted with permission from Elsevier).

Then equation (6.16), with $\xi_A = \xi_0$ from equation (6.20), can be used to calculate the SCF at the transition at the tubular weld:

$$SCF = 1 \pm \frac{6\delta}{t} \frac{1}{1 + (T/t)^{2.5}} e^{-\xi_0} \tag{6.21}$$

where the plus sign applies to the outside and minus to the inside for the geometry shown in Figure 6.4.

Stress concentrations at tubular butt weld connections are due to eccentricities that result from a variety of sources. These may be classified as differences in tubular diameters, differences in the thickness of joined tubular members, out-of-roundness, and center eccentricity. The resulting eccentricity may be evaluated conservatively by direct summation of the contributions from the different sources. The eccentricity due to out-of-roundness usually makes the largest contribution to the resulting eccentricity, δ, for tubular members with longitudinal weld seams welded together from rolled plates.

The transition of the weld to the base material on the outside of the tubular can normally be classified as E. If the welding is performed in a flat position, then it can be classified as D. Weld roots in tubular members that are made from one side are normally classified as F3. Good workmanship during construction is essential to ensure that full penetration welds are made, and this is controlled by non-destructive testing (NDT). However, documentation of a full penetration weld may be generally difficult due to limitations in the NDT technique for detecting defects in the root area. The F3 curve can account for some lack of penetration, but it should be noted that a major proportion of fatigue life is associated with the initial crack growth while the cracks are still small. This may be evaluated by fracture mechanics, as described in Chapter 16. Therefore, if lack of penetration is to be expected due to the fabrication method used, the design S-N curves should be adjusted to account for this by use of fracture mechanics.

6.3 SCFs for Girth Welds in Tubular Members

The SCF equation for misaligned girth butt welds for tubular members that was derived by Connelly and Zettlemoyer (1993) is often used in projects. This equation reads:

$$SCF = 1 + \frac{2.6(\delta_t + \delta_m)}{t} \frac{1}{1 + 0.7\left(\dfrac{T}{t}\right)^{1.4}} \tag{6.22}$$

where

δ_m = maximum misalignment in fabrication standard
$\delta_t = \frac{1}{2}(T - t)$; eccentricity due to change in thickness

This equation was derived from finite element analyses of tubular sections with diameter to thickness ratios of $D/t = 25$, and was mainly intended for use with girth welds in tubular members in jacket structures. This equation and equation (6.21) provide SCFs that correspond well with results of numerical analysis for the low diameter

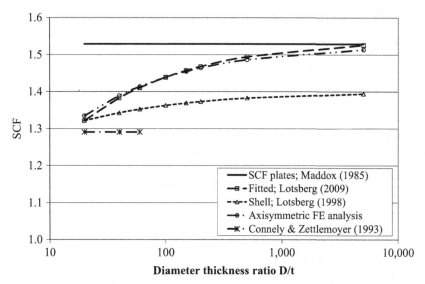

Figure 6.5. SCF as function of diameter-to-thickness ratio, with transition in thickness from 20 mm to 30 mm (Lotsberg 2009b) (reprinted with permission from Elsevier).

to thickness ratios typically used in the design of jacket structures, with a D/t ratio of between 20 and 40, as shown in Figure 6.5. However, it should be noted that the exponent on thickness ratio in equation (6.21) differs from that of equation (5.14) for plated structures. This indicates that the derived values for stress concentrations for tubular elements reach different values from plates when the diameter-to-thickness ratio approaches infinity. Thus, equation (6.21) provides non-conservative SCFs for large diameter to thickness ratios, as indicated in Figure 6.5 (Lotsberg 1998). For this reason, a more detailed assessment was performed, as is further described in this section, based on comparisons with numerical data that were derived from axisymmetric finite element analysis of tubular members with thickness transitions. The finite element modeling was performed with one isoparametric element over the thickness, representing a linear stress distribution. The element length at the weld region was equal to the smallest thickness, t. The transition from one thickness to another was modeled as 1:4, with the thickness transition modeled with two elements in the length direction. This was modeled in order that notch stress from the weld toe, which is assumed to be accounted for in the S-N curve for the butt welds, was not picked up. Stresses at the nodes at the transition from the thinnest member to the thicker section were derived as mean values from the adjacent elements. These stresses differed slightly from that of using only the stresses at the nodes calculated from the first element in the thinnest section. An alternative expression for SCF was then proposed that concurs with the numerical data from the finite element analyses performed (Lotsberg 2009b).

The exponent β in equation (6.23) was put as 2.5 for $D/t = 20$, where there is good agreement between numerical analysis and that derived from shell theory, and it approaches 1.5 for very large diameter thickness ratios (1.5 for $D/t = 1000$), which is the same as that used for plates. Calibration with numerical data is used as additional information for derivation of an expression for the β exponent shown in

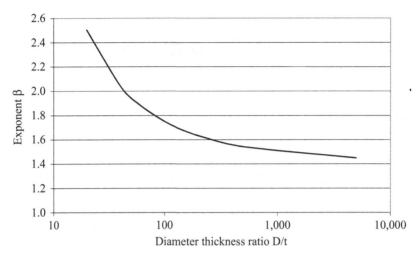

Figure 6.6. Exponent β as function of diameter thickness ratio from equation (6.23) (Lotsberg 2009b) (reprinted with permission from Elsevier).

equation (6.23). Based on this methodology, the design equation for SCFs for circumferential butt welds in shell structures reads:

$$SCF = 1 + \frac{6(\delta_t + \delta_m)}{t} \frac{1}{1 + \left(\dfrac{T}{t}\right)^{\beta}} e^{-\alpha}$$

where:

$$\alpha = \frac{1.82L}{\sqrt{Dt}} \cdot \frac{1}{1 + \left(\dfrac{T}{t}\right)^{\beta}} \qquad\qquad (6.23)$$

and

$$\beta = 1.5 - \frac{1.0}{\log\left(\dfrac{D}{t}\right)} + \frac{3.0}{\left(\log\left(\dfrac{D}{t}\right)\right)^{2}}$$

The exponent on thickness ratio as a function of diameter thickness ratio is shown in Figure 6.6. D = diameter measured to the mid-thickness of the shell. However, for practical purposes, the outer diameter, as shown in Figure 6.6, is often used as an approximation (D = d). This is because tubular sections are normally defined on design drawings by a geometry corresponding to the outer diameter. From Figure 6.5 it can be seen that the proposed formulation fits the numerical data well. Thus, the proposed equation can be recommended for use in fatigue design standards.

6.4 Recommended SCFs for Tubular Girth Welds

The S-N curve that should be used for fatigue design of butt welds in large tubular members depends on whether the welding is made from only one side or from both sides. Selection of the S-N curves to be used for the inside and the outside is also based on how the fabrication is performed. The stress concentration depends

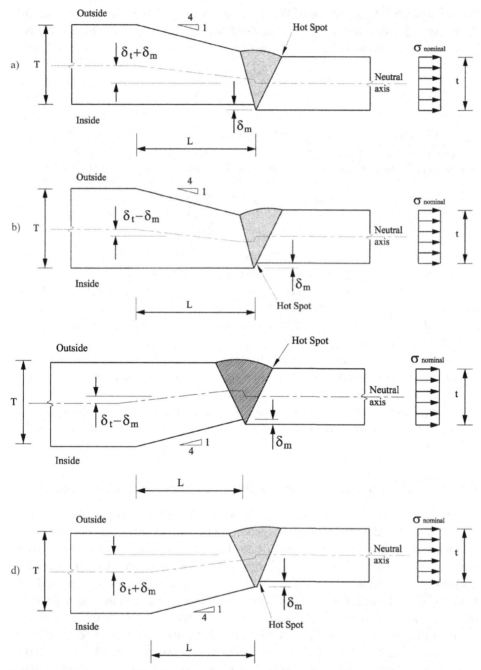

Figure 6.7. Different combinations of geometry and fabrication tolerances for girth weld in tubular members made from the outside (Lotsberg 2009b) (reprinted with permission from Elsevier).

on whether the slope transition at a thickness change is on the outside or inside of the tubular member. Therefore, equations for different geometries and different hot spots are required. Equations that are used for fatigue design of large tubular sections are presented below. For the hot spot shown in Figure 6.7a, the SCF expressed by equation (6.23) can be used. When considering local bending stress over the thickness, the following SCF can be derived for the inside of the tubular member at the weld root shown in Figure 6.7b, with $\delta_m = 0$:

$$\text{SCF} = 1 - \frac{6\delta_t}{t} \frac{1}{1 + \left(\dfrac{T}{t}\right)^{\beta}} e^{-\alpha} \tag{6.24}$$

By also including a value, δ_m, for fabrication tolerance that is larger than zero and that can be positive or negative relative to the neutral axis, the following equation is recommended as an SCF for the inside hot spot stress in Figure 6.7b:

$$\text{SCF} = 1 - \frac{6(\delta_t - \delta_m)}{t} \frac{1}{1 + \left(\dfrac{T}{t}\right)^{\beta}} e^{-\alpha} \tag{6.25}$$

If the transition in thickness is on the inside of the tubular member and the weld is made from the outside only, then equation (6.23) can be used for the inside and equation (6.25) for the outside; see Figure 6.7d and Figure 6.7c, respectively. These equations may also be used if the welding is performed from both sides. An example of a calculation of an SCF at a girth weld made from the outside in a pile supporting a jacket structure is presented in Section 16.3.4.

Section 7.1 was originally intended for fatigue assessment of risers and pipelines. However, it is considered to be of a similar relevance for fatigue assessment of girth welds in tubular structures. It is thus recommended that Section 7.1 is studied before the example in Section 6.6 is looked into. The example in Section 6.6 shows fatigue assessment of an attachment placed some distance from a girth weld in a tubular jacket leg.

6.5 Application of Eccentricity to Achieve an Improved Fatigue Strength

An engineer conducting fatigue design does not normally like to see eccentricities in a fabricated structure. However, circular closure welds made from the outside of a tubular member may be an exception to this, as explained in Section 6.2.2. They might also be used more actively to improve the fatigue strength of risers and tethers subjected to dynamic axial loading that require welding from the outside without the possibility of a final back welding. This can be achieved by installation of an additional tubular section, with thickness T, as shown in the sketch in Figure 6.8 where a section with thickness T is welded into the pipe. The design equations given in Section 6.2.2 may be used to determine the thickness, T, in relation to the diameter of the tubular member that gives a fatigue life on the inside that is equal to, or better than, that of the outside. The difference in S-N data on the inside and the outside can be taken into account as indicated in Section 6.2. The bending stress over the wall thickness in Figure 6.8 has been calculated based on the information in Section 7.1.4 in a similar way as in the example in Section 6.6. The bending stress has been calculated for transition in thickness only without including any fabrication tolerance in this example. Inserting a tubular section with an increased thickness, as shown, reduces the stress on the inside of the tubular member. At the same time, the stress on the outside is increased, which may be acceptable due to the higher-strength S-N curve

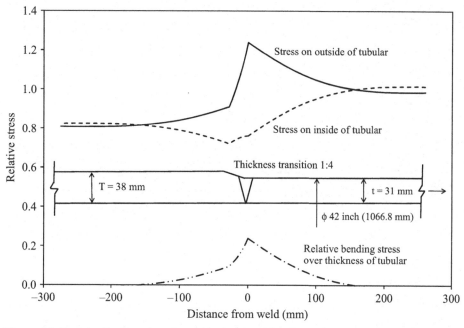

Figure 6.8. Longitudinal section through tubular, with schematic stress along the inside and the outside.

applicable to this side. Weld improvement is also possible for the outside. One of the girth welds on the tubular insert may be made from both sides in a shop, while the second weld is made onsite. The practical use of this concept depends on the actual design conditions, with a trade-off between weight and fabrication costs as compared with the benefits from an improved fatigue life. It might be added that good fatigue strength at the weld roots in single-sided girth welds in risers and pipelines has now been developed such that sufficient fatigue strength can be achieved without use of this methodology provided that there are no defects in the weld root; see, for example, Maddox and Johnston (2011). However, the methodology can be used at conical connections to reduce the stress concentration due to the conical geometry provided that the thickness transition is placed on the correct side (see Section 6.8.1). The calculation of stress is similar to that represented by equation (6.14).

6.6 Example of Fatigue Assessment of Anode Attachment Close to a Circumferential Weld in a Jacket Leg

Anode attachments are being welded to the legs in a jacket structure, and it is questioned what is the interaction between the stress concentrations due to the girth weld and due to the attachments shown in Figure 6.9a:

1. How to account for the effect of the stress at the girth weld due to the thickness transition and fabrication tolerance when performing fatigue assessment of the anode attachment?
2. How to consider the effect of stress increase due the anode attachment when performing fatigue assessment of the girth weld?

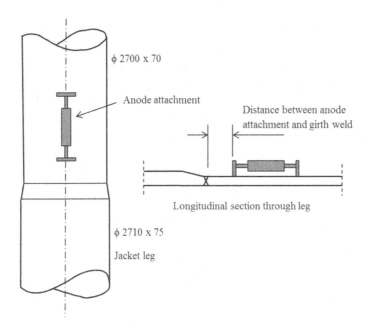

a) Geometry of jacket leg with anode attachment

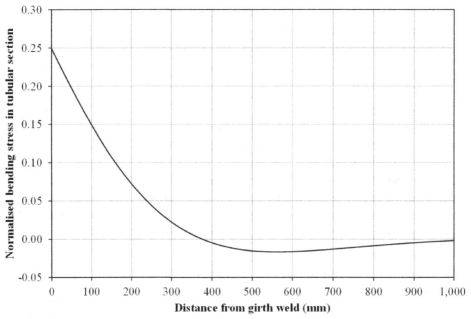

b) Relative bending stress in tubular due to stress increase from girth weld

Figure 6.9. Anode attachment at a girth weld in a jacket leg.

First the stress concentration due to the girth weld is calculated. The largest thickness of the leg is T = 75 mm and the smaller thickness of the leg is t = 70 mm. The outer leg diameter is D = 2700 mm, and the mean diameter $D_m = d - t = 2630$ mm. This gives $D_m/t = 2630/70 = 37.57$ (with large diameter ratios, normally no distinction is made between mean and outer diameters; see also equations used for calculation of SCFs in Section 6.3). The specified fabrication tolerance $\delta_m = 4.0$ mm. The thickness transition is made over a length $L = 4 \cdot (T{-}t) = 4 \cdot (75{-}70) = 20$ mm. The shift in neutral axis due to thickness difference is $\delta_t = 0.5 \cdot (75{-}70) = 2.5$ mm. This results in the following SCF for the girth weld from equation (6.23):

$$\alpha = \frac{1.82 \cdot 20}{\sqrt{2700 \cdot 70}} \frac{1}{1 + \left(\dfrac{75}{70}\right)^{2.06}} = 0.0389$$

$$\beta = 1.5 - \frac{1}{\log(2700/70)} + \frac{3}{(\log(2700/70))^2} = 2.06 \qquad (6.26)$$

$$SCF = 1 + \frac{6\,(2.5 + 4.0)}{70} \frac{1}{1 + \left(\dfrac{75}{70}\right)^{2.06}} e^{-0.0389} = 1.25$$

Then equations (7.8)–(7.11) are used to calculate the relative bending stress at a position x from the girth weld. From equation (7.11):

$$l_e = \frac{\sqrt{(2700/2) \cdot 70}}{\sqrt[4]{3(1 - 0.3^2)}} = 239.15\,mm \qquad (6.27)$$

The relative bending moment is derived as $M_0 = SCF{-}1$ and the relative bending moment at different positions is calculated from equation (7.8), and the result is shown in Figure 6.9b. It is observed that if the anode attachment is placed at a distance about 380 mm from the girth weld, one need not consider the effect of the girth weld when assessing the fatigue life of the anode attachment. If the anode attachment is closer to the girth weld, one should read out a relative bending stress value from Figure 6.9b at the actual position and add this to the relative nominal axial stress in the leg (=1.0), and use the resulting SCF value for fatigue design of the anode attachment, which is typically designed using nominal stress S-N curves E or F according to DNVGL-RP-C203 (2016) depending on plate thickness. For example, if the distance between the attachment and the girth weld is only 100 mm, a SCF = 1.15 should be applied as observed from Figure 6.9b. The bending moment contribution to the stress concentration factor in equation (6.26) is derived as $M_0 = 1.25 -1.0 = 0.25$. From equation (7.10): $\xi = 100/239.15 = 0.418$. From equation (7.8), the stress concentrate at position x = 100 mm is derived as:

$$SCF\,(x = 100\,mm) = 1 + (1.25 - 1)e^{-0.418} \cos 0.418 = 1.15 \qquad (6.28)$$

The effect of the attachment on the increased stress in the leg is considered to be small as the length of the attachment in the circumferential direction is short and does not result in any local bending of the wall thickness in the leg similar to that of a circumferential ring stiffener.

From Figure 9.14 it is observed that the membrane stress due to a weld decays rather early, such that two times the plate thickness away from the attachment, one do not see much stress increase due to a small attachment. Thus, one can say that the girth weld has influence on the stress condition at the attachment if it is placed close to the girth weld; however, the influence of the attachment on the stress at the girth weld is small.

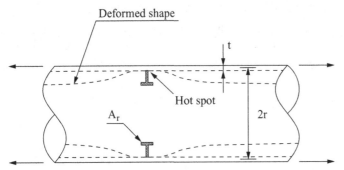

Figure 6.10. Ring stiffener in shell (Lotsberg 1998) (reprinted with permission from Elsevier).

6.7 Ring Stiffeners

Here, a ring stiffener in a shell, as shown in Figure 6.10, is considered. The inward displacement of a ring subjected to a radial force per unit length, $2Q_0$, on the outside can be calculated by use of Hooke's law as:

$$w_{ring} = -\frac{2Q_0 r^2}{EA_r} \tag{6.29}$$

where:

A_r = area of ring stiffener
r = radius
E = Young's modulus.

At a ring stiffener, due to symmetry, the derivative of the radial displacement with respect to the longitudinal direction at the shell is zero (see Figure 6.10), and from equation (6.8) for $\xi = 0$:

$$\frac{\partial w}{\partial x} = 0 \quad \Rightarrow \quad M_0 = -\frac{Q_0 l_e}{2} \tag{6.30}$$

The displacement of the ring is equal that of the shell at the circumferential weld connection between the stiffener and the shell.

Equations (6.29), (6.3) and (6.7) for $\xi = 0$ and $p = 0$ give:

$$-\frac{2Q_0 r^2}{EA_r} = \frac{l_e^2}{2D_k}(M_0 + Q_0 l_e) - \frac{vr}{Et}N \tag{6.31}$$

The moment in the shell at a ring stiffener is obtained by combining equations (6.30) and (6.31), and inserting values for D_k from equation (6.2) as:

$$M_0 = -\frac{vrNA_r t^2 l_e}{4r^2 t^3 + 6A_r l_e^3(1 - v^2)} \tag{6.32}$$

The bending stress in the shell is obtained by dividing by the section modulus, $W = t^2/6$, as:

$$\sigma_b = \pm\frac{M_0}{W} = \pm\frac{\sqrt{3}v}{\sqrt{1 - v^2}}\frac{1}{\beta}\frac{N}{t} \tag{6.33}$$

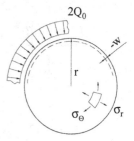

Figure 6.11. Stress distribution in bulkhead (Lotsberg 1998) (reprinted with permission from Elsevier).

where β is defined as:

$$\beta = 1 + \frac{2t\sqrt{rt}}{A_r\sqrt[4]{3(1-v^2)}} \tag{6.34}$$

With $v = 0.3$, the expression for β becomes:

$$\beta = 1 + \frac{1.56t\sqrt{rt}}{A_r} \tag{6.35}$$

Then the stress concentration at a ring stiffener can be presented based on its definition as:

$$SCF = 1 \pm \frac{\sqrt{3}v}{\sqrt{1-v^2}}\frac{1}{\beta} = 1 \pm \frac{0.548}{\beta} \tag{6.36}$$

where the minus sign applies to the inside and the plus sign to the outside. Due to the weld notch, the fatigue strength of the weld at the inside is lower than for the outer shell. As the stress on the inside is lower than on the outside, it is recommended that ring stiffeners are positioned on the inside, and not on the outside, of a shell structure that is subjected to axial loading.

In the following, a bulkhead in a shell is considered. The radial displacement of the weld connection between a bulkhead and a shell can be obtained from two-dimensional elastic theory as:

$$w = \varepsilon_r r = \frac{1}{E}(\sigma_r - v\sigma_\theta) \tag{6.37}$$

where:

$$\sigma_r = \sigma_\theta = -\frac{2Q_0}{t_b} \tag{6.38}$$

Here t_b is the thickness of the bulkhead; see also Figure 6.11. Then, from equations (6.37) and (6.38), the radial displacement is obtained as:

$$w = -\frac{2Q_0}{t_b}\frac{1-v}{E}r \tag{6.39}$$

Comparison of equations (6.29) and (6.39) shows that the same equations as those used for a ring stiffener can also be used for assessment of SCFs at a bulkhead by calculating an effective area as:

$$A_r = \frac{rt_b}{1-v} \tag{6.40}$$

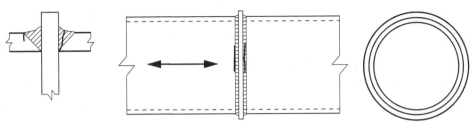

Figure 6.12. Tubular section welded against transverse plate.

6.7.1 Example: Assessment of Stress Concentration Inherent in Nominal Stress S-N curves

A tubular member is welded against a thick transverse plate (see Figure 6.12). The weld is to be classified in terms of nominal S-N curves to be used for the outside and the inside. A similar detail is found as construction detail no. 15 in table A-9 of DNVGL-RP-C203 (2016). This detail is classified as G; see S-N curve in Table 4.1. First the outside should be considered. Welding of one plate against another plate is considered as a cruciform joint and is classified as F from construction detail no. 1 in table A-8 in DNVGL-RP-C203 (2016). For a very stiff transverse plate at the tubular member, there will be a maximum stress concentration for the outside of 1.54, according to equation (6.36) when A_r is large. The F detail has an implicit stress concentration of 1.27 according to Table 4.1. By multiplying these two factors together, the hot spot stress concentration inherent for the detail for the outside can be derived as equaling 1.95. From Table 4.1 it can be seen that this is an SCF that is closer to W1 than to G; however, G may be considered reasonable for transverse plates that are not particularly stiff (the SCF for S-N curve G is 1.80). The stress concentration factor for the outside, SCF_0, can be derived from equation (6.36). Thus, the hot spot stress for the outside weld toe, relative to that of hot spot stress S-N curve D, can be calculated as $\sigma_{hot\ spot} = 1.27 \cdot SCF_0$. The lowest S-N curve that can be assumed for the root is W3 for construction detail no. 1 in table A-6 in DNVGL-RP-C203 (2016). W3 has an inherent SCF of 2.50, according to Table 4.1. The stress concentration factor for the inside, SCF_i, can be derived from equation (6.36). Thus, the hot spot stress for the inside weld root, relative to that of hot spot stress S-N curve D, can be calculated as $\sigma_{hot\ spot} = 2.50 \cdot SCF_i$. Now an equation for when the outside weld toe can be considered to be a more critical hot spot than the inside weld root can be estimated by inserting for SCFs from equation (6.36):

$$1.27 \cdot SCF_0 > 2.50 \cdot SCF_i$$

$$1.27 \left(1 + \frac{0.548}{\beta}\right) > 2.50 \left(1 - \frac{0.548}{\beta}\right) \tag{6.41}$$

$$\beta < 1.68$$

This corresponds to $SCF_0 = 1.326$ and $SCF_i = 0.674$ from equation (6.36). For actual geometries this can be further assessed using equations (6.40), 6.35) and (6.36); however, it is likely that the outside may be more critical than the inside if the transverse plate is relatively thick.

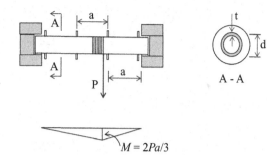

Figure 6.13. Drum for transportation.

$$M = 2Pa/3$$

6.7.2 Example: Fatigue Assessment of a Drum

A drum used to transport equipment is assessed with respect to fatigue (see Figure 6.13). The maximum allowable tension force in the wire on the drum needs to be determined. There are three different spaces for wire on the drum, separated by external ring stiffeners of 200×20 mm. The ring stiffeners are welded to the drum by double-sided fillet welds. The highest bending stress in the drum occurs when the wire is at one of the side plates at the center region of the drum. The reaction force at one support then equals $2P/3$, and the maximum bending moment at the highest stressed ring stiffener is $2Pa/3$. When the drum is rotated 180 degrees, the bending moment at the same position is reversed and the range in bending moment is derived as:

$$\Delta M = \frac{4}{3} Pa \tag{6.42}$$

The elastic section modulus for the drum is calculated as:

$$W = \frac{\pi}{32} \left(D^4 - (D - 2t)^4 \right) \frac{1}{D} \tag{6.43}$$

where:

D = diameter = 600 mm
t = thickness = 20 mm.

Then $W = 5114 \cdot 10^3$ mm^3.

For the outside of the drum, a SCF is calculated from equations (6.35) and (6.36):

$$\beta = 1 + \frac{1.56t\sqrt{rt}}{A_r} = 1 + \frac{1.56 \cdot 20\sqrt{300 \cdot 20}}{200 \cdot 20} = 1.60 \tag{6.44}$$

$$SCF = 1 + \frac{0.54}{\beta} = 1 + \frac{0.54}{1.60} = 1.34 \tag{6.45}$$

The nominal stress at the outside of the drum at the ring stiffener being considered is obtained as:

$$\Delta\sigma = \frac{\Delta M}{W} SCF = \frac{4Pa}{3W} SCF \tag{6.46}$$

The distance from the drum support to the ring stiffener, a = 1200 mm. A Design Fatigue Factor (DFF) of 2 is specified to be used for the design. The number of rotations of the drum is not specified and considered to be uncertain. Therefore, the aim

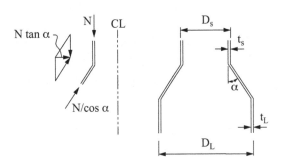

Figure 6.14. Conical transition (Lotsberg 1998) (reprinted with permission from Elsevier).

is for a stress range that is below the constant amplitude fatigue limit. The detail classification is found from table A-7 detail 8 in DNVGL-RP-C203 (2016), which gives S-N curve E for a T-joint plate connection. The allowable stress range is then obtained from Table 4.1 for an E –detail, and with modification of allowable stress range from Section 3.2.2 as

$$\Delta\sigma_{allowable\, t=25mm} = \frac{\Delta\sigma\, at\, 10^7 cycles}{DFF^{1/3.0}} = \frac{46.78}{2^{1/3.0}} = 37.13\, MPa \qquad (6.47)$$

Then the maximum tension force is derived from equation (6.45) as:

$$P = \frac{3\Delta\sigma_{allowable}W}{4aSCF} = \frac{3 \cdot 37.13 \cdot 5114 \cdot 10^3}{4 \cdot 1200 \cdot 1.34} = 88.5\, kN \qquad (6.48)$$

6.8 Conical Transitions

6.8.1 Weld at Conical Junction

SCFs for conical transitions may be deduced from the expression for local stress at welded junctions, as given by API RP2A-WSD (2014). This refers back to Boardman (1948), who used classical shell theory to derive stresses at conical junctions. A similar derivation is also shown in the following, in which a conical transition, as shown in Figure 6.14, is considered. At the junctions where the cone intersects with the tubular member, an unbalanced radial force per unit length around the junction occurs:

$$F = N \tan \alpha \qquad (6.49)$$

The radial stiffness of the cone is assumed to be equal to that of the shell. This implies that the thickness of the two members should be approximately equal and that the angle, α, should be limited. Normally α is less than 30° for practical design and fabrication of conical transitions, and this is considered to be within an acceptable validity for these equations. Equilibrium requires:

$$Q_0 = -\frac{F}{2} \qquad (6.50)$$

As for the ring stiffener, it can be assumed that the junction does not rotate if the thickness of the cone equals that of the tubular. Then, from equation (6.8):

$$M_0 = -\frac{Q_0 l_e}{2} = \frac{N l_e \tan \alpha}{4} \qquad (6.51)$$

The bending stress can then be obtained from equation (6.51) by inserting for the elastic section modulus and elastic length l_e from equation (6.4) as:

$$\sigma_b = \pm\frac{M_0}{W} = \pm\frac{3N\tan\alpha}{2t^2}\frac{\sqrt{rt}}{\sqrt[4]{3(1-v^2)}} \tag{6.52}$$

With $v = 0.3$:

$$\sigma_b = \pm 1.17\frac{N}{t}\sqrt{\frac{r}{t}}\tan\alpha \tag{6.53}$$

and the stress concentration is obtained as:

$$SCF = 1 \pm 1.17\sqrt{\frac{r}{t}}\tan\alpha \tag{6.54}$$

where the plus sign applies to the outside of the smaller junction ($r = (D_s - t_s)/2$, $t = t_s$) and to the inside of the larger junction ($r = (D_L - t_L)/2, t = t_L$). The minus sign applies to the inside of the smaller junction and to the outside of the larger junction. The calculated SCF should be used, together with the nominal stress in the tubular member in the calculation.

In API RP2A-WSD (2014), ISO 19902 (2007), and DNVGL-RP-C203 (2016) the difference in cone thickness from that of the connecting tubular member has been accounted for by inserting a mean thickness into the classical solution derived for equal thicknesses. These equations read:

$$SCF = 1 + \frac{0.6t_i\sqrt{D_i(t_i + t_c)}}{t_i^2}\tan\alpha \quad \textit{for the tubular side}$$

$$SCF = 1 + \frac{0.6t_i\sqrt{D_i(t_i + t_c)}}{t_c^2}\tan\alpha \quad \textit{for the cone side} \tag{6.55}$$

where:

D_i = cylinder diameter at junction (D_S, D_L)
t_i = tubular member thickness wall thickness (t_S, t_L)
t_c = cone thickness
α = the slope of the cone

By inserting numbers into these equations it can observed that equation (6.55) is slightly conservative due to differences in significant numbers and differences between radius to the mid-center and diameter measured to the outer shell in equation (6.55). More importantly, it should be noted that the neutral axis in the cone crosses the neutral axis in the tubular members at the junctions. Thus, it is unnecessary to consider shift in neutral axis due to thickness transitions and fabrication tolerances as is required for tubular structures in Section 6.4. However, in order that the equations for stress concentrations at the conical transitions are fully valid, any thickness transition should be placed on the outer side of the cone at the large diameter junction and on the inside of the smaller diameter junction (see Figure 6.15). This is of lesser importance for conical transitions at joints in jacket structures due to the large slope of the cones that are normally used. The slope angles in the cones used in large-diameter support structures for wind turbines are small, typically

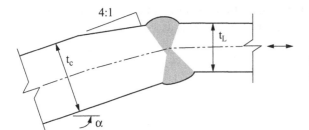

Figure 6.15. Thickness transition from cone to tubular at large diameter junction.

around 2°, and the calculated SCF then becomes more sensitive to change in direction of the neutral axis. If the transition in slope of the cone at the weld in Figure 6.15 is moved to the inside, it can be seen that the effective slope angle, α, increases and a larger stress concentration at the junction results. This is not recommended for achieving optimal design solutions.

6.8.2 Example of Conical Transition in Monopile for Wind Turbine Structure

An example of calculated SCFs at a conical transition in a monopile used for support of a wind turbine structure is shown in the following. The following geometry is assumed for this example:

Cone angle $\alpha = 2.0°$
$D = 5000$ mm
$t_L = 70$ mm
$t_c = 75$ mm

SCF due to conical transition equation (6.55):

$$SCF = 1 + \frac{0.6 \cdot 70\sqrt{5000 \cdot (75 + 70)}}{70^2} \tan 2° = 1.25 \tag{6.56}$$

$L = 4(75-70) = 20$ mm assuming thickness transition 1:4
$\beta = 1.83$ from equation (6.23)
$\alpha = 0.0288$ from equation (6.23)
$e^{-\alpha} = 0.97$
$\delta_t = 0.5(75-70) = 2.5$ mm

SCF due to thickness transition δ_t from equation (6.23):

$$SCF = 1 + \frac{6 \cdot 2.5}{70\left(1 + \left(\dfrac{75}{70}\right)^{1.83}\right)} 0.97 = 1.10 \tag{6.57}$$

The resulting stress concentration factor with thickness transition at inside at large diameter junction

$$SCF = 1.25 + (1.10 - 1.0) = 1.35 \tag{6.58}$$

The resulting stress concentration factor with thickness transition at outside at large diameter junction

$$SCF = 1.25 - (1.10 - 1.0) = 1.15 \tag{6.59}$$

This difference in calculated SCFs shows that it is important to consider where to place thickness transitions at conical transitions in these structural elements. It should be noted that the position of the thickness transition at the small diameter junction should be on opposite side of that of the large diameter junction.

6.8.3 Conical Transition with Ring Stiffeners at the Junctions

It is assumed in the following that a ring stiffener is placed at a junction without eccentricity. Reference is made to Section 6.8.4 for consideration of the effect of eccentricity. The radial displacement of a ring stiffener placed at the smaller junction is obtained as (see equations (6.29) and (6.49)):

$$w_{ring} = -\frac{r^2(2Q_0 - N\tan\alpha)}{EA_r} \tag{6.60}$$

The displacement of the ring equals that of the shell at the connection point between the ring and the shell, which, from equations (6.3), (6.7), and (6.60) for $\xi = 0$ and $p = 0$, gives:

$$-\frac{r^2(2Q_0 - N\tan\alpha)}{EA_r} = \frac{l_e^2}{2D_k}(M_0 + Q_0 l_e) - \frac{vr}{Et}N \tag{6.61}$$

This equation can be rewritten as:

$$-\frac{2Q_0 r^2}{EA_r} = \frac{l_e^2}{2D_k}(M_0 + Q_0 l_e) - \frac{r}{Et}\left(v + \frac{tr\tan\alpha}{A_r}\right)N \tag{6.62}$$

The similarity with equation (6.31) should be noted and the SCF at the smaller junction is derived as:

$$SCF = 1 \pm \frac{\sqrt{3}\left(v + \dfrac{tr\tan\alpha}{A_r}\right)}{\sqrt{1 - v^2}}\frac{1}{\beta} \tag{6.63}$$

With $v = 0.3$, equation (6.63) becomes:

$$SCF = 1 \pm \left(0.548 + \frac{1.82\,tr\tan\alpha}{A_r}\right)\frac{1}{\beta} \tag{6.64}$$

where plus applies for the outside and minus for the inside. β is obtained from equation (6.35). For the larger junction, the stress concentration is obtained correspondingly as:

$$SCF = 1 \pm \left(0.548 - \frac{1.82\,tr\tan\alpha}{A_r}\right)\frac{1}{\beta} \tag{6.65}$$

where minus applies for inside and plus for the outside. The radius, r, used in these equations should correspond to the junction being considered. The calculated SCF should then be used, together with the nominal stress in the tubular member with radius r, for calculation of hot spot stress.

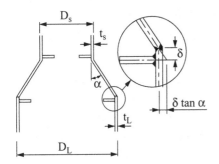

Figure 6.16. Conical transition with stiffeners at the junctions (Lotsberg 1998) (reprinted with permission from Elsevier).

6.8.4 Conical Transition with Ring Stiffener Placed Eccentrically at Junction

Ring stiffeners are often placed eccentrically with respect to the system lines of the cone and the shell in order to simplify fabrication. Such eccentricity, as shown in Figure 6.16, may lead to additional stresses to those given in Section 6.8.3. Assuming that the eccentricity moment is distributed in equal proportions to the cone and to the shell, an additional SCF is obtained as:

$$SCF = 1 + 3\frac{\delta}{t}\tan\alpha \qquad (6.66)$$

This SCF is conservative, as it does not include the bending stiffness of the ring. Therefore, the effect of this bending is considered with a bulkhead inserted at the same region. This design is sometimes used in crane pedestals.

A relationship between moment, M_s, and rotation, θ, around the circumference of a shell, as shown in Figure 6.17, can be derived from equation (6.3), with $w = 0$ ($Q_0 l_e = -M_0$), and equation (6.8):

$$M_s = -\frac{2D_k}{l_e}\frac{\partial w}{\partial x} = \frac{2D_k}{l_e}\theta \qquad (6.67)$$

where:

    D_k is the stiffness of the shell
    l_e is the elastic length as defined in Section 6.1 and equation (6.4).

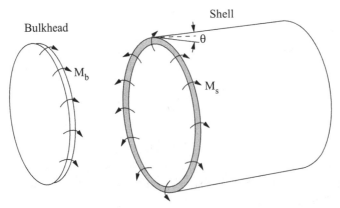

Figure 6.17. Moment loading on boundary of bulkhead and shell.

Following, for example, Roark and Young (1975), a similar relationship can be developed for the bulkhead:

$$M_b = \frac{D_b(1+v)}{r}\theta \tag{6.68}$$

where the bulkhead stiffness is defined as:

$$D_b = \frac{Et_b^3}{12(1-v^2)} \tag{6.69}$$

Assuming again that the stiffness of the cone is the same as that of the shell, the moment in the shell is obtained according to its contribution to the total rotation stiffness as:

$$M = \frac{2D_k/l_e}{4D_k/l_e + D_b(1+v)/r}N\delta\tan\alpha \tag{6.70}$$

The following stress concentration due to eccentricity, δ, is then obtained:

$$SCF = 1 + \frac{3}{1+\gamma}\frac{\delta}{t}\tan\alpha \tag{6.71}$$

with γ defined as:

$$\gamma = \frac{1+v}{4\sqrt[4]{3(1-v^2)}}\left(\frac{t_b}{t}\right)^3\sqrt{\frac{t}{r}} \tag{6.72}$$

For steel, γ becomes:

$$\gamma \cong 0.25\left(\frac{t_b}{t}\right)^3\sqrt{\frac{t}{r}} \tag{6.73}$$

It should be noted from this equation that the thickness of the bulkhead has to be significantly larger than the thickness of the shell in order to attract bending moment and thus to reduce the stress concentration in the shell significantly. (For a tubular member with a diameter over thickness of 40 and $t_b = t$, the value of γ is only 0.055 and it can be seen that the difference between equations 6.66 and 6.71 becomes small.)

6.9 Tethers and Risers Subjected to Axial Tension

Angle deviations between consecutive tubular segments in risers and tethers of tension leg platforms result in nonlinear stress concentrations being a function of axial pretension of tubular sections. Such effects can be assessed by nonlinear finite element analysis. More simply, however, they can be evaluated by analytical means, as shown in the following. The collinearity, with small angle deviation between consecutive fabricated tubular segments, in tethers and risers gives rise to increased stress due to a resulting global bending moment (see Figure 6.18). The eccentricity due to collinearity is a function of axial tension in the tubular member and is significantly reduced as the axial force increases by tension. Assuming that the moment, M, results from an eccentricity, δ_{max}, where pretension is accounted for in the analysis, the following derivation of a SCF is performed. It is assumed that the collinearity between

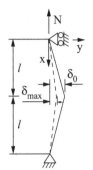

Figure 6.18. Model for analysis of effect of angle deviation between two tubular sections (Lotsberg 1998) (reprinted with permission from Elsevier).

two tubular segments results in an eccentricity, δ_0, without any axial tension force in Figure 6.18. The relationship between bending curvature and bending moment on a tubular element is expressed as (reference is made to textbook on structural mechanics):

$$\frac{\partial^2 y}{\partial x^2} = -\frac{M(x)}{EI} \tag{6.74}$$

where:

E = Young's modulus
I = moment of inertia of tubular section

The eccentricity along the tubular can be expressed as:

$$\delta(x) = \frac{\delta_0}{l}x \tag{6.75}$$

and the moment along the tubular is:

$$M(x) = -(\delta(x) + y)N \tag{6.76}$$

By combining equations (6.74) and (6.76), the following differential equation results:

$$\frac{\partial^2 y}{\partial x^2} - \frac{N}{EI}y - \frac{N\delta_0 x}{EIl} = 0 \tag{6.77}$$

The general solution of this equation is:

$$y = C_1 e^{\lambda x} + C_2 e^{-\lambda x} - \frac{\delta_0 x}{l} \tag{6.78}$$

where

$$\lambda = \sqrt{\frac{N}{EI}} \tag{6.79}$$

The boundary conditions are:

$$\begin{aligned} y &= 0 \quad for \quad x = 0 \\ \frac{\partial y}{\partial x} &= 0 \quad for \quad x = l \end{aligned} \tag{6.80}$$

Table 6.2. *SCF for a riser*

Number of elements	δ_0	l (m)	λl	δ_{max}/δ_0 for $N = 2$ MN	SCF for $N = 2$ MN	SCF for $N = 0$
2	$l' \cdot \alpha = 7.4$ mm	14.8	1.60	0.58	1.04	1.06
4	$4 \cdot l' \cdot \alpha = 29.6$ mm	29.6	3.20	0.31	1.08	1.24

These boundary conditions are used to determine the integration constants and the results are:

$$C_1 = \frac{\delta_0}{l} \frac{1}{\lambda(e^{\lambda l} + e^{-\lambda l})} \quad \text{and} \quad C_2 = -C_1 \tag{6.81}$$

Maximum eccentricity at x = *l* is then obtained as:

$$\delta_{max} = \delta_0 + y_{max} = \frac{\delta_0}{\lambda l} \tanh(\lambda l) \tag{6.82}$$

The maximum stress at x = *l* is obtained from:

$$\sigma = \frac{N}{A} + \frac{M}{W} = \frac{N}{\pi(d-t)t} + \frac{4N\delta_{max}}{\pi(d-t)^2 t} \tag{6.83}$$

This stress can also be presented as:

$$\sigma = \frac{N}{\pi(d-t)t} SCF \tag{6.84}$$

where the SCF is presented as:

$$SCF = 1 + \frac{4\delta_{max}}{d-t} \tag{6.85}$$

where:

δ_{max} is eccentricity as function of the axial force, N, as obtained from equation (6.82)
d is the outer diameter of the tubulars
t is the thickness of tubular sections.

6.9.1 Example: Pretensioned Riser

A riser with outer diameter d = 508 mm and t = 17.5 mm is considered. This gives a moment of inertia for the riser I = $8.12 \cdot 10^8$ mm^4. The riser is constructed from tubular elements of length l' = 14.8 m, and it is pretensioned to a quasistatic load of 2 MN.

E = $2.1 \cdot 10^5$ MPa is used.
Equation (6.79) results in $\lambda = 0.108$ m^{-1}.

The fabrication tolerance is given as $2\alpha = 1/1000$. The calculated SCFs for different numbers of riser strings are shown in Table 6.2. It should be noted from Table 6.2 that the pretension has a significant impact on the resulting eccentricity and on the SCF. It is, of course, conservative to assume that when a number of elements

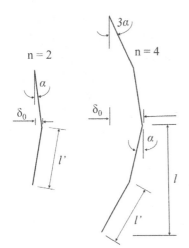

Figure 6.19. Addition of fabrication tolerances for a riser string (Lotsberg 1998) (reprinted with permission from Elsevier).

are welded together, all the tolerances accumulate in the least favorable direction as shown in Figure 6.19. For construction of many elements, some square root summation could be used if the fabrication deviation is not systematic. Engineering judgment is normally used for evaluation of such tolerances in an actual project. The expressions presented may be used for simple assessments of the effect on the resulting stress.

7 Stresses at Welds in Pipelines, Risers, and Storage Tanks

7.1 Stresses at Girth Welds and Ring Stiffeners due to Axial Force

7.1.1 General

Butt welds in pipelines are normally made from the outside during pipeline production (see Figure 6.3). This methodology is used for production of large-diameter pipelines, where fabrication is performed either by adding on pipe segments on a pipe-laying barge or by reeling of the pipeline on a large-diameter drum. The butt welds are the critical sections with respect to fatigue, and fatigue cracking may occur from the weld root or from the weld cap (weld toe). Welding these butt welds is time consuming and is a key issue when considering the efficiency and reliability of the final pipeline. In the production of seamless pipes, fabrication tolerances due to pipe ovality and scatter in thickness are important issues in pipeline design. Therefore, stress concentrations in butt welds in risers and pipelines due to tolerances are important.

In DNVGL-RP-C203 (2016), it is recommended to use a stress concentration factor (SCF) to account for the local bending over the tubular thickness, resulting from eccentricity or fabrication tolerance, δ_m, as shown in Figure 6.3. It is assumed that the local stress increase due to the weld notch itself is embedded in the S-N curve to be used. Especially for the weld toe (cap), accounting for the increased stress due to eccentricity, from tolerances or from a shift in neutral axis when going from one pipe segment to another, is recommended. In considering the axial stress in the pipe resulting from the end cap effect from internal pressure (or from other axial force in the pipes – this could also be due to the global bending of the pipes, which is normally the main contributor to stress range and fatigue damage), the following SCF can be used for the weld toe side in Figure 6.3.

$$SCF = 1 + 3\frac{\delta_m}{t} e^{-\sqrt{t/d}} \tag{7.1}$$

This equation corresponds approximately to equation (6.18) with L = t/0.91 for large d–t-ratios.

It is important to understand the physical basis of the effect of tolerances on fatigue capacity in order to derive reliable design S-N curves based on fatigue test data. Therefore, an axisymmetric finite element model of a butt weld in a pipe was

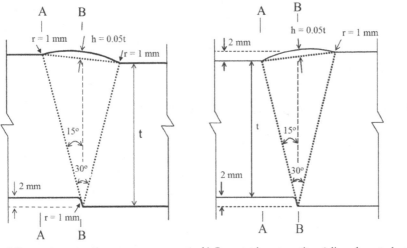

a) Eccentric connection b) Symmetric connection (aligned neutral axis)

Figure 7.1. Single-side welded connections (Lotsberg 2008a) (reprinted with permission from Elsevier).

made for the calculation of stress at the weld root, with and without eccentricity, for comparison with the analytical equation (7.1) (Lotsberg and Holth 2007). A tolerance implies a local notch at the weld root that may reduce the fatigue capacity, as compared with that obtained from fabrication with perfect tolerances. This should be accounted for in the fabrication of realistic test specimens, and for derivation of relevant fatigue test data and reliable design S-N curves for single-side welds for pipeline fabrication and in other tubular structures that are welded from the outside only. For the weld toe, in addition to the notch effect from the weld toe, a stress concentration due to bending effects in the pipe wall from tolerances should be accounted for in the design.

7.1.2 Circumferential Butt Welds in Pipes at Thickness Transitions and with Fabrication Tolerances

The bending stress at the root of the weld in Figure 7.1 is close to zero following the moment distribution shown in Figure 7.1b. An axisymmetric model of a pipeline section was made, with a diameter of 1066.8 mm (42 inches) and a wall thickness of 31.0 mm (1.220 inches), in order to assess this further (Lotsberg and Holth 2007). Two different geometries were analyzed, as shown in Figure 7.1:

- One single-side welded connection, with a fabrication tolerance of 2 mm as shown in Figure 7.1a, which also shows a shift in neutral axis of 2 mm from one pipe to the next.
- One analysis with the same geometry at the notch region, but with added material on the right pipe section, such that the connection is without shift in neutral axis, as shown in Figure 7.1b.

The notch at the weld root and at the two weld toes in Figure 7.1 was modeled by a radius of 1.0 mm. This requires modeling by using a fine mesh of elements, as typically

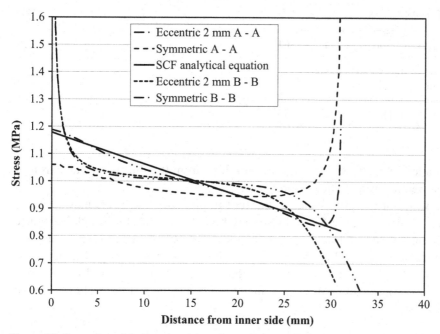

Figure 7.2. Stress in axial direction in sections A-A and B-B in Figure 7.1 (Lotsberg 2008a) (reprinted with permission from Elsevier).

recommended for the notch stress methodology; see Section 9.6. The connections were subjected to a unit axial load over the 31 mm-thick section. Stresses at sections A-A and B-B are presented in Figure 7.2.

First the stress in the direction normal to line A-A in Figure 7.2 is considered. The left part of Figure 7.2 shows the stress at the inner side of the pipe, while the right part shows the stress at the outer side. At the outer side there is a peak stress due to the notch at the weld toe. More interesting in this respect is the stress at the inner side, which is not affected so much by notches, as shown in Figure 7.1. The additional bending stress on this side is mainly due to bending over the thickness. The calculated stress is compared with that resulting from equation (7.1). It can be seen that the calculated stress for an eccentricity of 2 mm is slightly larger than that from the linear stress distribution from equation (7.1). The calculated stress distribution at the inner side seems to be slightly disturbed by the presence of the weld notch. The same disturbance also occurs for the situation without a shift in neutral axis, but with the presence of the root notch (symmetric geometry). Based on this, the global stress behavior that is represented by equation (7.1) is considered to concur with the numerical results from the finite element analyses. The results obtained using different equations for calculation of SCFs are shown in Table 7.1.

The stress distribution through section B-B, as shown in Figure 7.2, is then considered. The gradients of both the stress distributions for the two geometries can be seen to be small away from the notches. This indicates that a shift in neutral axis does not result in a significant local bending stress over the thickness at the root that must be included in a fatigue analysis. However, the notch stress due to the local root geometry needs to be accounted for. This effect is included in the S-N curves in DNVGL-RP-C203 (2016) for the root side. Using this calculation, it can be shown

Table 7.1. *Calculated SCFs*

Methodology used for calculation	SCF
Axisymmetric finite element model	1.19
SCF equation (5.2)	1.19
SCF equation (7.1)	1.18
SCF equation (6.21)	1.16

that the additional stress at the hot spot resulting from the shift in neutral axis may be accounted for by analytical expressions.

The width of the girth welds in the root in pipelines and risers may be wider than that shown in Figure 6.3 and also may be narrower on the outside to reduce the welding volume and increase fabrication efficiency. A more typical weld section through a girth weld is shown in Figure 7.3. For this geometry also, the stress due to local bending is less for the root than for the weld toe. The local bending stress due to a misalignment, δ_m, and membrane stress, σ_m, at the weld toe can be expressed as:

$$\sigma_{bt} = \frac{3\,\delta_m}{t}e^{-\alpha}\,\sigma_m \tag{7.2}$$

where α is derived from equation (6.23). The width of the weld at the root in Figure 7.3 is L_r and the width of the weld at the outside is L. Then the bending stress of the pipe wall at the transition from the weld to the base material at the root can be obtained from the linearized moment in Figure 7.3 as:

$$\sigma_{br} = \frac{3\,\delta_m\,L_r}{t\,L}e^{-\alpha}\,\sigma_m \tag{7.3}$$

Thus, for the weld root the effect of axial misalignment can be included by the following SCF for the weld root:

$$SCF = 1 + \frac{3\,\delta_m\,L_r}{t\,L}e^{-\alpha} \tag{7.4}$$

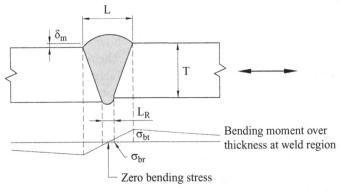

Figure 7.3. Stress distribution due to misalignment at single-sided weld in tubular member.

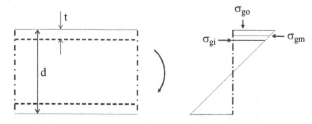

Figure 7.4. Membrane force is used for calculation of local bending stress over tubular thickness.

7.1.3 Nominal Stress in Pipe Wall and Derivation of Hot Spot Stresses

The stress condition in a tubular member subjected to a global bending moment is shown in Figure 7.4. The local bending stress over the thickness of a tubular member is due to the membrane stress over the thickness. Thus the membrane stress, σ_{gm}, should be used for calculating the additional stress resulting from a stress concentration at a weld. An SCF representative for the outside should be used for calculation of stress on the outside, and an SCF representative for the inside should be used for calculation of stress on the inside. The local bending stress over the thickness is derived as the stress concentration multiplied by the membrane stress:

$$\sigma_b = SCF\,\sigma_m - \sigma_m = (SCF - 1)\sigma_m \qquad (7.5)$$

where:

σ_m = membrane stress calculated at the middle of the thickness due to axial force and bending moment.

The local bending stress should be added to that from the global stress in the tubular member resulting from the axial force and bending moment. The stress on the outside should be used for fatigue assessment initiating from the outside, and the stress on the inside should be used for fatigue assessment initiating from the inside. With reference to Figure 7.4, the stress on the outside can be calculated as:

$$\sigma_{outside} = (SCF_{outside} - 1)\,(\sigma_{gm} + \sigma_a) + \sigma_{go} + \sigma_a$$
$$= SCF_{outside}(\sigma_{gm} + \sigma_a) + \sigma_{go} - \sigma_{gm} \qquad (7.6)$$

where:

$SCF_{outside}$ is SCF for calculation of stress on the outside, σ_a is stress due to axial force and the other parameters are defined in Figure 7.4 showing stresses due to global bending.

The stress on the inside of the tubular can, similarly, be derived as:

$$\sigma_{inside} = (SCF_{inside} - 1)\,(\sigma_{gm} + \sigma_a) + \sigma_{gi} + \sigma_a$$
$$= SCF_{inside}(\sigma_{gm} + \sigma_a) + \sigma_{gi} - \sigma_{gm} \qquad (7.7)$$

where:

SCF_{inside} is the SCF for calculation of stress on the inside, σ_a is stress due to axial force and the other parameters are defined in Figure 7.4.

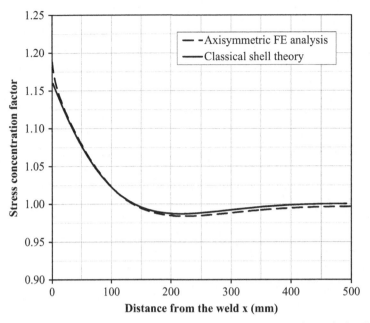

Figure 7.5. Stress concentration at distance from the weld toe (x in Figure 6.3) (Lotsberg 2009a) (reprinted with permission from Elsevier).

7.1.4 Stress Distribution in Pipe Away from a Butt Weld with Fabrication Tolerances

This section is included for information on the laboratory testing of fatigue strength of pipes. It may also be relevant for assessment of interaction between hot spots as shown by the example in Section 6.6. The recommended distance in the axial direction of the pipe, from the weld toe to the region where bending moments over the wall thickness due to fabrication tolerances at the weld is approaching zero, has been questioned. This is important information for the purpose of correct placement of strain gauges for measuring nominal strain/stress in fatigue tests.

Calculated stresses derived from the axisymmetric finite element model, as presented in Lotsberg and Holth (2007), are compared with those derived from classical shell theory in Figure 7.5. The geometry analyzed was a pipe with outer diameter 1066.8 mm (42 inches) and wall thickness 31 mm (1.220 inches). The bending moment at a distance x from the center of a weld can be calculated from shell theory based on the equations in Section 6.1. The bending moment over the wall thickness at a distance, x, from the center of the weld can be determined from equations (6.3) and (6.9) by requiring the radial displacement of the pipe due to moment loading to be zero in the center of the weld (asymmetry point). Equation (6.3) gives $Q_0 l_e = -M_0$. Then the following equation for the bending moment of the pipe wall at a distance x from the center of the weld is derived from equation (6.9), with definition of g-functions from equation (6.6):

$$M_x = M_0\big(g_3(\xi) - g_2(\xi)\big) = M_0(e^{-\xi}\cos\xi + e^{-\xi}\sin\xi - e^{-\xi}\sin\xi) = M_0 e^{-\xi}\cos\xi$$

$$(7.8)$$

where the eccentricity moment, M_0, is derived from Figure 5.1b as:

$$M_0 = \frac{\delta_m}{2}\sigma_a t \tag{7.9}$$

and

$$\xi = \frac{x}{l_e} \tag{7.10}$$

where the elastic length is calculated as:

$$l_e = \frac{\sqrt{rt}}{\sqrt[4]{3(1-v^2)}} \tag{7.11}$$

where:

   r = radius to the mid-surface of the pipe
   t = thickness of the pipe
   v = Poisson's ratio.

From Figure 7.5 it can be seen that there is a negligible difference in the calculated stresses using the two methods (within 1% of each other). The distance from the weld toe to the region where the effect from the bending moment is reduced to zero (or where the SCF approaches 1.0) is also similar (136 to 138 mm). Half the analyzed weld width is 8.3 mm. Thus, to obtain a general equation for the distance to the position where the moment is reduced to zero, it is practical to relate the distance from the center of the weld. This corresponds to $(138 + 8.3)/98.6 = 1.48\ l_e$, where l_e is the elastic length from equation (7.11); ($l_e = 98.6$ mm for the pipe being considered). Note that the distance to the region where the bending moment vanishes is now related to the center of the weld for reference, while Figure 7.5 shows an abscissa related to the distance from the weld toe. The distance from the weld to the right part of Figure 7.5, where the effect of the moment completely disappears, is approximately $4l_e$.

7.2 Stresses at Seam Weld due to Out-of-Roundness of Fabricated Pipes and Internal Pressure

Pipes that are used for transportation and storage of gas in pipelines, and cylinders used for transportation of compressed natural gas (CNG) on vessels, have a relatively large diameter. This requires the use of pipes that have been welded together from rolled plates and have a longitudinal weld seam. The longitudinal welds that are made during production of these pipes are made from both sides. The stress in the circumferential direction of the pipes due to the internal pressure is approximately twice that in the longitudinal direction. High-strength steel is used to reduce the weight of pipes used for transportation of CNG on ships. The fatigue strength of welded steel is largely independent of material yield strength, and due to the high stress range resulting from the load cycle of filling and emptying the pipes, fatigue of the weld seam is a challenging design issue, even when associated with relatively few load cycles. One main reason for this complication is the significant length of longitudinal welds that are mainly subjected to the same circumferential stress range. The fatigue strength of welded structures is reduced by increased thickness and weld length, as explained in Section 4.7.

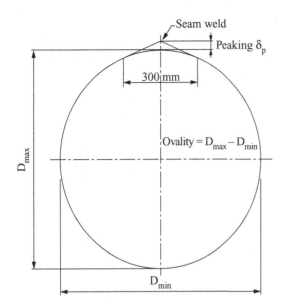

Figure 7.6. Location of seam weld in tubular.

In fabricated pipes there will always be some imperfections that are associated with fabrication. Residual out-of-roundness in the pipes is an important parameter in this respect as it introduces local bending stresses, together with the membrane stress in the circumferential direction in the pipe wall. The membrane stress and the local bending stress, which act normally to the longitudinal seam weld, may govern the fatigue lives of these connections due to the stress cycles from filling and emptying the pipes. During fabrication of cylinders or pipes for gas storage, out-of-roundness after welding is reduced to a minimum by plastic expansion of the pipe wall by mechanical forces in the radial direction. However, due to the "springback force" from elasticity, some out-of-roundness is likely to remain as shown in Figure 7.6. In the DNV pipeline rules, out-of-roundness is defined as $\delta_{OOR} = d_{max} - d_{min}$ (DNV-OS-F101, 2013). This means that the δ_0 in Figure 7.7 is $\delta_{OOR}/4$. δ_{OOR} is specified in the DNV rules as equalling 15 mm for "normal control" and 10 mm for "enhanced control."

An example of as-produced geometry is shown in Figure 7.6. The longitudinal weld seam is positioned at maximum ovality. In addition, there is an effect of peaking that results from the procedure used for pipe fabrication.

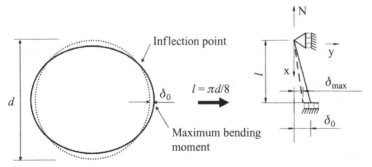

Figure 7.7. Model for analysis of out-of-roundness (Lotsberg 2008a) (reprinted with permission from Elsevier).

The out-of-roundness of fabricated pipe elements results in an increase in stress due to a bending moment over the wall thickness (see Figure 7.7). The eccentricity due to out-of-roundness is a function of the tension in the hoop direction of the pipe, which decreases as the internal pressure increases and the hoop tension rises. Thus, the bending stress over the wall thickness is a nonlinear function of the internal pressure.

Assuming that the moment, M, results from an eccentricity, δ, where the hoop tension is accounted for in the analysis, the following derivation of a SCF can be performed. It is assumed that the out-of-roundness results in an eccentricity, δ_0, without any hoop tension force from internal pressure. In order to simplify the calculation, the ring is transformed to an equivalent beam model, as shown in Figure 7.7. The circumference of a pipe between an inflection point, where the bending moments are zero, and a symmetry point with maximum moment is considered. Thus the length, l, in the beam model becomes $l = \pi d/8$. The relationship between bending curvature and bending moment on a beam can be expressed according to classical beam theory as:

$$\frac{\partial^2 y}{\partial x^2} = -\frac{M(x)}{EI} \tag{7.12}$$

where:

E = Young's modulus
I = moment of inertia of the pipe wall.

The eccentricity along the beam can be expressed as:

$$\delta(x) = \frac{\delta_0}{l}x \tag{7.13}$$

and the moment along the pipe wall beam is:

$$M(x) = -(\delta(x) + y)N \tag{7.14}$$

By combining equations (7.12)–(7.14), the following differential equation is obtained:

$$\frac{\partial^2 y}{\partial x^2} - \frac{N}{EI}y - \frac{N\delta_0 x}{EIl} = 0 \tag{7.15}$$

The general solution of equation (7.15) is:

$$y = C_1 e^{\lambda x} + C_2 e^{-\lambda x} - \frac{\delta_0 x}{l} \tag{7.16}$$

where:

$$\lambda = \sqrt{\frac{N}{EI}} \tag{7.17}$$

where:

N = membrane force in the circumferential direction for the unit length of the pipe.

The boundary conditions in the model shown in Figure 7.7 are:

$$y = 0 \quad for \quad x = 0$$
$$\frac{\partial y}{\partial x} = 0 \quad for \quad x = l \tag{7.18}$$

These boundary conditions are used to determine the integration constants and the results are:

$$C_1 = \frac{\delta_0}{l} \frac{1}{\lambda (e^{\lambda l} + e^{-\lambda l})} \quad and \quad C_2 = -C_1 \tag{7.19}$$

The maximum eccentricity at $x = l$ is then obtained as:

$$\delta_{max} = \delta_0 + y_{max} = \frac{\delta_0}{\lambda l} \tanh(\lambda l) \tag{7.20}$$

The maximum stress at $x = l$ (with $M = N \cdot \delta_{max}$) is obtained from:

$$\sigma = \frac{N}{A} + \frac{M}{W} = \frac{N}{t} + \frac{6 N \delta_{max}}{t^2} \tag{7.21}$$

This stress can also be presented as:

$$\sigma = \frac{N}{t} SCF \tag{7.22}$$

where the SCF is

$$SCF = 1 + \frac{6 \delta_{max}}{t} \tag{7.23}$$

where:

δ_{max} is eccentricity as a function of the axial force, N, as obtained from equation (7.20)
t is the thickness of the pipe.

In terms of out-of-roundness, the equation for SCF can now be derived from equation (7.20) and (7.23) and $\delta_0 = \delta_{OOR}/4$ as:

$$SCF = 1 + \frac{1.5 \delta_{OOR}}{t \lambda l} \tanh(\lambda l) \tag{7.24}$$

where the out-of-roundness is defined as $\delta_{OOR} = d_{max} - d_{min}$, $l = \pi d/8$ and with λ from equation (7.17), which is a function of the membrane hoop stress σ_m as follows:

$$\lambda = \sqrt{\frac{12 \sigma_m}{E t^2}} \tag{7.25}$$

Results from using this equation are derived for a pipe with a 1066.8 mm (42 inches) diameter and a wall thickness of 31.8 mm (1.252 inches) using tolerance requirements from DNV-OS-F101 (2013), which are also referred to in the DNV rules for compressed gas (DNV Rules for Classification of Ships 2012). For standard dimensional requirements for line pipes, the maximum allowable $\delta_{OOR} = d_{max} - d_{min} = 15$ mm. The maximum allowable tolerance for enhanced dimensional requirements for line pipes is $\delta_{OOR} = 10$ mm. The calculated bending stresses for this tolerance are shown in Figure 7.8. The results can be compared with finite element analysis using Abaqus, which accounts for nonlinear geometry. The analytical approach for derivation of

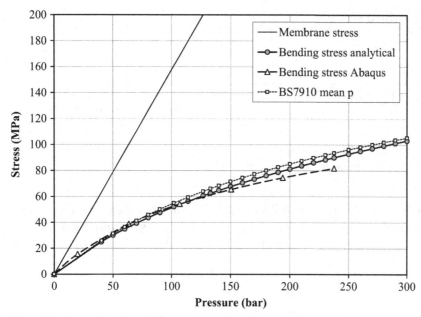

Figure 7.8. Example of membrane and bending stress in the hoop direction in a 42″ diameter pipe.

bending stress in the pipe wall can be seen to provide approximately the same results as those derived from the Abaqus analysis. The results can also be compared using the following equation from BS 7910 (2013):

$$\sigma_b = \sigma_m \frac{1.5\,(d_{\max} - d_{\min})}{t\left\{1 + 0.5\dfrac{p_m(1 - \nu^2)}{E}\left(\dfrac{d - t}{t}\right)^3\right\}} \tag{7.26}$$

where:

> p_m = maximum pressure at the operating condition being assessed; in the section for remarks it is stated that "if under fatigue loading p_m varies, use the mean value during the time interval considered"
>
> σ_m = membrane stress in the circumferential direction due to internal pressure.

The resulting SCFs can be used for fatigue assessment of pipelines, when the pipelines are subjected to variations in internal pressure, such as during the starts and stops of operation. The same SCFs can also be used for fatigue assessment of pipes for gas transportation, where each filling and emptying of the pipes gives a stress cycle that contributes to fatigue damage accumulation.

The internal pressure also results in longitudinal stress in the pipe wall, due to the end cap effect. This nominal longitudinal stress due to internal pressure is only half that of the circumferential stress.

7.3 Stresses at Ring Stiffeners due to Internal Pressure

In this section, a ring stiffener in a cylindrical shell, as shown in Figure 7.9, is considered. This may also apply to a short buckling arrestor in a pipeline or a bolted flange

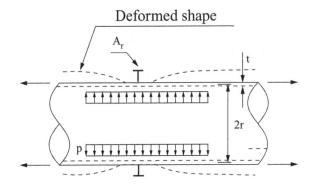

Figure 7.9. Ring stiffener on a pipe with internal pressure (Lotsberg 2008a) (reprinted with permission from Elsevier).

connection that is used in risers for oil and gas production, as indicated in Figure 7.10, where the most critical hot spot is found at point B. The inward displacement of a ring subjected to a radial force (per unit length) $2Q_0$ on the outside can be calculated as:

$$w_{ring} = -\frac{2Q_0\, r^2}{EA_r} \qquad (7.27)$$

where:

  A_r = area of ring stiffener (for a flanged connection, A_r = area of the two flanges).
  At a ring stiffener, due to symmetry the rotation at the shell is zero; see Figure 7.9.

From equation (6.8):

$$\frac{\partial w}{\partial x} = 0 \quad \Rightarrow \quad M_0 = -\frac{Q_0\, l_e}{2} \qquad (7.28)$$

The displacement of the ring equals that of the shell at the weld connection between the ring stiffener and the shell, which, from equations (6.3), (6.7), and (7.27), and for $\xi = 0$, gives:

$$-\frac{2Q_0\, r^2}{EA_r} = \frac{l_e^2}{2\, D_k}\,(M_0 + Q_0\, l_e) + \frac{pr^2}{Et}\left(1 - \frac{v}{2}\right) \qquad (7.29)$$

Combining equations (7.28) and (7.29), and inserting D_k from equation (6.2), the moment in the cylindrical shell at a ring stiffener can be obtained as:

$$M_0 = \frac{p\, r^2 t^2 A_r l_e (2 - v)}{8\, r^2\, t^3 + 12\, A_r\, l_e^3\,(1 - v^2)} \qquad (7.30)$$

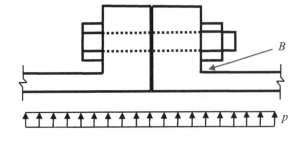

Figure 7.10. Section through a bolted flange connection (Lotsberg 2008a). (Reprinted with permission from Elsevier.)

By dividing by the section modulus the bending stress in the shell is obtained as:

$$\sigma_b = \pm \frac{\sqrt{3}\,(2-v)}{\sqrt{1-v^2}} \frac{1}{\beta} \frac{pr}{2t} \qquad (7.31)$$

where β is defined as:

$$\beta = 1 + \frac{2t\sqrt{rt}}{A_r \sqrt[4]{3(1-v^2)}} \qquad (7.32)$$

The stress concentration at a ring stiffener is obtained as:

$$SCF = 1 \pm \frac{\sqrt{3}\,(2-v)}{\sqrt{1-v^2}} \frac{1}{\beta} \qquad (7.33)$$

where the plus sign applies to the inner side and minus to the outer side. This stress concentration includes the effect from the internal pressure and the end cap pressure. It should be used together with the nominal stress acting in the axial direction of the pipe wall due to end cap pressure.

With $v = 0.3$ for steel the expression for β becomes:

$$\beta = 1 + \frac{1.56\,t\sqrt{rt}}{A_r} \qquad (7.34)$$

and the SCF for the inner side becomes:

$$SCF = 1 + \frac{3.087}{\beta} \qquad (7.35)$$

and the SCF for the outer side becomes:

$$SCF = 1 - \frac{3.087}{\beta} \qquad (7.36)$$

Due to the notch of the weld itself, the fatigue strength of the weld at the ring stiffener becomes lower than for the other shell side. As the stress is lower on the outside than the inside, it is recommended that ring stiffeners are positioned on the outside of a shell structure that is subjected to internal pressure. This is a different conclusion to that reached for ring stiffeners in tubular members that are subjected to pure external axial force, as shown in Section 6.7. From equation (7.35) it can be seen that the SCF at the inside of the pipe at a ring stiffener becomes significant.

An example of a pipe for CNG with outer diameter 1066.8 mm (42 inches) and wall thickness of 31 mm (1.220 inches) is considered. With a flat bar ring stiffener 30×200, the SCF equals 2.53 on the inside and 0.53 for the outside.

The same equations as those used for a ring stiffener can also be used for assessment of SCFs at an open bulkhead in a pipe subjected to internal pressure by calculating an effective area as:

$$A_r = \frac{r\,t_b}{1-v} \qquad (7.37)$$

where:

t_b is thickness of the bulkhead.

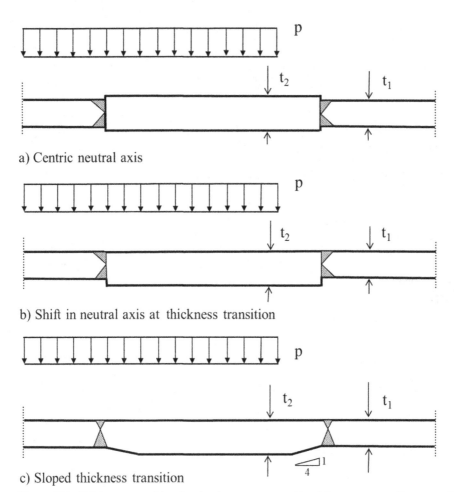

a) Centric neutral axis

b) Shift in neutral axis at thickness transition

c) Sloped thickness transition

Figure 7.11. Thickness transition in pipes at support and buckling arrestors (Lotsberg 2008a) (reprinted with permission from Elsevier).

7.4 Stresses at Thickness Transitions due to Internal Pressure

7.4.1 Circumferential Butt Welds in Pipes with Different Thicknesses

Local bending stresses in pipes will normally occur at connections welded together from pipes of different thicknesses, such as at pipe supports and buckling arrestors. A welded connection between pipes of different thicknesses, as shown in Figure 7.11, is considered below. The following derivation is based on an assumption of a centric neutral axis going from one thickness to the next, as shown in Figure 7.11a. A shift in neutral axis is then considered at the end of this section, with thickness transitions on the outer side (Figure 7.11b) or the inner side of the pipe. For derivation of an equation for SCF at the circumferential weld, the following requirements are defined at the welded connection:

- The radial displacement of pipe wall no. 1 is equal to that of pipe no. 2.
- The rotation of pipe wall no. 1 is the same as for pipe wall no. 2.
- The moments around the circumference are the same in the two pipes.
- There is continuity in shear force over the thickness between the two pipes.

Mathematically, these requirements can be expressed (in the same order) as:

$$w_1 = w_2 \tag{7.38}$$

$$\left(\frac{\partial w}{\partial x}\right)_1 = -\left(\frac{\partial w}{\partial x}\right)_2 \tag{7.39}$$

$$(M_0)_1 = (M_0)_2 \tag{7.40}$$

$$(Q_0)_1 = -(Q_0)_2 \tag{7.41}$$

The axial force in the pipes per unit circumferential length due to internal pressure can be calculated as end cap force divided by circumferential length:

$$N = \frac{p\,\pi\,r^2}{2\pi r} = \frac{pr}{2} \tag{7.42}$$

Then the particular part from equation (6.7) can be expressed as:

$$w_{part} = \frac{p\,r^2}{Et}\left(1 - \frac{\nu}{2}\right) \tag{7.43}$$

From equations (6.3), (7.38), (7.40), (7.41), and (7.43):

$$w = \frac{l_{e1}^2}{2D_{k1}}(M_0 + Q_0\,l_{e1}) + \frac{pr^2}{Et_1}\left(1 - \frac{\nu}{2}\right)$$
$$= \frac{l_{e2}^2}{2D_{k2}}(M_0 - Q_0\,l_{e2}) + \frac{pr^2}{Et_2}\left(1 - \frac{\nu}{2}\right) \tag{7.44}$$

From equation (6.8), (7.39), (7.40), and (7.41):

$$\frac{\partial w}{\partial x} = \frac{l_{e1}}{2D_{k1}}(2M_0 + Q_0\,l_{e1}) = -\frac{l_{e2}}{2D_{k2}}(2M_0 - Q_0 l_{e2}) \tag{7.45}$$

Then from equations (7.44) and (7.45):

$$M_0 = \frac{p\,r\,t_1^2\,(2-\nu)}{4\sqrt{3\,(1-\nu^2)}\,\gamma}\left(\frac{1}{t_1} - \frac{1}{t_2}\right) \tag{7.46}$$

and

$$Q_0 = -\frac{2\left(t_1^{2.5} + t_2^{2.5}\right)}{t_2^{2.5} - t_2^{0.5}t_1^2}\frac{M_0}{l_{e1}} \tag{7.47}$$

where:

$$\gamma = \frac{2\left(t_1^{2.5} + t_2^{2.5}\right)}{t_2^{2.5} - t_2^{0.5}t_1^2}\left(1 + \left(\frac{t_1}{t_2}\right)^{1.5}\right) + \left(\frac{t_1}{t_2}\right)^2 - 1 \tag{7.48}$$

The bending stress at the connection in the thinnest pipe is obtained as:

$$\sigma_b = \frac{M_0}{W} = \frac{6\,M_0}{t_1^2} = \frac{pr}{\gamma}\frac{2-\nu}{2}\sqrt{\frac{3}{(1-\nu^2)}}\left(\frac{1}{t_1} - \frac{1}{t_2}\right) \tag{7.49}$$

It is assumed that the radius $r \gg t_1$. The stress due to the end cap pressure is calculated as:

$$\sigma_a = \frac{pr}{2t_1} \tag{7.50}$$

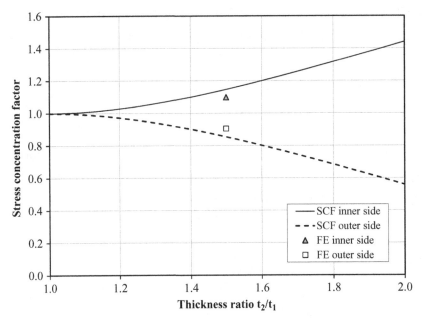

Figure 7.12. SCF for longitudinal stress from internal pressure disregarding thickness transition (or shift in neutral axis).

The total stress at the inner side and the outer side is calculated as:

$$\sigma_t = \sigma_a \pm \sigma_b \qquad (7.51)$$

This equation can also be written as:

$$\sigma_{tot} = \sigma_a \left(1 \pm \frac{\sigma_b}{\sigma_a}\right) = \sigma_a SCF \qquad (7.52)$$

where the SCF is:

$$SCF = 1 \pm \frac{\sigma_b}{\sigma_a} \qquad (7.53)$$

For the inner side of the pipe:

$$SCF = 1 + \frac{2-v}{\gamma}\sqrt{\frac{3}{1-v^2}}\left(1 - \frac{t_1}{t_2}\right) \qquad (7.54)$$

For the outer side of the pipe:

$$SCF = 1 - \frac{2-v}{\gamma}\sqrt{\frac{3}{1-v^2}}\left(1 - \frac{t_1}{t_2}\right) \qquad (7.55)$$

The equation for the moment in equation (7.46) is valid for all thickness relationships. The derivation of equations for SCFs (equations 7.54 and 7.55) is based on calculation of nominal stress in the pipe section with thickness t_1. It should be noted that equations (7.54) and (7.55) are based on a centric neutral axis, as shown in Figure 7.11a. SCFs that have been calculated using these equations are presented in Figure 7.12. The results from one geometry using finite element analysis are shown in Figure 7.12. It can be seen that the largest stress is found at the inner side of the pipe. An eccentric thickness transition as shown in Figure 7.11b also introduces local

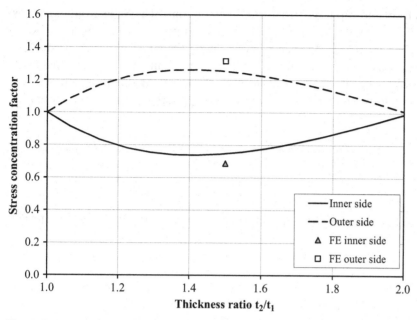

Figure 7.13. SCF with step in thickness on outside.

bending over the pipe wall when subjected to axial forces as explained in Section 6.2.2. The information from Section 6.2.2 can be used to derive SCFs for eccentric thickness transitions by using the superposition principle. For the inner side of the pipe with thickness transition on the outside:

$$SCF = 1 + \frac{2-\nu}{\gamma} \sqrt{\frac{3}{1-\nu^2}} \left(1 - \frac{t_1}{t_2}\right) - \frac{3(t_2 - t_1)}{t_1} \frac{1}{1 + (t_2/t_1)^{2.5}} e^{-\alpha} \quad (7.56)$$

The parameter α is defined in equation (6.23). For the outer side of the pipe with thickness transition on the outside:

$$SCF = 1 - \frac{2-\nu}{\gamma} \sqrt{\frac{3}{1-\nu^2}} \left(1 - \frac{t_1}{t_2}\right) + \frac{3(t_2 - t_1)}{t_1} \frac{1}{1 + (t_2/t_1)^{2.5}} e^{-\alpha} \quad (7.57)$$

The SCFs calculated using these equations are presented in Figure 7.13, together with results from a finite element analysis. It can be seen from Figure 7.13 that a transition in thickness on the outside of the pipe reduces the SCF at the inner side, as compared with that without shift in neutral axis shown in Figure 7.12. For the inner side of the pipe with thickness transition on the inner side:

$$SCF = 1 + \frac{2-\nu}{\gamma} \sqrt{\frac{3}{1-\nu^2}} \left(1 - \frac{t_1}{t_2}\right) + \frac{3(t_2 - t_1)}{t_1} \frac{1}{1 + (t_2/t_1)^{2.5}} e^{-\alpha} \quad (7.58)$$

The parameter α is defined by equation (6.23). For the outer side of the pipe with thickness transition on the inner side:

$$SCF = 1 - \frac{2-\nu}{\gamma} \sqrt{\frac{3}{1-\nu^2}} \left(1 - \frac{t_1}{t_2}\right) - \frac{3(t_2 - t_1)}{t_1} \frac{1}{1 + (t_2/t_1)^{2.5}} e^{-\alpha} \quad (7.59)$$

It should be noted that these SCFs apply only to the axial stresses that result from internal pressure. The SCFs should be used together with the nominal axial stress that results from end cap force or internal pressure only. The SCFs from equations (6.23) and (6.25) should be used for axial stress in the pipe wall resulting from external forces. Thus, for design it is necessary to divide the stress into these two stress components, axial stress due to internal pressure and additional external axial load, before the stress range is calculated. Therefore, the following superposition of stress conditions should be used for derivation of hot spot stress at the circumferential weld:

$$\Delta\sigma_{hot\,spot} = \Delta\sigma_{Internal\,axial\,pressure}SCF_{Internal\,pressure} + \Delta\sigma_{External\,axial}SCF_{External\,axial}$$

(7.60)

where:

$\Delta\sigma_{Internal\,axial\,pressure}$ = Stress in the axial direction of the pipe due to internal pressure

$\Delta\sigma_{External\,axial}$ = Stress in the axial direction of the pipe due to external axial force

$SCF_{Internal\,pressure}$ = SCFs from equations (7.56)–(7.59) depending on the position of the thickness transition and considered hot spot

$SCF_{External\,axial}$ = SCFs from equations (6.23) and (6.25) depending on the position of the thickness transition and considered hot spot.

In many practical cases, a slope transition of 1:4 is made for transition from one thickness to another, as indicated in Figure 7.11c. Finite element analyses of such geometries show that the effect of internal pressure on bending stress at the weld region decreases (see also equations 7.54 and 7.55 for effects from different pipe stiffness). The main contribution to local bending stress is then from the axial force at thickness transitions and fabrication tolerances represented by equations (6.23) and (6.25). In pipeline design, the thicker pipes are machined over a longer length, such that the two pipes are of similar thicknesses at the girth weld section and the thickness transition is sufficiently distant from the weld that it does not have a significant influence on the stress at the weld (see also Section 7.1.4).

7.5 Stresses in Cylinders Subjected to Internal Pressure

7.5.1 Classical Theory for Spherical Shells

For general theory on spherical shells, Flugge (1973) and Spence and Tooth (1994) are recommended readings for a more detailed background for this section. In the following, the relationship between the forces and rotations at the edge of the sphere shown in Figure 7.14 are presented. In Flugge (1973) and Spence and Tooth (1994), the following equation is derived for radial displacement:

$$w_s = \frac{2\,Q_{0s}\,a\,\lambda}{Et_s} + \frac{2\,M_{0s}\,\lambda^2}{Et_s}$$

(7.61)

and for rotation:

$$\frac{\partial w_s}{\partial x} = \frac{2\,Q_{0s}\,\lambda^2}{E\,t_s} + \frac{4\,M_{0s}\,\lambda^3}{E\,a\,t_s}$$

(7.62)

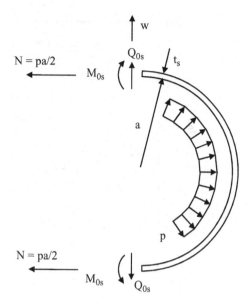

Figure 7.14. End sphere for gas storage pipe (Lotsberg 2008a) (reprinted with permission from Elsevier).

where:

$$\lambda = \sqrt[4]{3(1 - \nu^2)} \sqrt{\frac{a}{t_s}} \tag{7.63}$$

a = radius of sphere
t_s = thickness of sphere.

The radial displacement due to pressure, p, is:

$$w_p = \frac{p\,a^2}{2\,E\,t_s}(1 - \nu) \tag{7.64}$$

There is no rotation at the edge for a constant pressure, p.

7.5.2 Stresses at Girth Weld between Cylinder and Sphere in Storage Tank with Internal Pressure

The following derivation is based on an assumption of centric neutral axes going from the pipe to a sphere. An eccentric transition from a pipe to a sphere can be included similar to that derived for thickness transitions in pipes in Section 6.2.2. The methodology used to derive the stresses at the transition between a cylindrical pipe and the end spheres is similar to that used to derive the stress distributions at thickness transitions. This means that there is compatibility in displacements, rotations, bending moment, and shear force, with reference to equations (7.38)–(7.41). By putting a = r, and from equations (6.3), (6.7), (7.61), (7.64), (7.38), (7.40), and (7.41), the following equation can be derived:

$$
\begin{aligned}
w &= \frac{l_e^2}{2D}(M_0 + Q_0\,l_e) + \frac{pr^2}{Et_c}\left(1 - \frac{\nu}{2}\right) \\
&= \frac{2\lambda^2 M_0}{Et_s} - \frac{2\lambda\,rQ_0}{Et_s} + \frac{pr^2}{2\,Et_s}(1 - \nu)
\end{aligned} \tag{7.65}
$$

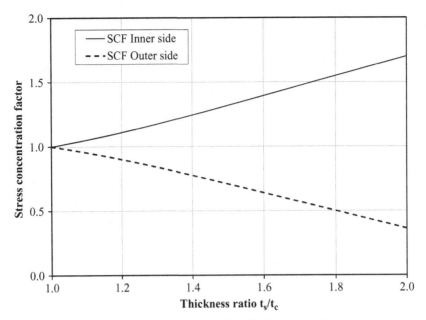

Figure 7.15. SCF at weld to sphere as function of thickness ratio.

From equations (6.8), (7.62), (7.39), (7.40), and (7.41), the following equation is derived:

$$\frac{\partial w}{\partial x} = \frac{l_{e1}}{2D_k}(2\,M_0 + Q_0\,l_e) = -\frac{4\,\lambda^3 M_0}{Et_s\,r} + \frac{2\,\lambda^2\,Q_0}{Et_s} \tag{7.66}$$

From equations (7.65) and (7.66), the bending moment at the junction is derived:

$$M_0 = \frac{p\,r^3}{4\sqrt{3(1-v^2)}}\,\frac{\psi}{\phi}\left(\frac{1-v}{t_s} - \frac{2-v}{t_c}\right) \tag{7.67}$$

and the SCF for the inner side is obtained as:

$$SCF = 1 + \sqrt{\frac{3}{1-v^2}}\,\frac{\psi}{\phi}\,\frac{r^2}{t_c}\left(\frac{1-v}{t_s} - \frac{2-v}{t_c}\right) \tag{7.68}$$

and for the outer side:

$$SCF = 1 - \sqrt{\frac{3}{1-v^2}}\,\frac{\psi}{\phi}\,\frac{r^2}{t_c}\left(\frac{1-v}{t_s} - \frac{2-v}{t_c}\right) \tag{7.69}$$

where:

$$\phi = r^2\psi\left(\frac{1}{t_c^2} - \frac{1}{t_s^2}\right) + 2\,r^{1.5}\left(\frac{1}{t_c^{1.5}} + \frac{1}{t_s^{1.5}}\right) \tag{7.70}$$

and

$$\psi = \frac{(t_c^2 - t_s^2)}{(t_c^{2.5} + t_s^{2.5})}\sqrt{\frac{t_s\,t_c}{r}} \tag{7.71}$$

where:

t_c = thickness of cylinder
t_s = thickness of end sphere.

The equation for the moment in equation (7.67) is valid for all thickness relationships. The derivation of equations for SCFs (equations 7.68 and 7.69) is based on calculation of nominal stress in the pipe section without shift in neutral axis going from pipe to sphere. The resulting SCFs are presented graphically in Figure 7.15.

The shear force at the connection between the cylinder and the sphere is obtained from equations (7.65) and (7.66) as:

$$Q_0 = \frac{2\sqrt[4]{3(1-\nu^2)}}{\phi r} M_0 \qquad (7.72)$$

For considerations of eccentric neutral axes, see Section 6.2.2. A similar term as the last terms in equations (7.56) and (7.57) may be added to equations (7.68) and (7.69), as approximations to include the effect of an eccentric thickness transition.

Most engineers will probably find it easier to make an axisymmetric finite element analysis model in order to establish the stress distribution in cylinders. The modeling should be made with elements that are representative for a linear stress distribution over the thickness in order to derive comparable results with these analytical equations.

8 Stress Concentration Factor for Joints

8.1 General

Many offshore steel structures are designed as truss frameworks in which tubular members are used as the structural elements, such as in the jacket structure in Figure I.5. For a more detailed background, see, for example, Marshall (1992).

Waves and currents generate relatively small loads on tubular members due to their low drag coefficients. However, the intersections between different members that are connected to the same joint may be rather complex, and this may lead to relatively high local stresses at the hot spot areas with correspondingly short fatigue lives. Thus, in order to design structures that meet the required fatigue life, it is necessary to have adequate knowledge about the stress condition at tubular joints.

Tubular joints may be classified into the following groups:

1. Simple tubular joints
2. Overlapping joints
3. Tubular joints with internal ring stiffeners
4. Heavy stiffened tubular joints
5. Grout reinforced joints
6. Cast steel joints.

A simple tubular joint, as shown in Figure 8.1, is understood to mean a joint other than a circumferential girth weld between tubular members, and that is not stiffened by internal or external stiffeners. Furthermore, the braces are welded into the chord without overlapping each other. Overlapping joints are understood to be overlapping braces at the intersection to the chord. Overlapping joints may be used when there is difficulty in placing the tubular members within a specific area, leading to acceptable eccentricities in the joint. These connections can show rather high capacity with respect to the Ultimate Limit State and the Fatigue Limit State. However, they may be more complex to fabricate than simple tubular joints are, and therefore other solutions are often preferred. For example, one solution would be to increase the chord diameter to allow for a larger space for brace intersections, using internal ring stiffeners to achieve the required chord ring stiffness and capacity. In some joints, such as K-joints, where the axial force in one brace is to be transferred to a second brace, it may also be efficient to use longitudinal stiffeners internally in

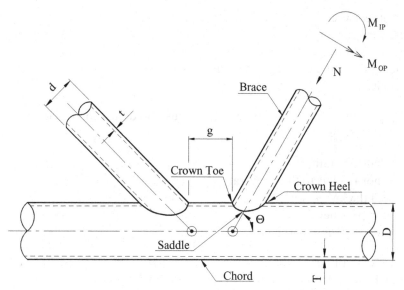

Figure 8.1. Geometrical definitions for tubular joints (DNVGL-RP-C203 2016).

addition to ring stiffeners to reduce the hot spot stress. These joints may be categorized as heavy stiffened joints. Heavy stiffened joints in jacket structures were more frequently designed during the 1980s than today; see, for example, Callan et al. (1981). By using a proper detailed design it was possible to keep the stress concentration factors (SCFs) in these joints at a low level. In special areas grout-reinforced joints may also be used. These joints are also attractive for reinforcement of older installed structures (Etterdal et al. 2001). Complex joints that are of significant importance for the integrity of the structure may be designed as cast joints. Cast joints provide the designer with good flexibility for determining a geometry that results in low stress concentrations; see also Section 4.7.5.

Joints in offshore steel structures are mostly multiplanar. Uniplanar joints are typically found in the bracing between the main legs in horizontal frames, such as used for sideways support of conductors. SCFs for tubular joints are normally presented in terms of uniplanar joints such as T-, Y-, X-, and K-joints, disregarding the effects of braces that are not lying in the planes under consideration.

8.2 Simple Tubular Joints

8.2.1 Definitions of Geometry Parameters and Stresses

Guidance on SCFs for simple tubular joints is included in some design standards for steel structures, such as ISO 19902 (2007), API RP2A (2014), and DNVGL-RP-C203 (2016). Simple tubular joints are defined by the main geometry parameters in Figure 8.1:

D = outside chord diameter
d = outside brace diameter
T = chord wall thickness

t = brace wall thickness
θ = brace inclination angle
g = gap between braces
L = total length of chord between side joints
e = eccentricity.

For establishing parametric equations for SCFs, various non-dimensional geometry parameters should be defined:

β = d/D = diameter ratio
γ = R/T = chord radius to thickness ratio
τ = t/T = wall thickness ratio
ς = g/D = gap parameter
α = 2L/D = chord length parameter.

γ represents the radial stiffness of the chord and, together with β and τ, is a significant parameter for SCFs for calculation of hot spot stress in tubular joints.

For fatigue analysis of tubular joints, it is convenient to divide the loading from a brace to the chord into three separate load cases: axial load, in-plane bending load, and out-of-plane bending load. Each of these load cases has its own particular stress distribution along the intersection line between the brace and the chord, and thereby its own particular influence on the fatigue life. The stress distribution at the intersection of tubular joints is rather complex and depends on the geometry of the joint. The locations at which the highest stresses occur are denoted hot spots. There are two different hot spots: one at the weld toe on the chord side and one at the weld toe on the brace side. Whether the largest hot spot stress is found on the chord side or the brace side depends on the geometry of the joints. Thus, the SCFs on the chord side and brace side must be considered individually.

The SCF is derived as:

$$SCF = \frac{\sigma_{hot\ spot}}{\sigma_{nominal}} \tag{8.1}$$

where the hot spot stress is derived from finite element analysis as described in Section 9.5. The nominal stress refers to the brace. The stresses considered in analysis of tubular joints are the principal stresses. At the hot spot region, the largest principal stress is normally close to perpendicular to the weld. The effect of the weld notch is assumed to be accounted for in the design S-N curve, in the way the hot spot stress is derived with extrapolation of stresses from a region outside that affected by the weld notch. Experimental determination of hot spot stress requires a similar extrapolation procedure of measured stress distribution to the weld toe from stresses outside the weld notch affected zone. During the 1980s it was standard practice to derive S-N curves for tubular joints based on uniaxial strain measurements, that is, from stresses derived as $\sigma = \varepsilon_1 E$, where ε_1 is the measured strain normal to the weld toe and E = Young's modulus; see chapter by Gibstein and Moe in Almar-Naess (1985). Assuming that the measured strain is in the direction of the largest principal stress, this approach will have some built-in safety compared with stresses derived from finite element analysis, due to a two-dimensional strain effect at the hot spot region. The relationship between stress and strains at a tubular joint hot spot area can be expressed by the following equations (reference is made to a textbook on elasticity theory):

$$\varepsilon_1 = \frac{1}{E}(\sigma_1 - \nu\sigma_2)$$

$$\varepsilon_2 = \frac{1}{E}(\sigma_2 - \nu\sigma_1)$$

(8.2)

where:

the strain ε_2 is normal to the strain direction ε_1, and σ_1 and σ_2 are the corresponding stresses
E = Young's modulus for steel
ν = Poisson's ratio.

Then by elimination of σ_2, the following principal stress is derived:

$$\sigma_1 = \frac{E(\varepsilon_1 + \nu\varepsilon_2)}{1 - \nu^2}$$

(8.3)

This gives a built-in factor equal to:

$$f = \frac{\sigma_1}{\sigma} = \frac{1 + \nu\varepsilon_2/\varepsilon_1}{1 - \nu^2}$$

(8.4)

At the majority of hot spot locations in ordinary tubular joints, the ratio between ε_2 and ε_1 is between 0.24 and 0.28; see Gibstein and Moe in Almar-Naess (1985). This results in a factor of f = 1.15 – 1.19. During later assessment of S-N curves for tubular joints, if only uniaxial strain had been measured, then the measured strain was increased by a factor of 1.17 before the data point was entered into an S-N diagram for regression analysis; see van Wingerde et al. (1997b).

Simple tubular joints are typically used in jacket structures. The first jackets installed in the North Sea were not analyzed in detail for fatigue. However, as fatigue assessment began to be included in normal design procedures, the need for SCFs for calculation of hot spot stresses in tubular joints also arose.

Due to the sensitivity of calculated fatigue life to hot spot stresses, it is important that the analysis methods are calibrated against measurement data. Such data were presented at two major conferences in 1981: "Fatigue in Offshore Structural Steels" in London in February and "Steel in Marine Structures" in Paris in October. See papers by Marsh (1981), Wylde and McDonald (1981), McDonald and Wylde (1981), Radenkovic (1981), and Pozzolini (1981), as well as an OTC paper by Dijkstra and de Back (1980) on fatigue strength of welded tubular T- and X-joints.

A number of different sets of parametric equations for estimating SCFs for tubular joints have been presented (e.g., Kuang et al. 1975; Potvin et al. 1977; Gibstein 1978; Wordsworth and Smedley 1978; Wordsworth 1981; UEG 1985; Efthymiou and Durkin 1985; Efthymiou 1988; Connolly et al. 1990; Hellier et al. 1990a, 1990b; Smedley and Fisher 1991a, 1991b; Brennan et al. 1999 and Smedley 2003). The performance of the various sets of SCF equations has been assessed in a number of studies, including Fessler et al. (1991), Dover et al. (1991), Lloyds Register for HSE (HSE OTH, 1997), and ISO 19902 (2007). It was determined that the parametric equations for SCFs for tubular joints provided by Efthymiou (1988) are preferable in comparison with other available equations when considering consistency and coverage. The

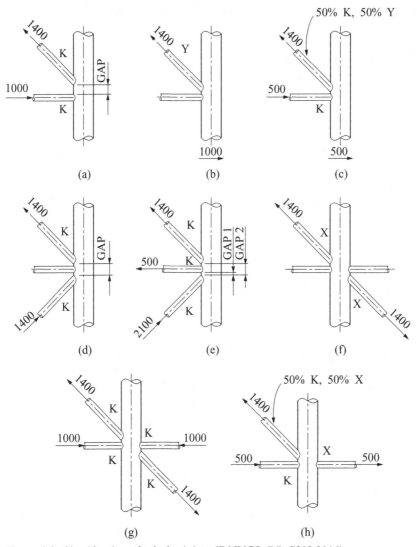

Figure 8.2. Classification of tubular joints (DNVGL-RP-C203 2016).

equations by Efthymiou (1988) are used in ISO 19902 (2007), and API RP2A (2014), and in recommendations in DNVGL-RP-C203 (2016). They are also referred to in BS 7910 (2013) for fracture mechanics. SCFs for axial load, in-plane bending, and out-of-plane bending have been presented, dependent on type of joint. The definition of joint type depends not only on geometry but also on force flow through the joints as that is important for derivation of hot spot stresses that reflect relevant physical behavior. Examples of joint classification are shown in Figure 8.2. The type of joint is thus a function of loading in addition to the geometry, and, as a result of this, the type of joint will alter as a wave passes through a jacket structure. Joint classification is the process by which the axial force in a given brace is subdivided into K, X, and Y components of actions that correspond to the three joint types for which stress concentration equations exist. Such subdivision normally considers all the members in a single plane at a joint. For the purposes of this provision, brace planes within ±15°

Table 8.1. *Calculated SCFs for axial load(s) in brace(s), based on equations from Efthymiou (1988)*

Joint geometry: $\beta = 0.6, \gamma = 15.88, \tau = 0.831, \theta = 60°, e = 0, \alpha = 25, \varsigma = 0.1$					
Joint type	X-joint	T-joint	Y-joint	K-joint (balanced)	K-joint analyzed by finite element method (Almar-Naess 1985)
SCF$_c$ at saddle	16.83	14.13	11.22	5.52	5.9
SCF$_b$ at saddle	9.13	10.12	7.50	4.01	4.3

of each other may be considered as being in a common plane. Each brace in the considered plane can have a unique classification that could vary with action condition. The classification can also be a mixture between the three joint types described earlier. Classification of joints can also be based on deterministic analysis, using a wave height corresponding to that contributing the most to calculated fatigue damage. A conservative classification may be used, keeping in mind that $SCF_X > SCF_Y > SCF_K$. There is one set of equations for SCFs on the brace side and another set of equations for the chord side. Examples of calculated SCFs for axial load(s) in brace(s) are shown in Table 8.1 for X-, T-, Y-, and K-joints. The values were calculated from the SCFs presented by Efthymiou (1988) and which also are included in DNVGL-RP-C203 (2016). The results from a finite element analysis for the same geometry (Gibstein and Moe, as cited in Almar-Naess 1985) is also included in Table 8.1. The validity range for the Efthymiou equations for SCFs are presented as follows:

$$0.2 \leq \beta \leq 1.0$$
$$0.2 \leq \tau \leq 1.0$$
$$8 \leq \gamma \leq 32$$
$$4 \leq \alpha \leq 40$$
$$20° \leq \theta \leq 90°$$
$$\frac{-0.6\beta}{\sin\theta} \leq \zeta \leq 1.0$$

Stress concentrations in multiplanar stiffened tubular KK joints for jack-up structures have been presented by Woghiren and Brennan (2009).

8.2.2 Influence of Diameter Ratio β on Stress Concentration

The influence of the diameter ratio β on SCF for different types of tubular joints is illustrated in Figure 8.3.

8.2.3 Influence of Radius-to-Thickness Ratio of Chord, γ, on Stress Concentration

SCF increases with increasing γ, as shown in Figure 8.4 for an X-joint with $\beta = 0.80$ and $\Theta = 60°$.

8.2.4 Influence of Thickness Ratio, τ, on Stress Concentration

The calculated SCF increases with increasing thickness ratio, τ, as shown in Figure 8.4 for an X-joint with $\beta = 0.80$ and $\Theta = 60°$.

Stress Concentration Factor for Joints

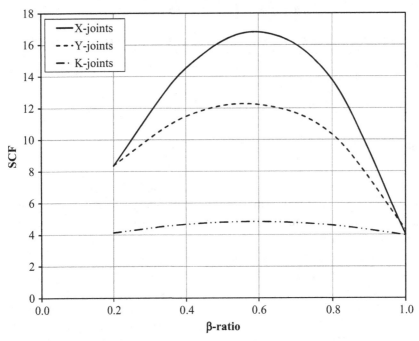

Figure 8.3. Effect of diameter ratio on calculated SCFs.

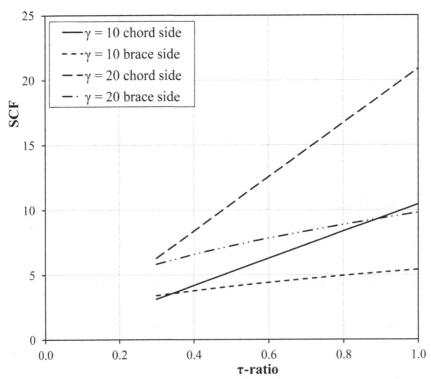

Figure 8.4. Effect of γ-ratio and τ-ratio on the chord side SCF and the brace side SCF in an X-joint.

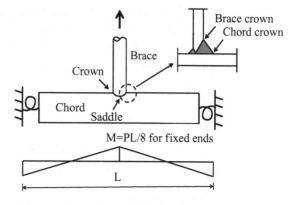

Figure 8.5. Illustration of a loaded brace giving bending moment in the chord (Lotsberg 2011) (reprinted with permission from Elsevier).

8.2.5 Influence of Chord-Length-to-Diameter Ratio, α, on Stress Concentration

The stress concentration at the crown point in T- and Y-joints increases with increasing chord-length-to-diameter ratio, α. In some special cases it has been found that the analysis programs do not provide physically correct SCFs for the considered problem at the crown points of tubular joints without some additional engineering assessment. The uncertainty related to correct hot spot stress methodology is enhanced when the α parameter exceeds its validity range ($\alpha = 2L/D$, where L = distance between supporting joints for the chord and D = chord diameter). The considered problem can also be related to geometric K-joints in special cases where, as a result of local loading, the K-joint becomes analyzed as a T-joint, due to force flow considerations in the joint when using the concept of influence functions (Efthymiou 1988). An influence function is an expression for the hot spot stress at a certain location around the chord-to-brace intersection that arises from a nominal stress of unit magnitude acting on any brace of the joint. The hot spot stress can be obtained by multiplying each influence function by its respective nominal stress and then superimposing the contributions from all braces of the joint, including the brace under consideration. Some examples of when this formulation requires special consideration are presented in greater detail in the following section.

To understand this problem fully, a short background on the equations used for calculation of stress concentrations at the crown points for T- and Y-joints is necessary. The equations for SCFs for T-joints and Y-joints in Efthymiou (1988) and ISO 19902 (2007) are made for a chord with a single load that is subjected to a loading similar to that shown in Figure 8.5. The axial loading in the brace in Figure 8.5 results in a bending moment in the chord. It is the bending moment below the brace that is the main reason for the SCF at the crown position for large chord-length-to-diameter ratios. Thus, it can be seen that the length of the chord is an important parameter here. It should also be noted that in these references, in order for the equations for the SCFs to be valid, $\alpha = 2L/D$ is limited to an upper value of 40. If the brace is not in the middle of the chord, then calculation of an equivalent chord length – that gives the same bending moment in the chord member as that of the actual bending moment in the chord at the considered brace – is recommended. When more attachments are added to the same chord, the moment in the chord will increase for loads transverse to the chord. This has to be considered in a fatigue analysis for derivation of the correct hot spot stresses at the crown points.

The equation for the SCF for the chord crown in Efthymiou (1988) and ISO 19902 (2007) reads:

$$SCF_{Cc} = \gamma^{0.2}\tau(2.65 + 5(\beta - 0.65)^2) + \tau\beta(C_2\alpha - 3)\sin\theta \tag{8.5}$$

For the brace crown it reads:

$$SCF_{Bc} = 3 + \gamma^{1.2}(0.12\exp(-4\beta) + 0.011\beta^2 - 0.045) + \beta\tau(C_3\alpha - 1.2) \tag{8.6}$$

where:

the geometric parameters are presented in Section 8.2.1

C = chord end fixity parameter, with C = 0.5 for fixed ends and C = 1.0 for free ends; thus, $0.5 \leq C \leq 1.0$.

C = 0.7 is typically recommended to be used in analysis. $C_1 = 2$ (C-0.5); $C_2 = C/2$; and $C_3 = C/5$. It should be noted that $C_3/C_2 = 0.4$. This means that from comparison of equations (8.5) and (8.6), 40% of the global bending stress on the chord side of the weld is transferred to the brace side of the weld at the crown position. For fixed ends, $C_2 = 0.25$. A chord with fixed ends, as shown in Figure 8.5, is considered in the following. When assuming a small wall thickness, as compared with the diameter of the tubular members, the relationship between the force, P, and the stress in the brace is derived as:

$$P = \pi \, dt\sigma_{Brace} \tag{8.7}$$

The elastic section modulus for the chord is calculated as:

$$W_{Chord} = \frac{\pi D^2 T}{4}. \tag{8.8}$$

The SCF at the chord crown due to global bending is derived from definition of the SCF as:

$$SCF_{Cc} = \frac{\sigma_{Chord}}{\sigma_{Brace}} \tag{8.9}$$

and

$$SCF_{Cc} = \frac{M_{Chord}}{\sigma_{Brace}W_{Chord}} = \frac{PL}{8\,\sigma_{Brace}W_{Chord}} \tag{8.10}$$

Further, from equations (8.7), (8.8), and (8.10):

$$SCF_{Cc} = \frac{\pi \, dt\sigma_{Brace}L}{8\,\sigma_{Brace}\pi D^2 T/4} = 0.25\frac{t}{T}\frac{d}{D}\frac{2L}{D} = 0.25\tau\beta\alpha \tag{8.11}$$

Note that the last part of equation (8.11) is the same expression as is included in the Efthymiou equation (8.5) when considering fixed ends with $C_2 = 0.25$ and C = 0.5.

From calculated SCFs as a function of the α parameter it can be seen that the stress at the chord crown position increases with increasing α due to increased global bending of the member, and for long members (and large α values) this stress becomes governing for the calculated fatigue life. Some examples in which the present formulation requires special consideration are presented below.

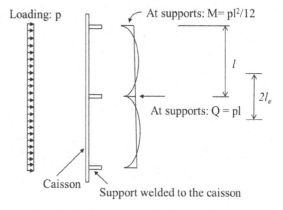

Figure 8.6. A caisson welded as T-joints to supports at horizontal frames in a jacket (Lotsberg 2011) (reprinted with permission from Elsevier).

Loads on Caissons

A caisson welded to stub supports, as shown in Figure 8.6, is considered. The actual moment at the support is derived for illustration as:

$$M_{actual} = \frac{pl^2}{12} \tag{8.12}$$

and the moment corresponding to the equation for the stress concentration is derived as:

$$M_{SCFequation} = \frac{Q2l}{8} = \frac{pl^2}{4} \tag{8.13}$$

It should be observed that the moments from these equations differ by a factor 3, and that the hot spot stress would be incorrect by the same factor. This would have a significant influence on the calculated fatigue life, bearing in mind that the stress is raised in power 3 to 5 for calculation of fatigue life. Thus, to derive a correct hot spot stress, an effective length between supports (L in Figure 8.5 is equal to an effective length, $2l_e$, as indicated in Figure 8.6) is obtained by requiring that a moment corresponding to the SCF equation (8.13) is equal to the actual moment at the support in Figure 8.6, as expressed by equation (8.12). The moment distribution in Figure 8.6 corresponds to that of fixed ends, and C = 0.5 applies. The effective length for calculation of moment can then be derived from the following relationship:

$$M = \frac{Q2l_e}{8} = \frac{pl^2}{12} \tag{8.14}$$

At the support the reaction force is Q = P and:

$$Q = pl \tag{8.15}$$

From equations (8.14) and (8.15), the following expression is derived for the effective length between horizontal supports:

$$l_e = \frac{1}{3}l \tag{8.16}$$

This means that a physically correct hot spot stress can be derived in this case when an effective length from equation (8.16) is used for calculation of SCF. While this is correct for a uniform loading, as shown, the real loading on structures is normally not

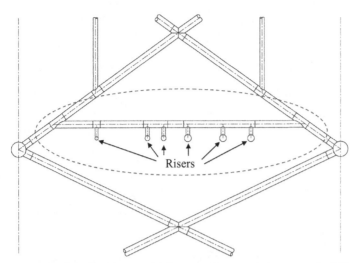

Figure 8.7. Detail of frame with beam with several risers attached (Lotsberg 2011) (reprinted with permission from Elsevier).

constant, and for more complex loads a frame analysis would be required to calculate the correct effective length.

Several Risers/Caissons Attached to the Same Chord

An example with a chord that is a support for several risers is shown in Figure 8.7. It is assumed that the wave loading acts in a direction normal to the chord. The bending moment in the chord in Figure 8.7 is approximately 3.3 times the bending moment of one single force, P, acting in the middle of the chord. Thus, in this case the SCF at the chord crown is 3.3 times higher than would be derived from the analytical equations for SCFs from equation (8.5), as illustrated in Figure 8.8.

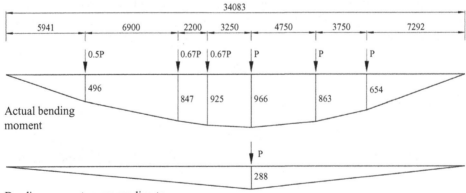

Bending moment corresponding to
that of SCF equation

Figure 8.8. Loading on beam with moment distribution for the considered loading as compared with a single load corresponding to that from using the SCF equation (Lotsberg 2011) (reprinted with permission from Elsevier).

K-Joint in Jacket Frame Analyzed with Influence Functions

During analysis of a geometric K-joint in a jacket at a position just below the waterline, it was realized that a correct bending moment in the chord was of significant importance for the fatigue analysis (Lotsberg 2011). For some positions of the wave, the vertical wave loading on the horizontal chord gave compressive forces in the two braces forming the K-joint, while for another wave position it produced tension forces in both braces. In reality, the braces are support points for the chord, but because the SCFs are calculated using the existing parametric equations, the brace loads simulate a large bending moment in the chord. This results in very large SCFs at the crown position, with correspondingly short calculated fatigue lives. Thus, a reduced effective length had to be used here also in order to simulate the correct bending moment in the chord when influence functions were used. Further details are provided in Lotsberg (2011). Based on this, an alternative formulation for SCFs at the crown points of T-joints and Y-joints has been proposed with the purpose of simplifying engineering work when performing fatigue analysis of frame structures. The proposed formulation involves only minor revision of parametric equations previously used in existing design standards. The bending moment in the chord is normally calculated in a computer program for frame analysis. Thus, the alternative formulation simply uses the calculated bending moment in the chord directly to derive the hot spot stress at the chord crown for calculation of fatigue damage. This can be performed by removing the term with bending moments in the existing equations for SCFs and then adding the actual calculated bending moment in the chord. Using this approach an alternative equation for SCF for the chord crown reads:

$$SCF_{Cc} = \gamma^{0.2} \tau (2.65 + 5(\beta - 0.65)^2) - 3\tau\beta\sin\theta + \frac{M_{Chord}}{W_{Chord}\sigma_{Brace}} \qquad (8.17)$$

Similarly, an alternative equation for SCF for the brace crown reads:

$$SCF_{Bc} = 3 + \gamma^{1.2}(0.12\exp(-4\beta) + 0.011\beta^2 - 0.045) - 1.2\beta\tau + \frac{0.4M_{Chord}}{W_{Chord}\sigma_{Brace}}$$

$$(8.18)$$

Equations (8.17) and (8.18) have been modified such that they provide the same SCFs as those achieved by equations (8.5) and (8.6) for the base case with axial loading in one brace, as shown in Figure 8.5 for the same end fixity. Using this formulation, the end fixity can be accounted for automatically through joint flexibility, provided that this is included in the computer program/analysis, and a C value to determine end fixity need not be specified for calculation of SCF. This formulation also does not require an upper limit on the α parameter. However, for long members with significant global bending of the chord, an additional SCF may be included to account for the difference between the T-curve and the F to F1 curves when considering the brace as an attachment to the chord. This will result in an additional SCF of approximately 1.27, according to DNVGL-RP-C203 (2016), and 1.25 as is proposed in ISO 19902 (2007).

Table 8.2. *Bias and uncertainty in SCFs for different simple tubular joints, depending on type of loading (based on assessment of data in HSE OTH 354 1997)*

Joint	Loading	Hot spot	Number of joints	Bias*	CoV (%)
T	Axial	Chord saddle	28	1.07	11
		Chord crown	9	1.12	26
		Brace saddle	8	1.29	25
		Brace crown	4	1.55	20
	Out-of-plane bending	Chord side	18	1.10	13
		Brace side	9	1.54	36
	In-plane bending	Chord side	21	1.09	17
		Brace side	24	1.22	20
X	Axial	Chord saddle	16	1.15	19
		Chord crown	3	1.04	6
		Brace saddle	7	1.22	13
		Brace crown	3	1.47	9
	Out-of-plane bending	Chord side	6	1.13	20
		Brace side	4	1.60	31
	In-plane bending	Chord side	12	1.33	38
		Brace side	6	1.33	26
K	Axial			1.19	19

* Bias is defined as the calculated value divided by the measured value.

8.2.6 Assessment of Accuracy of SCFs

The bias and uncertainty in calculated SCFs from parametric equations for different simple tubular joints, as presented in Table 8.2, should be included in further assessment of a reliable analysis procedure to determine the best possible estimates of fatigue lives. Uncertainties can be directly accounted for in the probabilistic analysis. It is more difficult to account for bias, as the fatigue damage is derived from fatigue analysis using influence functions where different SCFs are used to calculate fatigue damage. Thus, to account for different bias values associated with the different SCFs, these bias values would have to be considered before the fatigue damage is calculated. Bias in parametric equations for SCFs should not be used in design. Parametric equations for stress concentrations in tubular joints have also been assessed by Fessler et al. (1991), and it was concluded that the Efthymiou's equations provide values on the safe side, with a coefficient of variation (CoV) of around 0.20.

8.2.7 Combination of Stresses from Different Load Conditions

For combination of stresses due to axial force, in-plane bending, and out-of plane bending moments, reference is made to Gulati et al. (1982) and Buitrago et al. (1984). The resulting hot spot stresses at the eight positions around the circumference in Figure 8.9 can be derived by superposition of stresses due to axial force, in-plane

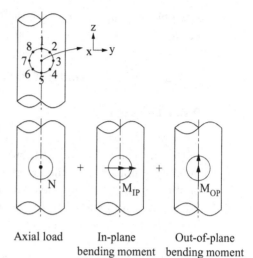

Figure 8.9. Superposition of stresses from different load components.

Axial load   In-plane   Out-of-plane
bending moment   bending moment

loading, and out-of plane loading, as shown by the following equations:

$$\sigma_1 = \mathrm{SCF_{AC}}\sigma_x - \mathrm{SCF_{MIP}}\sigma_{my}$$

$$\sigma_2 = \frac{1}{2}(\mathrm{SCF_{AC}} + \mathrm{SCF_{AS}})\sigma_x - \frac{1}{2}\sqrt{2}\,\mathrm{SCF_{MIP}}\sigma_{my} + \frac{1}{2}\sqrt{2}\,\mathrm{SCF_{MOP}}\sigma_{mz}$$

$$\sigma_3 = \mathrm{SCF_{AS}}\sigma_x + \mathrm{SCF_{MOP}}\sigma_{mx}$$

$$\sigma_4 = \frac{1}{2}(\mathrm{SCF_{AC}} + \mathrm{SCF_{AS}})\sigma_x + \frac{1}{2}\sqrt{2}\,\mathrm{SCF_{MIP}}\sigma_{my} + \frac{1}{2}\sqrt{2}\,\mathrm{SCF_{MOP}}\sigma_{mz}$$

$$\sigma_5 = \mathrm{SCF_{AC}}\sigma_x + \mathrm{SCF_{MIP}}\sigma_{my}$$

$$\sigma_6 = \frac{1}{2}(\mathrm{SCF_{AC}} + \mathrm{SCF_{AS}})\sigma_x + \frac{1}{2}\sqrt{2}\,\mathrm{SCF_{MIP}}\sigma_{my} - \frac{1}{2}\sqrt{2}\,\mathrm{SCF_{MOP}}\sigma_{mz}$$

$$\sigma_7 = \mathrm{SCF_{AS}}\sigma_x - \mathrm{SCF_{MOP}}\sigma_{mx}$$

$$\sigma_8 = \frac{1}{2}(\mathrm{SCF_{AC}} + \mathrm{SCF_{AS}})\sigma_x - \frac{1}{2}\sqrt{2}\,\mathrm{SCF_{MIP}}\sigma_{my} - \frac{1}{2}\sqrt{2}\,\mathrm{SCF_{MOP}}\sigma_{mz}$$

$$(8.19)$$

Here, σ_x, σ_{my}, and σ_{mz} are the maximum nominal stresses due to axial load and bending in-plane and out-of-plane, respectively. $\mathrm{SCF_{AS}}$ is the SCF at the saddle for axial load, and $\mathrm{SCF_{AC}}$ is the SCF at the crown. $\mathrm{SCF_{MIP}}$ is the SCF for the in-plane moment, and $\mathrm{SCF_{MOP}}$ is the SCF for the out-of-plane moment. The equation for combination of stresses is an approximation; the hot spots may be at different positions around the interface, and the stress distribution from moment loading is not necessarily a sinusoidal function around the circumference. However, this equation has been found practical and useful for more than 30 years (also used in DNV CN30.2 1984). In-service experience to date has shown that the fatigue analysis procedure used for jacket structures provides design solutions with acceptable fatigue behavior. Influence functions may be used to calculate hot spot stress as an alternative to the procedure given here (see, e.g., Buitrago et al. 1984; Efthymiou 1988). According to Efthymiou, the method using influence functions is preferable, as it was shown to provide more accurate results than other methods where the SCFs are a function of geometry and force flow through the joint. Another alternative that might be attractive for complex joints is to make a full finite element model of the joint and use the assembly as a super-element in the structural analysis model. Reference is made to Section 9.5 for finite element analysis of tubular joints.

8.3 Single-Sided Welded Tubular Joints

8.3.1 Background

Until recently, most of the jackets for use in the North Sea were designed and fabricated with stub sections at tubular joints, which make welding possible from the inside. The hot spots for fatigue assessment were for the outside of these joints. Use of stubs is understood to involve prefabrication of joints, with a short part of the brace (the stub) welded to the chord. The stub can be designed with a similar or larger thickness than the main part of the brace and is so short that the welding can be also done from the inside, such that a reliable double-sided weld is achieved in the joints between the brace and the chord. The stubs should be so long that there will not be significant interaction between stress concentrations from the tubular joint and the girth weld between the stub and the brace. The length needed for separation of these interaction effects depends on whether membrane stress or bending stress is considered. Reference is made to Section 7.1.3 regarding decay function for bending stress along a tubular member. A peak membrane stress is considered to become rater uniformly distributed over the circumference at a distance of one diameter from the hot spot. Thus, a requirement of a diameter or minimum 600 mm is required for stub lengths in tubular joints in ISO 19902 (2007), API RP2A (2014) and NORSOK N-004 (2013). As these tubular joints with stubs are prefabricated before the main construction of the jackets is started this requires also some chord length to be used. This chord length needs to be so long that the capacity equations for the tubular joints can be applied also when a thicker section for this part of the chord if used in these equations. Also here one should avoid significant interaction between the stress in the tubular joint and the girth welds in the chord. Therefore a free length outside the braces equal minimum of a quarter of the chord diameter and 300 mm should be used according to ISO 19902 (2007), API RP2A (2014), and NORSOK N-004 (2013). By this design principle there is flexibility to increasing the chord thickness locally at the joint and also to specify steel in this section with improved through thickness properties as explained in Section 11.2. During construction, the braces are welded to the stubs by closure girth welds made from the outside (see, e.g., Stacey et al. 1997). In other parts of the world where environmental conditions are milder than in the North Sea and fatigue design has been a less significant issue, single-sided welds have been more frequently used. The main tubular joints in jacket structures for the North Sea are currently also fabricated without stubs, where it is possible to document sufficient fatigue capacity with a simpler fabrication using single-sided welds. Such types of joints were previously only accepted in smaller diameter tubular joints, such as those used in conductor frames and flare tower structures. The use of single-sided welded tubular joints requires guidance on SCFs to be used for the inside, as well as information on S-N curves and the Design Fatigue Factors to be used. A procedure for fatigue assessment of fatigue cracking from single-side welds in tubular joints has been discussed in some literature, but recommendations have not been properly defined in design standards such as ISO 19902 (2007), although some information is provided in the informative section. Thus, although the existing literature may give the impression that the fatigue capacities of single-sided welded tubular joints are much the same as those of double-sided welded joints, in reality this is a more complex issue than reflected in existing design standards. The reason for this is explained in more detail in the next sections. Due to optimization of design, with possible post-weld improvement of the outside weld toes, the need for a reliable design procedure for the inside welds in single-sided welds in tubular joints is likely to increase in the future.

Qian et al. (2013) performed four fatigue tests on large-scale X-joints, each with different surface treatment at the weld toe. The joints were loaded by cyclic in-plane bending moment. The as-welded joints cracked from the weld toe. One joint was ground at the weld toe and this joint cracked also from the weld root.

8.3.2 Design S-N Curves

The girth welds between the brace and the stubs are classified as F3 according to DNVGL-RP-C203 (2016). It is considered more difficult to find defects in the root of single-sided welded tubular joints than in single-sided girth welds when using ultrasonic testing, due to the curvature of the steel surfaces at these hot spots. Mudge et al. (1996) assessed the detection performance of conventional Ultrasonic Testing for cracks initiation at the weld root of brace to toe chord welds in tubular joints and considered the detectability to be dependent on crack orientation. It is indicated that it might be possible to obtain responses from weld root cracks of 1–2 mm; however, discrimination of these responses from signals from the surrounding geometry may be difficult. Similar root defects in girth welds results in S-N curve F3, according to DNVGL-RP-C203 (2016). At some connections, like single-sided welded X-joints, there is a significant notch effect at the inside due to the weld profile not being as good as, for example, a butt weld. It is therefore optimistic to assume a D-curve in general for the inside, as indicated in HSE OTO 1999 022 (1999) document for such joints. Regarding in-service inspection, fatigue cracks initiating from the inside will be detected at a later stage than those initiating from an outside weld toe. Thus, the time interval from detecting a fatigue crack until it becomes critical for the integrity of the structure is significantly reduced, and therefore it may be prudent to have a larger Design Fatigue Factor for fatigue cracking from the inside than for cracking from the outside. See Table 4.1, showing that the stress concentration embedded in the S-N curve F3 is 1.61. With full weld improvement from grinding or hammer peening, the fatigue life for the outside can be increased by a factor of 3.5 for grinding and by a factor of 4 for hammer peening. An improvement factor of 2 is indicated for grinding in ISO 19902 (2007). If this factor can be allowed to be used for the outside, it is also accepted that the inside stresses increase by a factor of approximately equal $2^{1/3.0} = 1.25$. Thus, an S-N curve for the inside should be used that has an embedded SCF of $1.61 \cdot 1.25 = 2.01$; this corresponds to S-N curve W1 (from Table 4.1). If an improvement factor of 4 for hammer peening for the outside is accepted for use (or an improvement factor of 3.5, as allowed by DNVGL-RP-C203 2016), an increase in the inside stresses by a factor of $4^{1/3.0} = 1.59$ is also accepted. Thus, an S-N curve for the inside should be used that has an embedded SCF equal to $1.61 \cdot 1.59 = 2.55$; this corresponds to S-N curve W3. Thus, recommending use of the W3 curve for single-sided welds in tubular joints accounts for the possibility of improving the weld for the outside joint. However, it might still be debated whether this curve properly accounts for inferior non-destructive testing (NDT) in the root and the likelihood that in-service cracks will be detected at a later stage than those initiating from the outside, as discussed earlier in this section. The need for a larger Design Fatigue Factor should therefore be considered. There is also less certainty in fabrication of single-sided welded joints, as visual inspection cannot be used for quality control. With the exception of X-joints, fatigue cracking from the weld roots in single-sided welded joints has been infrequently reported, and this might indicate that strict design procedures for such joints are unnecessary. However, the reason for this lack of fatigue cracking from the weld roots in single-sided welded joints may be explained by the fact that strength considerations generally prevail over fatigue in structures that are located in milder environmental zones. In this respect, the possibility should be borne in mind of performing weld improvements on the outside to achieve acceptable calculated fatigue lives at the design stage or for life extension of structures. Relevant in-service experience on this has not yet been gained. Based on this, the W3 curve is generally recommended as a basic curve for single-sided welds in tubular joints. However, if the designer and the owner agree that it is acceptable to waive the possibility of weld improvement, then a higher S-N curve may be selected, like that of F3 for single-sided welds in tubular joints.

8.3.3 Design Fatigue Factor

As indicated in Section 8.3.2, there are several reasons why the use of a larger Design Fatigue Factor for the inside than for the outside may be considered. This becomes more important if weld improvement for the outside is used, such that the inside becomes more critical than that for which experience is currently available. Alternatively, in-service inspection intervals for fatigue cracks may be reduced for these joints.

8.3.4 SCFs for Inside Hot Spots

Validity Range for the Equations

SCFs for inside tubular joints have been published by Lee (1999) and HSE OTO 1999 022 (1999). The procedure for derivation of inside hot spot stress is based on calculation of a reduction factor, R, that is multiplied by the outside stress derived from Efthymiou's parametric equations. It should be noted that the validity range from Lee (1999) does not include a relevant range of the gamma term ($\gamma = D/2T$, where D = diameter of the chord and T = thickness of the chord) for X-joints and Y-joints. The HSE report OTO 1999 022 gives equations with valid gamma values for X-joints and Y-joints down to 10. A comparison of this term is shown in Figure 8.10 for X-joints and in Figure 8.11 for Y-joints. The differences between Lee (1999) and the OTO 1999 022 report are not very large for in-plane bending and out-of-plane bending for these joints, and therefore the validity range for the gamma term can be altered from 21 to 10 without further modification of the SCF equations. There are larger differences between Lee (1999) and the OTO report for axial forces for both X-joints and Y-joints, and it is therefore proposed that a term in the equation is used that is made linear-dependent on gamma. This modification is shown by the

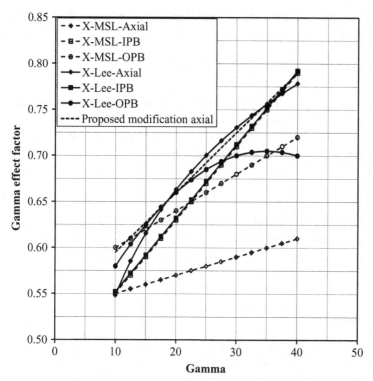

Figure 8.10. Comparison of gamma effect factor for X-joints.

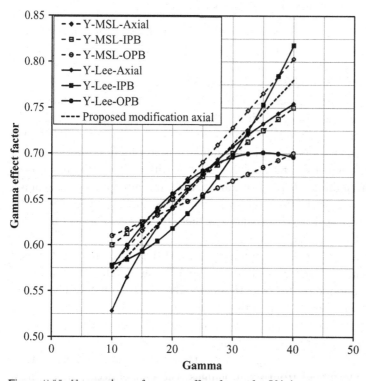

Figure 8.11. Comparison of gamma effect factor for Y-joints.

dotted line in Figure 8.10 for X-joints; a similar modification is proposed for Y-joints as indicated in Figure 8.11.

Proposed Modification of SCF Equations in Order to Broaden the Validity Range

Based on the considerations presented in this section, it is proposed that the SCF equations for single-sided tubular joints are modified to that shown in DNVGL-RP-C203 (2016). It could also be noted that the maximum reduction factor, R, has been set to 0.85, based on a finite element analysis of an X-joint.

8.4 Overlap Joints

Overlap joints are understood to be joints in which braces overlap at the connection to the chord. This means that the gap in Figure 8.1 becomes negative. Efthymiou's equations for SCF can be used for $g \geq -0.6d/\sin\theta$, where d = diameter of the brace and θ = brace angle, as shown in Figure 8.1.

8.5 Stiffened Tubular Joints

The effects of stiffeners on hot spot stress in large-diameter tubular joints have been reported by Callan et al. (1981). In these studies, SCFs were derived using finite element analysis and strain-gauged acrylic models for studying the effects of:

- chord ring stiffener sizes and location of peak stresses;
- longitudinal chord stiffeners on peak stresses.

In Callan et al.'s (1981) study it was observed that light and closely spaced stiffeners reduce the SCF at the saddle position but have little effect on the SCF at the brace toe and heel. In contrast, longitudinal stiffeners significantly reduced the toe SCF. Recommendations on stiffeners generally depend on the type of joint, the position of the considered hot spot, and the type of loading (axial, in-plane, and/or out-of-plane bending moments). Internal ring stiffeners are considered efficient when the largest stresses are at the saddle region. If the hot spot is closer to the crown toe position, as is normally seen in typical K-joints, it is considered efficient to add some longitudinal stiffeners that are welded to the ring stiffeners in order to increase the effective section modulus of the chord and reduce the hot spot stress. When a finite element analysis is performed, the principal stress direction at the hot spot region can be studied in order to assess how further stiffening can most efficiently reduce the stresses to an acceptable value.

Smedley and Fischer (1991b) presented SCFs for stiffened tubular joints based on strain measurements in acrylic models. A total of 328 ring-stiffened T-, Y-, X-, and K-joints were tested, and the results from these models corresponded well with data from steel joints and with results from finite element analyses.

Slater and Tubby (1996) reported fatigue testing of ten large-scale tubular joints, each with a different configuration of internal ring stiffeners. The specimens were tested either under axial brace loading or unbalanced out-of-plane bending. Stress distributions were determined from strain gauges. The conventional linear stress extrapolation procedure for determination of hot spot stress in unstiffened tubular joints was found to be applicable to the range of stiffened tubular joints investigated. The hot spot stresses derived from measurements were compared with that derived from the Smedley and Fischer's (1991b) parametric equations. The design equations were found to be always conservative. The fatigue test data were compared with the T-curve from Department of Energy (1984), and it showed a penalty of around 2.0 on life compared with unstiffened joints is needed to give a similar level of confidence for a through-thickness crack. A faster fatigue crack was observed in ring-stiffened joints than in simple tubular joints. This should be considered when planning in-service inspection for fatigue cracks.

Stiffened tubular joints with low stress concentrations tend to show a more uniform stress distribution around the brace/chord intersection than simple (unstiffened) tubular joints. This may lead to multiple crack initiation in the brace or the chord and subsequent crack coalescence, leading to a long crack when it has grown through the thickness (Nichols and Slater 1996). This means that with a through-thickness crack as a failure criterion, there is less static capacity left in a stiffened tubular joint than in simple joints. This also means that the inspection interval for stiffened joints should be shorter than for simple tubular joints with the same calculated fatigue life.

This indicates that it may be considered acceptable to use the parametric equations for fatigue assessment of stiffened joints together with the S-N curve established for simple tubular joints. However, the potential conservatism in the parametric equations is removed if direct finite element analysis is performed and the structural behavior of these joints is more similar to that of other types of welded joints where the hot spot S-N curve D applies. Thus, it may be recommended to apply the D-curve in fatigue design based on finite element analysis of stiffened tubular joints and the hot spot stress methodology presented in Section 9.1.

8.6 Grout-Reinforced Joints

8.6.1 General

In grouted joints the chord is either completely filled with grout (single-skin grouted joints), or the annulus between the chord and an inner member is filled with grout (double-skin grouted joints). The SCF of a grouted joint depends on load history and loading direction. The SCF is lower if the bond between the chord and the grout is unbroken. For testing of grouted joints the bond should be broken prior to SCF measurements. The tensile and compressive SCF may differ due to the bond. In order to achieve a fatigue design that is on the safe side, use of SCFs derived from tests where the bonds are broken and the joint is subjected to tensile loading is recommended. The bonds can be broken by a significant tension load, which may be determined during testing by an evaluation of the force displacement relationship (when incrementing the loading into a nonlinear behavior).

8.6.2 Chord Filled with Grout

Grouted joints should be treated as simple joints, except that the chord thickness in the γ-term for saddle SCF calculations for brace and chord should be substituted with an equivalent chord wall thickness given by:

$$T_e = (5D + 134T)/144 \qquad (8.20)$$

where D and T are chord diameter and thickness, respectively. The dimensions are in mm.

Grout has little effect on joints with high β-ratios or low γ-ratios, and the benefits of grouting should be neglected for joints with $\beta > 0.9$ or $\gamma < 12.0$, unless documented otherwise. This equation is used in ISO 19902 (2007) and DNVGL-RP-C203 (2016).

> Example: A similar geometry as analyzed in the example in Table 8.1 is considered, with D = 800 mm, d = 480 mm, T = 25.2 mm. This gives T_e = 51.2 mm and the SCF for the Y-joint is reduced to 3.02 on the chord side and to 3.41 on the brace side.

8.6.3 Annulus between Tubular Members Filled with Grout

For joints where the annuli between tubular members are filled with grout, such as in joints in legs with insert piles, the grouted joints should be treated as simple joints, with the exception that the chord thickness in the SCF calculation for brace and chord should be substituted with an equivalent chord wall thickness given by:

$$T_e = T + 0.45T_p \qquad (8.21)$$

where:

T = chord thickness
T_p = thickness of insert pile.

This equation is based on a few in-house laboratory test data.

8.7 Cast Nodes

Cast nodes are normally analyzed by finite elements in order to obtain a hot spot stress that is acceptably low, with an optimal geometry in terms of weight. Use of three-dimensional elements is required. The maximum surface stresses can be determined directly from such models and used together with a relevant S-N curve for such connections. See also Section 4.7.5.

8.8 Joints with Gusset Plates

Insert gusset plates are sometimes used in joints in topside structures to connect rectangular hollow sections (RHS) or tubular members to main girders; see Figures 8.12 and 9.17. Not many SCFs for gusset plates have been presented in design standards or in literature, probably because it is difficult to define geometries that are suitable under all conditions. Thus, in order to achieve reliable fatigue assessments, use of finite element analysis of gusset plate connections is recommended. When joints with gusset plates are subjected to dynamic loading, a full penetration weld between the member and the gusset plate is preferable. Without a full penetration weld, it is difficult to document the fatigue capacity for fatigue cracking starting from the weld

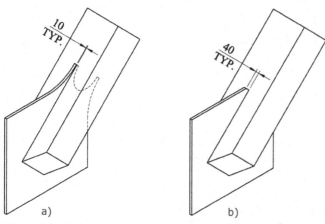

Figure 8.12. Joints with gusset plates. (a) Favorable geometry. (b) Simple geometry.

root. In dynamically loaded structures, shaping the gusset plate is also recommended in order to achieve a smooth stress flow from the member into the gusset plate, as indicated in Figure 8.12a.

Gusset plates in semi-submersible platforms may be placed on the inside of the brace, or also on the outside in special cases such as for repair and strengthening. The effect of different design solutions was investigated by Eide and Berge (1990) by fatigue testing of simplified models of stiffened brace/column connections. The local strains were measured for derivation of hot spot stress, and based on this the fatigue lives were found to be in good agreement with the UK Department of Energy T-curve from 1984 (which is comparable with the D-curve). The effect of adding on external gussets reduced the local bending stress over the plate thickness and increased the fatigue life by a factor of 1.3 as compared with internal gusset plates only. However, it should be noted that the relative number of cycles from a through-thickness crack in the brace to a large crack was significantly reduced (from a factor of 1.5 to 1.06). This should be kept in mind when planning in-service inspection for fatigue cracks.

Some guidance on S-N curves for slotted tubular connections is presented by Baptista et al. (2015). Where a reliable fatigue life must be documented, finite element analysis is recommended; see Section 9.3.

8.9 Rectangular Hollow Sections

Rectangular Hollow Sections (RHS) are typically used in deck structures and module support frames in different types of offshore structures. SCF for RHS joints can be found in ISO 14347 (2008). See van Wingerde et al. (1996, 1997a), with the advice that the T-curve can also be used for joints made with RHS. See also Zhao and Packer (2000).

8.10 Fillet-Welded Bearing Supports

Where support plating below bearings is designed with fillet-welded connections (see Figure 8.13), it should be verified that fatigue cracking of the weld will not occur.

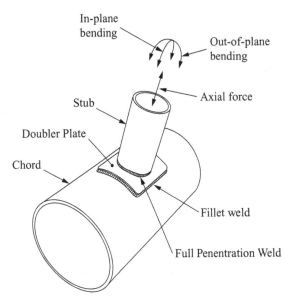

Figure 8.13. Fillet-welded plate on tubular member.

Although the joint may be required to carry wholly compressive stresses and the plate surfaces may be machined to fit, for fatigue assessment the total stress fluctuation should be considered to be transmitted through the welds. If it is assumed that compressive loading is transferred through contact, it should be verified that the contact will not be lost during welding. The actual installation conditions, including maximum construction tolerances, should be accounted for.

8.11 Cutouts and Pipe Penetrations in Plated Structures

There are hundreds of welded pipe penetrations in the plated structures of ships and floating production platforms. These details are important with respect to reliable fatigue design. The hydrophone holder welded into one of the important structural members of *Alexander L. Kielland* is an example of a welded pipe penetration in which fatigue failure became catastrophic for the integrity of the platform. The fatigue design procedure for such details in most design standards is rather imprecise, and therefore during the 1990s, DNV undertook work to improve the basis for a more reliable fatigue design procedure for welded penetrations in plated structures. This procedure accounted for one main stress direction, and was included in DNV CN 30.7 (2001) "Fatigue Assessment of Ship Structures" and DNV-RP-C203 (2001) "Fatigue Strength Analysis of Offshore Structures." Fatigue cracking from the weld toe and fatigue cracking from the root of fillet welds were considered. In recent years, further work has been performed to simplify the procedure for use together with calculated stresses from finite element analysis of two-dimensional stress states in plated structures, where the penetrations have not been part of the analysis model, see also Section 9.4. The procedure is based on calculated SCFs using finite element analyses for relevant geometries of penetration. SCFs were derived by certain requirements to finite element mesh and derivation of hot spot stress (see Lotsberg et al. 1997). The accuracy of this procedure was further investigated in the FPSO Fatigue Capacity Joint Industry Project (JIP) (Fricke 2001). This methodology was also used in the same JIP to derive a design procedure for fatigue design of manholes with different

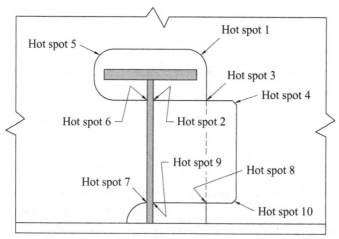

Figure 8.14. Hot spots at an intersection between a longitudinal stiffener and a transverse frame in a tanker.

reinforcements (Chen and Landet 2001). The numbers of cycles until failure, using the fatigue design procedure described, have been compared with fatigue test data; the calculated data and fatigue test data have been found to correspond well (Lotsberg 2004a), see also Section 5.5.

8.12 Details in Ship Structures

8.12.1 Lugs at Side Longitudinals

Damage statistics from tankers showed that 80% of the cracks that had to be repaired in the 1970s occurred at the connections between the longitudinal stiffeners and the transverse girders (see Haslum et al. 1976). Therefore, finite element analyses of these intersections were made to determine SCFs for alternative designs, together with practical design recommendations. These connections have also been considered susceptible to fatigue cracking more recently (see, e.g., Lotsberg et al. 1998; Ulleland et al. 2001; Lindemark et al. 2009). Depending on the geometry, there may be several hot spots that should be considered with respect to fatigue, as indicated in Figure 8.14. In-service experience shows that it is important to achieve designs with sufficient radius in the transverse plates to avoid fatigue cracking from the edge of the base material. Based on this experience, the geometry of intersections between longitudinals and transverse frames shown in Figures 2.22 and 2.23 was proposed and fatigue tested in the FPSO Fatigue Capacity JIP presented in Section 2.4. The lug or collar plates are placed eccentrically on the transverse plate and connected with fillet welds all around (see also Figure 8.15). This results in hot spot no. 3 in Figure 8.14 being susceptible to fatigue cracking, and requires particular consideration during design.

8.12.2 Asymmetric Sections Subjected to Dynamic Sideway Loading

The background for this section is from Holtsmark (1993). Refined versions of this procedure are also included in the DNV CN 30.7 (2014). The shear center in

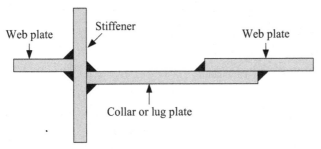

Figure 8.15. Eccentricity at collar or lug plates.

asymmetrical sections is outside the web centerline, which implies a skew bending when these sections are subjected to lateral pressure. The nominal bending stress at the top of a flange of a stiffener subjected to lateral pressure load p is derived as:

$$\sigma_{nominal} = \frac{psl^2}{12Z} \tag{8.22}$$

where

p = lateral pressure

s = plate width between stiffeners

l = effective length between supports (or transverse frames). Use of brackets at ends of stiffeners gives an effective length that is shorter than the distance between the transverse frames.

Z = elastic section modulus at the flange of the panel stiffener including plate flange with respect to neutral axis normal to the stiffener web.

The SCFs at the flange of asymmetrical stiffeners on laterally loaded panels as defined in Figure 8.16 are calculated as follows:

The following stress concentration factor can be applied at the flange edge

$$SCF_{n1} = \frac{1 + \gamma \beta}{1 + \gamma \beta^2 \psi} \tag{8.23}$$

and the following stress concentration factor can be used at the mid-thickness of the web

$$SCF_{n2} = \frac{1 + \gamma \beta^2}{1 + \gamma \beta^2 \psi} \tag{8.24}$$

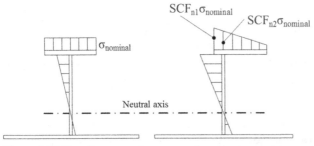

Figure 8.16. Bending stress in symmetric and asymmetric panel stiffeners with the same web and flange areas.

Table 8.3. *Characteristic flange data for HP bulb profile*

HP-bulb		Geometry of equivalent built-up flange		
Height (mm)	Web thickness t_w (mm)	b_f (mm)	t_f (mm)	b_g (mm)
200	9–13	$t_w + 24.5$	22.9	$(t_w + 0.9)/2$
220	9–13	$t_w + 27.6$	25.4	$(t_w + 1.0)/2$
240	10–14	$t_w + 30.3$	28.0	$(t_w + 1.1)/2$
260	10–14	$t_w + 33.0$	30.6	$(t_w + 1.3)/2$
280	10–14	$t_w + 35.4$	33.3	$(t_w + 1.4)/2$
300	11–16	$t_w + 38.4$	35.9	$(t_w + 1.5)/2$
320	11–16	$t_w + 41.0$	38.5	$(t_w + 1.6)/2$
340	12–17	$t_w + 43.3$	41.3	$(t_w + 1.7)/2$
370	13–19	$t_w + 47.5$	45.2	$(t_w + 1.9)/2$
400	14–19	$t_w + 51.7$	49.1	$(t_w + 2.1)/2$
430	15–21	$t_w + 55.8$	53.1	$(t_w + 2.3)/2$

where

$$\beta = 1 - \frac{2b_g}{b_f} \quad \text{for built-up sections}$$

$$\beta = 1 - \frac{t_w}{b_f} \quad \text{for rolled angles} \tag{8.25}$$

ψ = ratio between elastic section modulus of the stiffener web with plate flange, as calculated at the flange and the elastic section modulus of the complete panel stiffener = Z_w/Z.

$$\psi = \frac{(h - t_f)^2 t_f}{4Z} \tag{8.26}$$

Z_w = section modulus of the stiffener web with respect to the top surface of the stiffener flange

b_f = breadth of stiffener flange; for bulb profiles, see Table 8.3

t_f = flange thickness; for bulb profiles, see Table 8.3

b_g = distance from the nearest edge of the flange to the mid-thickness plane of the web (see Figure 8.17); for rolled profiles, see Table 8.3

h = stiffener height

t_w = stiffener web thickness

s = plate width between stiffeners

t_p = plate thickness

At the ends of continuous stiffeners:

$$\gamma = \frac{3\left(1 + \dfrac{\mu}{280}\right)}{\left(1 + \dfrac{\mu}{40}\right)} \tag{8.27}$$

where:

$$\mu = \frac{l^4}{b_f^3 t_f h^2 \left(\dfrac{4h}{t_w^3} + \dfrac{s}{t_p^3}\right)} \tag{8.28}$$

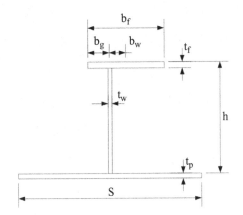

Figure 8.17. Asymmetrical profile dimensions.

The formulation is not directly applicable for bulb sections, for which the equivalent built-up section in Table 8.3 can be used.

8.12.3 Example of Calculated SCFs for an Asymmetric Section

An example with a Hollande Profile (HP) longitudinal with dimensions $340 \cdot 14$ is shown. The distance between transverse frames is $l = 3200$ mm, the distance between longitudinals s $= 800$ mm, and the plate thickness $t_p = 16$ mm. From Table 8.3 it can be seen that $b_f = 57.3$ mm, $t_f = 41.3$ mm, and $b_g = 7.85$ mm. The distance from the outside of the plate to the neutral axis is derived as 82.05 mm. The moment of inertia about the neutral axis, including the longitudinal and the plate, is $I = 2.827 \cdot 10^8$ mm^4. The section modulus at top of the flange $Z = 1.032 \cdot 10^6$ mm^3. This gives $\psi = 0.303$ from equation (8.26), $\mu = 168.962$ from equation (8.28), $\gamma = 0.921$ from equation (8.27), and $\beta = 0.726$ from equation (8.25) for built-up sections.

This gives SCF$_{n1}$ = 1.388 and SCF$_{n2}$ = 1.295 from equations (8.23) and (8.24), respectively. This is a typical stress concentration due to skew loading on a bulb section. The SCFs for angles due to skew loading are larger, being up to approximately 2.0.

9 Finite Element Analysis

9.1 Welded Connections in Plated Structures

9.1.1 General

Assessment of high cycle fatigue in marine structures is based on linear elastic structural finite element analysis (FEA). Assessment of low cycle fatigue can be based on nonlinear FEA; see also Section 3.1.3. Some basic knowledge about finite elements is recommended in order to prepare finite element models that are representative of the physical behavior of different structural details. Before starting to model a detail, it is necessary to have a clear view of what the outcome of the analysis should be and of how the analysis results should be used, together with S-N data, for assessment of calculated fatigue life. In principle, three different types of finite element models can be prepared for fatigue assessment:

1. Model for calculation of membrane stresses, to be used together with S-N curves for nominal stress for calculation of fatigue damage.
2. Model for calculation of structural stress or hot spot that represents the stress due to the considered geometric detail, which is entered into a hot spot stress S-N curve for calculation of fatigue damage.
3. Model that accounts for the considered geometric detail, including the weld toe, where the calculated notch stress is entered into a notch S-N curve for calculation of fatigue damage.

A rather coarse finite element model may be appropriate for analysis of membrane stresses in plated structures. The elements used should represent a linear membrane stress distribution within each element. This is achieved by using eight-node isoparametric shell elements and four-node shell elements with internal degrees of freedom. The ability to represent bending stress over the thickness is less important for calculation of membrane stresses. This means that if only the membrane stress is to be derived from the analysis, this model cannot be used for fatigue assessment of plated structures that are subjected to significant dynamic lateral pressure. This limitation is also considered a drawback to using the nominal stress approach for such loading conditions and is one reason why structural stress methods were introduced into fatigue design of marine structures during the early 1990s. Structural stress methods are also called hot spot stress methods, and the local geometry of a detail is accounted for in the stress calculation, in addition to plate bending due to lateral pressure. The finite element model here needs to represent linear stress behavior through the plate thickness. This can be achieved by using shell elements or one 20-node isoparametric element over the plate thickness. With a fine mesh model at a weld notch, the analysis starts to pick up a singularity at the weld toe if the weld is included in the finite element model. The nonlinear part of the stress distribution due to the weld notch is considered to disappear some distance away from the weld toe. The extent of this zone is less than 0.4t, where t is plate thickness. Therefore, the IIW (Hobbacher 2009) uses 0.4t as the closest extrapolation point for derivation of hot spot stress at weld toes based on measurements. When a stress distribution through the thickness of a plate, $\sigma(x)$, has been calculated (see also Figure 3.9), the stress components can be separated as described in the following. The membrane stress is derived as:

$$\sigma_m = \frac{1}{t} \int_{x=0}^{x=t} \sigma(x)\, dx \tag{9.1}$$

where x is measured from bottom of plate. The bending stress at the plate surface is derived as:

$$\sigma_b = \frac{6}{t^2} \int_{x=0}^{x=t} \left(\sigma(x) - \sigma_m \right) \left(\frac{t}{2} - x \right) dx \tag{9.2}$$

The nonlinear stress at position x from the bottom of the plate is derived as:

$$\sigma_{nl}(x) = \sigma(x) - \sigma_m - \left(1 - \frac{2x}{t} \right) \sigma_b \tag{9.3}$$

These equations may be used to estimate the relevant stresses to be entered into the different S-N curves in the three different fatigue analysis approaches listed above. The membrane stress is needed as input to the S-N analysis in the first approach. The membrane stress plus the linearized bending stress to the surface is needed as input to the S-N analysis in the second approach. In the third approach, the nonlinear stress is added to the membrane and bending stress, and the surface stress (notch stress) is used as input to the S-N analysis.

A number of different methods have been proposed using the finite element method for derivation of hot spot stress or structural stress in plated structures. The different methodologies are based on calibration with laboratory fatigue test data, as explained in Sections 2.2–2.7. This also includes calculation of hot spot stress from FEA results and links to S-N data. For methodologies based on simple surface stress calculation or surface stress calculation with stress extrapolation to the hot spot, see Fricke (2001), Lotsberg (2004b), and Storsul et al. (2004a). Alternatively, the structural stress can be calculated from the stress distribution through the plate thickness at the hot spot. This through-thickness methodology is used in the Battelle approach (Dong 2001). The stress distribution through the plate thickness can be derived directly from the nodal forces calculated by the finite element programs, and some equations for this are presented in BS 7608 (2014). Provided that the recommended finite element modeling at the hot spot areas is followed, the two methods provide similar calculated structural stresses (see, e.g., Poutiainen et al. 2004).

Similar comparisons of surface stress methods and the through-thickness method have been presented by Doerk et al. (2003) for the following details: plate lap fillet weld, one-sided doubling plate, bracket toe ending on a beam, and a fillet weld around a plated edge (Specimen 5 in Section 2.3.2). For simple two-dimensional geometries the through-thickness method may be claimed to be mesh-insensitive, but not so for three-dimensional problems. It was found that the results using the through-thickness method for the bracket toe were highly mesh-dependent. The surface extrapolation method also has problems in such cases, but they seem to be less severe according to Doerk et al. (2003).

Fricke and Kahl (2005) presented a comparison of three different hot spot methods for three different details. This included use of surface stress extrapolation methods as presented by IIW (Niemi et al. 2006), through-thickness method (Dong 2001), and the method proposed by Xiao and Yamada (2004) in which the hot spot stress is derived as the calculated stress in the main stress direction at a depth 1 mm below the weld toe. This method was found to show small scatter; it is considered to account directly for the thickness effect; however, a rather fine mesh model is required. The fatigue lives were predicted using the design S-N curves recommended within the different approaches and compared with the results of fatigue tests evaluated for a corresponding probability of survival. The following main conclusions were drawn based on the derived results: uncertainties in the computed structural stresses and predicted fatigue lives are mainly due to the element properties and sizes, to the stress evaluation, and in particular to the weld modeling in shell element models. In spite of the different structural stress definitions, the fatigue lives predicted with the three approaches are not too distant from each other.

The structural stress methodology has also been assessed by a number of other researchers, such as Aygül et al. (2012) on orthotropic bridge deck details and Aygül et al. (2013) on welded bridge details.

The structural hot spot stress methods are in general considered to be practical for fatigue analysis of welded structures when using finite element analysis for assessment. There are still questions that remain open when applying the methods to complex details such as at welds on thick members, such as bulb profiles, where it is difficult to select a thickness for stress extrapolation point along the surface or for the depth for through-thickness stress linearization. Furthermore, the fatigue life predictions may be strongly affected by residual stresses or large variations in local weld profile, which should be kept in mind when assessing the reliability of the different methods. In addition, the practicality is very important for industrial application.

9.1.2 Finite Element Modeling for Structural Stress Analysis

From detailed FEA of structures it may be difficult to assess the stresses that can be defined as "nominal stress" to be used together with the selected nominal stress S-N curves, as some of the calculated stress may be part of the stress that is accounted for in the nominal stress S-N curve for the considered detail. Therefore, in many cases it is more convenient to use an alternative approach for calculation of fatigue damage, when local stresses at a detail are obtained from FEA. There is normally significant scatter in weld notch geometry, and therefore it can be difficult to calculate the notch stress at a weld based on the local weld geometry. This scatter is normally accounted for more efficiently in an appropriate S-N curve. This leads to a compromise with respect to finite element modeling; the mesh must be fine enough to represent the stress concentration due to the analyzed detail, but it should not be so fine that it starts to pick up stress due to the weld notch. This is achieved by using the second

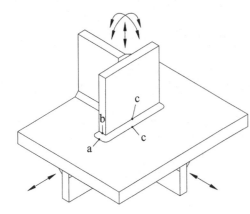

Figure 9.1. Different hot spot positions.

finite element model listed in Section 9.1.1, where a linear stress distribution over the plate thickness is calculated for derivation of the structural stress. This procedure is also denoted the hot spot method. The following guidance is made for computation of hot spot stresses with potential fatigue cracking from the weld toe, with local models using the finite element method. Hot spot stresses are calculated assuming linear material behavior and using an idealized structural model with no fabrication-related misalignments. The extent of the local model has to be chosen such that effects due to the boundaries and load application on the structural detail being considered are sufficiently small, and such that reasonable boundary conditions can be formulated. It is important to have a continuous, not too steep change in the density of the element mesh in the areas where the hot spot stresses are to be analyzed. The geometry of the elements should be evaluated carefully in order to avoid errors due to deformed elements. For example, corner angles between 60° and 120° and a length-to-breadth ratio of the elements below 5 are recommended. In plated structures, three types of hot spots at weld toes can be identified, as exemplified in Figure 9.1:

a) At the weld toe on the plate surface at an ending attachment.
b) At the weld toe around the plate edge of an ending attachment.
c) Along the weld of an attached plate (weld toes on both the plate and attachment surface).

Models with thin plate or shell elements, or, alternatively, with solid elements, are normally used for analysis. It should be noted that while on the one hand the arrangement and type of elements have to allow for steep stress gradients as well as for the formation of plate bending, on the other hand only the linear stress distribution in the plate thickness direction needs to be evaluated with respect to definition of the structural stress. The following methods of modeling are recommended. The simplest way of modeling is offered by thin plate and shell elements, which have to be arranged in the mid-plane of the structural components; see also Figure 9.2a. Use of eight-noded elements is particularly recommended in cases of steep stress gradients. Care should be given to possible underestimation of stress, especially at weld toes of type (b) in Figure 9.2a. Use of four-noded elements with improved in-plane bending modes is a good alternative. The welds are usually not modeled, except in special cases where the results are affected by high local bending, for example, due

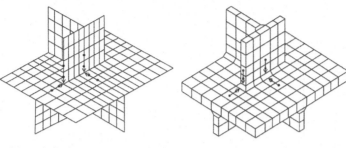

a) In a shell element model to the   b) In a model with three-dimensional
intersection line   elements to the weld toe

Figure 9.2. Stress extrapolation.

to an offset between plates or due to a small free plate length between adjacent welds, such as at lug (or collar) plates. Here, the weld may be included by transverse plate elements with appropriate stiffness or by introducing constrained equations for coupled node displacements. A thickness equal to twice the thickness of the plates may be used for modeling the welds by transverse plates. For four-node shell elements, with additional internal degrees of freedom for improved in-plane behavior, and for eight-node shell elements, a mesh size from $t \times t$ up to $2t \times 2t$ may be used. Larger mesh sizes at the hot spot region may provide non-conservative results. A mesh of $t \times t$ is generally preferable at the hot spot region for efficient readout of element stresses and hot spot stress derivation. An alternative modeling, particularly for complex cases, is offered by solid elements that need to have a displacement function, allowing steep stress gradients as well as plate bending with linear stress distribution in the plate thickness direction. This is offered by isoparametric 20-node elements (with mid-side nodes at the edges), which mean that only one element in plate thickness direction is required. It is then easy to evaluate the membrane and bending stress components if a reduced integration order, with only two integration points in the thickness direction, is chosen. A finer mesh subdivision is necessary, particularly if eight-noded solid elements are selected. Here, at least four elements are recommended in the thickness direction. Modeling the welds is generally recommended and easily achievable, as shown in Figure 9.2b. However, the analysis results may also be acceptable without including the weld in the model. For modeling with three-dimensional elements, the dimensions of the first two or three elements in front of the weld toe should be chosen as described in the following. The element length may be selected to correspond to the plate thickness. In the transverse direction, the plate thickness may be chosen again for the breadth of the plate elements. However, the breadth should not exceed the "attachment width," that is, the thickness of the attached plate plus $2 \times$ the weld leg length (in case of hot spot type c in Figure 9.1: the thickness of the web plate behind plus $2 \times$ weld leg length as shown in Figure 9.2b). The length of the elements should be limited to $2t$. In order to capture the properties of bulb sections with respect to St. Venant torsion, use of several three-dimensional elements for modeling of a bulb section is recommended. If, in addition, the weld from stiffeners in the transverse frames is modeled, the requirements with respect to element shape are likely to govern the finite element model at the hot spot region. Reference is made to local geometries of connections between longitudinal

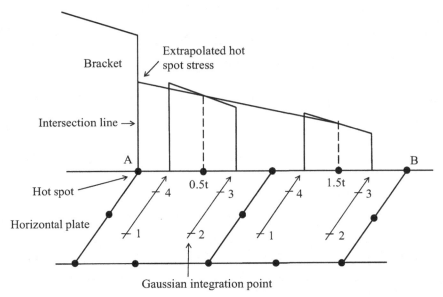

Figure 9.3. Example of derivation of hot spot stress.

sections and transverse frames presented in Section 2.4. If there is uncertainty about properties of finite elements to be used at a hot spot, assessment of the computer program used for analysis against calculated stress concentration factors (SCFs) presented for six different geometries in appendix D of DNVGL-RP-C203 (2016) is recommended.

9.1.3 Derivation of Hot Spot Stress from Finite Element Analysis

The stress components on the plate surface should be evaluated along the paths shown in Figure 9.2 and extrapolated to the hot spot. The average stress components between adjacent elements are used for the extrapolation. Recommended stress evaluation points are located at distances 0.5t and 1.5t away from the hot spot, where t is the plate thickness at the weld toe. These locations are also denoted as stress readout points. If the element size at a hot spot region of size t × t is used, then the stresses may be evaluated as follows.

In cases of plate or shell elements, the surface stress may be evaluated at the corresponding mid-side points. Thus, the stresses at the mid-side nodes along the line A–B in Figure 9.3 may be used directly as stress at readout points 0.5t and 1.5t. In cases of solid elements, the stress may first be extrapolated from the Gaussian points to the surface. These stresses can then be interpolated linearly to the surface center or extrapolated to the edge of the elements, if this is the line for hot spot stress derivation.

For meshes with four-node shell elements larger than t × t, fitting a second-order polynomial to the element stresses in the three first elements and deriving stresses for extrapolation from the 0.5t and 1.5t points is recommended. An example of this is shown schematically in Figure 9.4. This procedure may be used to establish stress values at the 0.5t and 1.5t points.

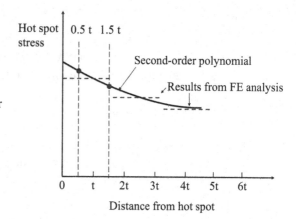

Figure 9.4. Derivation of hot spot stress for element sizes larger than t × t.

For eight-node elements, a second-order polynomial may be fitted to the stress results at the mid-side nodes of the three first elements, and the stresses can be derived at the readout points 0.5t and 1.5t.

The surface stress methodology in FEA follows similar principles as for structural stress derivation from measured data, in terms of the number and position of strain gauges, and for extrapolation of measured stresses to derive the structural stress (see Hobbacher 2009). Hobbacher (2009) also presents different types of meshing, defined as coarse mesh and fine mesh, where each type is linked to a procedure for derivation of structural stress. A different procedure is also presented for location b in Figure 9.1, where the stress distribution does not depend on the plate thickness. Therefore, the reference points are given as absolute distances from the weld toe. One method is to use a mesh with higher-order finite elements with a length of 10 mm. The hot spot stress is then derived from the stresses at the mid-side nodes of the first two elements by linear extrapolation to the intersection line (see Figure 9.3), or to the weld toe if the weld is included in the finite element model.

Two alternative methods, as described in the following sections, can be used for hot spot stress derivation in DNVGL-RP-C203 (2016).

Method A

For modeling with shell elements without any weld included in the model, linear extrapolation of the stresses to the intersection line from the readout points at 0.5t and 1.5t from the intersection line can be performed to derive hot spot stress (see Figure 9.3). For modeling with three-dimensional elements with the weld included in the model, linear extrapolation of the stresses to the weld toe from the readout points at 0.5t and 1.5t from the weld toe can be performed for derivation of hot spot stress. The effective hot spot stress range to be used, together with the hot spot S-N curve, is derived as (see also Section 2.7):

$$\Delta\sigma_{Eff} = \max \begin{cases} \sqrt{\Delta\sigma_\perp^2 + 0.81\,\Delta\tau_{//}^2} \\ \alpha\,|\Delta\sigma_1| \\ \alpha\,|\Delta\sigma_2| \end{cases} \qquad (9.4)$$

where:

> $\alpha = 0.90$ if the detail is classified as C2, with stress parallel to the weld at the hot spot; see table A-3 in DNVGL-RP-C203 (2016) for S-N classification
>
> $\alpha = 0.80$ if the detail is classified as C1, with stress parallel to the weld at the hot spot
>
> $\alpha = 0.72$ if the detail is classified as C, with stress parallel to the weld at the hot spot
>
> $\Delta\sigma_\perp$ = stress range normal to the weld toe
>
> $\Delta\sigma_{//}$ = stress range parallel with the weld toe
>
> $\Delta\tau_{//}$ = shear stress range parallel with the weld toe.

The principal stresses are calculated as:

$$
\begin{aligned}
\sigma_1 &= \frac{\sigma_\perp + \sigma_{//}}{2} + \frac{1}{2}\sqrt{\left(\sigma_\perp - \sigma_{//}\right)^2 + 4\,\tau_{//}^2} \\
\sigma_2 &= \frac{\sigma_\perp + \sigma_{//}}{2} - \frac{1}{2}\sqrt{\left(\sigma_\perp - \sigma_{//}\right)^2 + 4\,\tau_{//}^2}
\end{aligned}
\tag{9.5}
$$

The stress ranges in equation (9.5) are derived for each considered stress component as the difference between the maximum and the minimum stress. The first equation for effective stress (equation 9.5) is made to account for the situation with fatigue cracking along a weld toe as shown in Figure 2.62. The second and third equations are made to account for fatigue cracking when the principal stress direction is more parallel with the weld toe, as shown in Figure 2.62.

Method B

For modeling with shell elements without any weld included in the model, the hot spot stress is taken as the stress at the readout point 0.5t away from the intersection line. For modeling with three-dimensional elements with the weld included in the model, the hot spot stress is taken as the stress at the readout point 0.5t away from the weld toe.

The effective hot spot stress is derived as:

$$
\Delta\sigma_{Eff} = \max \begin{cases} 1.12\sqrt{\Delta\sigma_\perp^2 + 0.81\,\Delta\tau_{//}^2} \\ 1.12\,\alpha\,|\Delta\sigma_1| \\ 1.12\,\alpha\,|\Delta\sigma_2| \end{cases}
\tag{9.6}
$$

where the notations are as explained for Method A.

The first equation for effective stress (equation 9.6) is made to account for the situation with fatigue cracking along a weld toe, as shown in Figure 2.62. The second and third equations are made to account for fatigue cracking when the principal stress direction is more parallel with the weld toe, as shown in Figure 2.62.

Convergence of calculated hot spot stresses using different finite elements, mesh sizes, and methods for hot spot stress derivation has been reported by Fricke (2001), Lotsberg (2004b), and Storsul et al. (2004a). Examples of analyses of the hopper corner detail in Figure 2.16 is shown in Figure 9.5 for element width (B) equal the thickness of the main plate and different mesh sizes (L) measured normal to the weld toe. It is observed that different elements and different methods provide different results. The target value for finite element analysis is in the range 1.82–1.96

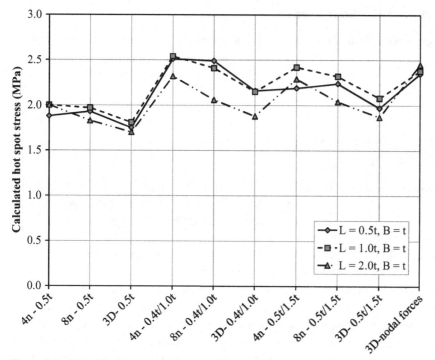

Figure 9.5. Calculated hot spot stress by different types of finite elements, element sizes, and methods for hot spot stress derivation (Storsul et al. 2004a). 4n: Four node shell element with internal degrees of freedom. 8n: 8 – node isoparametric shell element. 3D: 20 node isoparametric element.

according to Table 2.6. The lowest scatter is observed for methods that do not imply stress extrapolation as the readout of stress at 0.5t and stress calculation through the thickness. However, the last method provided a rather large hot spot value for mesh size similar to the largest values derived by extrapolation. Due to varying stress range along the weld toe, a wider element (larger B) would reduce the calculated hot spot stresses. This demonstrates the importance of calibrating the analysis methods to well-known cases before it is fully relied upon for fatigue assessment as also pointed out in Section 2.4.5.

For the purpose of fatigue damage calculation using rainflow counting, equation (9.6) may be written as:

$$\Delta\sigma_{Eff} = \max \begin{cases} 1.12\,\Delta\sigma_{\perp}\sqrt{1 + 0.81\left(\dfrac{\Delta\tau_{//}}{\Delta\sigma_{\perp}}\right)^2} \\ 1.12\,\alpha\,|\Delta\sigma_1| \\ 1.12\,\alpha\,|\Delta\sigma_2| \end{cases} \qquad (9.7)$$

and the $\Delta\sigma_1$ and $\Delta\sigma_2$ may first be calculated without use of absolute sign to achieve correct stress ranges. Furthermore, the rainflow counting need to be performed for each of the equations in (9.7) for calculation of fatigue damage before it can be decided on what of these equations is governing the design.

9.1.4 Effective Hot Spot Stress

At hot spots with significant plate bending, an effective hot spot stress for fatigue assessment can be derived that is based on the following equation:

$$\Delta\sigma_e = \Delta\sigma_a + 0.60\,\Delta\sigma_b \qquad (9.8)$$

where:

> $\Delta\sigma_a$ = membrane stress
> $\Delta\sigma_b$ = bending stress.

Load shedding during crack growth is the reason for this reduction factor on the bending stress, as explained in Section 2.3.5. The effect is limited to areas with a localized stress concentration, which may occur, for example, at a hopper corner in a ship. However, in cases where there is little stress variation along the weld, the difference in fatigue life between axial loading and pure bending is much smaller. Therefore, it should be noted that it is incorrect to reduce the bending part of the stress to 60% as a generalization; it must be restricted to cases with a pronounced stress concentration, where the stress distribution under fatigue crack development is more similar to a displacement-controlled situation than to that of a load-controlled development. Simple tubular joints are one example of where the bending stress cannot be reduced by this factor. A side longitudinal that is subjected to dynamic lateral pressure is another example where the bending stress in the plate cannot be reduced by this factor. This is because the crack-driving stresses along the longitudinals due to the local bending moments from the lateral pressure will not decrease during fatigue crack growth.

9.1.5 Hot Spot S-N Curves

It is generally recommended to link the hot spot stress derived from FEA to the D-curve in DNVGL-RP-C203 (2016) or FAT 90 in Hobbacher (2009). This applies to Method A. It also applies to Method B, when the derived stresses include the additional 1.12 factor in equation (9.6), as compared with equation (9.4). It should be noted that the definition of the stress field through the plate thickness in Section 9.1.3 implies that the hot spot stress methodology is not recommended for simple cruciform joints, simple T-joints in plated structures, or simple butt joints that are welded from one side only.

Calculation of hot spot stress at simple cruciform connections and butt welds needs some further consideration. The background for this is illustrated by considering a cruciform connection, as shown in Figure 9.6a. Only the mid-planes of the plates are modeled using shell elements, as shown in Figure 9.6b, in which the use of eight-node elements is illustrated. When the connection is loaded by a unit membrane load, it can be seen from Figure 9.6c that this stress of 1 MPa is calculated for all the shell elements in the horizontal plate, as the elements in the transverse direction do not represent transverse stiffness that generates local stresses at the crossing plate. Thus, the calculated hot spot stress equals the nominal stress, and a nominal stress S-N curve should be used for this connection. This is because the relevant nominal stress S-N curves for cruciform connections are lower than the general recommended hot

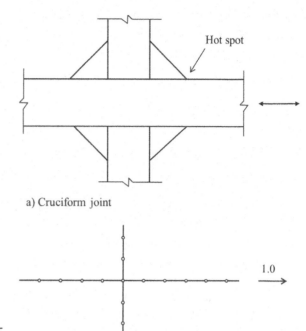

a) Cruciform joint

Figure 9.6 Finite element analysis of a simple cruciform joint.

b) Shell element model

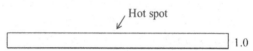

c) Calculated stress in shell element model

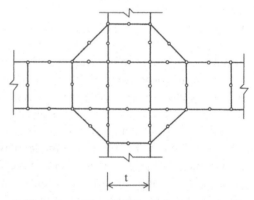

d) Element model with 20-node isoparametric elements

spot S-N curve D, as presented in appendix A of DNVGL-RP-0001 (2015). An alternative to using shell elements is to use one 20-node isoparametric element over the
thickness, as illustrated in Figure 9.6d. This element represents a quadratic displacement function, and thereby a linear stress distribution within each element as the
calculated stresses are derived from the derivative of the displacements. Due to the
symmetry of the connection, the calculated stresses at the top and bottom surfaces
of the connection must be the same. As this solid element represents a linear stress
distribution, the calculated stress will therefore be constant over the plate thickness

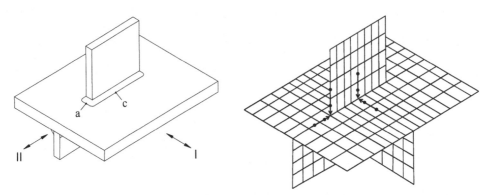

Figure 9.7. Different methods for analysis of normal stresses and in-plane stress in a shell finite element model.

and, based on equilibrium considerations, equal to the nominal stress. Thus, modelling with only one 20-node element over the thickness at a cruciform joint also implies use of a nominal stress S-N curve as a hot spot stress S-N curve. This is illustrated by the shell model in Figure 9.7. For stresses in the direction normal to the shell (direction I), there will be no stress flow into the transverse shell plating, which is represented by only one plane in the shell model. However, it attracts stresses for in-plane (direction II) shown in Figure 9.7. The described hot spot concept using the D-curve gives acceptable results when there is a bracket behind the transverse plate, as shown in Figure 9.2, acting with its stiffness in the direction I in Figure 9.7. As the nominal stress S-N curve for direction I is lower than that of the D-curve, it would be non-conservative to use hot spot stresses extracted from FEA for this connection for direction I at position "c" in Figure 9.7, whereas it would be acceptable for direction II at position "a". Therefore, the nominal stress approach is recommended for use for direction I at position "c" in Figure 9.7, which is typical for cruciform joints and T-joints. The nominal stress can be derived relatively easily from the analysis of these cruciform connections, and is recommended for use with the appropriate detail S-N class for nominal stress analysis, as given in appendix A in DNVGL-RP-C203 (2016). The calculation can simply use these nominal S-N curves as hot spot S-N curves for this detail. When the nominal stress S-N curves are used, the stresses extracted from the finite element model may normally use Method B, without inclusion of the 1.12 factor indicated in equation (9.6), as this nominal stress is not associated with a significant stress gradient at the considered detail for membrane load. Thus, for simple cruciform joints, simple T-joints in plated structures, or simple butt welds that are welded from one side only, using the calculated hot spot stress from FEA, together with the representative nominal S-N curves for these details, as presented in appendix A in DNVGL-RP-C203 (2016), is recommended.

A similar situation with respect to hot spot stress calculation and selection of recommended S-N curve as explained earlier in this section for a cruciform joint occurs when a brace with a simple ring stiffener on the inside or the outside is analyzed (see Figure 6.10). The hot spot stress will include the effect of decreased stress on the inside and increased stress on the outside due to the circumferential stiffness of the ring. However, as the ring will not attract stress normal to its plane, a lower S-N curve than D has to be used from the tables of S-N classification in appendix A of

DNVGL-RP-C203 (2016). This typically results in an E-curve or F-curve, depending on the thickness of the stiffener. The largest stress at the node at the shell ring stiffener interface in a shell element model should be used as nominal stress, as the stress distribution shows a significant gradient along the shell when approaching a ring stiffener due to the circumferential stiffness of the tubular (see also Section 6.7). This is important at an outside ring stiffener in order to include the bending stress in the shell due to its deformed shape (see Figure 6.10). Reference is also made to the example of fatigue analysis of a drum in Section 6.7.2. However, use of the S-N curve D is appropriate where a longitudinal stiffener ends at a circumferential ring stiffener. These considerations also apply to the assessment of fillet welds, and for connections such as gusset joints analyzed by finite element methods (see Section 9.3). Method A or Method B should be used for the root areas of gusset joints together with lower design S-N curves that are representative for the actual fatigue capacity of these areas.

The C-curve in AWS standard (2010) is similar to the D-curve in DNVGL-RP-C203 (2016) and FAT 90 in IIW (Hobbacher 2009) and is recommended as hot spot stress S-N curve by Dexter et al. (1994).

Fabrication tolerances are considered to be most important for butt welds and cruciform joints, and need to be considered in fatigue assessments of these connections. When misalignments exceed the values explicitly included in the S-N curve, appropriate SCFs should be applied to the analysis results, or the analysis should explicitly model misalignments in a conservative way. Reference is also made to Section 5.3.

9.1.6 Analysis Methodology for Fillet Welds

The limitations of the structural stress analysis methodology are partly explained in Section 9.1.5. However, by following the guidance on selection of S-N curves in Section 9.1.5, the structural stress methodology can be followed for all welded connections except for fatigue assessment of fillet welds.

Fillet welds at doubling plates will be subjected to bending stress over the throat thickness when the doubling plate is loaded by an axial force or by a bending moment, and performing an FEA for assessment of fatigue of these connections is recommended. Then the fatigue life can be assessed using the nominal stress analysis procedure presented in Section 3.2.1 or a hot spot stress approach as illustrated in Figure 9.8.

For fatigue assessment of fillet welds, the nominal analysis procedure presented in Section 3.2.1 is rather robust as it is not very sensitive to the element mesh in the fillet weld, and is thus recommended. The stresses in the doubling plate some distance from the fillet weld is integrated for calculation of the force P (see Figure 9.8a). Then the relevant stress in the fillet weld for fatigue assessment can be calculated as described in Section 3.2.1.

The bending stress in the fillet weld can be calculated using at least two second-order solid elements over the throat thickness, where each element represents a linear stress distribution. The calculated stress components at positions 0.25a and 0.75a, where a is throat thickness, can be extrapolated to the weld root where these stresses are used to calculate the principal stress, see Figure 9.8b (Dalsgaard Sørensen et al.

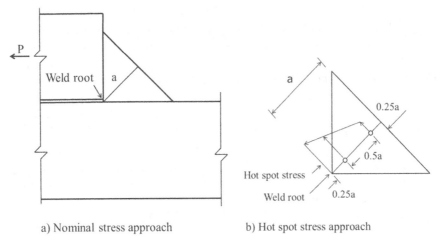

a) Nominal stress approach   b) Hot spot stress approach

Figure 9.8 Methods for analysis of stresses to be used for fatigue assessment of fillet welds.

2006). The range of the maximum principal stress can then be used, together with the F3-curve, for calculation of fatigue damage. The S-N curve for air environments can be used as it is assumed that potential fatigue cracking will initiate from the weld root. Reference is also made to Fricke (2006) for fatigue from the fillet weld root and to Fricke and Feltz (2009) for fatigue from the fillet weld toe.

Alternatively, the notch stress methodology described in Section 9.6 can be used for fatigue analysis of the weld root.

9.1.7 Verification of Analysis Methodology

The analysis methodology may be verified based on analysis of details with derived target hot spot stress. Such details with target hot spot stress are shown in appendix D in DNVGL-RP-C203 (2016). From the hopper corner detail no. 4 from Sections 2.3.1 and 2.3.3 it is likely that a higher hot spot stress than the target value will be calculated from FEA due to the reasons explained in Section 2.3.5.

9.1.8 Examples of Finite Element Models in Ship Structures

Some examples of finite element models used at critical areas in ship structures are shown in Figures 9.9 and 9.11. The hopper corner in Figure 9.9 was modeled by shell elements. In a project in which too short fatigue lives at a similar hopper corner were calculated, the use of 20-node isoparametric elements was suggested in order to obtain a more accurate analysis; the weld geometry could then be included in the model. During this work it was found that it was easier to make a finite element model using two 20-node elements over the plate thickness. This resulted in a higher calculated surface stress at the hot spot than that derived from the shell finite element model, as indicated in Figure 9.10. The explanation for this is that a bilinear stress distribution over the thickness had been calculated when using two elements over the plate thickness, and the element stress at the surface was higher than when a single element over the thickness was used. Thus, when using more

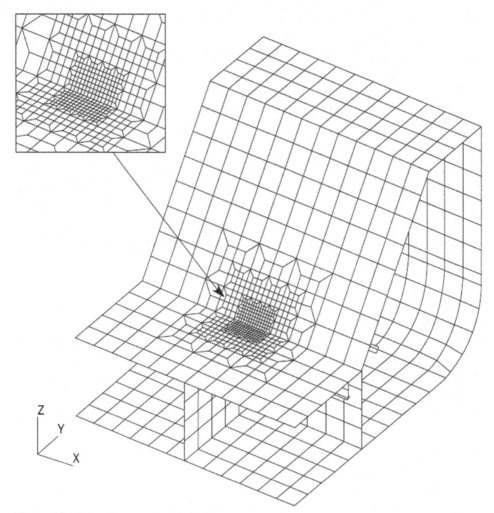

Figure 9.9. Finite element model of a hopper corner in a tank.

elements over the thickness, higher surface stresses will be calculated and the surface stresses approach infinity at a singularity, like a weld toe notch with no modeled
notch radius, if the element size approaches zero. In cases where a nonlinear stress
distribution through the thickness has been calculated, equations (9.1) and (9.2) can
be used to establish a linearized hot spot stress. When making finite element models
based on three-dimensional computer drawings, it may be convenient to use linear
tetrahedral elements (10 nodes) in the finite element model, as these elements readily adopt to the actual geometry. Five of these elements constitute a similar volume as
that of a 20-node isoparametric element, and this volume is considered to represent
a similar element stiffness and ability to represent a linear stress distribution. As it
may be difficult to restrict the modeling at more complex details, such that a linear
stress distribution over the thickness is achieved, equations (9.1) and (9.2) may again
be used to derive a representative linearized hot spot stress for use in fatigue life
calculations.

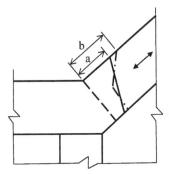

a: Calculated hot spot stress with a shell element or one 20-node isoparametric element over thickness

b: Calculated hot spot stress with two 20-node isoparametric elements over thickness

Figure 9.10. Illustration of calculated hot spot stress at a hopper corner using different finite element models.

9.2 Alternative Procedure for Analysis of Web-Stiffened Cruciform Connections

9.2.1 General

A number of FEAs using models with three-dimensional elements and models with shell elements of web-stiffened cruciform joints were performed in order to establish a simplified analysis procedure using shell elements of hopper connections, stringer heels, and joints connecting deck structures to vertical members in ship structures, as shown in Figure 9.12. Weld leg length is a parameter that was included in these analyses. Based on the results, a methodology for derivation of hot spot stress at

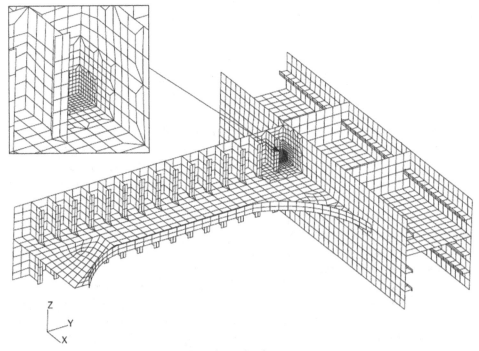

Figure 9.11. Finite element model of a stringer heel area.

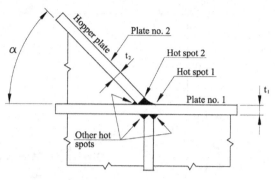

a) Hopper knuckle in tanker

Figure 9.12. Examples of web stiffened cruciform joints (Lotsberg et al. 2008c) (with permission from Taylor & Francis Ltd., http://
www.tandfonline.com).

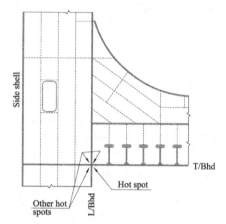

b) Heel of stringer in tanker

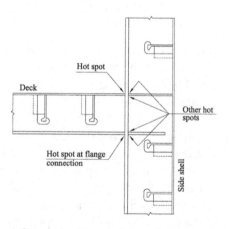

c) Connection between deck web frame and
side web frame in vehicle carrier

welded connections using shell finite element models was developed (Lotsberg et al.
2008c). It should be noted that the procedure described in the following is limited to
the plate flange connection. The "other hot spots," as indicated in Figure 9.12, should
be checked according to the general analysis procedure given in Section 9.1.

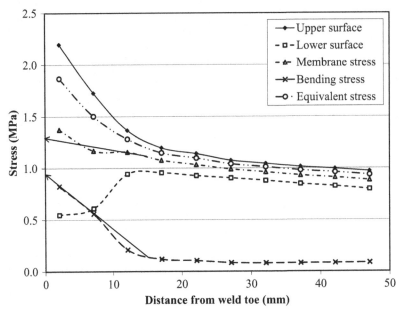

Figure 9.13. Stress distribution in plate no. 1 without an additional weld (Lotsberg et al. 2008c) (with permission from Taylor & Francis Ltd., http://www.tandfonline.com).

9.2.2 Plate Thickness to Be Used in Analysis Procedure

The main calibration of the procedure in Section 9.2.3 was performed on fatigue-tested specimens with $t_1 = t_2 = 10$ mm. The readout position has been made dependent on the plate thickness, t_1, in Figure 9.12a. It could be questioned whether the readout point should have rather been made as a function of plate thickness, t_2. With respect to addressing this question, the following points should be considered:

- It is assessed that it is the stress in plate no. 1 that is governing for the fatigue capacity at the weld toe on plate no. 1 side of Figure 9.12a, at hot spot no. 1, and that it is the stress in the inclined or vertical plate no. 2 that governs for the fatigue capacity of the weld toe on the transition from the weld to the plate no. 2 at hot spot no. 2. Thus, for finite element modeling of the hot spot regions and readout of hot spot stress, it would be the thickness, t_1, that is governing for the stress at weld toe to plate no. 1, and the thickness, t_2, that is governing for the stress at weld toe to plate no. 2.

- The stress distribution at a 45° hopper connection in Figure 9.12a is shown in Figure 9.13 without an additional weld and in Figure 9.14 with an additional weld. The local bending stress in plate no. 1 is the major contributor to the increase in hot spot stress, as compared with nominal membrane stress. The membrane stresses and the bending stresses are extrapolated back to the weld toe in the figure for illustration (extrapolation from stresses at $t_1/2$ and $3t_1/2$ to the weld toe is shown).

- The stress distribution at the hot spot no. 1 is not expected to change significantly, even if the thickness t_2 is increased to a large value. However, the readout point would be shifted to the right in Figure 9.15, resulting in a corresponding reduction in readout stress from the shell analysis model. Thus, this would provide a

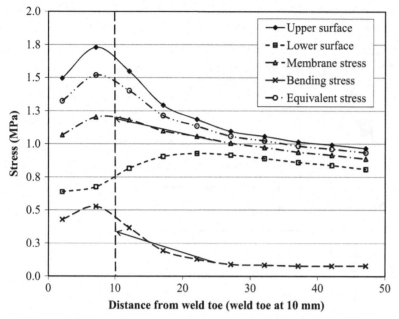

Figure 9.14. Stress distribution in plate no. 1 with an additional weld leg at 10 mm from weld toe (Lotsberg et al. 2008c) (with permission from Taylor & Francis Ltd., http://www .tandfonline.com).

non-conservative hot spot stress. Based on this, the thickness, t_2, is not a relevant parameter governing the distance from the intersection line to the readout point of stress in a procedure for direct readout of hot spot stress from a shell element analysis for hot spot no. 1.

9.2.3 Procedure for Analysis Using a Shell Element Model

In the procedure for analysis of hot spot number 1 in Figure 9.12a, it is assumed that the weld is not included in the shell FEA. The procedure is calibrated such that surface stress can be read out from readout points shifted away from the intersection line, at a position of the actual weld toe. The distance from the intersection line to the weld toe is obtained as:

$$x_{shift} = \frac{t_1}{2} + x_{wt} \qquad (9.9)$$

where:

t_1 = plate thickness of plate welded to the plate no. 1 in Figure 9.12a
x_{wt} = weld leg length.

The stress at the shift position (x_{shift}) is derived directly from the analysis (without any extrapolation of stresses). The surface stress (including membrane and bending stresses) is denoted $\sigma_{surface}(x_{shift})$. The membrane stress and the bending stress are denoted $\sigma_{membrane}(x_{shift})$ and $\sigma_{bending}(x_{shift})$, respectively. The derived stress range is then used as described in Section 9.1.3 before the hot spot stress S-N curve D (or FAT 90) is entered for fatigue life assessment.

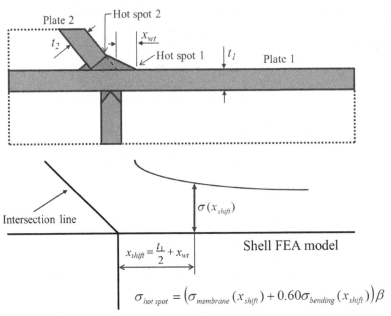

Figure 9.15. Procedure for derivation of hot spot stress using shell finite element model (Lotsberg et al. 2008c) (with permission from Taylor & Francis Ltd., http://www.tandfonline.com).

The hot spot stress is derived as:

$$\sigma_{hot\,spot} = \left(\sigma_{membrane}(x_{shift}) + 0.60\sigma_{bending}(x_{shift})\right)\beta \tag{9.10}$$

where:

$$\sigma_{bending}(x_{shift}) = \sigma_{surface}(x_{shift}) - \sigma_{membrane}(x_{shift}) \tag{9.11}$$

For connections with $\alpha = 45°$ (Figure 9.12a), a correction factor, β, is derived as:

$$\beta = 1.07 - 0.15\frac{x_{wt}}{t_1} + 0.22\left(\frac{x_{wt}}{t_1}\right)^2 \tag{9.12}$$

For connections with $\alpha = 60°$, a correction factor, β, is derived as:

$$\beta = 1.09 - 0.16\frac{x_{wt}}{t_1} + 0.36\left(\frac{x_{wt}}{t_1}\right)^2 \tag{9.13}$$

For connections with $\alpha = 90°$, a correction factor, β, is derived as

$$\beta = 1.20 + 0.04\frac{x_{wt}}{t_1} + 0.30\left(\frac{x_{wt}}{t_1}\right)^2 \tag{9.14}$$

The procedure is calibrated for $0 \le x_{wt}/t_1 \le 1.0$ and is thus considered to be applicable for this geometry. The derived hot spot stress is entered into the hot spot S-N curve for welded connections. The analysis procedure is illustrated in Figure 9.15. The same procedure can also be used to analyze hot spot no. 2 in Figure 9.12a (now with t_2 as reference thickness in the equations). The "other hot spots" located at the web, as indicated in Figure 9.12, can be checked by the following procedure: the maximum principal surface stress, defined at a distance $x_h = t_3/2 + x_{wt}$ (where t_3 is web thickness and x_{wt} is the smallest value for the two welds meeting in the corner) from

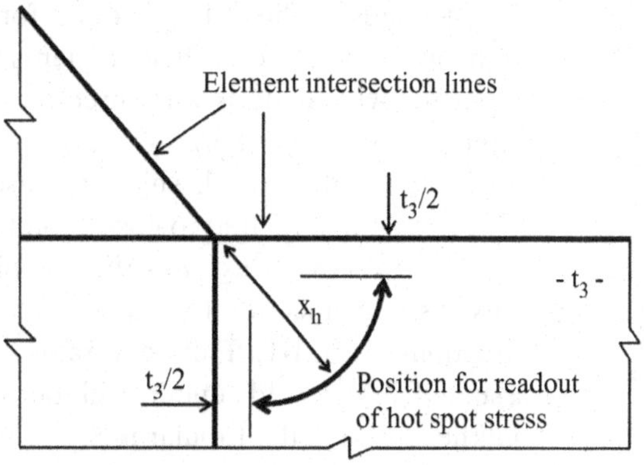

Figure 9.16. Readout of hot spot stress for other hot spots in Figure 9.12.

the crossing intersection lines to the hot spot, and not closer to the intersection lines than $t_3/2$ (see Figure 9.16), should be used for fatigue evaluation. This stress should be combined with nominal stress S-N curves E or F for cruciform joints, as presented in table A.7, details 8–10, in DNVGL-RP-C203 (2016).

9.3 Joint with Gusset Plates

Insert gusset plates are typically used in joints in topside structures to connect Rectangular Hollow Sections (RHS) or tubular members to main girders (see Figures 8.12 and 9.17). When such connections are subjected to dynamic loading, a full penetration weld between the member and the gusset plate is preferred; otherwise it can be difficult to document the fatigue capacity for fatigue cracking starting from the weld root. In dynamically loaded structures, shaping the gusset plate is also recommended to achieve a smooth stress flow from the member into the gusset plate. Where a reliable fatigue life of gusset plates is to be documented, performing an FEA is recommended to determine the hot spot stress, as there is not much guidance on SCFs for such details in the literature or design standards. Hot spot stresses at weld toes derived by FEA can be combined with the D-curve. The F3-curve or W3-curve should be used for the hot spot on the inside of the tubular members if the connection is made without any back weld, as is generally the case. The selection of S-N curve

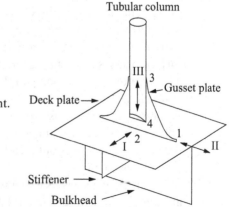

Figure 9.17. Gusset and cruciform joint.

then depends on the difficulty of performing reliable non-destructive testing (NDT) of the weld root. Use of shell elements for such analysis provides conservative SCFs compared with analysis with three-dimensional elements that includes modeling an external fillet weld. There should normally be a fillet weld on the outside of a full or partial penetration weld, which increases the effective area.

An example related to gusset joints and cruciform joints is considered in Figure 9.17; this example illustrates the complexity of fatigue design of such connections. This gusset connection to a deck may have varying cyclic stresses in three different directions (I, II, III). Full penetration welds between the gusset and the deck are assumed, in the welds under deck between the deck plate and the stiffener and the bulkhead, and in the tubular member that is slotted over the gusset, which has been profiled for 'favorable' geometry. Stress peaks may be expected at the points indicated as ①, ②, ③, and ④. Fabrication misalignments are not usually included in finite element models and fabrication misalignments are not modeled between the gusset and the underlying bulkhead. Furthermore it is assumed that the stresses are extracted from plate/shell elements, while the hot spot stress is derived using Method B in Section 9.1.3.

Considering the stresses in the deck plate for direction I, Section 9.1.5 with table A-7 (8) in DNVGL-RP-C203 (2016) leads to the use of the E-curve (FAT 80) for a gusset plate thickness of 25 mm or below, and F-curve (FAT 71) for thicknesses greater than 25 mm (hot spot 2). This would apply all along the gusset and to stresses at the top and bottom surfaces of the elements in the deck plate adjacent to the weld. Stresses are not factored by 1.12, because the stresses in the plate can be considered as nominal, and this stress is now used together with a nominal stress S-N curve.

Considering stresses in the deck plate for direction II, Section 9.1.5 leads to the use of the D-curve when extrapolating stresses to the plane of the gusset plate, and further extrapolation of stresses to the hot spot are made as shown in Figure 9.3. This principally applies at point ①. However, with Method B, stresses are typically extracted at the center of the elements located on either side of the axis of the gusset toe. Extrapolation of stresses to the gusset plane should then be performed at 0.5t, before this stress is multiplied by 1.12 for calculation of hot spot stress.

Considering stresses in the gusset (and in the bulkhead below), a peak stress may be found at point ② because of the presence of the stiffener for stress direction III. Here, the hot spot stress method can be used together with S-N curve D (due to the presence of the stiffener below the deck plate). For a region some distance away from the stiffener and stress direction III, Section 9.1.5 with table A-8 (1) in DNVGL-RP-C203 (2016) leads to the use of the F-curve (without the 1.12 factor) if it is assumed that the deck plate is thicker than 25 mm. This approach should be used for all gusset surface stresses extracted from the bottom row of gusset plate elements. In order to have a better understanding of this, a simple shell element model could be visualized in which the transverse deck plating does not attract additional stress, due to the lack of transverse stiffness, as explained in Section 9.1.5. If the fabrication eccentricity has not been modeled, it is unnecessary to apply an additional correction factor to either the deck or the gusset/bulkhead stresses, provided that the defined tolerance is less than that accounted for in the design S-N curve being used.

Considering stresses extracted from the tubular column, a stress concentration at point ③ may be expected. From the root of the single-sided weld, the stresses extracted from the inner surface of tubular elements adjacent to the gusset can be

Table 9.1. *SCFs for joints with gusset plate*

Geometry	SCF
RHS 250 × 16 with favorable geometry of gusset plate	2.9
RHS 250 × 16 with simple shape of gusset plate	3.8
Ø250 × 16 with favorable geometry of gusset plate	2.3
Ø250 × 16 with simple shape of gusset plate	3.0

Table 9.2. *Calculation of hot spot stress at gusset plate*

Analysis model	Calculated stress at 3t/2	Calculated stress at t/2	Extrapolated stress to the weld toe
Full penetration weld and axial load in RHS	1.62	1.97	2.15
6 mm gap in root of weld and axial load in RHS	1.69	2.11	2.32

used, and the F3-curve (without the 1.12 factor) should be used. For stresses extracted from the external surface of the tubular elements, the D-curve can be used (as the general hot spot stress methodology analysis applies here). A concentration of stress in the gusset at point ④ may also be found, where the welding is usually quite complex due to the presence of a cover plate. The structural stress methodology can be applied for the gusset plate surface stresses in this area, and W3 should be used to assess cracks from the root.

An example of SCFs for gusset plate joints analyzed in a project (Lotsberg et al. 2005b) is shown in Table 9.1. Examples of analysis results using 20-node isoparametric elements are shown in Figure 9.18 and Table 9.2. Due to the lack of penetration, using only one element over the thickness of the RHS was difficult, but this was resolved by disconnecting the degrees of freedom at the inside nodes (one 20-node element over the thickness).

9.4 Welded Penetrations in Plates

9.4.1 General

There are a large number of pipe penetrations in plated structures in semi-submersible platforms. These structures are normally analyzed with finite element programs. The welded penetrations are not included in the finite element models

Figure 9.18. Detail at end of gusset plate.

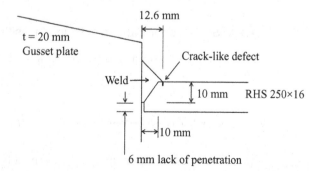

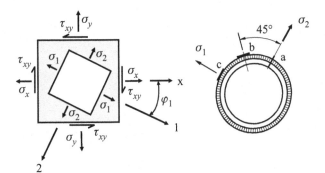

Figure 9.19. Hot spot locations around penetrations, relative to the direction of principal stress.

due to modeling time and analysis capacity. At the positions of the penetrations, a two-dimensional stress field is calculated using shell elements for modeling the plated structures. This implies stresses in the x direction and the y direction in the coordinate system of the element, and a shear stress. From these stresses, the principal stresses at the penetrations can be calculated as:

$$\Delta \sigma_1 = \frac{\Delta \sigma_x + \Delta \sigma_y}{2} + \frac{1}{2}\sqrt{\left(\Delta \sigma_x - \Delta \sigma_y\right)^2 + 4\,\Delta\,\tau_{xy}^2}$$

$$\Delta \sigma_2 = \frac{\Delta \sigma_x + \Delta \sigma_y}{2} - \frac{1}{2}\sqrt{\left(\Delta \sigma_x - \Delta \sigma_y\right)^2 + 4\,\Delta\,\tau_{xy}^2}$$

(9.15)

It can then be assumed that the axis of maximum principal stress is placed relative to the penetration in Figure 9.19. The fatigue design is performed for all three positions, a, b, and c, as shown in Figure 9.19. At each position, the first and the second principal stresses are accounted for by using the superposition principle, as explained in the following Sections 9.4.2–9.4.4.

9.4.2 Stresses for Fatigue Design at Position a

For a single stress state, the hot spot stress at position a in Figure 9.19 can be derived from the graph for SCFs shown in Figure 5.20. This graph can be used to calculate the stress range due to the first principal stress. The first principal stress will also create a parallel stress at position c, as shown in Figure 9.19. This parallel stress at position c (that is due to the first principal stress) can be derived from Figure 5.21. This graph can also be used to calculate the hot spot stress at position a (in Figure 9.19) for the second principal stress. The resulting hot spot stress at position a, due to the first and the second principal stresses, is derived by summation of the derived hot spot stresses (see Figure 9.20). Finally, this hot spot stress can be entered into the relevant S-N curve for calculation of fatigue damage.

9.4.3 Stresses for Fatigue Design at Position b

The principal stresses at position b in Figure 9.19 can be calculated as:

$$\Delta \sigma_{1b} = \frac{\Delta \sigma_{np} + \Delta \sigma_{//p}}{2} + \frac{1}{2}\sqrt{\left(\Delta \sigma_{np} - \Delta \sigma_{//p}\right)^2 + 4\,\Delta\,\tau_{//p}^2}$$

$$\Delta \sigma_{2b} = \frac{\Delta \sigma_{np} + \Delta \sigma_{//p}}{2} - \frac{1}{2}\sqrt{\left(\Delta \sigma_{np} - \Delta \sigma_{//p}\right)^2 + 4\,\Delta\,\tau_{//p}^2}$$

(9.16)

where the stress components can be derived from Figures 5.24, 5.25, and 5.26.

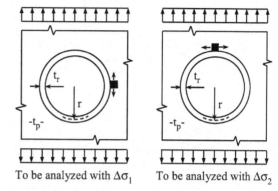

Figure 9.20. Parallel stresses at position a.

To be analyzed with $\Delta\sigma_1$ To be analyzed with $\Delta\sigma_2$

The stress components are evaluated once for the first principal stress and once for the second principal stress shown in Figure 9.19. The stress components are then added together, and equation (9.16) is used to calculate the principal stresses, $\Delta\sigma_{1b}$ and $\Delta\sigma_{2b}$. Finally, the effective hot spot stress at position b is calculated as:

$$\Delta\sigma_{Eff} = \max \begin{cases} \sqrt{\Delta\sigma_{np}^2 + 0.81\,\Delta\tau_{//p}^2} \\ \alpha\,|\Delta\sigma_{1b}| \\ \alpha\,|\Delta\sigma_{2b}| \end{cases} \qquad (9.17)$$

where α is presented in Section 9.1.3. This stress range is then entered into the D-curve in DNVGL-RP-C203 (2016) or the FAT 90 curve (Hobbacher 2009).

9.4.4 Stresses for Fatigue Design at Position c

Graphs for SCFs for the conditions shown in Figure 9.21 are required for calculation of the hot spot stress at position c in Figure 9.19. One SCF from Figure 5.23 is combined with the first principal stress, and another SCF from Figure 5.22 is combined with the second principal stress. These hot spot stresses are then added together, similarly to that in Section 9.4.2, before the D-curve (or the FAT 90 curve) is entered for calculation of fatigue damage.

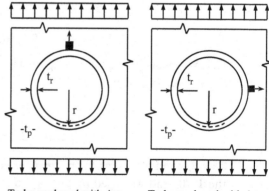

Figure 9.21. Stresses required for derivation of hot spot stress at position c.

To be analyzed with $\Delta\sigma_1$ To be analyzed with $\Delta\sigma_2$

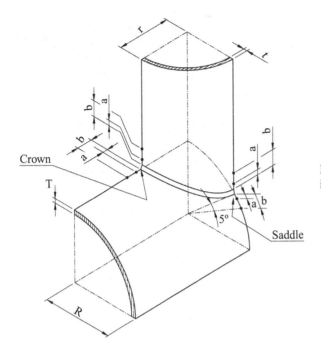

Figure 9.22. Derivation of hot spot stress in tubular joints.

9.5 Tubular Joints

The stress range at the hot spot of tubular joints should be combined with the T-curve. Analysis based on thick shell elements may be used. In this case, the weld is not included in the model. The hot spot stress may be determined as for welded connections. More reliable results are obtained by including the weld in the model by using three-dimensional elements.

The hot spot stress or geometric stress at tubular joints can be obtained by linear extrapolation of the stresses obtained from analysis at positions at distances a and b from the weld toe, as indicated in Figure 9.22. For extrapolation of stress along the brace surface that is normal to the weld toe:

$$a = 0.2\sqrt{rt}$$
$$b = 0.65\sqrt{rt} \tag{9.18}$$

For extrapolation of stress along the chord surface that is normal to the weld toe at the crown position:

$$a = 0.2\sqrt{RT}$$
$$b = 0.4\sqrt[4]{rt\,RT} \tag{9.19}$$

For extrapolation of stress along the chord surface that is normal to the weld toe at the saddle position:

$$a = 0.2\sqrt{RT}$$
$$b = 2\pi R\frac{5}{360} = \frac{\pi R}{36} \tag{9.20}$$

This methodology was based on American and British standards from the end of the 1970s and was used by the various test laboratories in the Common Research

Programme on Offshore Structures presented at the conference on Steel in Marine Structures in Paris in 1981 (see Pozzolini 1981; de Back 1981). It has later been denoted the ECSC procedure for hot spot stress derivation (see, e.g., Berge et al. 1994).

An alternative to the procedure presented above is to place the Gaussian points, where stresses are calculated, at a distance $0.1\sqrt{rt}$ from the weld toe (r = radius of considered tubular member and t = thickness). This procedure was calibrated with measurement data during the 1980s (DNVGL-RP-C203 2016). The stress at this point may be used directly in the fatigue assessment. For analysis of cast nodes, see Section 8.7.

Gibstein (1981) recommended derivation of hot spot stress by a linear extrapolation through the nearest point from the weld toe of 4 mm as this was considered to provide consistent results.

Berge et al. (1994) pointed out that different procedure for derivation of hot spot stresses may be needed to represent the physical behavior depending on type of joint. Furthermore, the hot spot stress is defined as a surface stress at the intersection between the brace and the chord in simple tubular joints. No information is contained about the stress gradient through the thickness or the gradient along the intersection from the hot spot. During crack growth in these joints there will also be a load shedding or redistribution of the stress as the crack grows. This can only be properly considered with a crack included in the analysis model for analysis based on fracture mechanics. With reduced stress concentration factor due to stiffened joints by, for example, internal ring stiffeners, the possibility for redistribution of stresses during crack growth is reduced and the crack growth is faster for the same hot spot stress indicating that a lower S-N curve should be used for stiffened tubular joints as indicated in the test data by Slater and Tubby (1996). See also Section 8.5 for hot spot stress analysis of stiffened tubular joints.

It is important to use a proper link between method for derivation of hot spot stress and the design S-N curve – that is, the design S-N curve should be established based on the same methodology for hot spot stress derivation from measured data as later used in design analysis.

9.6 Notch Stress Method

9.6.1 General

There are welded details where the structural stress method will not predict accurately the fatigue behavior of a joint by use of stress extrapolation along a plate surface or from the stress distribution through the plate thickness as shown by Tovo and Lazzarin (1999). A load-carrying fillet-welded cruciform joint is used as an example where different local stress fields at a weld toe are shown for a full penetration weld and a connection with a fillet weld. This difference in local stresses adds to the general problem with calculation of the hot spot stress (or structural stress) by finite element analysis as explained in Section 9.1.5.

Fatigue failures may be explained by high-peak stress ranges at geometric notches. For the base material also, the stress field (or stress gradient) at the notch is considered to influence the fatigue strength in addition to the peak stress range. In welded connections the local notch radius and weld flank angle is considered to govern the maximum peak stress range; see, for example, formulas for stress concentration factors for notches by Iida and Uemura (1996). Measurement of weld toe radius and weld flank angle show significant scatter (Engesvik 1981), which makes a direct relationship between a local stress and a design S-N curve difficult. Tovo and Lazzarin (1999) proposed a mathematical relation between notch stress intensity factor and the most significant stress parameters at the notch based on the classical stress solution by Williams (1957) at a notch in a connection with zero radius. Wathne Tveiten and Moan (2000) used a similar approach to derive improved hot spot stresses at welded connections in aluminum structures where the actual notch geometry could be accounted for based on measurements.

Chattopadhyay et al. (2011) proposed a methodology for calculation of peak stress ranges at the weld toe based on different stress concentration factors for axial stress and bending stress. The fatigue process for crack initiation at the weld toe is assumed to be nonlinear and is proposed to be based on peak stresses determined by the Neuber or Glinka methods as presented in Section 3.1.4 and a material cyclic stress–strain relationship as presented in Section 3.1.2. The fatigue damage can be calculated from the Coffin-Manson equation (3.11). Here the size of the initial crack is needed for further crack growth analysis based on fracture mechanics. An initial size in the range of 0.5–0.8 mm is indicated. A significantly lower initial crack size has lately been derived from fatigue tests by Lassen and Recho (2015). This makes a two-step approach somewhat uncertain and complex at present, as it is observed that significant fatigue life is associated with growth of small cracks.

An overview of fatigue strength assessment of non-welded and welded structures based on local parameters has been provided by Radaj (1996). Radaj concludes that there are quite a number of local approaches available for assessing the fatigue strength and service life of welded structures based on structural stresses, notch stresses, and fracture mechanics. The state of development, the contents, and the range of applicability of the approaches are rather different. Around 1996, Radaj also considered that it was difficult to standardize and verify them for industrial application and inclusion in design standards. However, Petershagen (1991) had successfully showed that the effective notch stress approach correctly classifies simple welded joints due to the IIW FAT value recommendations of the nominal stress approach and the notch stress method proposed by Radaj was included in the IIW guideline for fatigue assessment of structures (Hobbacher 1996). A few years later it was included also in DNV standards and it has later been used in a number of research and industry projects. It has been found attractive because it is simple and easy to use as shown in the following sections.

9.6.2 The Notch Stress Method

The notch stress is the total stress at the root of a notch, obtained assuming linear-elastic material behavior. In order to take into account the statistical nature and scatter of weld shape parameters, as well as of nonlinear material behavior at the notch

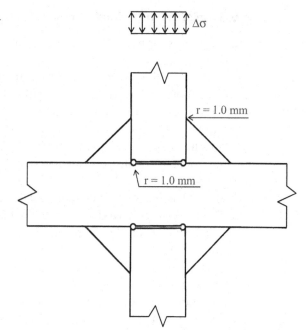

Figure 9.23. Analysis of notch stress.

root, the real weld is replaced by an effective one, as shown in Figure 9.23. For structural steels, an effective notch root radius of $r = 1.0$ mm has been verified to give consistent results (Fricke 2013). Reference is also made to a German research project on comparison of the notch stress method, the structural stress method, and the nominal stress method with fatigue test data on a number of different details. The thickness of the test specimens was in the 1–20 mm range. This work is presented by Bruder et al. (2012).

For fatigue assessment, the calculated notch stress is compared with a notch stress S-N curve, as shown in Section 4.1.9. The notch stress S-N curve corresponds to the proposed notch radius. The background for the notch stress method is explained in Radaj et al. (2006). A guideline for the assessment of weld root fatigue using this method has been presented by Fricke (2013). The method is restricted to welded joints that are expected to fail from the weld toe or from the weld root. Other causes of fatigue failure, such as from surface roughness or from embedded defects, are not covered. It is also not applicable where there are considerable stress components parallel to the weld toe or parallel to the root gap. The method is suitable for comparing alternative geometries. Unless otherwise specified, flank angles of 30° for butt welds and 45° for fillet welds are suggested. The method is limited to thicknesses $t \geq 5$ mm; for smaller wall thicknesses, a method with a smaller notch radius that is linked to a different S-N curve may be used, as shown by Sonsino (2009). In cases where a mean geometrical notch root radius can be defined; for example, after certain post-weld improvement procedures, such as grinding, the actual geometrical radius may be used in the stress analysis. The calculated surface stress should then be entered into a relevant S-N curve, such as the C-curve or C1-curve. For potential fatigue crack growth from ground areas at welded regions, see table 5 in appendix A in DNVGL-RP-C203 (2016).

9.6.3 Calculation of Notch Stress

The notch stress or notch SCF can be calculated by parametric formulas, taken from diagrams or calculated from FEA. The effective notch radius is introduced, such that the tip of the radius touches the root of the real notch, for example, the end of an unwelded root gap. Calculation of effective notch stress by the finite element method requires the use of a fine element mesh around the notch region. The effective notch stress to be used, together with the recommended S-N curve, is the maximum calculated surface stress in the notch. This maximum surface stress may be obtained directly from the nodal stress calculated at the surface, or from extrapolation of element stresses to the surface. In some finite element programs it may also be efficient to add bar elements with a negligibly small area along the notch surface. The surface stress is then finally derived as force in the bar divided by area. The requirement for mesh size depends on the elements used. If elements with quadratic displacement functions are used, then a minimum of four elements should be used along a quarter of the circle circumference. If simpler elements are used, the mesh refinement should be improved correspondingly. The mesh should be made with regular elements without transition to reduce refinement within the first three element layers from the notch surface. An element shape that is close to "quadratic" is preferred. In cases of uncertainty about the elements' ability to provide reliable surface stresses, validation of the methodology against a well-known case is recommended. The notch stress methodology is frequently used in research studies, but has also been used by the industry in a number of cases where it has been found difficult to decide on an agreed and documented S-N curve.

9.6.4 Example of Validation of Analysis Methodology

The notch stress concept using FEA can be validated against a well-tested detail that can also be assessed using the nominal stress approach. A cruciform joint may be selected for analysis and the F-curve may be used for fatigue assessment of the weld toe using the nominal stress approach. A target notch stress at the weld root is then 3.17 times greater than the nominal stress in the plate. This ratio between notch stress and nominal stress is denoted f, and is defined as:

$$f = \frac{S_{notch}}{S_{nominal}} \tag{9.21}$$

The S-N curve for the nominal stress can be presented as:

$$\log N = \log a_{d,\,nominal} - m \log S_{nominal} \tag{9.22}$$

and the S-N curve for the notch stress can be presented as:

$$\log N = \log a_{d,\,notch} - m \log S_{notch} \tag{9.23}$$

By requiring the same number of cycles from notch stress analysis as from nominal stress fatigue analysis, the following relationship between the two S-N curves is derived from equations (9.21)–(9.23):

$$\log a_{d,\,nominal} = \log a_{d,\,notch} - m \log f \tag{9.24}$$

The ratio on stress f may then also be derived from this equation as:

$$f = \left(\frac{\log a_{d,\,notch}}{\log a_{d,\,nominal}} \right)^{1/m} \tag{9.25}$$

Reference is made to Table 4.1 and Table 4.6 for parameters describing the S-N curves needed for calculation of f in this equation.

A fillet-welded cruciform joint may be selected for analysis, as shown in Figure 9.23. The W3-curve may then be used for fatigue assessment of the weld root using the nominal stress approach. The nominal stress in the weld is derived as:

$$\sigma_w = \sigma_{Nominal} t / 2a \tag{9.26}$$

where:

t = thickness of member
a = throat thickness.

A target notch stress in the weld root is then obtained as 6.25 times greater than the nominal stress in the fillet weld for an example with plate thicknesses of 16 mm and weld throat thickness of 8 mm. If the calculated stress from validation analyses is distant from the target values, the accuracy of the methodology used should be considered, such as type of element in relation to mesh refinement and readout of notch surface stress. The toe of the same cruciform fillet-welded detail would be classified as F3 according to DNVGL-RP-C203 (2016) or FAT 56 according to IIW (Hobbacher 2009). This results in a target notch stress that is 4.02 times larger than the nominal stress.

10 Fatigue Assessment Based on Stress Range Distributions

10.1 Weibull Distribution of Long-Term Stress Ranges

For fatigue assessment of floating structures it is often assumed that the long-term stress range distribution can be adequately represented by a two-parameter Weibull distribution (see, e.g., Guedes Soares, and Moan 1991). The probability density function for a two-parameter Weibull distribution reads:

$$f(\Delta\sigma) = \frac{h}{q}\left(\frac{\Delta\sigma}{q}\right)^{h-1} e^{-\left(\frac{\Delta\sigma}{q}\right)^h} \tag{10.1}$$

where:

  h = shape parameter
  q = scale parameter.

The scale parameter is related to the maximum stress range, $\Delta\sigma_0$, during n_0 stress cycles by:

$$q = \frac{\Delta\sigma_0}{(\ln n_0)^{1/h}} \tag{10.2}$$

The following cumulative distribution for stress ranges is derived by integration of equation (10.1) from 0 to $\Delta\sigma$:

$$F(\Delta\sigma) = \int_0^{\Delta\sigma} f(\Delta\sigma)d\Delta\sigma$$

$$= \int_0^{\Delta\sigma} \frac{h}{q}\left(\frac{\Delta\sigma}{q}\right)^{h-1} e^{-\left(\frac{\Delta\sigma}{q}\right)^h} d\Delta\sigma \tag{10.3}$$

$$= \left| -e^{-\left(\frac{\Delta\sigma}{q}\right)^h} \right|_0^{\Delta\sigma}$$

$$= 1 - e^{-\left(\frac{\Delta\sigma}{q}\right)^h}$$

For engineering purposes, the scale parameter, q, can be eliminated by introducing the maximum stress range, $\Delta\sigma_0$, during n_0 number of stress cycles. This can be

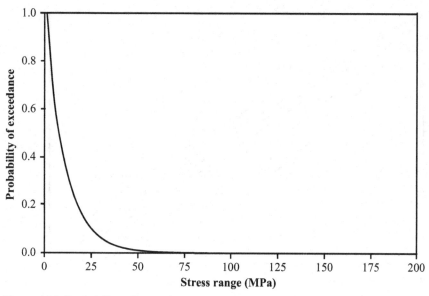

Figure 10.1. Probability of exceedance Q.

performed by using the complementary function, Q, which expresses the probability that the stress range, $\Delta\sigma$, is exceeded:

$$Q(\Delta\sigma) = 1 - F(\Delta\sigma) \tag{10.4}$$

The largest stress range occurs only once during n_0 cycles, and the probability that this occurs is:

$$Q(\Delta\sigma_0) = \frac{1}{n_0} = e^{-\left(\frac{\Delta\sigma_0}{q}\right)^h} \tag{10.5}$$

The number of cycles, n, that exceeds the stress strange, $\Delta\sigma$, can then be expressed as:

$$Q(\Delta\sigma) = \frac{n}{n_0} = e^{-\left(\frac{\Delta\sigma}{q}\right)^h} \tag{10.6}$$

This function is illustrated in Figure 10.1 for h = 1.0 and q = 10.86 MPa.

Taking the logarithm on both sides of equation (10.6) results in:

$$\ln\frac{n}{n_0} = -\left(\frac{\Delta\sigma}{q}\right)^h \tag{10.7}$$

Then, by inserting for q from equation (10.2), the equation becomes:

$$\ln\frac{n}{n_0} = \ln n - \ln n_0 = -\left(\frac{\Delta\sigma}{\Delta\sigma_0}\right)^h \ln n_0 \tag{10.8}$$

And by rearranging equation (10.8) gives:

$$\frac{\Delta\sigma}{\Delta\sigma_0} = \left(1 - \frac{\ln n}{\ln n_0}\right)^{1/h} = \left(1 - \frac{\log n}{\log n_0}\right)^{1/h} \tag{10.9}$$

In the last part of this equation, the relation log n = Constant · ln n is used.

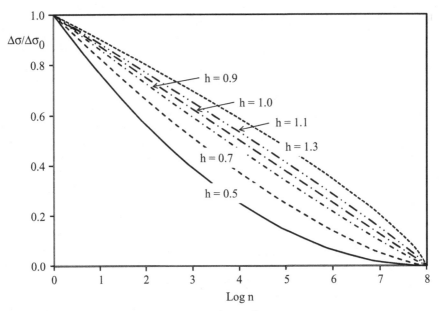

Figure 10.2. Long-term stress range exceedance diagram.

The result from this equation for different values of the shape parameter, h, is shown in Figure 10.2.

10.2 Closed-Form Expressions for Fatigue Damage Based on the Weibull Distribution of Stress Ranges

The gamma function (see, e.g., Ambramowitz and Stegun 1970) is defined as:

$$\Gamma(z) = \int_0^\infty e^{-t} t^{z-1} dt \qquad (10.10)$$

The incomplete gamma function is defined as:

$$\gamma(a, x) = \int_0^x e^{-t} t^{a-1} dt \qquad (10.11)$$

(a in this expression is a different parameter from that used for description of S-N curves.)

The complementary incomplete gamma function is defined as:

$$\Gamma(a, x) = \Gamma(a) - \gamma(a, x) = \int_x^\infty e^{-t} t^{a-1} dt \qquad (10.12)$$

The two-parameter Weibull distribution is presented in equation (10.1). For a one-slope S-N curve, the fatigue damage can be calculated based on integration of the

Palmgren-Miner damage equation (3.40) and the S-N curve in equation (4.1) as:

$$D = \int_0^\infty \frac{dn}{N(\Delta\sigma)} d\Delta\sigma$$

$$= n_0 \int_0^\infty \frac{f(\Delta\sigma)}{N(\Delta\sigma)} d\Delta\sigma \qquad (10.13)$$

$$= \frac{n_0}{a_d} \int_0^\infty \Delta\sigma^m \frac{h}{q} \left(\frac{\Delta\sigma}{q}\right)^{h-1} e^{-\left(\frac{\Delta\sigma}{q}\right)^h} d\Delta\sigma$$

The following variable is then introduced:

$$t = \left(\frac{\Delta\sigma}{q}\right)^h \qquad (10.14)$$

The result below is derived by further differentiation of equation (10.14):

$$\frac{dt}{d\Delta\sigma} = \frac{h}{q}\left(\frac{\Delta\sigma}{q}\right)^{h-1} \qquad (10.15)$$

Then, by combining equations (10.13), (10.14), and (10.15), the following expression for calculated fatigue damage is derived (to get it comparable with equation 10.10):

$$D = \frac{n_0}{a_d} q^m \int_0^\infty e^{-t} t^{(1+\frac{m}{h})-1} dt \qquad (10.16)$$

By comparing equations (10.16) and (10.10) it can be seen that the fatigue damage can be calculated from the following closed-form expression:

$$D = \frac{n_0}{a_d} q^m \Gamma\left(1 + \frac{m}{h}\right) \qquad (10.17)$$

With q from equation (10.2), the fatigue damage can be calculated as:

$$D = \frac{n_0}{a_d} \frac{\Delta\sigma_0^m}{(\ln n_0)^{m/h}} \Gamma\left(1 + \frac{m}{h}\right) \qquad (10.18)$$

For calculation of crack growth by fracture mechanics when neglecting threshold values it may be convenient to derive an equivalent constant stress range that gives the same calculated fatigue damage as from a Weibull long-term stress range distribution for the same number of stress cycles. This can be achieved by requiring that the fatigue damage under constant amplitude loading, considering one block with n_0 cycles in equation (3.40), gives the same fatigue damage as by equation (10.17):

$$D = \frac{n_0}{a_d} \Delta\sigma_{eq}^m = \frac{n_0}{a_d} \frac{\Delta\sigma_0^m}{(\ln n_0)^{m/h}} \Gamma\left(1 + \frac{m}{h}\right) \qquad (10.19)$$

Solving this equation for the constant equivalent stress range gives:

$$\Delta\sigma_{eq} = \frac{\Delta\sigma_0}{(\ln n_0)^{1/h}} \sqrt[m]{\Gamma\left(1 + \frac{m}{h}\right)} \qquad (10.20)$$

It is noted that with q from equation (10.2) this equation can also be presented as:

$$\Delta\sigma_{eq} = q \sqrt[m]{\Gamma\left(1 + \frac{m}{h}\right)} \tag{10.21}$$

For two-slope (bilinear) S-N curves, the fatigue damage can similarly be calculated from the Palmgren-Miner rule by integration of fatigue damage below each part of the bilinear S-N curves presented in Section 4.1. The number of cycles in the nominator of the damage rule is derived from the expression for the Weibull long-term stress range distribution. Thus, the fatigue damage is calculated as:

$$D = \frac{n_0}{a_{d1}} \int_{S_1}^{\infty} \Delta\sigma^{m_1} \frac{h}{q}\left(\frac{\Delta\sigma}{q}\right)^{h-1} e^{-\left(\frac{\Delta\sigma}{q}\right)^h} d\Delta\sigma + \frac{n_0}{a_{d2}} \int_{0}^{S_1} \Delta\sigma^{m_2} \frac{h}{q}\left(\frac{\Delta\sigma}{q}\right)^{h-1} e^{-\left(\frac{\Delta\sigma}{q}\right)^h} d\Delta\sigma \tag{10.22}$$

Again, the variables in equations (10.14) and (10.15) are introduced, and the fatigue damage is calculated as:

$$\begin{aligned}
D &= \frac{n_0}{a_{d1}} \int_{(S_1/q)^h}^{\infty} q^{m_1} t^{\frac{m_1}{h}} \frac{dt}{d\Delta\sigma} e^{-t} d\Delta\sigma + \frac{n_0}{a_{d2}} \int_{0}^{(S_1/q)^h} q^{m_2} t^{\frac{m_2}{h}} \frac{dt}{d\Delta\sigma} e^{-t} d\Delta\sigma \\
&= \frac{n_0}{a_{d1}} q^{m_1} \int_{(S_1/q)^h}^{\infty} e^{-t} t^{\frac{m_1}{h}} dt + \frac{n_0}{a_{d2}} q^{m_2} \int_{0}^{(S_1/q)^h} e^{-t} t^{\frac{m_2}{h}} dt \\
&= \frac{n_0}{a_{d1}} q^{m_1} \int_{(S_1/q)^h}^{\infty} e^{-t} t^{\left(1+\frac{m_1}{h}\right)-1} dt + \frac{n_0}{a_{d2}} q^{m_2} \int_{0}^{(S_1/q)^h} e^{-t} t^{\left(1+\frac{m_2}{h}\right)-1} dt
\end{aligned} \tag{10.23}$$

where S_1 is stress at transition from the left to the right part of the S-N curve. Comparing this expression for fatigue damage with the definition of the incomplete gamma functions in equations (10.11) and (10.12), the following closed-form equation for fatigue damage using a bilinear S-N curve is derived:

$$D = \frac{n_0}{a_{d1}} q^{m_1} \Gamma\left(1 + \frac{m_1}{h}; \left(\frac{S_1}{q}\right)^h\right) + \frac{n_0}{a_{d2}} q^{m_2} \gamma\left(1 + \frac{m_2}{h}; \left(\frac{S_1}{q}\right)^h\right) \tag{10.24}$$

10.3 Closed-Form Expressions for Fatigue Damage Based on the Rayleigh Distribution of Stress Ranges

The Rayleigh distribution is normally used for calculation of fatigue damage within a sea state when a frequency response fatigue analysis is performed. The Rayleigh distribution reads:

$$f(\Delta\sigma) = \frac{\Delta\sigma}{4\sigma_x^2} e^{-\frac{\Delta\sigma^2}{8\sigma_x^2}} \tag{10.25}$$

where:

$\Delta\sigma$ = stress range

σ_x = standard deviation in the stress response spectrum.

By comparison of distributions, it can be observed that the Rayleigh distribution is a special case of the Weibull distribution, with:

$$h = 2$$
$$q = 2\sqrt{2}\sigma_x \tag{10.26}$$

Thus, the closed-form expression for calculated fatigue damage can be derived directly from equation (10.24) as:

$$D = \frac{n_0}{a_{d1}} \left(2\sqrt{2}\sigma_x\right)^{m_1} \Gamma\left(1 + \frac{m_1}{2}; \left(\frac{S_1}{2\sqrt{2}\sigma_x}\right)^h\right)$$
$$+ \frac{n_0}{a_{d2}} \left(2\sqrt{2}\sigma_x\right)^{m_2} \gamma\left(1 + \frac{m_2}{2}; \left(\frac{S_1}{2\sqrt{2}\sigma_x}\right)^h\right) \tag{10.27}$$

10.4 Example of Use of Closed-Form Expressions for Fatigue Damage in Calculation Sheets Based on a Bilinear S-N Curve

In the following example, it is shown how the derived closed-form equations for fatigue damage based on a bilinear S-N curve can be used in a practical calculation sheet. The gamma function may be calculated directly from an available function in a calculation sheet. In an Excel sheet, the EXP(GAMMALN(a)) function can be used for calculation of $\Gamma(a)$.

The gamma distribution (see, e.g., Abramowitz and Stegun 1970; Gran 1992) may be used for calculation of the incomplete gamma functions in equation (10.24). This function is defined as:

$$P(a, x) = \frac{1}{\Gamma(a)} \int_0^x e^{-t} t^{a-1} dt \quad a > 0 \tag{10.28}$$

and

$$\gamma(a, x) = P(a, x)\Gamma(a) = \int_0^x e^{-t} t^{a-1} dt \quad a > 0 \tag{10.29}$$

The complementary incomplete gamma function is derived as:

$$\Gamma(a, x) = \Gamma(a) - \gamma(a, x) = \int_x^\infty e^{-t} t^{a-1} dt \tag{10.30}$$

It is observed that:

$$x = \left(\frac{S_1}{q}\right)^h$$
$$a = 1 + \frac{m_1}{h} \quad \textit{for left part of S-N curve} \tag{10.31}$$
$$a = 1 + \frac{m_2}{h} \quad \textit{for right part of S-N curve}$$

Table 10.1. *Calculation sheet for fatigue damage with bilinear S-N curves*

Cell		D	E	Comments
3	Hot Spot no.	1	2	Input in cells no. 4-17
4	S-N curve	B2	F1	Reference is made to DNVGL-RP-C203
5	Nominal stress range [MPa]	131.61	136.75	Maximum allowable stress range during n_0 cycles
6	Hot spot stress factor	3.00	1.15	Additional SCF
7	Weibull: h	1.10	1.10	Weibull shape parameter
8	Cycles at knee in S-N curve N_1	1.E+07	1.E+07	Number of cycles at knee in S-N curve N_1
9	m_1 (N = <N_1 cycles)	4.0	3.0	Input design S-N data from DNVGL-RP-C203
10	$\log a_{d1}$ (N = <N_1 cycles)	14.885	11.699	
11	m_2 (N>N_1 cycles)	5.0	5.0	
12	$\log a_{d2}$ (N>N_1 cycles)	16.856	14.832	
13	Years in service	20.0	20.0	
14	Zero up-crossing frequency: v_0	0.159	0.159	Inverse of the mean load response period
15	Effective thickness [mm]	30.0	30.0	Effective plate thickness for calculation of size effect
16	Reference thickness	25.0	25.0	
17	Thickness exponent k	0.00	0.25	
18	T_d = Time in service (in years) $\cdot 60 \cdot 60 \cdot 24 \cdot 365$	6.31E+08	6.31E+08	Service life in seconds
19	Calculated number of cycles: n_0	1.00E+08	1.00E+08	Based on service life and the mean load response
20	Calculated Weibull scale parameter: q	27.932	11.126	
21	Thickness or size correction	1.000	1.047	Calculated size correction
22	Gamma(1+m_1/h)	14.089	4.306	
23	Gamma(1+m_2/h)	56.331	56.331	
24	Stress at knee in S-N curve: S_1	93.594	36.841	
25	$(S_1/q)^h$	3.781	3.733	
26	Gamma distribution $P((1 + m_1/h) , (S_1/q)^h)$	0.395	0.570	
27	Gamma distribution $P((1 + m_2/h) , (S_1/q)^h)$	0.242	0.233	
28	Calculated fatigue damage: D	1.000	1.000	
29	Calculated life time T [years]	20.006	20.002	

In an Excel sheet, the GAMMA.DIST function can be used for calculation of the gamma distribution. This function is defined as GAMMA.DIST(x; alpha; beta; cumulative), and has the following arguments:

- x: the value at which the distribution is calculated;
- Alpha: parameter of the distribution;
- Beta: parameter of the distribution: If Beta = 1, then GAMMA.DIST will return the standard gamma distribution.
- Cumulative: a logical value that determines the form of the function. If cumulative is TRUE, then GAMMA.DIST will return the cumulative distribution function; if FALSE, it will return the probability density function.

An example of a calculation for two different details in a deck structure is shown in Table 10.1. The variables in the cells no 4-17 are input to the analysis and the

Table 10.2. *Detailed content of cells in calculation sheet*

Cell	Description of content	Cell E
18	$T_d =$ Service life $\cdot 60 \cdot 60 \cdot 24 \cdot 365$	=60*60*24*365*E13
19	Calculated number of cycles: n_0	=E14*E18
20	Calculated Weibull scale parameter: q	=E5*E6/(LN(E19)^(1/E7))
21	Thickness or size correction	=IF(E15<=E16;1;(E15/E16)^E17)
22	Gamma$(1+m_1/h)$	=EXP(GAMMALN(1+E9/E7))
23	Gamma$(1+m_2/h)$	=EXP(GAMMALN(1+E11/E7))
24	Stress at knee in S-N curve: S_1	=10^((E10-LOG(E8))/E9)
25	$(S_1/q)^h$	=(E24/E20)^E7
26	Gamma distribution $P((1+m_1/h)\,,(S_1/q)^h)$	=GAMMA.DIST(E25;(1+E9/E7);1;TRUE)
27	Gamma distribution $P((1+m_2/h)\,,(S_1/q)^h)$	=GAMMA.DIST(E25;(1+E11/E7);1;TRUE)
28	Calculated fatigue damage: D equation (10.24)	=E21^E9*E19/(10^E10)*E20^E9*(1-E26)*E22+ E21^E11*E19/(10^E12)*E20^E11*E27*E23
29	Calculated life time T [years]	=20/E28

calculated values are shown in the other cells. Two hot spot areas are considered: one in base material at a circular hole with a stress concentration factor (SCF) of 3.0, and the other at a doubling plate some distance away from the hole, such that the stress concentration due to the hole is reduced to 1.15. A preliminary design is considered, such that a Weibull shape parameter h = 1.10 is assumed. The maximum nominal stress range for the deck structure is requested. The plate thickness is 30 mm and the design life is assumed to be 20 years. A mean zero up-crossing period of 6.3 seconds is assumed. S-N curves in air are used. The details of the main cells used for the calculation of fatigue damage are shown in Table 10.2. When this calculation sheet has been prepared, it can be used to estimate maximum allowable nominal stress range in cells D5 and E5, based on trial and error, until a fatigue life of 20 years has been calculated in cells D29 and E29.

The reader may find it useful to prepare such a calculation sheet for simplified assessment of fatigue of marine structures.

10.5 Probability of Being Exceeded

The concept of probability level is used in the design of ship structures. It is also used in fatigue analysis related to calculation of side pressure loads on floating production vessels (DNV-RP-C206 2012). This notation has a different meaning from that of annual probabilities of failure as described in Sections 12.1.2 and 12.1.3. The notation of probability level or probability of being exceeded is used with reference to a Weibull description of a long-term distribution of stress ranges or wave heights. The probability level, as used here, is defined as the probability that a value is exceeded. For example, if there are $n_0 = 10^8$ load cycles during 20 years, the probability that the maximum wave height or a loading is exceeded during these load cycles in these 20 years is $1/n_0 = 10^{-8}$. Thus, it can be seen that this is rather different from that of an annual probability, which is a more frequently used expression in assessing the safety levels of marine structures. Assuming that the long-term wave height distribution has a Weibull distribution, a similar equation for the relationship between wave height and probability of being exceeded, as for stress range in equation (10.9), can be derived as:

$$H = H_{max}\left(1 - \frac{\log n}{\log n_0}\right)^{1/h} \tag{10.32}$$

A probability level referred to in DNV CN 30.7 (2014), "Fatigue Assessment of Ship Structures," of 10^{-4} means that the wave height corresponding to this probability level for a Weibull shape parameter of 1.0 is:

$$H = H_{max}\left(1 - \frac{\log 10^4}{\log 10^8}\right)^{1/h} = \frac{1}{2}H_{max} \tag{10.33}$$

where H_{max} is the maximum wave height during 20 years.

Thus, from this equation it can be seen that a probability level of 10^{-4} corresponds to half a maximum wave height during 20 years, with a Weibull shape parameter, h, of 1.0. Thus, this wave height is exceeded 10,000 times in the course of 20 years. The shape parameter for the stress range distribution can differ from that for wave height, depending on the relationship between wave height and load on the structure as well as on the possible dynamic response of the structure.

Equation (10.2) may be used directly for calculation of stress ranges at different probability levels, as the scale parameter is a given value within a specified Weibull distribution. Thus, the relationship between two different probability levels can be presented as:

$$q = \frac{\Delta\sigma_0}{(\ln n_0)^{1/h}} = \frac{\Delta\sigma}{(\ln n)^{1/h}} \tag{10.34}$$

This gives:

$$\Delta\sigma = \Delta\sigma_0\left(\frac{\ln n}{\ln n_0}\right)^{1/h} = \Delta\sigma_0\left(\frac{\log n}{\log n_0}\right)^{1/h} \tag{10.35}$$

This equation is practical for transformation of stress ranges between different probability levels. Thus, considering, for example, an offshore structure subjected to wave loading during 20 years, the number of cycles is typically 10^8. The largest stress range during 20 years can be denoted as $\Delta\sigma_{20}$. The corresponding number of cycles over 100 years is $5 \cdot 10^8$. The largest stress range during 100 years can be denoted as $\Delta\sigma_{100}$. Equation (10.35) can be used to establish the relationship between these two stress ranges.

$$\begin{aligned}
\Delta\sigma_{20} &= \Delta\sigma_{100}\left(\frac{\log n_{20}}{\log n_{100}}\right)^{1/h} \\
&= \Delta\sigma_{100}\left(\frac{\log 10^8}{\log 5 \cdot 10^8}\right)^{1/h} = \Delta\sigma_{100}(0.92)^{1/h}
\end{aligned} \tag{10.36}$$

This equation is presented in some design standards for marine structures.

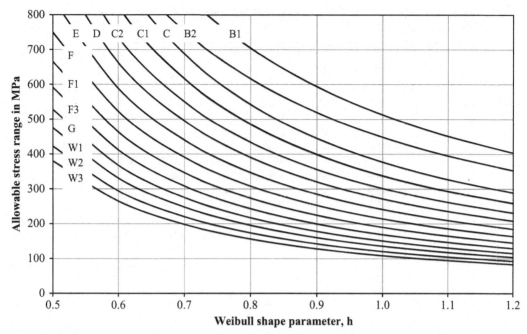

Figure 10.3. Allowable extreme stress range during 10^8 cycles for components in air (DNVGL-RP-C203 2016).

10.6 Maximum Allowable Stress Range

10.6.1 Design Charts

Equation (10.24) or the calculation sheet from Section 10.4 can be used to calculate the maximum allowable stress range during a considered lifetime period by requiring that the fatigue damage is limited to D = 1.0. Based on this, a design chart with allowable stress ranges for a design life of 20 years with 10^8 stress cycles is derived, as shown in Figure 10.3. The allowable stresses are derived for different Weibull shape parameters, and the S-N curves in air shown in Table 4.1. Numerical values for the allowable stress ranges are shown in Table 4.3.

10.6.2 Effect of Design Fatigue Factor and Other Design Lives

The allowable maximum stresses in the design chart in Section 10.6.1 are derived for a Design Fatigue Factor (DFF) of 1.0. For further description of DFF, see Section 12.7. Larger DFFs imply a larger number of stress cycles, which reduces the maximum allowable stress range to achieve the same acceptable calculated fatigue damage. Furthermore, the in-service life may not necessarily be 20 years. Thus, to use the design chart in Section 10.6.1, the allowable stress range can be decreased by a reduction factor that has been calculated for different shape parameters in Table 10.4. In-service lives other than that of 20 years can be accounted for by including the utilization factors in Table 10.5. These factors are also related to the value of the DFF, or, more directly, to the inverse of the DFF. The resulting fatigue damage utilization factor in Table 10.4 is then used to obtain a reduction factor for fatigue assessment of welded

Table 10.3. *Allowable extreme stress range in MPa during 10^8 cycles for components in air (DNVGL-RP-C203 2016)*

S-N curves	Weibull shape parameter, h							
	0.50	0.60	0.70	0.80	0.90	1.00	1.10	1.20
B1	1,449.3	1,092.2	861.2	704.7	594.1	512.9	451.4	403.6
B2	1,268.1	955.7	753.6	616.6	519.7	448.7	394.9	353.1
C	1,319.3	919.6	688.1	542.8	445.5	377.2	326.9	289.0
C1	1,182.0	824.0	616.5	486.2	399.2	337.8	292.9	258.9
C2	1,055.3	735.6	550.3	434.1	356.3	301.6	261.5	231.1
D	949.9	662.1	495.4	390.7	320.8	271.5	235.4	208.1
E	843.9	588.3	440.2	347.2	284.9	241.2	209.2	184.9
F	749.2	522.3	390.8	308.2	253.0	214.1	185.6	164.1
F1	664.8	463.4	346.7	273.5	224.5	190.0	164.7	145.6
F3	591.1	412.0	308.3	243.2	199.6	169.0	146.5	129.4
G	527.6	367.8	275.2	217.1	178.2	150.8	130.8	115.6
W1	475.0	331.0	247.8	195.4	160.4	135.8	117.7	104.0
W2	422.1	294.1	220.1	173.6	142.5	120.6	104.6	92.5
W3	379.9	264.8	198.2	156.0	128.2	108.6	94.2	83.2

connections in an air environment. The factor from Table 10.4 is multiplied by the allowable stress ranges from Table 10.3 to derive an allowable maximum stress range that corresponds with the planned in-service life and the required DFF.

10.6.3 Some Guidance on Selection of a Weibull Shape Parameter

The calculated fatigue damages are sensitive to the value selected for the Weibull shape parameter, and this makes it somewhat difficult to use in several situations. This concept may be used for design assessment at an early design stage, or for screening purpose of significant areas for more detailed fatigue assessment. In such

Table 10.4. *Reduction factor on stress to correspond with utilization factor, η, for C–W3 curves in an air environment (DNVGL-RP-C203 2016)*

Fatigue damage utilization, η	Weibull shape parameter h							
	0.50	0.60	0.70	0.80	0.90	1.00	1.10	1.20
0.10	0.497	0.511	0.526	0.540	0.552	0.563	0.573	0.582
0.20	0.609	0.620	0.632	0.642	0.652	0.661	0.670	0.677
0.22	0.627	0.638	0.648	0.659	0.668	0.677	0.685	0.692
0.27	0.661	0.676	0.686	0.695	0.703	0.711	0.719	0.725
0.30	0.688	0.697	0.706	0.715	0.723	0.730	0.737	0.743
0.33	0.708	0.717	0.725	0.733	0.741	0.748	0.754	0.760
0.40	0.751	0.758	0.765	0.772	0.779	0.785	0.790	0.795
0.50	0.805	0.810	0.816	0.821	0.826	0.831	0.835	0.839
0.60	0.852	0.856	0.860	0.864	0.868	0.871	0.875	0.878
0.67	0.882	0.885	0.888	0.891	0.894	0.897	0.900	0.902
0.70	0.894	0.897	0.900	0.902	0.905	0.908	0.910	0.912
0.80	0.932	0.934	0.936	0.938	0.939	0.941	0.942	0.944
1.00	1.000	1.000	1.000	1.000	1.000	1.000	1.000	1.000

Table 10.5. *Utilization factors, η, as a function of design life and DFF (DNVGL-RP-C203 2016)*

DFF	Design life in years						
	5	10	15	20	25	30	50
1	4.0	2.0	1.33	1.00	0.80	0.67	0.40
2	2.0	1.0	0.67	0.50	0.40	0.33	0.20
3	1.33	0.67	0.44	0.33	0.27	0.22	0.13
5	0.80	0.40	0.27	0.20	0.16	0.13	0.08
10	0.40	0.20	0.13	0.10	0.08	0.07	0.04

cases, shape parameters that are on the safe side should be assumed. For example, if the shape parameter for a floating structure is expected to be approximately 1.0, a value in the range of 1.05–1.10 may be used in order to be more on the safe side. Some guidance on selection of shape parameters for fatigue assessment of ship structures can be found in DNV CN 30.7 (2014), where the shape parameter is presented as a function of ship length. It is also a function of position in the ship under consideration. The shape parameter in the Weibull distribution was formerly approximately 1.0 for floating structures and for structures where the load response is governed largely by the mass term in the Morison equation for load calculation (see Naess and Moan 2013). Where the loading is governed more by the drag term, such as for structures with slender elements – for example, in jackets and flare towers – providing advice on the appropriate shape parameter is more difficult, as it can be in a larger range, from 0.5 to 0.9. Thus, a fatigue assessment based on a 100-year load, for example, becomes rather uncertain. This uncertainty would be significantly reduced if a load response corresponding more to that giving the largest contribution to fatigue damage is also calculated. If a stress range around the region where the largest contribution to fatigue damage is known, it is less important to know the largest stress ranges accurately. This can be assessed based on the figure for contribution to fatigue damage, as shown in Figure 1.2. The following guidance on shape parameter is found in literature (see, e.g., API RP2A-WSD 2014):

- 0.5 for static base shear in Gulf of Mexico jackets and truss spars
- 0.7 for waterline braces and dynamic shear in Gulf of Mexico and also tension leg platform pontoons
- 1.0 for North Sea, South China Sea, and Southern California (static shear)
- 1.3 for North Sea, South China Sea, Southern California (dynamic), and West Africa (persistent swell).

10.6.4 Example of Use of Design Charts

Assume that a topside stool is welded to a deck structure on a floating production vessel. The aim is to perform assessment for a fatigue design life of 25 years. A DFF of 2 is used in the design. The stool is so large that the length in the stress direction implies that the S-N curve F3 should be used for the fatigue assessment, according to appendix A of DNVGL-RP-C203 (2016). A Weibull shape parameter h = 1.0 is assumed. From Table 10.3, this gives an allowable stress range of 169 MPa during 20 years. A utilization factor of 0.40 for 25 years' service life and DFF = 2 is derived from Table 10.5, and a reduction factor of 0.785 on allowable stress range is derived from Table 10.4. Thus, the allowable stress range is reduced to $169 \cdot 0.785 = 132.67$ MPa for a plate thickness of 25 mm. The allowable stress range should be further reduced if the plate thickness, t_{actual}, is larger than 25 mm by a factor $(25/t_{actual})^k$, with k = 0.25, due to the thickness effect for the F3-curve from Table 4.1.

10.7 Combined Load and Response Processes

10.7.1 General

A combined load and response process is understood to mean that two load processes that occur at the same time are superimposed, resulting in one long-term distribution of stress ranges for calculation of fatigue damage. Calculation of fatigue damage for each of these processes is relatively straightforward, but the fatigue damage for the combined process can be more difficult as the stress ranges are raised in the inverse negative slope of the S-N curve for calculation of fatigue damage. Therefore, it becomes non-conservative to simply add together the calculated fatigue damage from each of these processes. Fatigue damage can be summed if the processes are sequential in time, such as for calculation of fatigue damage due to the dynamics from pile driving combined with the following in-service accumulation of fatigue damage due to wave action on the installed structure. Similarly, transport of a structure to the installation field followed by in-service accumulation of fatigue damage due to wave action on the structure is an example of sequential processes where the fatigue damage from each process can be added together. The more complex dynamic analysis under simultaneous load action is described by Naess and Moan (2013). Examples of combined processes include:

- dynamic loading of a flare tower subjected to wind dynamics and that is supported on a jacket structure subjected to wave action;
- wind turbine structures that are subjected to loading from both wind, rotor actions and wave actions;
- floating production vessels that are subjected to loading from both wave and swell actions (see DNV-RP-C206 2012; DNVGL-RP-C203 2016).

Combined loading may also result from operational loads that occur on a structure in addition to the response from environmental loading, as presented in the following section.

10.7.2 Example of Fatigue Analysis of Pipes on a Floating Production Vessel

Pipes are installed onboard a floating production vessel for operation of some machinery that is a component of the production equipment in the topside structure. The pipes are fabricated with welding from the outside only, which implies single-sided welds for S-N classification. It is assumed that the F3-curve is applicable for fatigue assessment of these connections. The pipes are subjected to an internal pressure of 20 bars during 1,000 cycles in the 20 years' service life of the vessel. The temperature inside the pipes increases from ambient to 310°C during each load cycle. The question then arises regarding how to calculate the fatigue life of the butt welds in these pipes. It is assumed in this example that the calculated fatigue damage, D_w, due to wave action is based on detailed fatigue analysis or on a Weibull long-term stress range distribution, and is calculated using S-N curves for the ambient temperature. Next, it is assumed that the stress cycles due to operation of the machinery are distributed evenly over the vessel's service life. It is assumed that the stress ranges due to operation of the machinery is added to the Weibull long-term stress range distribution, which corresponds to load cycles over 20 years. The fatigue damage due to this new load distribution, D_{w+o}, is then calculated numerically, using an S-N curve that is representative for a temperature of 310°C. By this analysis, too many operational load cycles result, such that the calculated fatigue damage in this example must be reduced by a factor $n_{operation}/n_{waves} = 1000/10^8 = 10^{-5}$. The resulting fatigue damage, D_{tot}, can then be derived as a sum of the two calculated fatigue damages (fatigue damage due to waves only plus fatigue damage due to wave and operational response). For operational load cycles, the effect of the wave load response is included twice. This can be corrected for, even if it is of negligible significance for the numerical results in this example. The total fatigue damage is then derived as:

$$D_{tot} = D_w \left(1 - \frac{n_{operation}}{n_{waves}}\right) + D_{w+o}\frac{n_{operation}}{n_{waves}} \tag{10.37}$$

In this section, this concept is now illustrated by an example with numerical values. The stress range in the pipes due to operational loads is 140 MPa, and it is assumed that there are 1,000 operational cycles during the lifetime. This stress range is assumed to include the effects due both to internal pressure and to temperature loads. The 100-year loading condition due to wave action on the vessel gives 150 MPa as nominal stress range in the pipes. It is assumed that a Weibull shape parameter, $h = 1.0$, can be used to transfer the load to 20 years by using equation (10.36). This gives $\Delta\sigma_{20} = 137.95$ MPa. It is assumed that there are 10^8 cycles due to wave loading in the course of 20 years. S-N curve F3 from Table 4.1 in the air environment is used for the fatigue analysis. By using the closed-form equations from Section 10.2, fatigue damage of $D = 0.468$ is calculated for 20 years of service life. In order to add on the stress ranges for 1,000 operational stress cycles, a numerical procedure is required. First, the accuracy for the same Weibull long-term stress range distribution is calculated. The results are shown in Table 10.6. The long-term stress range distribution is divided into 20 blocks, where the stress values in the blocks are calculated for the number of cycles being exceeded in the left column of the table. The corresponding stress ranges, as derived from equation (10.9), are listed in the second column of the table. Within each block a mean stress range is calculated, as shown in the third column of the table, while calculated fatigue damage within each block is shown in the fourth column. It can be seen that the calculated fatigue damage obtained using this procedure is slightly larger than that obtained by the analytical approach. The S-N data that correspond to a temperature of 310°C are then calculated. From equation (4.14), this gives $\log a_{d1} = 11.306$ and $\log a_{d2} = 14.176$. The fatigue damage for this S-N curve is then calculated using the closed-form equation (10.24), with the result

Table 10.6. *Calculated fatigue damage for long-term stress distribution*

Number of cycles	Stress range	Mean stress range in block	Calculated fatigue damage at ambient temperature	Calculated fatigue damage at 310°C	Additional fatigue damage due to 1,000 cycles at stress range 140 MPa
1	137.95	131.92	2.61E-05	4.54E-05	3.97E-09
5	125.90	123.30	2.67E-05	4.63E-05	4.51E-09
10	120.71	114.68	0.000171	0.000298	3.26E-08
50	108.65	106.06	0.000167	0.000295	3.68E-08
100	103.46	97.44	0.00105	0.001830	2.64E-07
500	91.41	88.82	0.00100	0.001732	2.96E-07
1,000	86.22	80.19	0.00587	0.010201	2.11E-06
5,000	74.17	71.57	0.00521	0.009065	2.34E-06
10,000	68.98	62.95	0.02838	0.049346	1.65E-05
50,000	56.93	54.33	0.02280	0.039653	1.81E-05
100,000	51.74	45.71	0.10866	0.188909	0.00013
500,000	39.68	37.09	0.07256	0.126140	0.00014
1,000,000	34.49	28.47	0.19822	0.456317	0.00094
5,000,000	22.44	19.85	0.04080	0.102556	0.0010
10,000,000	17.25	11.23	0.01889	0.047489	0.0068
50,000,000	5.20	2.60	1.58E-05	3.98E-05	0.0072
100,000,000	0.01				
Calculated fatigue damage			0.5039	1.0340	0.0163

$D = 0.936$. It can be seen that the numerical value is slightly conservative, with calculated $D = 1.034$. Finally, the damage due to 1,000 constant stress ranges at 140 MPa, in addition to wave load response, is calculated in the last column in Table 10.6. It can be seen that this additional fatigue damage is relatively small. Equation (10.37) can now be used in the calculation of the total fatigue damage:

$$D_{tot} = 0.936 \left(1 - \frac{1000}{10^8}\right) + 0.0163 = 0.952 \qquad (10.38)$$

10.8 Long-Term Loading Accounting for the Mean Stress Effect

Methods for derivation of long-term loading are needed that include the mean stress level, in which the mean stress effect is accounted for. This can be relevant for the following examples:

- Fatigue assessment of details in the base material fabricated without significant residual stress, where part of the stress cycle is compressive (see Section 3.3.3).
- Fatigue assessment of welded details where the residual stresses have been removed by post-weld heat treatment and where part of the stress cycle is compressive (see Section 3.3.5).
- Planning of in-service inspection for fatigue cracks in floating production vessels (see Section 3.3.6).
- Assessment of fatigue crack growth using fracture mechanics, when crack closure due to partly compressive load cycles is included in the analysis (see Section 16.9).

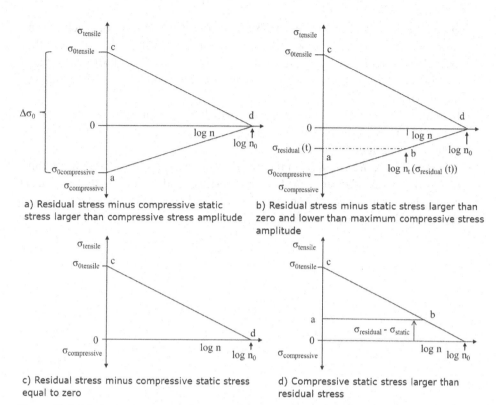

a) Residual stress minus compressive static stress larger than compressive stress amplitude

b) Residual stress minus static stress larger than zero and lower than maximum compressive stress amplitude

c) Residual stress minus compressive static stress equal to zero

d) Compressive static stress larger than residual stress

Figure 10.4. Methodology for calculation of equivalent fatigue driving stress as a function of loading.

- Assessment of structural capacity of grouted connections, where the capacity is significantly reduced if reversed sliding occurs, as compared with loading in one direction only (see Section 17.2.8).
- The effect of mean stress can be illustrated using the Weibull distribution of the long-term stress (or force, for grouted connections) as indicated in Figure 10.4. This figure is made for the purpose of analysis of fatigue in a connection with residual stress from fabrication. However, a similar approach can also be used for assessment of the structural capacity of grouted connections as shown in Section 17.2.8. Figure 10.4 illustrates a long-term stress range exceedance diagram that is similar to a Weibull two-parameter distribution, with shape parameter $h = 1.0$ and scale parameter $q = \Delta\sigma_0/(\ln(n_0))^{1/h}$. Here, $\Delta\sigma_0$ is the largest stress range out of n_0 stress cycles. In Figure 10.4a, it is assumed that the tensile residual stresses are so large that the mean stress from the wind and wave loading and the static load from self-weight do not influence the effective stress range that is to be used for fatigue damage calculation. The loading consists of a tensile part and a compressive part, such that $\Delta\sigma_0 = \Delta\sigma_{0tensile} + \Delta\sigma_{0compressive}$. Thus, the full stress range will contribute to the calculated fatigue damage.
- In Figure 10.4b it is assumed that the largest compressive stress exceeds that of the residual stress, such that the effective long-term stress range distribution that tends to drive crack growth is within the area a-b-d-c. Integration of effective stress range for analysis of crack growth can be performed by integration

of crack-driving stresses from a to b, and another integration of crack-driving stresses from b to d. Crack closure is assumed for stress cycles in compression that exceed the compressive stress indicated by line a-b. Thus, crack-driving stresses are understood to refer to the stress ranges above this line that tend to open the fatigue crack.

- In Figure 10.4c, a situation is indicated in which the compressive part of the load equals the tensile residual stress, such that only the tensile stresses should be included for fatigue life calculation.
- In Figure 10.4d, the stress from the mean static load and from wind, waves, and the vertical load is larger than the residual stress, and only the long-term stress range distribution within a-b-c is required for calculation of the effective stress range that drives crack growth.

11 Fabrication

11.1 General

It is important that designers work in close cooperation with fabricators to achieve structures that are suitable for welding. Simple fabrication and the possibility of access for welding and non-destructive testing (NDT) during fabrication should be planned in order to realize structures that fulfill specified requirements. In order to weld braces to chords without an overlap, there should be a gap between the braces not less than 50 mm, and the brace angle should not be below 30° to allow for proper welding at the heel area (see Figure 8.1). Seam welds and girth welds in tubular sections are usually placed outside the main hot spot areas in tubular joints. Recommendations regarding this can be found in design and fabrication standards such as ISO 19902 (2007), API RP 2A (2014), and NORSOK M-101 (2011).

11.2 Selection of Material

Material should be selected to meet requirements in material and fabrication standards, in addition to requirements for yield strength with respect to design for the Ultimate Limit State, as presented in Section I.4. Where the Fatigue Limit State governs the design, using material with a very high-yield strength is not recommended, as the fatigue strength of most steels does not particularly depend on the yield strength when it is welded, as explained in Section 4.4. It is important to use materials with documented properties showing good weldability. Where significant stresses from in-service loads or from fabrication are going to be transferred in the thickness direction, it is important to use material with documented through-thickness properties. It is also important to use material with sufficient ductility and fracture toughness for the lowest in-service temperature that can be expected during the lifetime of the structure. Thus, the grade of steel to be used in different structural parts is generally related to service temperature and thickness. Fracture toughness becomes reduced after welding and this should be accounted for when material is selected. For marine structures it is important to consider the actual environment, potential corrosion, and corrosion protection for material selected for a specific project.

11.3 Welding

Welding is a process in which notch-like imperfections and deviations in geometry from nominal are difficult to avoid completely. The degree of severity of such imperfections depends primarily on the level of control of the welding process, but also on other factors such as type of welding process, the materials, the joint geometry, and the availability of access for welding. The size of imperfection that can be tolerated without compromising fatigue life, based on classified details, decreases as the requirement for using higher S-N classes increases. The S-N class C (or FAT 125 in IIW)) is the highest class in DNVGL-RP-C203 (2016) for butt welds with a cyclic stress range acting normally to the weld direction. The use of this S-N class involves removal of all local stress raisers, such as weld toes, and welding access from both sides is normally required, together with high-quality NDT; see also Section 11.4. When tubular girth welds are made, welding from the inside for small-diameter tubular members may be difficult, and for such cases achieving a C class detail is more difficult. Welding can be performed from the outside, provided that there is some excess thickness (1–2 mm) on the inside that can be removed by machining after welding in order to achieve the required flush surface, provided there is access for this work to achieve the highest S-N classes. If the component is more complex and the welding is different from that of a flat position, welding access is reduced and there is a greater likelihood of producing larger imperfections. Thus, in order to achieve a correspondingly high S-N classification for details that are not welded in flat position, the NDT should be significantly enhanced. Otherwise, the practice is to recommend a reduced S-N class for joints that are not made in a flat position. The requirement for local geometry decreases for the lower S-N classes, such as for the W3-curve as presented in Section 4.1.3. This also relates to requirements for the remaining defects in the welds after fabrication. This can be clearly seen from the requirements for classification of details in appendix A of DNVGL-RP-C203 (2016). The direction of stress fluctuations with respect to the weld axis is also an important factor for classification of details with respect to fatigue strength. Experience indicates that the more severe weld imperfections, such as cracks, lack of fusion, lack of penetration, undercut, and slag inclusions, are oriented parallel to the weld axis and are therefore much more severe, in terms of fatigue strength, when the stress is transverse, rather than parallel, to the weld axis. Local stress raisers, such as weld toes, are similarly more unfavorably oriented with respect to transverse stresses. This is reflected in the tables for joint classification where, for example, a continuously fillet-welded T-joint, without start and stop positions, and with longitudinal stress ranges, may be acceptable for a C class. However, if this detail is stressed transversely through the base plate in a cruciform joint, then it would be classified as E or F, depending on the thickness of the transverse plate.

The control of welding in production is generally achieved by the following measures:

a) Use of a relevant welding procedure specification (WPS). The specified welding procedure is normally qualified through relevant mechanical qualification testing (WPQT) for significant details.

b) Welding being conducted by welders who are qualified to perform the work according to the specified or qualified welding procedure.

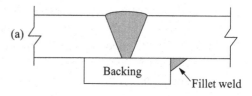

Figure 11.1. Welding on backing bars.

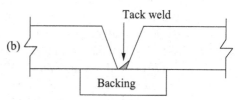

c) Following an appropriate welding quality management system.
d) Acceptable practice in shop for storage and handling of welding consumables to prevent dampness and contamination that may lead to hydrogen cracking in the weld.
e) Appropriate inspection and acceptance criteria for the production of the welds.

These measures are normally specified in a fabrication specification or a fabrication standard that is referred to, such as ISO 5817 (2014). It is important that there is equivalence between the assumptions made for the S-N classification and the actual quality that is achieved during fabrication.

In some cases, welding is performed from one side against a backing bar. Such welding may be performed, for example, in girth closure welds, as presented by Stacey et al. (1997). A methodology is presented that allows the use of backing bars without fillet-welding the backing bar to any of the plates outside the main weld groove. It should be noted that welding the backing bar outside of the weld groove by fillet welds as shown in Figure 11.1a typically reduces the detail classification from F to G. Remaining fillet welds on the backing plates should be avoided. Instead, some tack welds may be used in the main groove as shown in Figure 11.1b if the subsequent welding will melt the tack welds such that they become part of the main weld. These tack welds should then be part of the welding procedure. Use of backing bars enables use of a wide weld groove, such that a full penetration in the root can be achieved. However, a permanent backing bar also reduces the possibility of detecting defects in the root area by NDT. Different types of backing bars may be used depending also on the geometry of the parts to be joined.

11.4 Defects

NDT is performed to remove initial unacceptable defects in the welds. The type of inspection method used depends on the type of joint and defect to be detected. Furthermore, the amount of testing depends on the consequences of fatigue failure. Some testing may efficiently reduce the probability of systematic errors in the fabrication method. Surface-breaking defects can significantly reduce the fatigue life and should be avoided, and can be efficiently detected by magnetic particle inspection (MPI) for ferromagnetic materials. Normally 100% of inspection of welded details subjected to dynamic loading is performed by MPI, as this is a rapid testing method and not particularly expensive. Ultrasonic testing (UT) is used for detection of internal flaws, and these may also be detected by radiographic testing (RT). Acceptance criteria are needed for assessment of the inspection results, and whether repair of the weld is necessary is a frequent question. The acceptance criteria are strongly dependent on the S-N classification used in design. However, this is normally not well reflected in existing fabrication standards. When using ISO 5817 (2014) as the fabrication standard it is recommended to refer to its annex C for requirements to components subjected to dynamic loading. This is recommended for the higher S-N classes (see Hobbacher and Kassner 2012); thus, reference is made to annex C of ISO 5817 (2014) and to quality level ISO 5817-B125 for the highest S-N class for welded connections in IIW, which is FAT 125 (Hobbacher 2009), which is the same as S-N curve C in DNVGL-RP C203 (2016). Planar imperfections are the most severe in terms of reduction of fatigue strength, due to their sharp boundaries. These imperfections include cracks, lack of fusion, incomplete penetration, sharp undercuts, and overlapping. The latter involves lack of fusion and a profile that hinders reliable NDT. In general, the acceptance sizes for imperfections for the highest S-N classes are so small that the probability of detection and sizing with the necessary accuracy is low. It is also difficult to distinguish between planar and volumetric imperfections in such cases. Planar imperfections at weld toes are the most critical as they are located in regions of high stress concentrations. Shallow surface imperfections can simply be removed by surface grinding, and further rewelding is usually unnecessary provided that a smooth profile is achieved.

The requirements for acceptance criteria and the relationship between fabrication and S-N curves vary somewhat among the different standards. In BS 7608 (2014), the acceptance criteria are related to the S-N curves such that the strictest criteria are associated with the highest S-N curves. In the NORSOK M-101 (2011), the acceptance criteria are related to the consequence of failure. Although these methodologies are different, both seem logical. A further refinement might be to combine these two approaches. In both standards some porosity and slag inclusions are accepted; however, planar imperfections, such as lack of fusion and cracks, are not allowed.

The fabrication tolerances in terms of geometry required in offshore projects today are rather similar to those used about 40 years ago; see NORSOK M101 (2011), NPD (1977), and DNV (1977).

11.5 Fabrication Tolerances

Control of fabrication tolerances is important, especially for butt welds and cruciform joints. Effects of misalignment are considered in greater detail in Section 5.3.3. There are also requirements regarding the shape of the weld; for example, the height of the weld convexity in butt welds should not be greater than 10% of the weld width, with a smooth transition to the plate surface in order that the D-curve (FAT 90) can be applied.

The effect of weld shape on fatigue strength is indicated in Figure 11.2, which is based on Gurney (1979). There are rather few fatigue test data points behind this figure. However, this figure has sometimes been used to estimate the relative reduction in fatigue strength when standard fabrication criteria with respect to geometry have not been met.

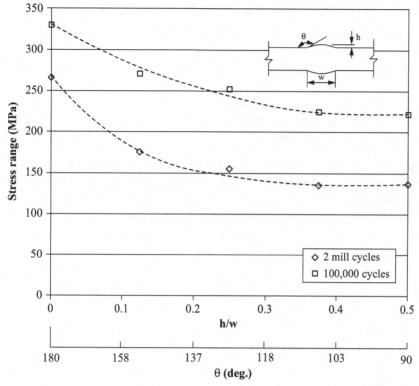

Figure 11.2. Effect of weld shape on fatigue strength, based on Gurney (1979).

Engesvik and Moan (1983) presented fatigue test data with measured weld toe radius and weld toe angles, and based on probabilistic fracture mechanics analysis they found that the weld shape and especially the toe radius contributed significantly to the scatter in the fatigue test data. Thus, a nice weld profile should be achieved during fabrication; otherwise, post-weld improvement might be recommended if a high fatigue strength is aimed for. The local weld shape is considered to depend on the welding method.

Studies on the significance of weld profile in tubular joints have been presented by Marshall (1993) and Maddox et al. (1995).

11.6 Non-Destructive Testing for Defects

11.6.1 General

Different methods of NDT are used to ensure that there are no significant defects in a structure after fabrication. NDT methods are understood to refer to inspection and testing or measuring weld geometry and potential weld defects without damaging the material or connection. These methods are also used for detection of potential fatigue cracks in structures that are subjected to significant dynamic loading during in-service life. All inspection methods and techniques have limitations with respect to their effectiveness and accuracy. This should be considered when the testing methods are selected and when the resulting data are assessed. The capabilities of the various individual inspection methods and techniques generally depend on:

- metallurgy of the inspected area;
- NDT technique and procedure employed;
- capabilities of the equipment used;
- flaw type and orientation of the flaws;
- local geometry and access for inspection;
- working conditions, including lighting;
- operator training and experience.

For all techniques, it is assumed that they are performed by suitably trained and certified operators and that there are well-designed testing procedures (DNVGL-OS-401 2015 and NORSOK M-101 2011). However, experience shows that two different operators using UT may detect and size defects rather differently from each other at more complex connections. Thus, according to the situation, independent testing is often recommended. As NDT methods differ in their capabilities of detecting different types of defects, combining a range of methods may improve the probability of detecting potentially harmful defects. The following NDT techniques are used most frequently in the fabrication and maintenance of marine structures:

- Visual Inspection (VI)
- Magnetic Particle Inspection (MPI)
- Eddy Current (EC)
- Penetrant Testing (PT)
- Ultrasonic Testing (UT)
- Radiographic Testing (RT)
- Alternating Current Field Measurement (ACFM)

NDT of welds should normally not be performed until 48 hours has elapsed since completion of welding, due to possibility for hydrogen cracking. This is because, with hydrogen cracking, it takes some time before the cracks develop. MPI or EC are the preferred techniques for detection of surface imperfections in ferromagnetic materials. For detection of surface imperfections in non-magnetic materials, either liquid PT or EC is preferred. UT and/or RT are used for detection of internal imperfections. RT may be used to supplement UT, or vice versa, in order to increase the probability of detection, or for characterizing/sizing the type of flaws that can be inspected. UT is required for detection of planar imperfections, while RT is preferred for detection of volumetric imperfections.

Planar defects include:

- lack of fusion between weld and base material;
- lack of fusion between welding passes;
- lack of penetration;
- cracks.

Volumetric defects include:

- pores;
- clusters of pores;
- slag.

Volumetric defects are not regarded as critical as planar defects with respect to fatigue and unstable fracture. Planar defects are not normally accepted in structural connections that will be subjected to significant dynamic loading.

11.6.2 Visual Inspection

VI is the most important inspection method for assessment of surface geometry of a weld, such as excess height of the weld and undercut, that is, the shape of the transition between the weld and the base material at the weld toe. Most fatigue cracks initiate from these areas, and therefore they need to be properly inspected and assessed during fabrication. VI may also include use of magnifying glasses for detection of defects and use of gauges, rulers, and other measuring tools for control of weld geometry. Assessment of surface geometry and detection of potential surface defects must be done under good lighting. It is easier to detect fatigue cracks in a white-coated surface than in a dark or corroded surface. VI can be carried out on a wide range of objects. The condition should be documented by photography or video recording.

11.6.3 Probability of Detection by Visual Inspection

Little information is available regarding probability of detection (PoD) data for close VI, although Fujimoto et al. (1996, 1997) have presented some curves based on experience and judgment without testing. Assuming that access for VI is moderate, cracks will be rather deep before they can be detected by VI. Where plate thicknesses are moderate, this implies that the cracks will have grown through half the plate thickness before they are likely to be detected. There may then be only a short period before the cracks grow through the entire thickness. Thus, the probability of detecting a crack by VI alone while it is so small that it can be repaired by grinding is very low.

11.6.4 Magnetic Particle Inspection

MPI is used to detect microcracks and surface cracks at weld toes. The probability of detecting such cracks by MPI depends on the depth and size of the crack, as indicated for a typical case in Figure 11.3. MPI has been used in air and underwater for many years. It is the most commonly used NDT method for detecting surface-breaking defects in welds, and is easily carried out using equipment that is well proven. MPI uses the magnetic properties of iron particles that are held in a suspension fluid that is used to cover the area to be inspected. Contrast paint is normally applied to the inspected area first to enable visualization of the pattern of the iron particles. The area for inspection is then subjected to a magnetic field, and the flaws are detected by concentration of the iron particles in the region of magnetic flux leakage around the flaw.

Use of magnetic yokes, using alternating current, is recommended for MPI inspection. As the defects do not have to be completely open to the surface, MPI is considered to be the primary method for detection of surface-breaking flaws in ferromagnetic materials, as it is simple and rapid. One limitation is that it can only

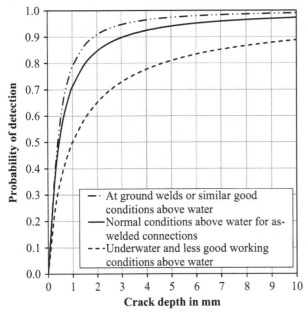

Figure 11.3. Probability of detection curves EC, MPI, and ACFM (DNVGL-RP-0001 2015).

be used on ferromagnetic materials and is not applicable to austenitic stainless steel. Furthermore, MPI cannot be performed on coated surfaces.

11.6.5 Penetrant Testing

Liquid PT methods include the various different techniques in which a liquid is placed on a dry surface and is given time to soak into surface-breaking cracks and cavities by capillary action. After removal of excess liquid, the dye in the cracks and cavities is visualized through the application of a developing suspension that extracts the penetrant such that the defect becomes visible. The advantage of dye PT is that it is simple to use and particularly suitable for fieldwork. It is the primary technique for detecting surface-breaking defects in non-magnetizable materials (like aluminum, copper, magnesium, and titanium) and is also applicable on non-ferromagnetic materials. PT is regarded as a good alternative to MPI, but the defect must be open to the surface. It is considered a simple, rapid method for detecting surface defects in non-ferromagnetic materials, and the equipment required is simple. However, it requires thorough preparation with respect to cleaning and is considered time-consuming compared with MPI.

11.6.6 Ultrasonic Testing

Different UT methods are used for detection of planar internal defects. The testing is performed by transmission and detection of ultrasonic waves. The reflected signal is presented on a cathode ray screen and records any deviations in the material, either through back wall reflections or from reflections of buried or surface flaws. The methodology needs to be carefully calibrated against reference blocks before use.

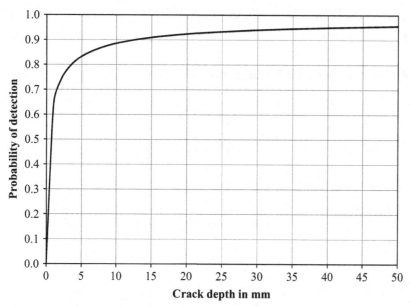

Figure 11.4. PoD curve for UT inspection (DNVGL-RP-0001 2015).

Welded connections in structures are often tested by a handheld probe. For other connections, like girth welds in pipelines, an automated ultrasonic testing (AUT) device is used. UT techniques are well known, and there are many publications on this subject; they are good methods for detecting crack-like defects, but one disadvantage for manual systems is the lack of a permanent electronic or photographic record for retention. The methodology is considered to be very suitable for detection of planar defects below the surface and gives a good cross-sectional positioning of a defect. It may also be used for thickness measurements. Furthermore, the use of UT is not hazardous for other activities. However, the methodology can only be used for thicknesses larger than 10–12 mm; it is not applicable for partial penetration/fillet welds and shows less sensitivity to volumetric defects. Use of UT requires very experienced and well-trained operators.

There are also more refined methods that are based on the same principles, such as Time Of Flight Diffraction (TOFD) technique. This is a UT technique that relies on measurement of the time taken for an ultrasonic wave to be diffracted from a crack tip as it travels from the transmitter to the receiver. TOFD is particularly suitable for sizing defects. A permanent record of an inspected weld can be obtained and can be compared with radiographic testing, particularly for thicker sections. It may also be useful for testing of weld roots. Human factors are a recognized aspect of all manual inspection methods, but even AUT systems require interpretation, and thus operator performance should be regularly checked. Where the consequences from a defect are large, independent testing may be recommended. A PoD curve for UT is shown in Figure 11.4. It is assumed that the inspection is performed by well-trained and qualified personnel under good conditions. When defects are detected by UT, different techniques may be used to size the defects. There is some sizing uncertainty associated with these techniques, and this should be considered if there is to be an engineering assessment of the consequences of a defect. Some guidance on this can be found in BS 7910 (2013).

11.6.7 Probability of Detection for Ultrasonic Testing

The PoD curve for UT is defined by:

$$P(a) = 1 - \frac{1}{1 + (a/X_0)^b} \tag{11.1}$$

where:

    a = crack depth in mm
    X_0 = distribution parameter (= 50% median value for the PoD)
    b = distribution parameter.

The parameters X_0 and b are calculated by curve fitting to experiments: $X_0 = 0.410$ and $b = 0.642$. The PoD curve depends on qualification and execution of work. If no other documentation is provided, the PoD curves in Figure 11.4 can be used for inspection planning. It should also be noted that different versions of UT exist and some of these are more reliable than that represented by this PoD curve.

11.6.8 Radiographic Testing

RT is used to detect volumetric flaws such as gas pores, inclusions, and undercuts. However, it has a generally low capability for detecting planar defects. In this technique, radiation is used to capture an image on film of the defects in a weld or in a parent material. A defect absorbs less radiation than the surrounding material and is therefore recognized as a line on the film. The source of radiation is generally either an X-ray tube or gamma rays from a radioactive isotope. Special precautions are required to protect inspectors from radiation hazards. For material thicknesses above 25 mm radiographic testing should always be supplemented by UT.

11.6.9 Eddy Current

EC detection of defects is based on the principles of electromagnetic induction and is concerned with the interaction of defects in metallic components with the magnetic field generated by a coil carrying an alternating current. Eddy currents are induced in the conductor material due to the alternating flux produced by the coil when an eddy current inspection probe carrying an alternating current is placed close to, or on the surface of, a conductor (such as steel). The induced eddy currents produce, in turn, an alternating magnetic flux that opposes the field produced by the current-carrying coil; this effect is detected as a change in the electrical impedance of the coil that can be measured electronically. Alternatively, the effect of the flux produced by the eddy current is detected by monitoring the voltage induced in a second coil similar to the excitation coil. The magnitude of the eddy current (and hence of the response of the instruments) will be affected by cracking, surface pitting, inclusions, and microstructure – that is, all discontinuities. EC can be used without removal of coating, making it attractive for in-service inspection method for surface cracks. When MPI is performed, the coating must be removed first, and it can be difficult to return it to a similar condition; corrosion of areas inspected using MPI may occur. However, EC may give spurious indications, and MPI is then performed for ferromagnetic materials for further assessment. The physics for applying EC above water is only marginally different from underwater applications and, although working conditions can be more challenging underwater, these are compensated for by special quality assurance measures. A similar performance as underwater is thus also expected above water, and the PoD curve generated is regarded as being representative for above-water applications also.

11.6.10 Alternating Current Field Measurement and Alternating Current Potential Drop Methods

Other methods referred to in literature are the Alternating Current Field Measurement (ACFM) method and the Alternating Current Potential Drop (ACPD) method. ACFM is used for detecting and sizing surface-breaking flaws. ACFM was developed as an extension of the successful ACPD technique. It was initially conceived for use underwater to detect flaws in offshore structures and to overcome the unsuitability of ACPD for such applications because of the need for good electrical contact between probes and the structure's surface. Presently, however, ACFM is applied to structures both in and out of the water. One advantage of ACFM over some other techniques is that the structure requires minimal cleaning and that the method can be applied over paint and other coatings up to several millimeters in thickness.

ACFM is an electromagnetic technique. A sensor probe is placed on the surface to be inspected, and an alternating current is induced into the surface. When there are no defects, the alternating current produces a uniform magnetic field across the surface. Any defect will perturb the current, forcing it to flow around and underneath the defect. This causes the magnetic field to become non-uniform, and sensors in the ACFM probe measure these field variations.

ACPD is understood to refer to measurement of the change in electrical resistance across a flaw when an alternating current potential is applied at either side of the flaw. When an alternating electric current flows between two electrodes connected to the surface of a metal, it tends to flow in a thin layer close to the surface. This current must also follow the profile of a surface-breaking crack. This will result in a voltage drop across the crack that can be measured by a suitable probe. The voltage drop is proportional to the depth of the crack and the current in the test piece. By comparing the voltage drop across a crack and across a similar uncracked area, the crack depth can be assessed. ACPD is mostly used for sizing cracks that have been detected by other methods.

11.6.11 Probability of Detection Curves for EC, MPI, and ACFM

The PoD curve for detection of defects is often quoted, along with a specified confidence limit. PoD curves are normally derived from testing combined with engineering judgment to relate the test data to the performance on actual structures. It is

Table 11.1. *PoD curves for EC, MPI, and ACFM (DNVGL-RP-0001 2015)*

Description	X_0	b
At ground welds or similar good working conditions above water	0.40	1.43
Normal working conditions above water	0.45	0.90
Underwater and less good working conditions above water	1.16	0.90

assumed that the inspection has been performed without hindrance and with good surface conditions. If not, a lower PoD curve is recommended for use. The test data are derived for a relevant crack opening. The inspection reliability may be poorer if the crack opening is smaller or if the structure was subjected to compressive stress at the time of inspection. The distribution functions for PoD for EC, MPI, and ACFM are assumed to be similar and can be presented in the form of equation (11.1). The PoD curves are dependent on qualification and execution of work. If no other documentation is provided, the PoD curves in Figure 11.3 with parameters in Table 11.1 can be used.

11.6.12 Methodology to Provide Reliable PoD Curves for Other Inspection Methods

The reliability of an inspection process depends on:

- the capability of the actual technique;
- the degree of reliance on operator skill;
- the inspection procedure used;
- Auditability.

In its simplest form, PoD is a percentage of cracks detected. However, in practice, this must refer to cracks of a particular size; that is, cracks must be grouped together within certain size ranges. The PoD can refer to crack length or crack depth, and the relevance depends on the inspection technique. PoD provides a basis by which inspection methods can be compared. PoD performance relates only to the trial from which it was derived, and the defects must be real cracks – not artificial defects or slots. Furthermore, the samples must be representative for the components to be inspected in the field (shape, size, and material properties), and there must be sufficient defects to make the trial representative as it is not satisfactory to rely on repeat inspections of the same crack. The way in which the PoD is reported must refer to the way in which the trial was conducted, and is often presented as a PoD curve. The PoD is also a function of the surface of the hot spot areas and depends on the amount of corrosion and cleaning. It is easier to detect fatigue cracks in a white-coated area by VI than in an area that is corroded. The test data should be subject to an engineering assessment before a PoD curve is recommended. The data should be assessed in terms of requirements for cleaning and preparation of areas to be inspected. The working conditions at relevant hot spot areas should be considered. This also means that actual PoD curves cannot be expected to approach 100% detection probability, even for deep cracks, if conditions at the inspection areas are unreliable. Reference is made to Visser (2002) for further reading on this topic.

11.7 Improvement Methods

11.7.1 General

Many fatigue design standards, such as API RP2A-WSD (2014) and ISO 19902 (2007), recommend an improvement in fatigue life by a factor of 2 when the weld toe has been ground. For hammer peening, an improvement factor of 4 is recommended in these standards. Some other standards, such as Department of Energy (1984), HSE (1990), HSE OTH 92390 (1999), BS 7608 (1993), and IACS (2013), recommend an improvement factor on calculated fatigue life of 2.2 for burr toe grinding. DNVGL-RP-C203 (2016) allows for a larger improvement factor, depending on material yield strength, up to a maximum factor of 3.5 for burr grinding and a maximum of 4 for hammer peened joints. It also allows for the use of S-N curves for improved joints that have a less steep slope of the S-N curve when the fatigue strength is assessed, based on a long-term stress range distribution. IIW recommends a factor of 1.3 on a stress range at $2 \cdot 10^6$ cycles for grinding the weld toe (Haagensen and Maddox 2013). A factor of 1.5 on stress range can be used for welds treated with hammer peening. The literature indicates that weld improvement leads to more horizontal S-N curves, and therefore the improvement factor depends on the position considered in the S-N curve. Thus, to present the improvement using factors on fatigue life is not always meaningful unless this factor is simultaneously linked to some long-term stress range distribution. Less improvement is associated with low cycle fatigue (large stress ranges) than for more high cycle loading, such as from wind or wave action. A negative inverse slope, m = 3.5, is used for ground details in DNVGL-RP-C203 (2016), and this is the same as in BS 7608 (2014). Based on constant amplitude test data, Yildirim and Marquis (2012a, 2012b) determined that the slope of the S-N curve m = 5.0 can be used for weld toes improved by ultrasonic peening. It should be kept in mind that the efficiency of this method can be significantly reduced under spectrum loading (Huo et al. 2005), where the compressive residual stresses are lost during large loads (see BS 7608 2014). It should be noted that improvement of the toe does not improve fatigue life if fatigue cracking from the weld root is the most probable failure mode. The considerations in the following sections are for conditions where the weld root is not considered to be a critical initiation point. An example of a welded detail that is well suited for post-weld improvement is shown in Figure 11.5, where a potential fatigue crack is expected to grow from the weld toe into the base material. Where the forces are transferred directly through the weld, the possibility for fatigue cracking from internal defects needs to be assessed as this may put a limitation on the actual effect of post-weld improvement.

Except for weld profiling (Section 11.7.2), the effects from different improvement methods as given in the following sections cannot be added. Reference is made to IIW Recommendations (Haagensen and Maddox 2013) for post-weld improvement with respect to execution. For background, see also Haagensen (1994, 1997), Zhang et al. (2003), and Zhang and Maddox (2009b).

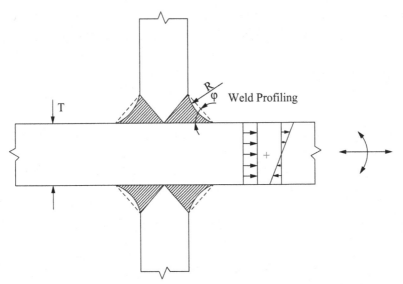

Figure 11.5. Weld profiling of cruciform joint (DNVGL-RP-C203 2016).

11.7.2 Weld Profiling by Machining and Grinding

Weld profiling is understood to refer to profiling by machining or grinding, as profiling using only welding is not considered to be an efficient means of improving
fatigue life. In design calculations, the thickness effect may be reduced to an exponent
0.15, provided that the weld is profiled by either machining or grinding to a radius
of approximately half the plate thickness, $(T/2)$, with the stress direction as shown
in Figure 11.7 (B). Where weld profiling is used, fatigue life can be increased, taking
account of a reduced local stress concentration factor (SCF) that can be obtained by
calculating a reduced hot spot stress as follows:

$$\sigma_{Local\ reduced} = \sigma_{Membrane}\alpha + \sigma_{Bending}\beta \tag{11.2}$$

where:

    α and β are derived from equations (11.3) and (11.4), respectively.

$$\alpha = 0.47 + 0.17(\tan\varphi)^{0.25}(T/R)^{0.5} \tag{11.3}$$

$$\beta = 0.60 + 0.13(\tan\varphi)^{0.25}(T/R)^{0.5} \tag{11.4}$$

For description of geometric parameters, see Figure 11.5. The membrane part and
the bending part of the stress must be separated from the local stress as:

$$\sigma_{Local} = \sigma_{Membrane} + \sigma_{Bending} \tag{11.5}$$

where:

    $\sigma_{Membrane}$ = Membrane stress
    $\sigma_{Bending}$ = Bending stress.

If a finite element analysis (FEA) of the connection under consideration has been
performed, the results from this can be used directly to derive membrane stress and

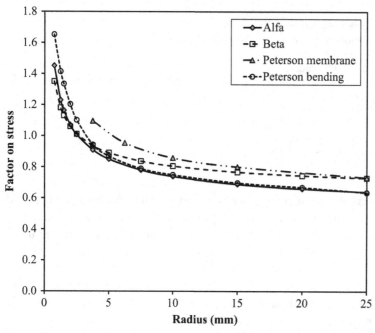

Figure 11.6. Effect of radius on stress at hot spot.

bending stress. For cruciform joints and heavy stiffened tubular joints, it may be assumed that the hot spot stress is mainly due to membrane stress. For simple tubular joints with a large stress concentration factor it may be assumed that the hot spot stress in the chord is mainly due to bending. The reduced local stress in equation (11.2) should be used, together with the same S-N curves for which the detail is classified without weld profiling. It is assumed that $R/T = 0.1$ without weld profiling for a plate thickness $T = 25$ mm. In addition, the fatigue life can be increased by taking account of local toe grinding; reference is made to Section 11.7.3. The grinding radius should still be tangential to the plate surfaces, as indicated in Figure 11.5, with grinding below any indication of undercut. Equations (11.2)–(11.5) were based partly on FEAs and by comparison of SCFs from Peterson's SCFs (1974). The format used for the equations was based on a similar expression for stress concentrations at weld toes, as presented by Ho and Lawrence (1984). A comparison of equations (11.3) and (11.4), with SCFs from diagrams in Peterson (1974) for membrane loading and bending loading, is shown in Figure 11.6. The curve forms are similar, but there are some differences in the absolute values. In order to achieve results that are on the safe side, it was decided to begin from $R/T = 0.1$ without weld profiling for a reference thickness of 25 mm. This corresponds to $R = 2.5$ mm. Measurement of weld toe radius shows mean values that are typically around 1–2 mm; however, the scatter in toe radius is normally significantly larger (Engesvik 1981). The SCFs from Peterson (1974) are related to a smooth transition surface corresponding to S-N curve C or FAT 125. The corresponding S-N curve for the same detail without grinding is classified as F or FAT 71. This information, together with a negative inverse slope of the considered S-N curves of $m = 3.0$, is used to derive the curves in Figure 11.6 in the

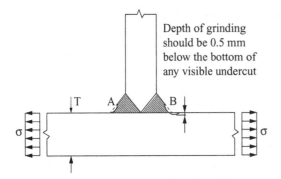

Figure 11.7. Grinding of welds (DNVGL-RP-C203 2016).

same format. With good workmanship, resulting in a smooth transition from the weld to the base material, a further reduction in calculated fatigue damage by a factor of 2 may be accepted. This is based on comparison of calculated fatigue damage using S-N curves that are made especially for ground connections; see Section 11.7.13.

Addition of some weld leg length will move the hot spot to a region with reduced stress; see, for example, Maddox et al. (1995) for the effect of weld profile in tubular joints and Section 9.2.3 for methodology to assess the effect of increased weld leg length in cruciform joints.

It takes much longer to perform weld profile grinding than to perform just toe grinding. The time needed for grinding a weld length of 1 m to achieve a radius of 25 mm has been indicated to be 8 hours, and therefore this methodology should be used only in special situations.

11.7.3 Weld Toe Grinding

Where local grinding of weld toes below any visible undercuts is performed, the fatigue life may be increased by a factor given in Table 11.3 (DNVGL-RP-C203, 2016). In addition, the thickness effect may be reduced to an exponent k = 0.20 for S-N curves F and lower, and reduced to k = 0.15 for S-N curves D and E. Grinding a weld toe tangentially to the plate surface, as at A in Figure 11.7, will produce only a slight improvement in fatigue strength. In order to be efficient, grinding should extend below the plate surface, as at B, so that toe defects are removed. Grinding is normally carried out by a rotary burr. The treatment should produce a smooth, concave profile at the weld toe, with the depth of the depression penetrating into the plate surface at least 0.5 mm below the bottom of any visible undercut (see Figure 11.7). The grinding depth should not normally exceed 2 mm, or 7% of the plate thickness, whichever is smaller. Grinding has been used as an efficient method for reliable improvement in fatigue life during fabrication in a number of projects. Grinding also improves the reliability of the NDT during fabrication and during service life. However, it may be a good design practice to exclude this factor at the design stage. The designer has been advised to improve the details locally by other means, or to reduce the stress range through design, and retain the possibility of fatigue life improvement as a reserve to allow for a possible increase in fatigue loading during the design and fabrication process. It should also be noted that if grinding is required to achieve a specified fatigue life, then the hot spot stress range is rather high. Due to grinding, more of the fatigue life occurs during the initiation of fatigue cracks, and the crack grows faster after initiation. This indicates that inspection intervals should be shorter during in-service life such that potential fatigue cracks can be detected before they become critical to the integrity of the structure. Use of a rotary ball–shaped burr or cylindrical-shaped burr, with a diameter of approximately 12 mm, is recommended for grinding weld toes. If the intention is to achieve a larger ground radius, a small-radius rotary burr should be used initially, and the burr diameter should be increased during the course of the work, as recommended in Haagensen and Maddox (2013).

Haagensen (1994) reported weld toe grinding on T-joints with fillet welds joining the attachment to the main plate by a rotary burr of 12 mm in diameter using an electric grinder at 15,000 to 20,000 rpm with grinding to a depth 0.5 to 1.0 mm below the deepest undercuts. The increase in stress range at 2 million cycles was 94% when compared with as-welded condition. It is observed that the scatter in the fatigue test data for ground specimens is larger than for test data from as-welded specimens (see also Haagensen 1997).

As the grinding should be made 0.5 mm below the bottom of any visible undercut, a target grinding depth of 1 mm may be recommended for plates thicker than 15–20 mm.

The effect of toe grinding the weld in tubular joints is normally considered to be large. However, there are geometries where fatigue cracking has occurred from the weld inter-run positions. This was observed in fatigue tests on welded tubular T-joints with equal brace and chord diameters (Wylde 1983). Therefore, when performing weld toe grinding of tubular joints with large β-ratios to improve fatigue life, it is recommended to perform grinding of the full weld.

Weld toe grinding is being used not only on steel structures but also on aluminum structures (see, e.g., Ye and Moan 2008).

Grinding of weld toes is a comparatively fast operation; typically 3–5 m of weld per hour can be ground. When planning for weld toe grinding, it should be remembered that the burrs wear out; a single burr typically does not last for more than approximately 10 m of grinding.

11.7.4 Workmanship

In order to achieve the intended surface finish of ground and machined areas, proper workmanship is essential. Girth welds in tethers and risers that are subjected to large dynamic loading have to meet the requirements of a high S-N curve when installed in the North Sea. This means grinding flush on both the inside and outside. However, some girth welds can only be made from the outside. In such cases, the connection should be fabricated with some additional thickness such that the weld root can be ground smooth in order to achieve a similar surface as when the weld is made from both sides, and then made flush by machining of both sides. The weld overfill may be removed by a coarse grinder tool, such as a coarse grit flapper disk, with grit size 40–60. The final surface should be achieved by fine grit grinding below that of weld toe defects. The surface should show a smooth or polished finish, with no visible score marks. The roughness should correspond to $R_a = 3.2$ μm or better; it should be remembered that if coating of the area is planned, then a roughness larger than $R_a = 3.2$ μm is often recommended (see also Section 11.9 for standards for coating being referred to). The surface should be checked by MPI. Grinding should be performed until all indications of defects are removed. The possible presence of internal defects in the weld may then be a limitation for use of a high S-N class, and it is important that a reliable NDT examination is performed and that acceptance criteria are used that correspond with the relevant S-N classification.

11.7.5 Example of Effect of Grinding a Weld

In this example, grinding of a cruciform joint in a floating platform in air environment, as shown in Figure 11.5, is considered. It is assumed that the plate thickness T = 25 mm. The connection is subjected to pure membrane loading. Furthermore, it is assumed that a full penetration weld is made, with added leg length of 0.71T to enable the possibility of making a smooth transition by grinding from the weld to the plate. It is assumed that a fatigue life of 10 years has been calculated. This calculated fatigue life does not meet the design requirements, and grinding of the weld according to the methodology presented in Section 11.7.2 is being considered. A calculated service life of 10 years corresponds to approximately $0.5 \cdot 10^8$ stress cycles. Grinding by a radius of 10 mm is suggested. Three different methods may then be used for fatigue assessment:

1. Calculation of reduced stress at the weld toe using the methodology in Section 11.7.2. The calculation is based on nominal stress and S-N curve F, which is also used for calculation of fatigue life for the as-welded condition.
2. Calculation of a stress concentration due to the ground radius at the weld toe combined with S-N curve C (FAT 125), which is the highest S-N curve that can be applied for welds that are machined smooth. The calculation is made based on hot spot stress, with SCF included.
3. Calculation based on nominal stress and S-N curve F for ground welds.

The methodology for fatigue life calculation using a two-parameter Weibull distribution, as presented in Chapter 10, is used in the following. A shape parameter h = 1.0 is assumed. The final result is not very dependent on this assumption, as it is first used to calculate the scale parameter in the Weibull distribution or the maximum allowable stress range for the F-curve in the air environment for the actual number of load cycles. This gives $\Delta\sigma_0 = 250.89$ MPa, as shown in Table 11.2. A stress reduction factor of 0.739 is next derived from equation (11.3) with a radius R = 10 mm and a fatigue damage of 0.339 is calculated using the original S-N curve F that also was used for the as-welded detail. The calculated fatigue damage can be reduced further by a factor of 2 when following the procedure in Section 11.7.2. Thus, the resulting calculated fatigue damage becomes 0.169.

Considering the alternative approach using S-N curve C, an SCF of 1.51 is estimated for a radius of 10 mm from graph (figure 66) for SCFs in Peterson (1974). A fatigue damage of 0.583 is calculated, which is on the high side in comparison with the other approaches. The negative inverse slope of the C-curve with m = 3.0 is considered to be one reason for this, as the slope of the ground S-N curves is presented with m = 3.5. By a smooth grinding of the weld into the base material the hot spot is at the transition from the weld into the base material in Figure 11.7. Thus, also S-N curve B2 may be relevant for the fatigue behavior and the calculated fatigue

Table 11.2. *Calculated fatigue damage using different assessment methods*

	S-N curve in air	Stress range (MPa)	Stress concentration or reduction factor	Calculated fatigue damage
As-welded condition	F-curve for as-welded condition	Nominal stress range: 250.89 MPa	1.000	1.000
Ground weld with radius r = 10 mm using S-N curve for as-welded condition	F-curve for as-welded condition	$250.89 \cdot 0.739 =$ 185.36 MPa	Reduction factor 0.739 from equation eq. (11.3)	0.339
Ground weld with radius r = 10 mm using SCF	C-curve B2-curve	$250.89 \cdot 1.51 = 378.84$ MPa	1.51 from Peterson (1974)	0.583 0.284
Ground weld using S-N curve for ground welds	F-curve for ground welds (Table 11.4)	250.89 MPa which equals the nominal stress range	1.00	0.328

damages using this S-N curve is also shown in Table 11.2 for information. Using the nominal stress, together with the F-curve for a ground connection, results in a calculated fatigue damage of 0.328. The results are summarized in Table 11.2. For the connection being considered, it may be possible to increase the radius slightly and achieve a longer calculated fatigue life. However, that will require correspondingly good workmanship. The calculated fatigue damages show that grinding of the weld and weld toes is an efficient improvement method.

11.7.6 TIG Dressing

Fatigue life may be improved by using Tungsten Inert Gas (TIG) dressing. The improvement factors are given in Table 11.3. TIG dressing is understood to refer to remelting of the weld toe region in order to achieve a smooth transition from the weld to the base material. Uncertainties in the quality assurance of this improvement method suggest that it probably should not be recommended for general use at the design stage.

11.7.7 Hammer Peening

Fatigue life may be improved by hammer peening by the factor given in Table 11.3. However, the following limitations apply:

- Hammer peening should only be used on members for which failure will not result in substantial consequences.
- Overload in compression must be avoided, because the compressive residual stress setup by hammer peening will be significantly reduced.
- Grinding a steering groove by use of a rotary burr of a diameter suitable for the hammerhead to be used for the peening is recommended. The peening tip must be small enough to reach the weld toe (radius 3–9 mm is recommended).

Table 11.3. *Improvement in fatigue life by different methods*
(DNVGL-RP-C203 2016)

Improvement method	Minimum specified yield strength	Increase in fatigue life (factor on life)[1]
Grinding	Below 350 MPa	$0.01\sigma_y^2$
	Above 350 MPa	3.5
TIG dressing	Below 350 MPa	$0.01\sigma_y^2$
	Above 350 MPa	3.5
Hammer peening[3]	Below 350 MPa	$0.011\sigma_y^2$
	Above 350 MPa	4.0

[1] The maximum S-N class that can be claimed by weld improvement is C2 to C depending on NDT and quality assurance for execution; see table 5 in appendix A in DNVGL-RP-C203 (2016).
[2] σ_y = characteristic yield strength for the actual base material.
[3] The improvement effect depends on the tools used and workmanship. Therefore, if the fabricator is inexperienced in hammer peening, fatigue testing of the relevant detail (with and without hammer peening) should be performed before a factor on improvement is decided.

Uncertainties regarding workmanship and quality assurance of the process suggest that this method should not be recommended for general use at the design stage. The effect of improvement in the low cycle range is less than in the high cycle range, as demonstrated by test data in Section 2.6.4.

Haagensen (1994) reported fatigue testing of similar test specimens as ground with the weld toes improved by hammer peening. Four runs were made with an air hammer with an air operating pressure of approximately 6 bars. The solid tool was a 12 mm diameter hemispherical tip. At 2 million cycles, the improvement represented an increase over as-welded fatigue strength of 142%, or 25% above that for ground specimens. Variable fatigue tests indicated that some effect of the improvement was lost in the short-life, high-stress region of the S-N diagram. However, in the long-life region, the fatigue performance under variable loading was similar to that of constant amplitude loading.

The effect of hammer peening performed underwater is considered to be similar to that performed above water, provided that proper workmanship and recommended procedures are followed (Buitrago and Zettlemoyer 1997).

The time needed for hammer peening in air, assuming four passes, is similar to that of grinding a weld toe. The time needed to grind a steering groove – approximately 20 m per hour – is added to that. The peening tips or hammer bits become dull after some use; thus, sufficient number of bits should be planned for if long welds are going to be hammer-peened.

For example, hammer peening was used to upgrade the fatigue strength of the ship sides at the waterline area in a conversion from a sailing ship to an FPSO (floating production, storage, and offloading unit) in the North Sea. The thickness of the ship side plates was not large enough that sufficient fatigue life at the fillet weld toes along the side longitudinals could be documented. A somewhat similar fatigue analysis example of a ship is shown in Appendix A. Improvement of these details by other methods is considered too complex, and therefore it was decided to hammer peen approximately 3 km of the welds along the side longitudinals. First a grinding of a steering groove along the weld toes was made, then photos of the ground tracks were made, then peening was performed, and finally photos of the full peened length were made for quality assurance and documentation.

11.7.8 High-Frequency Mechanical Impact Treatment

In the past decade or so, a number of papers on fatigue strength improvement of welded joints treated by High Frequency Mechanical Impact (HFMI) have been published (e.g., Weich et al. 2007; Lopez Martinez 2010; Yildirim and Marquis 2012a, 2012b). HFMI refers to methods based on ultrasonic impact treatment of the weld toes by impacting the material at a high frequency, 100–400 Hz, such that local yielding occurs and compressive stresses are introduced at the surface similar to that by hammer peening. The radius of the impacting tool is in the 3–5 mm range. Compared to hammer peening, it is more user-friendly as it does not imply significant noise or vibration transferred to the operator. Also, the action force required from the operator to the impacting tool is small compared to hammer peening. Various names have been used in literature to describe the devices: Ultrasonic Impact Treatment (UIT), Ultrasonic Peening (UP), Ultrasonic Peening Treatment (UPT), high-frequency impact, and Ultrasonic Needle Peening (UNP).

The improvement under constant amplitude fatigue testing is found to be large. The efficiency is increasing with material yield strength. The improvement is mainly due to the compressive stresses introduced at the weld toe down to a depth of 1 mm and exceeding the material yield strength at the surface (Cheng et al. 2003). Large compressive stress cycles lead to further yielding at the surface and the shakedown of the compressive stresses, which reduce the improvement effect (see Polezhayeva et al. 2014). For this reason it is also limited how large a benefit of HFMI can be counted on in fatigue design standards when the details are subjected to large compressive or tensile stress amplitudes (see BS 7608 2014). The effect of spectrum loads is being tested and it is expected that further recommendations on this will be provided in the near future by the IIW (Marquis et al. 2013).

11.7.9 Post-Weld Heat Treatment

It is well known that the fatigue behavior of welded parts can be improved significantly if the residual stresses are removed by means of post-weld heat treatment (PWHT). Equation (2.4) can be used to calculate a reduced effective stress range after a post-weld heat treatment where part of the stress range is compressive. Thermal stress relieving can be costly in the case of treatment of large specimens, and it has been suggested to replace PWHT with vibration stress relieving. An investigation of this, reported by Sonsino et al. (1996), showed that a vibration relief could not act as a substitute for thermal stress relief, since the residual stresses in the welded specimens could not be removed by this method. Improvement of the specimen was only achieved where the tensile residual stresses were reduced during the initial loading cycles due to local yielding at the weld toe (see Section 3.3.2 with respect to shakedown).

11.7.10 Extended Fatigue Life

An extended fatigue life is considered acceptable and within normal design criteria if the calculated fatigue life is longer than the total design life multiplied by the Design Fatigue Factor. Otherwise, an extended life may be based on results from inspections performed during the prior service life. Such an evaluation should be based on:

1. Calculated crack growth.
   Crack growth characteristics; that is, crack length/depth as a function of time/number of cycles (this depends on type of joint, type of loading, and possibility for redistribution of stress). See also Chapter 18.
2. Reliability of inspection method used.
3. Time elapsed from the last inspection. Use of EC or MPI is recommended for inspection of surface cracks starting at hot spots (in ferromagnetic materials).

The following procedure may be used for calculation of an elongated fatigue life for welded connections that are ground and inspected for fatigue cracks during in-service life. Provided that grinding below the surface to a depth of approximately 1.0 mm is performed, and that fatigue cracks are not detected by a detailed MPI of the considered hot spot region at the weld toe, the fatigue damage at this hot spot may be considered to start again at zero. If a fatigue crack is detected, further grinding should be performed to remove any indication of this crack. If more than 10% of the thickness is removed by grinding, the effect of this on increased stress should be included when a new fatigue life is assessed. In some cases, as much as 30% of the plate thickness may be removed by grinding before resorting to a weld repair. This depends on the type of joint, loading conditions, and accessibility for repair work. It should be noted that it can be difficult to detect fatigue cracks growing from the weld root of fillet welds by NDT. Furthermore, the fatigue life of such regions cannot be improved by grinding the surface. It should also be remembered that if renewal of one hot spot area is performed by local grinding, then there are probably other areas close to the hot spot region being considered that are not ground and that also experience significant dynamic loading. The accumulated fatigue damage at this region is the same as previously. However, this fatigue damage may also be reassessed taking into account:

- the correlation with a ground neighbor hot spot region that has not cracked;
- an updated reliability assessment, taking the reliability of performed in-service inspections into account; see also Sections 18.6 and 18.9.

11.7.11 Stop Holes

Stop holes may be drilled at crack ends to stop or retard further crack growth due to cyclic loading. The stop holes should be placed such that the crack tips are removed or, if there is uncertainty about where the cracks end, the holes can be placed in front of the visible crack tips. By making a hole at the crack tip, the stress intensity here is removed; however, a significant stress concentration remains that must be assessed. If crack growth initiates from the stop hole, the resulting stress intensity factor is due to the full crack length and crack propagation occurs as earlier. The stress concentration due to the holes at the ends of an internal crack can be calculated from the following expression:

$$SCF = 1 + 2\sqrt{\frac{a}{r}} \qquad (11.6)$$

where:

r = radius of the hole
a = half crack length.

From equation (11.6) it can be seen that a large radius relative to the crack length should be used in order to keep the stress concentration as low as possible. A fatigue assessment can then be performed based on this stress concentration, an S-N curve for the base material, and a relevant long-term stress range distribution. Making stop holes is sometimes an attractive option underwater, and tools are available for grinding large holes in plated structures. Stop holes may be made for temporary repairs, while more permanent repairs are planned and executed. Stop holes may also be used as part of a repair solution, where the crack-driving stresses are reduced by other means. The hole-drilling technique can be improved by use of high-strength bolts that are inserted in the holes and pretensioned to introduce compression stresses around the hole (Haagensen 1994). Cold working of the stop holes by expansion, involving forcing an oversized conical mandrel through the hole, can, depending on loading, significantly increase fatigue life, as has been demonstrated by Chandawanisch (1979). The mandrel is used to expand the hole such that compressive residual stresses around the hole remain when the mandrel is removed. Another method to improve the fatigue capacity in cracked plates is by making a hole inside the cracked area, but some distance away from the crack, and inserting a stud into the hole that enlarges the material such that crack closure occurs at a distance close to the crack tip (Mori and Arakawa 2013).

The following equation can be used when a crack at an edge is ground to a semi-elliptic shape:

$$SCF = 1 + 2\sqrt{\frac{a}{c}} \qquad (11.7)$$

where:

a = half axis into the material
c = half axis along the surface in the stress direction.

11.7.12 Grind Repair of Fatigue Cracks

A significant amount of fatigue life of tubular joints can still be documented, even if a considerable proportion of material is removed by grinding away a fatigue crack. An example of this is shown in Figure 11.8. Before starting grinding, a detailed FEA should be performed for documentation of stresses that will be used for fatigue assessment of a grind repair.

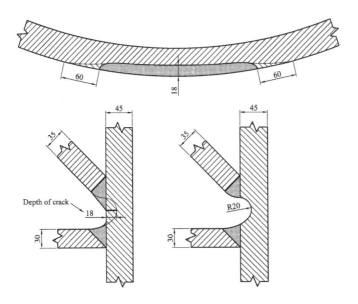

Crack at crown position in K-joint   Joint after grinding

Figure 11.8. Example of a grind repair of a deep fatigue crack in a tubular K-joint.

11.7.13 S-N Curves for Improved Areas

New S-N curves for ground weld toes have been included in DNVGL-RP-C203 (2016) based on the same improvement factor on stress range at $2 \cdot 10^6$ cycles as IIW and assuming an inverse slope of the ground hot spot S-N curve m = 3.5 for fewer than 10^7 cycles in air environment. A similar principle is followed for deriving design S-N curves for weld toes that have been peened. Here, a single-slope S-N curve with m = 5.0 is proposed. The design hot spot S-N curves for ground and peened weld toes are shown in Figure 11.9, together with the as-welded curve (S-N curve D). For the left part of the S-N diagram, it is assumed that S-N curves for weld-improved areas are no lower than for the as-welded condition. It should be emphasized that the S-N curves for improved areas should only be used when there is proper workmanship of the improvement and attention has been paid to the possibility of fatigue crack initiation from areas other than those that have been improved, such as from internal defects in the welds. In addition, it is important to evaluate the maximum stresses from static and dynamic loading in order to assess the feasibility of the improvement method; that is, it must be evaluated whether the beneficial compressive stresses at the hot spot are lost during early service life if peening is performed. If the connection can be subjected to significant compressive stresses during in-service life, it is recommended that the improvement factor for the peening method is reduced. Fatigue testing of improved details using a relevant spectrum loading is recommended when there is uncertainty about the improvement.

11.8 Measurement of Surface Roughness

Measurement of the surface roughness is important for evaluating whether the surface is smooth enough to correspond to a high S-N curve, such as the C-curve or the

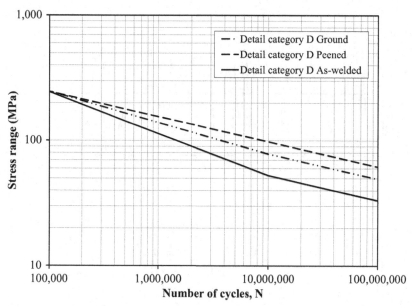

Figure 11.9. Example of design S-N curves (D-curve) for a butt weld in as-welded condition and improved by grinding or peening (DNVGL-RP-C203 2016).

high strength S-N curve in DNVGL-RP-C203 (2016). On the same surfaces, application of a durable coating to ensure long-term corrosion protection may be desirable. Some surface roughness is necessary to fix the coating to the steel. Requirements for maximum surface roughness are presented in fatigue design standards, while minimum values for surface roughness are provided in standards for coatings. These requirements are therefore somewhat contradictory to each other, and knowledge on how to define and measure surface roughness becomes an important issue. Different methods may be used in the same company standards. The R_a notation is frequently used for fatigue assessment, while R_z is used for the surface requirement for coating.

Table 11.4. *S-N curves for improved details by grinding or hammer peening (DNVGL-RP-C203 2016)*

	Improvement by grinding		Improvement by hammer peening
S-N curve	$N \leq 10^7$ cycles $m_1 = 3.5$ $\log a_{d1}$	$N > 10^7$ cycles $m_2 = 5.0$ $\log a_{d2}$	For all N m = 5.0 $\log a_d$
D	13.540	16.343	16.953
E	13.360	16.086	16.696
F	13.179	15.828	16.438
F1	12.997	15.568	16.178
F3	12.819	15.313	15.923
G	12.646	15.066	15.676
W1	12.486	14.838	15.448
W2	12.307	14.581	15.191
W3	12.147	14.353	14.963

See equation (4.1) for notation.

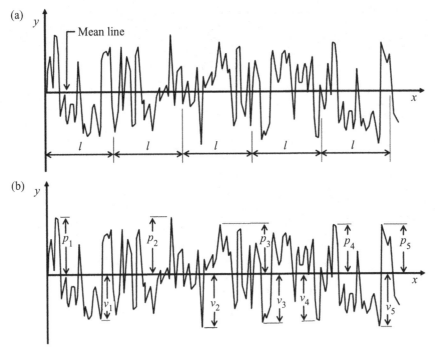

Figure 11.10. Definition of R_a and R_z for characterization of surface roughness.

The implication is that this results in rather different numbers, and may be misunderstood if these numbers are not clearly linked to the definition of the method used. Many roughness parameters are published in the literature (see, e.g., Gadelmawla et al. 2002). Some of the most frequently used parameters are defined later in this section. The most common method of measuring surface roughness uses a mechanical device that works by dragging a probe with a measuring tip across the surface. As the probe is drawn slowly across the surface, it moves up and down over the contours of the surface. This motion is recorded. The instrument is limited by the radius of the probe tip; if the radius is too large, it may be unable to penetrate the finer cracks or scratches, resulting in an incorrect measurement.

The arithmetic average height parameter, R_a, is a frequently used parameter for assessment of fatigue strength. This parameter is the average deviation from the mean line over a sampling length, as indicated in Figure 11.10a. This is derived as:

$$R_a = \frac{1}{l} \int_0^l |y(x)| dx \qquad (11.8)$$

R_q or RMS is the root mean square roughness, which is the standard deviation of the distribution of surface heights, which is another common parameter. R_q is more sensitive to a large deviation from the mean line than R_a. This parameter is expressed mathematically as:

$$R_q = R_{RMS} = \sqrt{\frac{1}{l} \int_0^l \{y(x)\}^2 dx} \qquad (11.9)$$

This measurement parameter is used by Buitrago and Zettlemoyer (1999) in their presentation of fatigue testing of tendon welds with internal defects. The girth welds in the tendons were ground to a surface roughness, R_q, not greater than 3 μm related to marks normal to the main stress direction.

The parameter R_z, also known as the ten-point height, is defined as the difference in height between the average of the five highest peaks and the five lowest troughs, as indicated in Figure 11.10b. R_z is considered more sensitive to occasional peaks and troughs than R_a is. It seems strange that R_z is not used more frequently in relation to fatigue, as the largest irregularities would be expected to have the greatest influence on the fatigue strength. R_z is defined mathematically by:

$$R_z = \frac{1}{n} \left(\sum_{i=1}^{n} p_i + \sum_{i=1}^{n} v_i \right) \qquad (11.10)$$

R_z is the same as R_{y5} that is referred to in the DNV-RP-B101 (2007) standard for corrosion protection of floating production and storage units with respect to measurements of surfaces before coating. Where larger roughness may be wanted for surface preparation for coating than recommended for fatigue, due emphasis should be made on achieving adequate protection against corrosion.

11.9 Effect of Surface Roughness on Fatigue Strength

In design standards such as NORSOK M-101 (2011) grit blasting of the surfaces is recommended prior to coating to Sa 2½ in order to achieve a surface roughness, $R_{y5} = 50$–85 μm. This corresponds to Grade G in ISO 8501 2012, Part 1. R_{y5} has the same definition as R_z, as explained in Section 11.8. The definition of the R_a parameter used to describe the acceptance criterion for grinding thus differs from that of R_{y5}. This means that it is difficult to make a direct comparison of requirements for surface finish, because a well-defined ratio between R_a and R_{y5} is lacking. Some measurements on steel surfaces indicate that $R_z/R_a = R_{y5}/R_a \approx 6$ may be used as an approximate value. However, a somewhat smaller value has also been reported from measurements of machined surfaces. Thus, $R_{y5} = 50$–85 μm corresponds approximately to $R_a = 8$–14 μm. This means that where surface preparation for coating is specified according to NORSOK M-501 (2012), the benefits from grinding that are obtained in laboratory testing with specimens ground to $R_a = 3.2$ μm are unlikely to be achieved.

In the FKM guideline (2003), a reduction factor on fatigue strength relative to a polished surface is presented as a function of material ultimate strength, R_m, and surface roughness, R_z, as:

$$K_r = 1 - 0.22 \log R_z \log \left(\frac{R_m}{200} \right) \quad for \quad 1 \mu m < R_z < 200 \mu m \qquad (11.11)$$

This reduction factor is plotted as a function of R_m for some R_a values in Figure 11.11 for $R_z/R_a = 6$. The graphs in this figure may be used to establish a reduction factor on fatigue strength when going from $R_a = 3.2$ μm to $R_a = 12.8$ μm, where surface preparation is performed as a basis for coating according to NORSOK. For

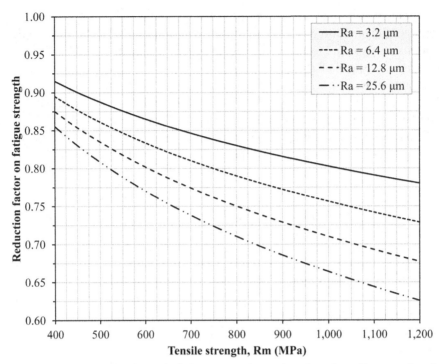

Figure 11.11. Reduction factor in fatigue strength as a function of tensile strength and surface roughness relative to a polished surface.

steel materials with a tensile strength of 600 MPa, this gives a reduction factor of 0.93 on fatigue strength, which corresponds to approximately a factor of 1.27 on fatigue life; for a tensile strength of 1,000 MPa this gives a reduction factor of 0.90, which corresponds to approximately a factor of 1.45 on fatigue life.

12 Probability of Fatigue Failure

12.1 Failure Probability at the Design Stage

12.1.1 General

The probability of fatigue failure is typically time-dependent due to the time-dependent nature of fatigue damage accumulation over service life, and possibly also due to time-dependent loading (see Section 18.10). A limit state function for a fatigue failure as function of time t can be defined as:

$$M(t) = R(t) - S(t) \tag{12.1}$$

where:

R is a function modeling fatigue capacity
S is a function modeling the load effect.

Both R and S are defined from underlying stochastic variables. If the loading, S, is larger than the capacity, R, the limit state function is negative, which means fatigue failure. The capacity indicated in Figure 12.1 can most often be considered as a distribution independent of time in an S-N diagram provided that the structure is properly corroded protected during the life time.

The accumulation of load cycles, S(t), can be illustrated by another distribution in the same diagram that moves to the right as more cycles are accumulated with time in service. Marine structures are subjected to variable long-term loading; however, here an equivalent distribution of the load effect is shown for illustration purpose. The distribution denoted $S(t_1)$ in Figure 12.1 does not result in any calculated probability of failure. However, the distribution $S(t_2)$ has so many accumulated load cycles that there is a probability that part of the accumulated load history is larger than the lowest values for fatigue capacity. The probability of this occurring defines the probability of fatigue failure during service life and can be expressed as:

$$P_F(t) = P(M(t) \leq 0) \tag{12.2}$$

This equation expresses the probability that the accumulated fatigue damage exceeds the fatigue capacity at time t. The calculated probability of failure can be derived from direct integration of a function expressing the probability that the accumulated

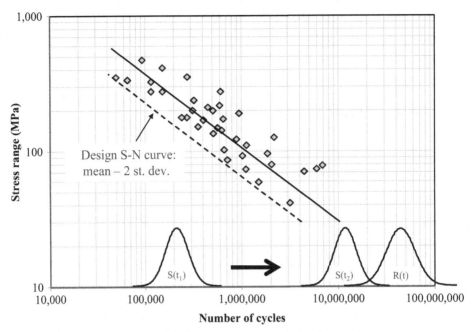

Figure 12.1. Illustration of accumulated fatigue damage with time.

load effect is larger than the fatigue capacity (see also Naess and Moan 2013; Madsen et al. 1986). In the derivation of the failure probabilities, special attention should be paid to the type of probability being derived. These can be an accumulated or time-limited failure probability or an annual failure probability, as explained in the following sections. A more detailed description of calculated probability of failure is presented in Section 18.7.

12.1.2 Accumulated and Annual Failure Probability

For structures that are subjected to dynamic loading, fatigue damage accumulates over the service life, and it is practical to relate this damage to an accumulated failure probability. This means that the calculated failure probability is the probability that the structure fails during the time period *prior* to the time considered. The annual failure probability is obtained by subtracting the accumulated failure prior to the year considered from the accumulated failure probability at the end of the year being considered (this is an acceptable approximation when small probabilities are considered; otherwise, the hazard function described in books on statistics should be used). The annual failure probability from year t_i to t_{i+1} is then derived as:

$$\Delta P(t_i \text{ to } t_{i+1}) = P(t_{i+1}) - P(t_i) \qquad (12.3)$$

Examples of accumulated probability of failure and annual probability of failure with respect to fatigue are presented in Section 12.2.

12.1.3 Time-Limited Failure Probability

The type of failure probability derived for structures that do not degrade over time depends on the formulation of the time-varying loading. For example, if the loading

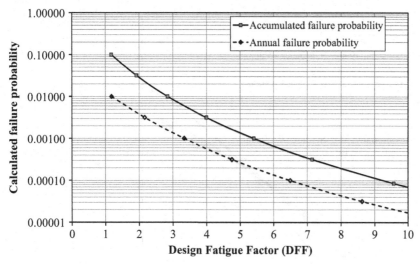

Figure 12.2. Fatigue failure probability as a function of DFF (DNVGL-RP-C203 2016).

is expressed as the annual largest load, then the annual failure probability is derived when a probability that the load exceeds the capacity is calculated. Expressing the same calculation using the 100-year largest load enables determination of the 100-year failure probability (given that the structure does not degrade over time). This corresponds to a probability of being exceeded equal to 10^{-2} on an annual basis. This type of probabilistic analysis is relevant for calculation of collapse of the structure, given that a fatigue crack is present.

12.2 Uncertainties in Fatigue Analysis

Fatigue life assessments are normally associated with large uncertainties. Reliability methods, such as presented in Section 12.1.1, may be used to illustrate the effect of uncertainties on the calculated probability of fatigue failure. Using this methodology, the probability of fatigue failure as a function of Design Fatigue Factor (DFF) can be calculated, as shown in Figure 12.2. This figure shows the accumulated probability of fatigue failure and the probability of fatigue failure during the last year in service, when a structure is designed with a 20-year design life. Uncertainties in the most important parameters in the fatigue design procedure were accounted for when these probabilities of fatigue failure were calculated. This figure was derived by uncertainty in loading, in addition to uncertainties in the S-N data and in the Palmgren-Miner damage accumulation rule. The loading, which includes all the structural stress analyses, was assumed to be normally distributed with a coefficient of variation (CoV) = 25%, a typical standard deviation in S-N data of 0.20, assuming a normal distribution in the logarithmic scale, and the Palmgren-Miner sum was also assumed to be log normally distributed with median 1.0 and CoV = 0.3. CoV is a normalized standard deviation or, as defined in statistics textbooks, a standard deviation divided by the mean value. Thus, using mathematical symbols, it can be expressed as CoV $= \sigma / \mu$, where σ = standard deviation and μ = mean value.

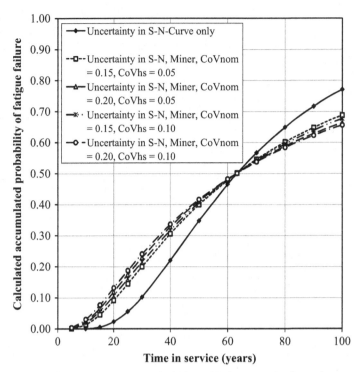

Figure 12.3. Accumulated probability of fatigue cracks occurring as a function of service life for a 20-year design life (CoV_{nom}: uncertainty in nominal stress range; CoV_{hs}: uncertainty in calculated hot spot stress given no uncertainty in nominal stress range; see also the main text for explanation of abbreviations in the legend) (DNVGL-RP-C203 2016).

The accumulated probability of failure is independent of design life and need not be linked to a 20-year service life. However, for calculating the annual probability of failure, a service life must be specified (here 20 years).

An expected long-term stress range is typically used in fatigue design of marine structures. However, some assumptions that tend toward the safe side are often made to ensure proper documentation of fatigue analysis. If uncertainties in parameters other than the S-N data are neglected, a DFF of 1.0 implies a probability of a fatigue crack occurring during service life of 2.3% due to the safety in the design S-N curve (see Section 4.1.1). Figure 12.3 shows the accumulated probability of fatigue failure as a function of years in service for different assumptions of uncertainty in the input parameters. This figure shows the results for DFF = 1.0. A more detailed left part of this figure, corresponding to the first 20 years of service life, is shown in Figure 12.4. Results for other DFF values may be obtained as approximations (for bilinear S-N curves) by simple multiplication of the time scale on the abscissa axis by the actual DFF that is under consideration. Figures 12.3 and 12.4 show the accumulated probability of fatigue failure for uncertainty in S-N data corresponding to a standard deviation of 0.20 in log N scale. A normal distribution of the S-N data in logarithmic scale is assumed. The uncertainty in Palmgren-Miner summation is described as log normal, with median of 1.0 and CoV of 0.30 (Wirsching 1984; Wirsching and Chen 1988). Other uncertainties, such as load and response, are assumed to be normally distributed with CoV of 15–20% and the hot spot stress derivation is also assumed to be normally distributed with CoV of 5–10%. The calculated fatigue life forms the

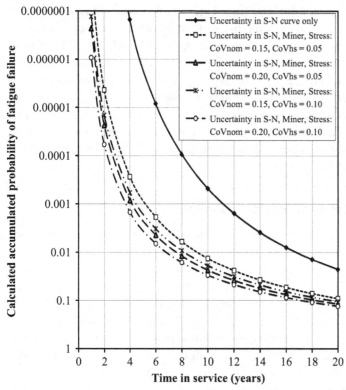

Figure 12.4. Accumulated probability of occurrence of through-wall fatigue cracks as a function of service life for a 20-year calculated fatigue life (left part from Figure 12.3); (see also caption text in Figure 12.3 for abbreviations in the legend) (DNVGL-RP-C203 2016).

basis for assessment of the probability of fatigue cracking during service life. Thus, it implicitly forms the basis for the requirements for in-service inspection (see also Section 12.3 and Chapter 18). For details showing a short fatigue life at an early design stage, the details being considered should be evaluated in terms of improvement of local geometry in order to reduce their stress concentrations. At an early design stage it is more cost-efficient to make minor geometric modifications than to rely on methods for fatigue improvement under fabrication and construction, such as grinding and hammer peening. DFFs for different areas of the structures should be defined in specifications in the contract document for design.

12.3 Requirements for In-Service Inspection for Fatigue Cracks

Uncertainties associated with fatigue life calculation normally imply that some in-service inspection for fatigue cracks will be required during service life, depending on the consequences of fatigue failure and calculated fatigue life. Figure 12.4 may be used for an initial estimate of time to first inspection of marine structures, based on the lower graph in this figure if normal uncertainties are associated with the fatigue life calculation. Figure 12.4 is derived for a calculated fatigue life of 20 years. For other calculated fatigue lives (L_{calc}), the numbers on the abscissa axis can be scaled by a factor $f = L_{calc}/20$ for estimating the time until the first required inspection. If a fatigue crack is without substantial consequences, an accumulated probability of

10^{-2} may be considered acceptable, and from Figure 12.4, inspection is not required during the first 6 years (for a calculated fatigue life of 20 years). If the consequences from a fatigue crack occurring might be substantial, the accumulated probability of fatigue failure should be less than 10^{-4}, and from Figure 12.4 the first inspection would be already required after 2 years. A calculated fatigue life significantly longer than 20 years would normally be required during design for a connection with the potential for substantial consequences should failure occur. After the first inspection, the time interval until the next inspection can be estimated on the basis of fracture mechanics and probabilistic analysis, taking the uncertainty in the inspection method into account, as presented in Chapter 18

12.4 Target Safety Level for Structural Design

Caution is needed when relating calculated probabilities of failure to acceptance criteria or the target safety level for structures. This is because calculated values are sensitive to the design format used and the input parameters to the analysis. A design format is understood to refer to how the design equations are formulated in terms of design parameters and safety factors. Therefore, the target safety level should be assessed in a similar way as used for calculation of probability of failure. This is reflected in the text of some design standards that state that reliability methods may be used in design, provided that they can be shown to document a similar safety level as the design standard in well-known cases. A "well-known case" can, for example, be a requirement for a DFF that has been used for a prolonged period and that has been shown to provide a sound structural design in terms of reliability, and acceptable fabrication and maintenance costs. Based on the specified design formulation in the standard, and upon certain assumptions regarding the parameter distributions used as input, a probability of failure can be calculated that now corresponds to the DFF that is used. It is assumed that the structure is fully utilized in terms of calculated Palmgren-Miner fatigue damage. The calculated probability of failure depends on the distributions of the parameters used. Thus, it is termed a "calculated probability of failure" as it is not normally related directly to any statistical data regarding failure in similar structures. This calculated probability of failure is also denoted as a "nominal" failure probability in some literature. If this failure probability is later used as a target safety value, the same distributions should be used for the parameters as input to the further analysis as far as relevant. In this way, the analysis procedure is accepted as not being perfect, but the error is reduced as the same error is included in both the calculation of the implicit target value and in the analysis for calculating the probability to be compared with the target value. Thus, there is a similar error in the calculated probability on the left-hand side of a design equation as in the target value on the right-hand side, and therefore this error can be eliminated. Examples of calculated accumulated probabilities of fatigue failure for different uncertainties in dynamic loading as a function of DFF are shown in Figure 12.5 with uncertainty in S-N data and Palmgren-Miner sum as presented in Section 12.2. The analyses are based on 20 years' service life with a mean zero up-crossing frequency equal 0.13 sec^{-1} and a Weibull long term stress range distribution with shape parameter h = 1.0. The uncertainty in loading is described by the CoV. The corresponding calculated annual probabilities of fatigue failure as a function of uncertainty in load for a 20-year service life are shown in Figure 12.6.

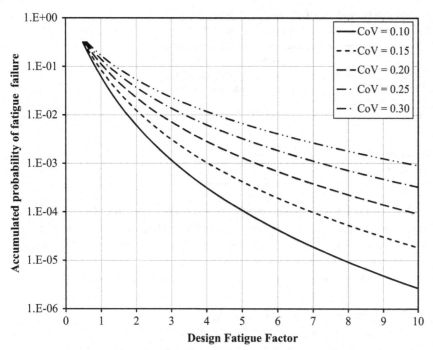

Figure 12.5. Calculated accumulated probability of fatigue failure as a function of uncertainty in load (DNVGL-RP-0001 2015).

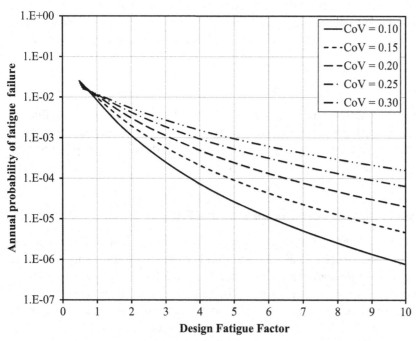

Figure 12.6. Calculated annual probability of fatigue failure as a function of uncertainty in load for a 20-year service life (DNVGL-RP-0001 2015).

Table 12.1. *Relationship between DFF and probability of failure for CoV = 0.20 on the load effect (DNVGL-RP-0001 2015)*

DFF	Accumulated probability of fatigue failure	Annual probability of fatigue failure the last year in a 20-year service life
1	$1.1 \cdot 10^{-1}$	$1.0 \cdot 10^{-2}$
2	$2.2 \cdot 10^{-2}$	$3.0 \cdot 10^{-3}$
3	$7.1 \cdot 10^{-3}$	$1.1 \cdot 10^{-3}$
5	$1.3 \cdot 10^{-3}$	$2.4 \cdot 10^{-4}$
10	$9.1 \cdot 10^{-5}$	$2.0 \cdot 10^{-5}$

The recommended DFF in design standards are normally determined on the basis of a relative large uncertainty in the dynamic loading, as it shall apply to several different details. Thus, if knowledge of less uncertainty in load is available, if, for example, monitoring of response is performed, then these figures may be used to estimate a lower calculated probability of failure. However, it should not be forgotten that measurement data are also associated with some uncertainties. Thus, some uncertainty should still be included in the dynamic loading. See also Section I.4 for assessment of absolute values for target safety levels in marine structures.

Assuming that a CoV = 0.20 is representative for mean uncertainty in load effect for typical offshore structures and with other uncertainties as used for derivation of Figure 12.5 and Figure 12.6, the target probabilities of failure are given in Table 12.1. These values are also proposed as acceptance criteria for establishing the inspection intervals for non-destructive testing (NDT) in Chapter 18. The annual probabilities of failure listed in Table 12.1 are presented for the last year in a service life of 20 years. For another service life (derived by x years) an annual probability of failure can be derived as:

$$p_{f\ annual\ x} = p_{f\ annual\ 20} \frac{20}{x} \tag{12.4}$$

The wind and wave actions may vary from one year to another. This also has influence on the calculated accumulated fatigue damage on a short-term basis. The effect of this on probability of failure from one year to the next year depends on type of marine structures, as explained in Moan et al. (2005).

12.5 Residual Strength of Structures with a Fatigue Crack

Fatigue crack growth through the thickness of a joint or connection does not necessarily mean that the structure being considered is close to collapse. This is because jacket structures, semi-submersibles, and floating production vessels are relatively redundant structures. One reason for this is that the requirements imposed in design related to the Accidental Limit State (ALS) in design standards for offshore structures, such as NORSOK N-001 (2012) and ISO 19902 (2007).

Residual capacity in structures with a crack present is considered in Section 12.6, along with some simplified guidance regarding system failures of jacket structures. This is important for assessment of the target safety level. Simplified procedures are assessed as normally being sufficient for documentation of the safety level. It is likely

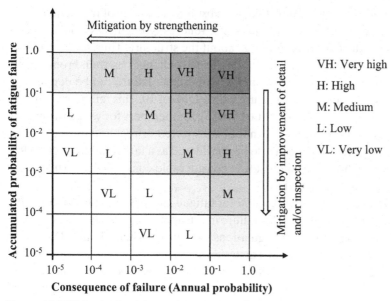

Figure 12.7. Risk matrix with consequence of failure and probability of failure.

that a limit on the number of fatigue cracks in the structures will govern the assessment of safety level more than an advanced assessment of system reliability. In this respect, selection of connections for inspection, such that a possible progressive failure path is avoided, becomes important. Progressive failure is understood to refer to failure occurring in several joints, such that a failure mechanism is established and the structure collapses. A risk matrix, with the relative size of the consequences of failure along the horizontal axis and the probability of failure along the vertical axis, is shown in Figure 12.7.

The probability of fatigue failure increases over time because of the time-dependent accumulation of fatigue damage during cyclic loading. Inspection increases the available knowledge of potential fatigue cracks in the structure, and may reduce the estimated risk by decreasing the calculated probability of failure. The consequence of fatigue failure is considered to be less dependent on time. However, if there are many hot spots with potential fatigue cracks in the structure, the consequences of a fatigue failure may also increase as more fatigue damage is accumulated. In order to reduce the consequences of fatigue failure, mitigating measures may be introduced, such as strengthening or improvement of details. However, such mitigations may also include the inspection of surrounding structures for assessment of whether they have a sound capacity without any deficiencies.

In general, the target level of the structural failure probability is defined as being dependent on the consequence and nature of failure. Evaluation of the consequences of failure comprises an assessment regarding human life, environmental impact, and economics, as described in NORSOK N-001 (2012). It must be kept in mind that the structural reliability analysis described here does not include gross errors that need to be analyzed by other techniques, such as traditional risk analysis (Lotsberg et al. 2005a). Possible sources of gross errors and evaluation of their probabilities must be considered separately for each structure. Thus, it should be noted that calculated probabilities of failure should not be taken as absolute measures of the frequency of failure. Rather, they are nominal measures that reflect the engineers' confidence in the reliability, given the current knowledge about the structure. Thus, if the available information changes, the estimated reliability usually changes as well. From Figure 12.6 and equation (12.4) an annual probability of fatigue failure can be derived for a connection for a nonredundant structure with a DFF of 10. This failure probability is accepted, according to NORSOK N-001 (2012), for members for which the consequences of failure are large and that cannot be inspected or repaired during service life. Thus, this failure probability can also be considered as a target failure probability for members, for which failure will result in considerable consequences. This target failure probability can be denoted as $P_{\text{f annual x} - \text{Target with DFF} = 10}$.

DFF = 10 has been required since 1977 for fatigue analysis of details with significant consequences in terms of a fatigue failure in offshore structures in the North Sea (Norwegian Petroleum Directorate regulations). When assessing whether DFF = 10 is a sound requirement for such details in design standards, experiences with DFF = 1.0 for similar details in structures with smaller potential consequences should first be examined, as a design with a low DFF implies a larger probability of fatigue failure during its service life than a detail with a larger DFF. Provided that a design with DFF = 1.0 results in an acceptable safety level, the safety level for DFF = 10 may be argued for as being appropriate, based on the calculated relative differences in failure probabilities, such as shown in Figure 12.5.

The probability that a structure fails, given that there is a fatigue crack failure at the connection being considered, is described by failure probability P_{SYS}. The following equation can then be used to derive target failure probabilities for such connections, in which the consequences of a fatigue failure have been accounted for:

$$P_{f \text{ annual Target}} = \frac{p_{f \text{ annual x} - \text{Target with } DFF = 10}}{P_{SYS}} \tag{12.5}$$

where:

$p_{\text{f annual x} - \text{Target with DFF}=10}$ is derived from equation (12.4) for DFF = 10

P_{SYS} is the failure probability of the structure, given that there is already a fatigue failure at the considered hot spot.

12.6 System Reliability Method

12.6.1 Robustness

Design standards require robustness and capacity in the damaged condition, as was shown in Section I.4 (ALS). These parameters also have an effect on inspection requirements for fatigue cracks, as a fatigue crack at a hot spot should not normally lead to catastrophic failure. If a potential fatigue crack has the potential to result in catastrophic failure, then, according to NORSOK N-001 (2012), it should be designed with a large DFF. The degree of robustness or redundancy is a parameter that influences the DFF requirement at the design stage. This also indicates that robustness (or redundancy) is a significant parameter for planning the requirements for in-service inspection. In order to assess the consequences of a fatigue crack, it is necessary to evaluate the probability of further progressive collapse, given the presence of a fatigue crack at a specific location. The probability of further collapse may also be denoted as the probability of a system collapse. The probability of collapse is an important measure when assessing the requirements for target reliability. Thus, the purpose of this section is to describe methods for assessing the probability of further collapse, given that there is fatigue cracking at a considered hot spot.

12.6.2 Assessment of Collapse Capacity in Jacket Structures

The structural behavior near to total collapse failure can be very complex, and is expensive to assess fully by computer analysis. This complexity can be due to the nonlinear mechanical behavior of the structure and the applied loading and load distribution close to failure. The structural behavior beyond the first-member failure depends not only on the degree of static indeterminacy but also on the ability of the structure to redistribute the load and on post-failure behavior, such as the ductility of individual members and joints. For jacket structures, collapse is generally load driven, and the variability in loads is greater than the uncertainties in the collapse capacity. Simulation studies of jacket structures have shown that the ultimate structural capacity, or collapse capacity, can be directly related to the total base shear force. The load pattern also has a minor effect on collapse capacity as calculated in a simulated collapse analysis, also often referred to as pushover analysis. The term "pushover analysis" is used by the industry for illustrating the reserve capacity of structures. For example, the vertical load may be kept constant, and then the environmental load is increased until collapse, using a model that accounts for nonlinearity in terms of material and geometry behavior. A complete reliability analysis of a real multi-leg jacket structure with respect to structural collapse is very complicated, and simplifications (approximations) are required. Section 12.6.3 describes a simplified methodology for calculating the probability of collapse as a function of the reserve strength ratio (RSR), which is defined as the ratio between the load-carrying capacity and the corresponding load effect, such as extreme load effect with known return period. (By return period of n years is understood a maximum load that is expected to occur within the considered time of n years; for example, 100-year or 10,000-year wave heights.) Comparison of the estimated collapse probability obtained by this simplified approach with results from more advanced models shows acceptable agreement for typical jacket structures.

12.6.3 Simplified Method for Estimation of Probability of System Failure

A fatigue crack in a structure does not necessarily imply a significant risk of collapse of that structure. The probability of collapse can be formulated as the probability of a fatigue crack failure multiplied by the probability of collapse, given that there is a fatigue crack in the structure, P_{SYS}. The resulting probability should be lower than a target value, P_t:

$$P_F P_{SYS} \leq P_t \tag{12.6}$$

Various refinements of this methodology have been presented in the literature. A simplified method for determining P_{SYS} has been used since the 1980s for inspection planning using probabilistic methods such as presented by Stahl (1986). For jacket structures, a methodology in which one member is removed has been used to assess the residual capacity. From this, a safety index can be estimated (Naess and Moan 2013) as:

$$\beta = \frac{\ln \left(\psi \dfrac{Q_{RSYS}}{Q_{SSYS}} \right)}{\sqrt{CoV_{RSYS}^2 + CoV_{SSYS}^2}} \tag{12.7}$$

where:

> Q_{RSYS} = structural capacity obtained from the analysis of the structure with the considered element removed from the analysis model
> Q_{SSYS} = load on the structure for the considered environment and return period
> CoV_{RSYS} = coefficient of variation for the structural capacity
> CoV_{SSYS} = coefficient of variation for the load
> Ψ = factor accounting for modeling error.

It is normal practice to assume input values for this equation on the safe side. P_{SYS} can be calculated from $P_{SYS} = \Phi(-\beta)$. It should be noted that the calculated probability shown here is related to the probability level used in the load calculation. Thus, if the derivation of β is based on a 100-year loading, the calculated P_{SYS} should be multiplied by 10^{-2} for estimating the annual probability of failure. The following values may be used (see, e.g., Naess and Moan 2013):

- Proposed value for CoV for the structural capacity: $CoV_{RSYS} = 0.10$.
- Proposed value for CoV for the load: $CoV_{SSYS} = 0.15 - 0.30$.

The uncertainty in a maximum load resulting in structural collapse is considered to be larger than in long-term fatigue loading. This way of calculating system reliability may be interpreted as system analysis, considering a failure path with a full correlation with respect to the resistance of the remaining elements. The methodology used here should be considered together with requirements for the target safety level of the structures, as this requires that there are only a few fatigue cracks in members that are significant for the structural integrity.

According to Stacey et al. (1996), the static capacity of simple tubular joints fabricated with ductile material with cracked areas up to 20% is not significantly less than uncracked joints. The static capacity of ring-stiffened tubular joints is considered to be less when there is a through-thickness crack as explained in Section 8.5.

12.7 Design Fatigue Factors

The recommended DFF for different structural connections depend on:

- the consequences of failure;
- crack growth characteristics to assess whether it is possible to detect fatigue cracks within a relevant inspection period;
- possibilities for inspection and, more importantly, whether in-service inspection for fatigue cracks is planned;

Table 12.2. *Design Fatigue Factors for structures from NORSOK N-004 (2013)*

Classification of structural components based on damage consequence	Inaccessible for inspection and repair or in the splash zone	Accessible for inspection, maintenance, and repair, and where inspections or maintenance are planned	
		Below splash zone	Above splash zone
Substantial consequences	10	3	2
Without substantial consequences	3	2	1

- the reliability of the inspection method for the considered hot spot;
- uncertainties associated with the calculated fatigue damage;
- the definition of long-term loading and design format.

Thus, a recommended DFF depends on the type of structure being considered. It also depends on the formulation of the safety format to be used in the design and on definitions of characteristic values used in the design. This is illustrated with some examples in the next three subsections.

12.7.1 Structures

According to NORSOK N-004 (2013) for design of offshore steel structures, the design fatigue life for the structural components should be based on the structural service life that is specified by the operator. If a structural service life has not been specified by the operator, a service life of no less than 15 years should be used. A short design fatigue life will imply shorter inspection intervals for the same DFF. The number of load cycles for fatigue assessment should be multiplied by the appropriate factor in Table 12.2 before the fatigue analysis is performed.

"Substantial consequences" in Table 12.2 means that failure of the joint will entail one or more of the following:

a) danger of loss of human life;
b) significant pollution;
c) major financial consequences.

"Without substantial consequences" is understood to mean failure, but where it can be demonstrated that the structure satisfies the requirement for the damaged condition according to the ALS, with failure in the actual joint as the defined damage. For further information on ALS, see also Section I.4. Welds in joints below 150 m water depth should be assumed inaccessible for in-service inspection. Furthermore, it is stated in NORSOK N-004 (2013) that in project phases where it is possible to increase the fatigue life by modifying structural details, grinding of welds should not be assumed to provide a measurable increase in the fatigue life at the design stage, as explained in Section 11.7.3.

It is normal practice to estimate the long-term loading for marine structures based on expected values for the installed in-service condition and relate the DFF to

Table 12.3. *Uncertainties in load effects during pile driving (Lotsberg et al. 2008b)*

Parameter	CoV before pile driving	CoV after pile driving
Model uncertainty for calculation of stress range during pile driving	0.17	0.17
Stress range in pile due to uncertainty in soil data	0.16–0.20	
Resulting CoV on load effect	0.23–0.26	0.17
CoV on load effect used for analysis	0.26	0.17

this. Many structures may be transported to the installation site over a short period. In such a situation it is difficult to define a DFF that is related to the expected loading. Here a dynamic loading that is associated with a rather small probability of being exceeded can be used, together with an associated DFF that need not be much larger than 1.0. However, the probability level related to environmental conditions should be considered, together with the value of DFF, such that the probability of accumulation of significant fatigue damage during transportation is kept sufficiently low.

If potential fatigue damage occurs at the same hot spot during transportation as during in-service life, then the damage must be summed together. This damage may be derived using different DFF values from the two phases. However, it should be kept in mind that it is the accumulated fatigue damage that governs the probability of fatigue failure such that control on this must be maintained in both the phases in order to achieve a sound basis for planning in-service inspection.

12.7.2 Piles

During in-service life the piles in a jacket structure are subjected to much the same loading as the structure itself. The connections in the piles are butt welds, and these may be considered to be somewhat simpler than the tubular joints in the structure. However, inspection of these during in-service life is not possible, and therefore a high reliability of these is aimed for during design with a high DFF. The piles are subjected to a different load history during pile driving; therefore, the required DFF for pile driving are further assessed in the following section. The different uncertainties associated with pile driving were assessed in Lotsberg et al. (2008b), and the main uncertainties are summarized in Table 12.3. A bias predicted/measured ratio of 1.03 was also included. A bias in the engineering context is understood to refer to a ratio between the mean value of prediction and that measured. As there is no general definition related to this expression, understanding how it is defined is recommended when a number related to a bias is presented. Assessment of soil data is associated with significant uncertainty, such that after pile driving, when the number of blows is known, the uncertainty related to accumulated fatigue damage is significantly lower than before pile driving. This also influences the requirements for DFF, as shown in Figure 12.8, for thickness of the pile equal to 100 mm. Based on this study it was assessed that the required DFF for life extension of piles in the NORSOK N-006 (2015) standard for life extension of platforms could be reduced to 3, where the stress cycles have been well controlled and documented during pile driving.

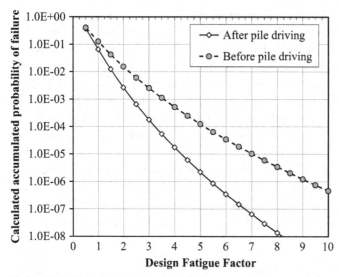

Figure 12.8. Calculated accumulated probability of fatigue failure as a function of Design Fatigue Factor for piles with thickness of 100 mm (Lotsberg et al. 2008b) (reprinted with permission from ASME).

12.7.3 Example of Design Methodology for Storage Pipes for Compressed Gas

Relationships between Analysis Models, Design Format, and Reliability

A design format for fatigue design of seam welds in pipes used for storage of compressed gas is considered below. These pipes are subjected to gas filling and emptying, and a design format can be written as:

$$n(\Delta\sigma_{Sd})DFF \leq N_{Rd}(\Delta\sigma_{Sd}) \qquad (12.8)$$

where:

> n ($\Delta\sigma_{Sd}$) = number of cycles at characteristic stress range at $\Delta\sigma_{Sd}$
> DFF = Design Fatigue Factor required to achieve a required target safety level
> N ($\Delta\sigma_{Sd}$) = number of cycles to failure at characteristic stress range $\Delta\sigma_{Sd}$. This can also be denoted as characteristic fatigue strength.

The basic safety format in equation (12.8) may look simple in principle; however, several questions are raised when establishing the recommendations to be included in design standards. These questions include:

- How should characteristic stress ranges and the number of stress cycles to be used in design be defined?
- Should additional bending stress range from fabrication tolerances be accounted for in derivation of the characteristic stress range, or should this stress be included in the values for the characteristic fatigue strength? See Section 7.2 for bending stress in pipes due to ovality.
- How is the weld length accounted for in the fatigue test data and the design procedure that is being used?

- Most often in design rules the characteristic fatigue strength is defined as 97.7% probability of survival. The S-N curves are established on the basis of mean minus two standard deviations for relevant experimental data in a logarithmic format (see Section 4.11). However, a standard that requires a high safety level may also define the characteristic fatigue strength as mean minus three standard deviations (see, e.g., DNV Rules for Classification of Ships 2012). How should the characteristic fatigue strength based on fatigue tests be defined?
- How are correlation effects in loading and fatigue strength accounted for in the design procedure?
- If the pressure in one cylinder is independent from another cylinder, there is no correlation effect in loading and there is full correlation if there is a connection between the cylinders, such that the pressure in these is equal. How will these differences in correlation influence on requirement to DFF in the design procedure?
- Can correlation effects on fatigue strength be considered to be accounted for through the "weld length effect" when this is included in the design procedure?
- Which DFF is recommended for achieving the required safety level?

It may be possible to define different design procedures that may lead to designs with an acceptable safety level. However, it is important that all required parameters in the design procedure are well defined and that each procedure is calibrated based on the actual safety format and appropriate definitions of the parameters involved. Thus, different safety formats or different definition of the parameters involved in the actual design procedure will also result in different requirements for DFFs. This means that use of design parameters from different standards should not be mixed as they are linked together in the way that the safety format and safety level are established within each standard. In comparison with other structural components, there are two important issues that are significant for the design format to be decided here; these are:

- the effect of fabrication tolerances on the design load effect or characteristic stress range;
- the effect of weld length on fatigue strength.

These issues are considered in further detail in the following.

Alternative Design Approaches
Three alternative approaches are considered:

- Alternative approach A: Use of DNV CNG rules with DFF = 10, and with the design S-N curve established as mean minus three standard deviations. The pipes are assumed to have been fabricated with enhanced tolerance control. The out-of-roundness is not specifically considered in design and the weld length is not assumed to have been accounted for in the design analysis. (The safety level is calibrated for a long weld.)
- Alternative approach B: Use of equation (4.12) for the length effect with the design S-N curve established as mean minus two standard deviations. The pipes are assumed to have been fabricated with enhanced tolerance control. Out-of-roundness is not specifically considered in design.

- Alternative approach C: Use of equation (4.12) for the length effect with the design S-N curve established as mean minus two standard deviations. It is assumed that the out-of-roundness has been specifically considered in the design. This does not necessarily require enhanced control in pipe fabrication as the actual out-of-roundness is assumed to have been accounted for in the design.

The requirements for values of DFFs can be assessed such that the safety levels obtained using the three different approaches become similar.

S-N Data and Fabrication Tolerances

A standard deviation in S-N data of 0.20 is considered relatively large. This is a typical value representative for the S-N curves for butt welds in DNVGL-RP-C203 (2016) and IIW (Hobbacher 2009). Therefore, this scatter represents more than that of the S-N data alone in the way they have been derived. The same may also be said about scatter in the S-N data for pipes, and this includes scatter due to local bending resulting from out-of-roundness in addition to that of the fatigue test data, given that the hot spot stress is measured. The stress in the S-N curve depends on how the stress range is measured during testing. The S-N curve will be different if it is based on measurement of internal pressure only, as compared with measurement by strain gauges close to the longitudinal weld.

The standard deviation in the S-N data used for design may be separated into scatter in test data, when the hot spot stress is known, and local bending stress by assuming that $\delta_{OOR} = 10$ mm corresponds to 2.5% of being exceeded. δ_{OOR} may be assumed to be normally distributed (see Section 7.2 for definition of δ_{OOR}). A revised standard deviation representative for the S-N data can then be calculated. The standard deviations are separated, based on the following relationship:

$$X = aY_1 + bY_2 \tag{12.9}$$

where Y_1 and Y_2 are independent variables.
Then:

$$VarX = a^2 VarY_1 + b^2 VarY_2 \tag{12.10}$$

and

$$St.dev. = \sqrt{VarY_1} = \sqrt{(VarX - b^2 VarY_2)/a^2} \tag{12.11}$$

Results for different tolerances and different standard deviations are shown in Figure 12.9. This figure may be used to read out separated standard deviations. A total standard deviation of 0.18 is assumed. For enhanced control of $\delta_{OOR} = 10$ mm, a standard deviation of 0.16 is derived for the S-N data in this figure. It is important that the test data are representative for the actual production weld. Fatigue cracking in small-scale test specimens will usually occur from the weld toes and not from internal defects. It then becomes essential to achieve the same quality for the kilometers of longitudinal welds that are made during pipe production. This involves certain requirements regarding quality control, NDT, and acceptance criteria.

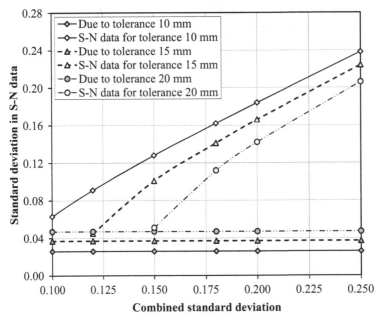

Figure 12.9. Relationship between separate standard deviations for tolerances and S-N data.

Design Methodology and Basic Assumptions for Analyses

A design case of long pipes has been considered for assessment of sensitivity and the significance of different parameters. The following assumptions are made:

- The S-N data are assumed to have been derived from fatigue of ring tests, where the hot spot stress is measured by strain gauges. The width of test specimens is assumed to be w = 120 mm.
- The length of each cylinder is assumed to be l = 36 m, welded together from three pipe sections. The number of specimens in each cylinder = l/w = 36000/120 = 300.
- Pipe geometry: Ø = 1066.8 mm (42 inches) and t = 32 mm (1.26 inches).
- Young's modulus: E = 2.1 · 10^5 N/mm².
- Assume n = 2000 load cycles from filling and emptying the pipes.

The number of load cycles to be used for design is the number of cycles expected during the design life multiplied by a DFF in order to achieve an appropriate safety level. In the following it is assumed that the main contribution to the loading is from filling and emptying the pipes. A larger DFF may be required to achieve sufficient reliability in cases of loads with larger uncertainties.

- Internal pressure range Δp. This gives $\Delta\sigma_m = \Delta p/(r_i, t)$ (from simple equilibrium consideration of half a pipe circumference) and $\Delta\sigma_b = f(\delta_{OOR}, \Delta\sigma_m)$.
- Assume the loading is normally distributed, with CoV = 0.05 as baseline case.
- Consider fabrication tolerances corresponding to "enhanced control" (DNV-OS-F101, 2013): $\delta_{OOR} = d_{max} - d_{min} = 10$ mm

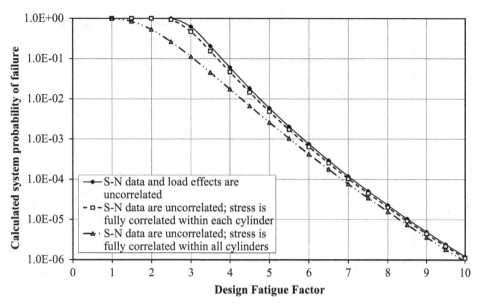

Figure 12.10. Effect of correlation in pressure on total probability of fatigue failure for 100 cylinders, each 36 m long.

- Assume that this tolerance corresponds to 2.5% of being exceeded. The tolerances are assumed to be normally distributed for the example analysis. However, the actual distribution will probably depend on how the pipes are fabricated.

Effect of Correlation

The effect of correlation with respect to loading or internal pressure is shown in Figure 12.10, using 100 cylinders as an example. A load effect with CoV = 0.05 is used. A standard deviation in the S-N curve of 0.15 is used for the analyses. The design S-N curve is established as mean minus two standard deviations. The uncertainty in load effect is low compared with that of the S-N data. Therefore, the effect of the correlation is not highly significant for the assessment, provided that the S-N data are not assumed to be correlated.

System Effect

The system effect in terms of probability of fatigue failure shown in Figure 12.11 is derived from minima distributions, considering welds in one cylinder of assumed length of 36 m (for minimum distributions, see also background for design equation for long welds in Section 4.7.6). The system effect for a number of cylinders is then derived from assumption of a series system:

$$p_{snc} = 1 - (1 - p_{1c})^n \tag{12.12}$$

Figure 12.11 is derived for CoV = 0.05 on loading, and standard deviation = 0.15 inherently in the S-N curve. It is also based on tolerance "Enhanced control," with maximum $\delta_{OOR} = 10$ mm. It is assumed that the tolerances have not been accounted for in the design analysis. The design is assumed to have been performed according to the DNV CNG rules (2012), with characteristic S-N curve of mean minus three standard deviations, and the length effect included simply as a system effect according to

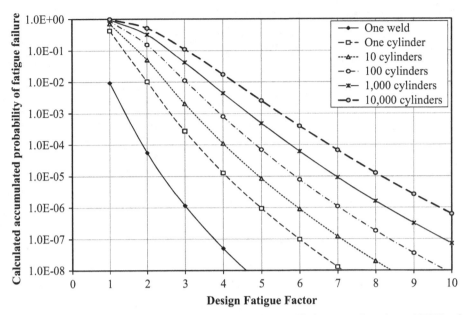

Figure 12.11. Accumulated probability of failure during lifetime as a function of DFF using design S-N curves derived as mean minus three standard deviations equal to 0.15.

equation (12.12). Full correlation of internal pressure within each cylinder, L = 36 m, is assumed. The use of equation (12.12) implies no correlation of internal pressure (or S-N data) between the cylinders. Based on these results, it is likely that a safety level in terms of accumulated probability of a fatigue failure is less than 10^{-5} during design life for pipes that can be stored on board one vessel when assuming a DFF of 10, by using enhanced control on pipe tolerances and sufficient control of internal pressure in the pipes. This is one of the alternative design procedures included in the DNV Rules for Classification of Compressed Natural Gas Carriers (2012).

Use of a DFF = 10 for structural components, where fatigue failure at a single hot spot may have significant consequences, has a long tradition in the offshore industry (see, e.g., DNV 1977). However, when the design of very long pipes subjected to the same stress range along the weld toes is considered, it can be seen from Figure 12.11 that the length effect, in addition to the DFF, is a key parameter governing failure probability. Thus, alternative formulations should be provided in design standards for these elements so that a more uniform safety level can be achieved for different weld lengths.

Assessment of Recommended Design Fatigue Factors

Assessment of the different approaches is based on the assumption that mean value is the same and the standard deviation in S-N data of 0.15 is the same. It is assumed that the loading is normally distributed, with CoV = 0.05.

A fabrication tolerance corresponding to "enhanced control": $\delta_{OOR} = d_{max} - d_{min} = 10$ mm is assumed. It is further assumed that this tolerance corresponds to 2.5% of being exceeded.

The assessments have been performed for the same number of load cycles in the alternative approaches, with n = 2000 cycles from filling and emptying of tanks. This corresponds to one cycle per week during the course of 20 years. The results

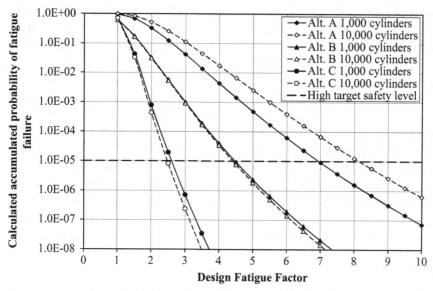

Figure 12.12. Calculated accumulated probability of failure for the three alternative procedures for 1,000 and 10,000 cylinders of 36 m in length.

from the probabilistic analyses for 1,000 cylinders and 10,000 cylinders are shown in Figure 12.12. The length of each cylinder is 36 m. The calculated probabilities of failure for alternatives B and C can be seen to be not significantly affected by the number of cylinders; this might be expected, as the weld length is a parameter in the design procedure. Alternative A does not include a weld length parameter, and the calculated probability of fatigue failure increases as the weld length or the number of cylinders increases. Figure 12.12 presents the procedures in terms of accumulated probabilities of failure. The calculated annual probability of failure is lower than the accumulated probability of failure by an order of magnitude. This means that a probability of failure below 10^{-5} in these figures corresponds to a high safety level. Based on this, the following recommendations can be made for the alternative approaches described above:

Alternative approach A. A DFF = 10 can be used together with characteristic S-N curves, established as mean minus three standard deviations from fatigue test data and presented as normally distributed in a logarithmic format for pipes fabricated with enhanced control. Out-of-roundness is not specifically considered in the design, and weld length is not assumed to have been accounted for in the design analysis.

Alternative approach B. A DFF = 5 can be used together with characteristic S-N curves established as mean minus two standard deviations in a logarithmic format for pipes fabricated with enhanced control. Out-of-roundness is not specifically considered in design, but weld length is assumed to have been accounted for according to equation (4.12).

Alternative approach C. A DFF = 3 can be used together with characteristic S-N curves, established as mean minus two standard deviations in a logarithmic format when out-of-roundness is specifically considered in design. This does not necessarily require enhanced control in pipe fabrication, as the actual out-of-roundness is assumed to have been accounted for in the design. The weld length is assumed to have been accounted for according to equation (4.12).

The main contribution to the loading is assumed to be due to filling and emptying of the pipes. A higher DFF may be required to achieve similar reliability in cases of loading with larger uncertainty, such as loads resulting from ship movements during transport of compressed gas. Uncertainty in Palmgren-Miner's rule has not been included in the present analyses, as filling and emptying of pipes has been assumed to imply a constant stress range. However, for probabilistic analysis, including long-term stress ranges from wave action, inclusion of uncertainty in the Palmgren-Miner rule is recommended. However, environmental loading is not considered relevant for the considered failure mode here.

There are various different ways to formulate a design procedure that may result in designs with an acceptable safety level. However, it is important that all the required parameters in a design procedure are well defined, and that final calibration of each procedure is based on the actual safety format and appropriate definitions of the parameters involved. Thus, different safety formats or different definition of the parameters involved in the actual design procedure will also result in different requirements for DFFs. This demonstrates that assessment of the safety level in a design procedure should not only be based on the DFF; the full procedure, including proper definition of each significant parameter, requires evaluation. Calibration of a sound design procedure should include consideration of the length effect of the longitudinal welds in the pipes The effect of correlation in fatigue strength is considered accounted for in the way the length effect in equation (4.12) is derived; see Section 4.7.6. This has resulted in alternative design procedures, for design of cylinders for transportation of compressed natural gas that is aimed for by the industry, which include the effect of weld lengths and require lower DFF values in order to achieve a high target safety level. All three alternative design procedures, A, B, and C, have been included in the DNV Rules for Classification of Compressed Natural Gas Carriers (2012) (see also Valsgård et al. 2010).

12.8 Example of Calculation of Probability of Fatigue Failure Using an Analytical Approach

As indicated by Naess and Moan (2013), there are some cases where reliabilities can be calculated by analytical approaches. Gran (1992) has also presented some expressions for reliability analysis of fatigue, and a similar analysis approach is followed here. For simplicity, it is assumed that calculation of fatigue damage is based on a single-slope S-N curve and that the fatigue damage can be calculated according to equation (10.17). A single-sided butt weld between plates of equal thickness, $t = 25$ mm, is selected as an example. This detail is assumed to be classified as F1.

This gives $\log a_d = 11.699$ for the characteristic or design S-N curve from Table 4.1.

The mean S-N curve is derived $\log a = \log a_d + 2 \cdot s_{\log N}$.

With $s_{\log N} = 0.20$: $\log a = 11.699 + 0.20 \cdot 2 = 12.099$.

Fatigue failure is assumed to occur when the Palmgren-Miner fatigue damage equals or exceeds 1.0. The limit state function for this can be written as in equation (12.1), with R and S as indicated in Figure 12.1.

The limit state function can also be written as $g = 1 - D$, where failure is defined for $g \leq 0$ where D is accumulated fatigue damage. For $g = 0$ the natural logarithm of this function can be used for definition of a new limit state function, such that it reads $g = -\ln D$.

From equation (10.17) and by using mean S-N data (log a), the following limit state function is derived:

$$g = -\ln n + \ln a - m \ln q - \ln \Gamma \left(1 + \frac{m}{h}\right) \tag{12.13}$$

Here the parameters n, a, and q can be considered as random variables. For simplicity, the negative inverse slope, m, of the S-N curve and the shape parameter in the Weibull long-term stress range distribution are kept constant; $m = 3.0$ and $h = 1.0$. These parameters are normally kept constant in reliability calculations in order to be able to estimate sound values of the associated parameter distributions for log a and the scale parameter, q.

The random variables in equation (12.13) are termed z_n, z_a, and z_q and are supposed to be normally distributed. $z_n = \ln n$ represents uncertainty in the number of actual stress cycles. This uncertainty is usually considered to have a relatively minor impact on the end result compared with uncertainties in the load and response and the S-N data. There is also uncertainty in the Palmgren-Miner damage accumulation rule, and this uncertainty may be represented by the same parameter. The expected value $\mu_{\ln n} = \ln n_0$ and the standard deviation is $\sigma_{\ln n}$. According to, for example, Wirsching and Chen (1988), uncertainty in the Palmgren-Miner rule is difficult to assess. In the following $\sigma_{\ln n} = 0.30$ is used.

z_a represents uncertainty in S-N data. This uncertainty is normally presented in a logarithmic format with base equal 10. Thus, the expected value reads $\mu_{\ln a} = E(\log 10a) \cdot \ln 10 = 2.30 \cdot E(\log 10a)$ and the standard deviation is $\sigma_{\ln a} = \sigma_{\log a} \cdot \ln 10 = 2.30 \cdot \sigma_{\log a}$.

z_q represents uncertainty in load and load response data. This uncertainty can be represented by a mean value $\mu_{\ln q} = \ln q$ and a normalized standard deviation as $CoV = \sigma/\mu$. The following approximation is then made for further analysis, $\sigma_{\ln q} \approx CoV$. The limit state function can then be written as:

$$g = -z_n + z_a - m z_q - \ln \Gamma \left(1 + \frac{m}{h}\right) \tag{12.14}$$

As all variables are normally distributed, then g is also normally distributed, and the mean value is derived as:

$$\mu_g = -\mu_{\ln n} + \mu_{\ln a} - m\mu_{\ln q} - \ln \Gamma \left(1 + \frac{m}{h}\right) \tag{12.15}$$

The variance is derived as the sum of the individual variances as:

$$\sigma_g^2 = \sigma_{\ln n}^2 + \sigma_{\ln a}^2 + m^2 \sigma_{\ln q}^2 \tag{12.16}$$

The reliability index can then be calculated as:

$$\beta = \frac{\mu_g}{\sigma_g} = \frac{-\mu_{\ln n} + \mu_{\ln a} - m\mu_{\ln q} - \ln \Gamma \left(1 + \frac{m}{h}\right)}{\left(\sigma_{\ln n}^2 + \sigma_{\ln a}^2 + m^2 \sigma_{\ln q}^2\right)^{1/2}} \tag{12.17}$$

and the probability of fatigue failure calculated as:

$$P_f = P_f(g \leq 0) = \Phi(-\beta) \tag{12.18}$$

Some numerical examples are provided below:

Assume a service life of 20 years with $n_0 = 10^8$ cycles.

$\mu_{lnn} = \ln n_0 = 18.4207.$

$\mu_{lna} = 12.099 \cdot \ln 10 = 27.8590.$

A Weibull long-term stress range distribution with shape parameter h $= 1.0$ is assumed.

The scale parameter, q, is determined from equation (10.17) as:

$q = 4.368$ MPa for DFF $= 10; \mu_{lnq} = \ln q = 1.4743.$

$q = 6.525$ MPa for DFF $= 3.0$

$q = 9.411$ MPa for DFF $= 1.0$

$\Gamma(1 + 3.0/1.0) = 6.0. \ln(6) = 1.7918$

$\sigma_{lna} = 0.20 \cdot \ln 10 = 0.4605. \sigma_{lnq} = 0.25$ is assumed

For DFF $= 10$, this gives from equation (12.15): $\mu_{lng} = 1.473 \, \mu_g = 3.2236$; from equation (12.16): $\sigma_g = (0.30^2 + 0.4505^2 + 3.0^2 \cdot 0.25^2)^{1/2} = 0.928$; from equation (12.17): $\beta = 3.4669$ and from equation (12.18): $P_f = 0.000263$

For DFF $= 3$, this gives $\beta = 2.1720$ and $P_f = 0.0149$

For DFF $= 1.0$, this gives $\beta = 0.9905$ and $P_f = 0.1610$.

When these results are compared with the numerically derived results in Figure 12.2, it can be seen that these values that have been calculated using an analytical approach are somewhat on the safe side. For further derivation of analytical expression for design points and sensitivity factors, see Gran (1992), as this may be of interest for further assessment of the contributions from different factors to the probability of failure.

13 Design of Bolted and Threaded Connections

13.1 Introduction

The methodology used for fatigue design of bolted connections that are subjected to dynamic loading is rather old. This becomes clear when studying the third edition of the book on design of bolt connections by Wiegand and Illgner (1962); this book was first issued in 1940, and in the 1962 third edition, among the 254 references provided, the oldest one dates back to 1908. However, the main span of references is from 1930 to 1960.

In the *Handbook of Bolts and Bolted Connections*, edited by Bickford and Nassar (1998), it is stated that: "The bolted joint is a surprisingly complicated affair." Although many factors affect its behavior and life, the design of most joints is based on "feel," supported by "past experience" or "custom." Because of this, most joints are overdesigned. Nevertheless, due to the uncertainties, it may be considered sound practice to make the joints stronger than the members. Many books and standards on bolted connections have been published; compared with some of these weighty books, the quantity of recommendations that can be written into this single chapter in this book is limited. Thus, this chapter should be considered more as providing an overview, and for more detailed studies, the reader should see other literature, such as the *Handbook* mentioned earlier, as well as Kulak et al. (1987). A number of ISO and ASTM standards can also be consulted regarding notations, production of bolts, tolerances of bolts, washers, and nuts, mechanical properties, testing, and documentation.

The terminology used for fasteners is presented in ISO 1891 (2009). Fasteners are understood to encompass bolts, and include head bolts that have threads for a nut at one end and stud bolts with threads at both ends and an unthreaded shank between them. The shank between the head and the threaded section of a stud bolt is also called a grip length, and the bolt diameter refers to this part of the bolt.

A number of bolt failures occurred during the first years of offshore activity in the North Sea. This resulted in stricter requirements for bolt design, fabrication, installation, maintenance, traceability, and quality assurance. The design of bolts in marine structures was found to require special attention regarding corrosion protection and dynamic loading. Due to the dynamic loading, it is recommended that bolts are pretensioned, as these bolts are most effective with respect to fatigue capacity.

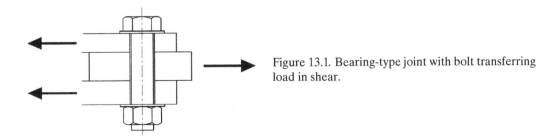

Figure 13.1. Bearing-type joint with bolt transferring load in shear.

When more than one bolt is used in a connection, due to fabrication tolerances, it is impossible to achieve an even distribution of load transfer between the different bolts in the connection if they are not pretensioned. Therefore, for bolted connections subjected to load reversals, use of a design solution with pretensioned bolts is normally recommended due to fabrication tolerances, or one may use tightly fitted bolts. It is important to maintain the pretension of the bolts during their service life.

Fasteners are used to connect different types of joints, as follows:

1. Bearing type or shear connections. Plated connections mainly subjected to in-plane loading, where the load between the plates is transferred as shear load in the bolts (see Figure 13.1). Preloading and special provisions for contact surfaces are not provided.
2. Slip-resistant connections. The in-plane loading between the plates is transferred through friction between the plates, due to the coefficient of friction and pretension bolt force (see Figure 13.2). Preloaded 8.8 or 10.9 bolts should be used (see also Eurocode EN 1993–1–8 2009).
3. Tension-type connections. Clamped connection where the external load direction is in line with that of the bolts (see Figure 13.3). Preloaded 8.8 or 10.9 bolts should be used (see also Eurocode EN 1993–1–8 2009).

There are also connections that are subjected to a combination of in-plane load, as described under point 2 in the list above, and external load, as described under point 3, which reduces the friction force between the plates. If relevant, this needs to be considered in design (see Eurocode EN 1993–1–8 2009).

For aluminum structures similar requirements are presented in Eurocode EN 1999-1-1 2009; however, experience has shown that design of slip resistant connections with use of stainless bolts is more challenging than in steel structures with loss

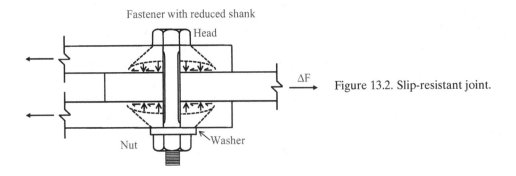

Figure 13.2. Slip-resistant joint.

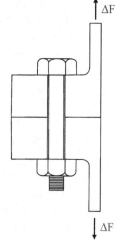

Figure 13.3. Tensioner-type joint with fasteners subjected to dynamic axial stress

of pretension of the bolts. Therefore fitted bolts have been used in some later design of helideck structures.

Various factors should be considered when designing a bolted connection subjected to dynamic loading:

- selection of relevant standards for specification of materials for bolts and nuts, design, and installation of the bolts for the actual type of connection;
- steel grade or property class, and selection of fastener, nut, and washer material for the relevant environment and functional requirements;
- geometry of bolts, nuts, and washers, including thread design for control of stress distribution along the threads when also considering fabrication tolerances;
- selection of corrosion protection;
- fabrication tolerances and surface finish of parts to be connected by the fasteners;
- pretension level and consideration on how to maintain long-term pretension of the bolted connection.

These items are considered in greater detail in the following sections.

13.2 Failure Modes of Bolts and Bolted Connections Subjected to Dynamic Loading

The most common places for fatigue failure in bolts are in the regions of highest stress concentrations, such as:

- where the head joins the shank of the bolt, due to a small transition radius;
- at the thread run-out (see Figure 13.7);
- at the first or second thread of engagement in the nut;
- at any change in diameter of the shank.

The dynamic loading to be transferred through the bolts in a connection depends on the design. In tension connections, like bolted flanges in risers, some dynamic loading in the bolts cannot be avoided. In slip-resistant bolted connections subjected to shear loading, the dynamic load in the bolt is negligible, and when the bolts are pretensioned and the load transfer is made by friction between the plates, a fatigue assessment of the bolts is unnecessary. Bolted joints subjected to load reversal in shear should be designed as slip-resistant connections, otherwise they should be designed with use of fitted bolts and correspondingly tight tolerances.

Fatigue failure of bolts subjected to dynamic axial load often initiates from the first bolt thread in the nut. In simple bolts that have been made without any special consideration for stress reductions, like stress relief grooves, the largest stresses are at the first thread. It is possible to provide a design solution that gives a more uniform stress distribution along the different threads, as, for example, shown by Wiegand and Illgner (1962) and Peterson (1974). This may be achieved by using special geometries of the nut or by using a nut with lower yield strength than the bolt material. However, if the hardness of the nut is insufficient, it may result in thread stripping due to the tensile load. This failure mode should be avoided by ensuring that the nut has a design capacity that is equal to, or larger than, the bolt capacity (Kulak et al. 1987; Kirkemo 2014; see also Section 13.4.3).

Nishida et al. (1997) tested bolts with different types of threads, different root radius, and different bolt and nut materials and found that the fatigue strength could only be marginally improved by changing these parameters. The fatigue strength was found to be a larger function of nut material than of bolt material. The traditional triangular thread geometry was found to show a good balance when considering static strength, machinability, and fatigue strength. The largest improvement in fatigue strength was achieved by shaping the threads with a reduced thread gradient against the thread run-out and making it possible to introduce a larger radius to the bolt at the tread run-out region. (Instead of tapering the nut as indicated in Figure 13.7, the tapering is made on the bolt side.) This reduces the contact area at the first threads and also the moment arm on the threads with respect to the force on the first threads. Thus, an optimal distribution of the force transfer along the threads can be achieved with a corresponding high bolt fatigue strength.

The likelihood of failure at the bolt head can be avoided by using stud bolts that are threaded at both ends.

An example of a large-diameter bolt that failed due to fatigue is shown in Figure 13.4. A fatigue crack initiated at the thread root (at the arrow) and propagated through most of the bolt area, before a final fracture occurred. This indicates that the force in the bolt due to pretension and external loading has not been very high, as the crack growth area is a relatively large fraction of the total cross section.

Loss of pretension force due to short grip length to bolt diameter ratio and loosening of bolts due to vibrations, where proper locking devices have not been used, is an observed failure mode in some structures.

13.3 Stress Corrosion and Embrittlements

In addition to the mechanical aspects for design of bolted assemblies, fastener materials must be selected to be compatible with the relevant environmental conditions with regards to corrosion and embrittlement. Guidelines for such materials selection can be found in, for example, ISO 21457 (2010).

To design against fastener corrosion, the following approaches, or a combination of these, are usually selected: (1) surface treatments, such as galvanizing (hot-dip

Figure 13.4. Photograph of bolt failure due to dynamic axial loading.

or electroplating), phosphatizing, or organic coatings; (2) selection of stainless steel, titanium, nickel alloy, or other material sufficiently corrosion resistant to the relevant environmental conditions and operating temperatures; and (3) cathodic protection (CP). CP is only relevant for fasteners fully submerged in an electrolyte, such as seawater or possibly buried. Electrical continuity to the cathodic protection system (sacrificial anodes or impressed current) must be ensured.

Fasteners must be selected not to be brittle for the relevant service conditions; it is a good practice to specify impact testing at minimum design temperature. For fasteners to be used in high-temperature service, precautions must be taken to select materials not expected to suffer detrimental effects from the high-temperature exposure, for example from temper embrittlement.

Hydrogen is in general detrimental to steel structures (see, e.g., Vander Voort 1990). A large fraction of fastener failures can be contributed to hydrogen damage, or more specifically, hydrogen-induced stress cracking (HISC). HISC is a delayed cracking mechanism that may occur when the three following factors coincide: (1) high level of tensile utilization; (2) presence of atomic hydrogen in the metal lattice; and (3) susceptible material.

Hydrogen can be introduced either during the manufacturing process of the fastener or from the environment during operation. Steelmaking, pickling processes (e.g., prior to coating/galvanization), and electroplating are examples of possible hydrogen sources from the manufacturing process. Failures often occur within a few days after assembly. Such sources must be avoided or hydrogen removed for fasteners susceptible to hydrogen, for example by performing mechanical cleaning instead of

pickling or by baking the hydrogen out of the material. When the hydrogen is introduced from the environment, failures may occur long after assembly, depending on
the rate of hydrogen absorption into the material. For subsea fasteners connected to
a cathodic protection (CP) system, hydrogen must always be considered, as H^+ will
be generated on the exposed steel surfaces from the cathodic reaction. It is therefore critical that hydrogen-resistant material is selected for subsea fasteners under
cathodic protection.

Carbon steel fasteners will have increasing susceptibility to hydrogen with
increasing strength. DNV-RP-B401 (2011) recommends a hardness of 350 HV as
an absolute upper limit for subsea fasteners exposed to CP. Bolts with SMYS up to
720 MPa, such as ASTM A193 grade B7 and ASTM A320 grade L7 (equivalent to
ISO strength class 8.8), have well-documented compatibility with CP; however, failures due to fasteners exceeding the specified hardness requirements are not uncom-
mon. For all practical applications, austenitic stainless steels and nickel-based alloys
are generally considered immune to HISC in the solution-annealed condition. With
the exceptions of UNS S30200 (AISI 302) and UNS S30400 (AISI 304) stainless
steel, moderate cold work does not induce HISC sensitivity of these materials. Bolts
in AISI 316 stainless steel manufactured according to ISO grade A4, property class
80, have proven compatibility with CP. For certain nickel-based alloys (i.e., austenitic
alloys, including, e.g., UNS N05500 and N07750), precipitation hardening may induce
high sensitivity to HISC.

For topside applications, it is common to allow up to ISO grade 10.9 fasteners.
However, as ISO 898 specifies hardness between 320 and 380 HV, 10.9 fasteners may
be susceptible to hydrogen cracking if hydrogen is not avoided. Baking is therefore
normally specified for 10.9 fasteners if hydrogen may have been introduced during
manufacturing.

A variant of hydrogen damage is sulfide stress-cracking (SSC). This must be considered for fasteners directly exposed to H_2S containing production fluids. H_2S corrosion may cause high levels of hydrogen absorbed on the surface. For carbon steel
fasteners to be SSC resistant, ISO 15156-2 requires the hardness to be limited to maximum 250 HV. ASTM A193 grade B7M and ASTM A320 grade L7M are examples of
SSC resistant grades. However, selecting these grades will have design implications
due to lower strength. Alternatively, fasteners of corrosion resistant alloy (CRA)
may be selected in accordance with ISO 15156-3, for example, ASTM A453 grade
660D.

13.4 Fatigue Capacity of Bolts

13.4.1 Geometry

Designers normally prefer to use bolts with a high capacity relative to the bolt stress
area. The reasons for this are:

- Large holes reduce the capacity of the plated structure to be connected when
  using larger bolts.
- The eccentricity of the connection, such as in a flanged connection for transfer of load, will increase with the use of larger bolt areas. This may lead to an

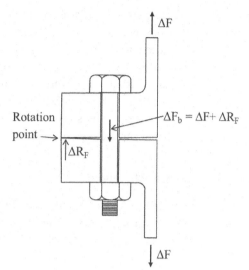

Figure 13.5. Prying action in bolted connection.

increased prying effect that reduces the capacity of the connection. A prying action is understood to refer to an additional reaction force between the connected plates as a result of moment due to transfer of eccentric force. This can significantly increase the force in the bolt(s), as indicated in Figure 13.5, due to the reaction force, ΔR_F, that is added to the external force, ΔF.

The maximum recommended pretension depends on the material in the fasteners. It is important to be within a minimum specified pretension level during service life if there are no plans for inspection and retensioning of the bolts. In order to achieve a long-term pretension in a bolted connection, it is recommended to:

- Fabricate the parts to be connected with accurate tolerances, to minimize local yielding and creep during long-term cyclic loading.
- Use a bolt geometry in which, if possible, the length of the bolt in tension is larger than approximately five times the diameter.
- Use bolts with a reduced shank, as indicated in Figure 13.2, in order to reduce the stiffness ratio between the bolt and the contact surfaces, if necessary.
- Use nuts with larger capacity than the fastener, such that stripping of the threads in the nuts will not occur.
- Limit the external load to approximately 75–85% of the fastener capacity, as larger loads tend to reduce the pretension of the bolts due to creep from local yielding and tolerances.

Increasing the bolt length and reducing the shank area reduces the bolt stiffness, which, in turn, reduces the stress cycling in the bolt, as explained in Section 13.6.2. Reduced shank diameters for different bolt sizes are given in DIN 2510 (1971). In general, the geometry of the bolts is a significant parameter with respect to fatigue capacity. More detailed guidelines on design of bolted connections can be found in the VDI 2230 (2003) Part 1 Guideline. The bolt length shall be chosen such that after tightening the following requirements are met for bolt and protrusion beyond the nut face and the thread length:

- The length of protrusion shall be at least the length of one thread pitch measured from the outer face of the nut to the end of the bolt.
- According to EN-1090-2:2008 at least four full threads (in addition to the thread run out) shall remain clear between the bearing surface of the nut and the unthreaded part of the shank for pretensioned bolts.
- For bolts that are not pretensioned, at least one full thread (in addition to the thread run out) shall remain clear between the bearing surface of the nut and the unthreaded part.

13.4.2 Chemistry

Please see ISO and ASTM standards for description of chemistry and requirements for alloying elements in the bolts. In general, the impurity elements should be kept at a low level to achieve sufficient fracture toughness at low temperatures.

13.4.3 Material Strength

The most commonly used grades or property classes in offshore structures are 8.8 and 10.9, where the first number in these notations shows the minimum specified tensile strength (SMTS) of the bolts in MPa divided by hundred, and the second number shows the minimum specified yield strength (SMYS) to tensile strength (including the dot in front of the number). Thus, the property class 8.8 is understood to refer to a bolt with a specified minimum tensile strength of 800 MPa and a minimum specified yield strength of $800 \cdot 0.8 = 640$ MPa, while the corresponding values for a 10.9 bolt is specified minimum tensile strength equal to 1000 MPa and a minimum specified yield strength of $1000 \cdot 0.9 = 900$ MPa. Elongation of the bolt at ultimate load decreases with increasing yield strength, while the risk for embrittlement and fracture increases with increased tensile stress in the bolt. Therefore, 10.9 grade is considered the upper limit of high-capacity bolts recommended for use in marine environments, but it is not recommended to be used in seawater. The 8.8 bolt shows significant elongation at ultimate load capacity; it is considered to be less sensitive to the pretension method and pretension force than a 10.9 bolt is. Thus, an 8.8 bolt is considered relatively robust in terms of installation, and is normally the first choice when designing a highly loaded, pretensioned bolted connection. A number of different standards within ISO and ASTM can be used for specification of the requirements for bolts and nuts. The majority of fasteners used in offshore structures are specified according to ASTM A193 (2014) Grade B7 or ASTM A 320 (2015) Grade L7 for studs M36 through to M56, and A320 Grade L43 for studs M64 through to M80. The nut material should comply with the requirements of ASTM A 194 (2014) Grades 4 and 7.

Mechanical properties for the bolts should be specified in accordance with ISO 898-1 (2013) or ASTM standards:

- ultimate strength
- proof stress
- elongation at fracture
- reduction of area
- impact strength (Charpy V-notch) at a defined minimum service temperature
- hardness of bolt and nut (minimum and maximum values to be specified).

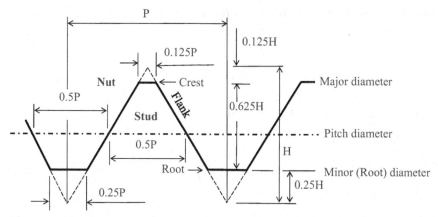

Figure 13.6. Basic profile of UN, UNR, and metric M threads (UN: Unified National Threads; UNR: Unified National Threads with controlled root to assure a proper radius in the thread root).

Mechanical properties for the nuts should be specified in accordance with ISO 898-1:2012 or ASTM standards. The nut should meet the proof load requirements to avoid thread stripping. For heavy hex ASTM A194 (2014) nuts, the proof load of the nut should be a minimum of 1.4 times the ultimate tensile capacity of the high-strength stud, and a minimum of 1.5 times the minimum ultimate capacity of the low-strength stud; see ASTM A320 (2015). These requirements imply that the strongest stud in a stud-nut assembly will always break prior to the nut failing by stripping.

When nuts are proof-loaded, they should be tested on the same type of fastener as planned for use in the connections. The nut should be loose after testing, and it should be possible to turn by hand.

Originally, a high-strength structural bolt assembly included a bolt with a nut and two hardened washers (Kulak et al. 1987). The washers were considered necessary for the following purposes:

- to protect the outer surface of the connected material from damage or galling as the bolt or nut was torqued or turned;
- to assist in maintaining a high clamping force in the bolt assembly;
- to provide surfaces of consistent hardness, so that variation in the torque-tension relationship could be minimized.

It has since been shown that washers are not always required. However, for lower yield strengths of the plates to be connected, hardened washers are necessary in order to avoid galling during the nut being turned when high-strength bolts are used. When bolts pass through a beam or flange that has a sloped interface, a bevel washer is often used to compensate for the lack of parallelism.

13.4.4 Effective Bolt Area

The effective bolt stress area depends on the failure mode to be analyzed. For shear loading, the section of the root area at the minor diameter in Figure 13.6 is used. The diameter at this root area is derived as:

$$d - 0.75H \cdot 2 = d - P\cos 30°2 = d - 1.3P$$

And the section at the root for metric series is derived as:

$$A_r = \frac{\pi}{4}(d - 1.3P)^2 \qquad (13.1)$$

where:

    d = nominal bolt diameter in mm
    P = thread pitch in mm (see Figure 13.6).

The root area for inch series reads:

$$A_r = \frac{\pi}{4}\left(d - \frac{1.3}{n}\right)^2 \qquad (13.2)$$

where:

    n = number of threads per inch.

When the tensile strength of threaded fasteners is tested, they exhibit strengths that are greater than predicted by the material strength and the root area, as if the fastener had a larger cross-sectional area. This may be explained by the development of tri-axiality in the thread root, and an empirical formula has been developed to account for this behavior. For metric threads, the tensile stress area (or bolt stress area) is calculated as:

$$A_s = \frac{\pi}{4}(d - 0.9382P)^2 \qquad (13.3)$$

and for the inch series:

$$A_s = \frac{\pi}{4}\left(d - \frac{0.9743}{n}\right)^2 \qquad (13.4)$$

13.4.5 Fitted Bolts

Fitted bolts are used in connections with very accurately machined tolerances. For example, these bolts are used to fix machinery in ships to the hull structure, but are not normally used in structural design. However, fitted bolts have recently been used in helideck structures. Due to Poisson's ratio, the fitting is reduced with significant pretension of these bolts. Thus, less pretension is recommended for fitted bolts than for bolts used in slip-resistant connections. This shows that it is important to decide on a suitable design approach at an early stage, as the design should either be made slip-resistant or be using fitted bolts. A combination of these two approaches is not recommended.

13.4.6 Thread Forming

Fatigue strength depends on the fabrication of the bolt threads. The fatigue strength for cut threads is significantly lower than for rolled threads. The fatigue strength is reduced if residual stresses due to the rolling process are removed, for example by hot galvanizing (see also DNVGL-RP-C203 2016). It is therefore recommended that, if possible, all heat treatment should be performed prior to rolling the threads. Cold rolling after heat treatment is specified in ISO 261 (1998). For external threads, it

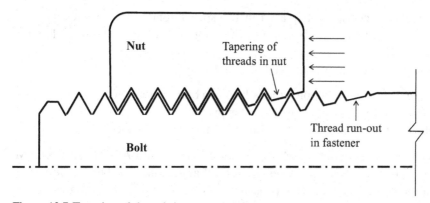

Figure 13.7. Tapering of threads in nut and thread run-out in fastener.

is normal to specify ISO metric coarse threads in accordance with ISO 68–1 (1998), as coarse threads are considered to be more robust with respect to handling and installation. Majzoobi et al. (2005) tested ISO bolts with coarse and fine threads, and nominal diameters in the 10–24 mm range, and found that coarse threads were superior to fine threads for the same core diameters. However, the same nominal bolt diameter bolts with coarse threads show a similar fatigue capacity as for fine threads. The thread type and diameters for the nuts should be the same as for the bolts.

An example of thread tapering of the nut and thread run-out in a fastener in order to reduce the stress concentration at the first thread is shown in Figure 13.7. However, design of bolted connections is normally performed without such specific requirements. To ensure that the threads in the roots of fasteners are made with a proper radius, specification of UNR threads is recommended (see Figure 13.6).

13.4.7 Tolerances

General thread tolerances for bolts and nuts are defined in ISO 965-1 (2013); however, the class of tolerances must be specified.

13.4.8 Surface Treatment

The surface roughness of the machined surfaces of the bolts should not exceed $R_a = 0.8\,\mu m$; see Section 11.8 for definition of surface roughness. The bearing surfaces of nuts and washers should not exceed $R_a = 1.6\,\mu m$. Furthermore, it is important that the surfaces are without crack indications or notches that can lead to fatigue crack initiation. It is recommended that 100% magnetic particle inspection (MPI) be performed on all procured bolts prior to application of corrosion protection (MPI can be used on ferromagnetic materials, but can not be applied on austenitic stainless steel, as explained in Section 11.6.4). Proper marking and identification of bolts is recommended as part of quality control.

Zinc or aluminum is used for metallic coating of fasteners for corrosion protection. Different types of other coatings can also be used as corrosive protection for atmospheric service, and cathodic protection can be used for installation below seawater surface. Bolt heads and nuts in cups filled with grease are also used in bolted riser flange connections placed in the splash zone. The type of corrosion protection may influence the selection of thread geometry and tolerance class for the nut.

Corrosion protection of bolts can be achieved by:

- hot-dip galvanizing, by dipping the bolts into a bath of hot metal; iron-zinc alloys or pure zinc alloys are often used for this process
- metallizing, by spraying hot zinc or aluminum over the surface.

Hot-dip galvanizing should be performed in accordance with ASTM A153/A153M (2009). The hot-dip galvanizing process requires that the mill scale be removed prior to application of the coating. This is usually achieved by pickling the member in a bath of acid. Galvanizing is subsequently performed by dipping the bolts into a bath of hot metal. The maximum temperature in the zinc bath should not exceed 460°C, unless special acceptance is approved. The zinc coating should have a minimum thickness of 80 μm at the threaded part and 100 μm on the shank. The properties of the substrates prior to the dip should be such that hydrogen embrittlement of the bolts is reduced by baking the bolts at around 180°C for 2–24 hours depending on type and size of fastener and other variables. Adhesion of zinc to the bolt surfaces should be verified on one representative dummy bolt specimen before commencing production. The results from the test should meet the requirements of ASTM A153/A153M (2009). During arc spraying the temperature in the bolt should not exceed 250°C, to avoid changes in microstructure.

Bolted connections used to connect aluminum plates to steel plates need special consideration, as bimetallic corrosion is liable to occur when dissimilar metals are in electrical contact in any electrolyte, including rainwater, condensation, and so forth. If an electric current flows between the two plates, the less noble metal becomes an anode and corrodes more rapidly than would have occurred if the metals were not in contact. The corrosion rate depends on the area relationship between the anode and the cathode. Adverse area ratios are likely to occur for fasteners at joints, such that electrical isolation of the dissimilar parts is recommended if water cannot be excluded from the connection. Electrical isolation may be achieved by painting the contact surfaces of the plates. Isolation around bolted connections may be achieved by using non-conductive plastic or rubber gaskets and nylon or Teflon washers and bushes (see Eurocode EN 1993–1–4 2006). This also relates to the use of stainless steel bolts for connecting steel plates.

Bolted connections in offshore structures are typically used between aluminum helideck structures and carbon steel support structures. Experience from in-service life shows that special precautions are needed to avoid corrosion when designing these connections. Aluminum alloy ALMg4.5Mn is frequently used in plates with thickness less than 16 mm and AlMgSi1Mn is frequently used in thicker plates and sections in aluminum structures. These alloys are providing good corrosion resistance in marine environments and are listed as recommended materials in NORSOK M-121 (1997) for aluminum structures for offshore applications. Other alloys listed in that document may be used for less severe environments.

A sketch of a bolted connection between aluminum and carbon steel in a severe marine environment is shown in Figure 13.8. Isolation between aluminum and carbon steel is recommended to avoid corrosion. A thin stainless steel plate (1 mm) may be used as isolation. Stainless steel bolts should be used, as experience shows that

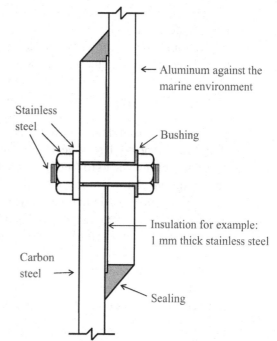

Figure 13.8. Example of a bolted connection between aluminum and carbon steel structure.

Aluminum against the marine environment

Stainless steel

Bushing

Insulation for example: 1 mm thick stainless steel

Carbon steel

Sealing

galvanized bolts may corrode even if they have been insulated from the plates. Also, the stainless steel bolts should be insulated by a bushing. Nylon bushings have been used in some projects; however, it is indicated that these do not show necessary long-term durability and may need to be replaced during service life. Other bushing material is preferred for the purpose of long-term pretension of the bolts. In areas with salt not removed by rain, it is observed that aluminum may corrode where a seal around the connection is not used or where the sealing is lost due to surface oxidation of aluminum, in which case the adhesion between the sealing and aluminum is lost. It is therefore recommended to perform the design such that there is proper space to add a sealing with proper adhesion to avoid split corrosion from aluminum. For example, where horizontal beams are bolted together, one of the flanges should be wider than the other so that a sealing can be properly placed. Experience also shows that there may be corrosion of aluminum close to the stainless steel bolt.

13.4.9 Effect of Mean Stress

The fatigue strength of bolts decreases with increasing mean tensile load. However, this decrease is not considered important relative to that of losing pretension in the bolted connections. Thus, this is not considered a significant issue when determining the pretension level for the material grades considered in Section 13.4.3.

13.5 Slip-Resistant Connections

A slip-resistant joint is understood to mean plated connections that are clamped together by pretension bolts, so that forces can be transferred through friction between the plates for forces acting in the planes of the plates (see Figure 13.2).

These joints are also denoted friction-type joints, due to the way in which the forces are transferred through the connection. Frictional resistance depends on the bolt pretension load and slip resistance at the faying surfaces. The maximum capacity is considered to have been exceeded when slip occurs that brings the plates into bearing against the bolts. Slip-resistant joints are considered to be efficient for transfer of dynamic loads, provided that the loads are below the load level that results in slip between the plates. High contact stresses between the joined plates will only occur in a small area around the bolt-holes. Fatigue testing of slip-resistant connections has shown that some small relative movement will be present at the ends of the plate connections. This may lead to fretting of the steel surfaces and fatigue initiation at the ends of the lap plates. If more of the load is transferred by shear or contact to the bolts, then the region around the holes may be most critical with respect to fatigue cracking. However, provided that the connection is designed so that slip will not occur, the gross plate area can be used for fatigue assessment of the plates. Otherwise, the net area at the holes including relevant stress concentration factors due to the geometry should be used for fatigue assessment.

When bolts are pretensioned, the force transfer is made by friction between the contact surfaces where the friction resistance force can be calculated as:

$$F = \mu \, F_0 \, n \qquad\qquad (13.5)$$

where:

μ = friction coefficient or slip coefficient
F_0 = pretension force in the bolt(s)
n = number of bolts.

The slip coefficient depends on fabrication of the plates and the type and thickness of the corrosion protection (see Kulak et al. 1987). Testing is recommended if there is uncertainty about the slip coefficient to be used in an actual design. Testing has shown that the friction coefficient is dependent on pressure and that contact pressure reduces the friction coefficient. Therefore, testing should be performed under relevant conditions such as surface roughness and contact pressure. A proper material coefficient is also recommended for determining design resistance capacity; see design standards such as EN 1993–1–8 (2009).

Most joints use a single fastening system to connect plates to members for transfer of forces acting on the joint. In land-based structures that are subjected mainly to static loads, it may be acceptable that the ultimate capacity of slip-resistant joints is increased by addition of fillet welds, provided that the last pretension of the bolts is made after the welding has been completed (Eurocode EN 1993–1–8 2009). However, these welds may reduce the fatigue capacity, as fatigue cracking can initiate more easily at the weld toes than at bolt-holes; thus, this solution is not recommended for fabrication of marine structures.

13.6 Tension-Type Connections

13.6.1 Application

Tension-type joints are used in flanged connections in risers. Joints with a somewhat similar behavior are also used to connect modules to deck support structures.

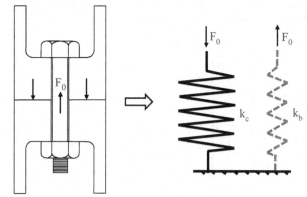

Figure 13.9. Analysis model for calculation of displacements during pretension of a bolted connection.

Elastomeric pads are being applied between accommodation modules (or living quarters) and platforms to prevent noise. Elastomeric pads might reduce the effect of the preloading as the stiffness of the contact surface for pads is less than for steel against steel and more of the dynamic load is transferred into the bolts as expressed by equation (13.13).

13.6.2 Structural Mechanics for Design of Bolted Connections

First, the simple connection in Figure 13.9 is considered for pretension of the bolt. Both the bolt and the connected plates are elastic and deform during pretension loading. This flexibility is illustrated in this figure by the springs. The bolt length is increased due to pretension force, F_0, and the plates are compressed due to the reaction force from the bolt, $-F_0$. The total displacement due to the pretension force can be calculated as:

$$\delta_{tot} = \delta_{0b} + \delta_{0c} \tag{13.6}$$

where:

δ_{0b} = elongation of the bolt
δ_{0c} = compression of the clamped plates.

Here, the elongation of the bolt for a constant bolt area can approximately be calculated as:

$$\delta_{0b} = \frac{F_0}{E\,A_s}l_b = \frac{F_0}{k_b} \tag{13.7}$$

where:

E = Young's modulus for the bolt material
l_b = effective length of the loaded bolt
A_s = bolt stress area
k_b = spring stiffness for the bolt.

When a bolt is made with different diameters along its length, the resulting elongation can be calculated as the sum of the separate elongations within each section.

The compression of the plates can be approximately calculated as:

$$\delta_{0c} = \frac{F_0}{E\,A_c}l_c = \frac{F_0}{k_c} \tag{13.8}$$

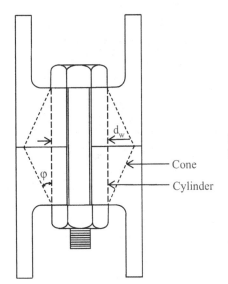

Figure 13.10. Models for derivation of stiffness from plates to be joined.

where:

E = Young's modulus for the plate material
l_c = sum of thickness of the loaded plates
A_c = effective plate area
k_c = spring stiffness for the joined plates.

Here it is a challenge to determine accurately the effective area, A_c. If this is significant for the fatigue assessment, a finite element analysis may be performed to determine k_c. Otherwise, k_c can be estimated by calculating the area of a cylinder with a diameter equal to the bolt head/nut, minus the area of the bolt-hole. In reality, a larger area of the plates will be subjected to prestress farther away from the bolt head and nut. Thus, this assumption leads to a conservative assessment of the dynamic loading in the bolt. An alternative to this is to assume that the stress flow in the plates lies within a cone-shaped geometry, as indicated in Figure 13.10. Here, an effective stiffness may be derived by integration over the joined plates. The results depend on angle φ, which is presented in the literature in the 30–45° range; see, for example, VDI 2230 (2003) Part 1 Guideline for a more detailed assessment. The following relationship between force and displacement can then be expressed by the resulting spring stiffness, k_s:

$$F_0 = k_s \, \delta_{tot} \tag{13.9}$$

The resulting spring stiffness for a series system is then derived from equations (13.6)–(13.9) as:

$$\frac{1}{k_s} = \frac{1}{k_b} + \frac{1}{k_c} \tag{13.10}$$

By pretension of the bolt equal to F_0, there will be an elongation of the bolt of δ_{0b} and a compression of the connected plates of δ_{0c}, as indicated in Figure 13.11. After pretension, the bolt and the clamped plates act together in a parallel system, as shown in Figure 13.12. This means that the resulting spring stiffness, k, is the sum of k_b and k_c.

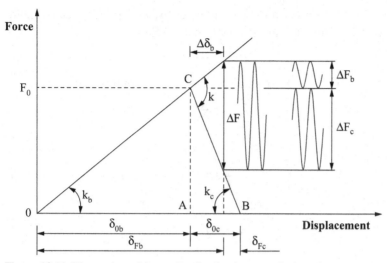

Figure 13.11. Illustration of force distribution between bolt and contact.

This resulting stiffness is illustrated at point C in Figure 13.11. This figure can be denoted as a "bolt diagram," as it illustrates the physical behavior of a pretensioned bolted connection subjected to dynamic loading in the axial direction of the bolt(s). The bolt pretension gives an elongation 0-A, and the axial stiffness of the bolt is shown in the figure as k_b. The reaction force from the bolt head and nut results in compression of the connected plates of B-A and to the compressive stiffness that is now plotted as k_c, which is the angle A-B-C. The resulting stiffness, k as shown at position C in Figure 13.11, can now be considered to be the stiffness condition after pretension of the bolt. By further loading of the connection by an external force, there is reaction or stiffness from both the bolt and the clamped plates, as illustrated in Figure 13.12. An external force, ΔF, will increase the force in the bolt by ΔF_b and

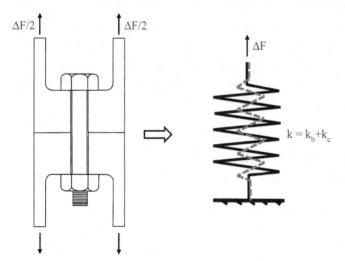

Figure 13.12. Analysis model for calculation of displacements for external load on a bolted connection.

decrease the contact stress between the joined plates by ΔF_c, as indicated in Figure 13.11. The bolt load after increase in external load by ΔF is:

$$F_b = F_0 + \Delta F_b \qquad (13.11)$$

The increase in bolt elongation due to the increase in load can be calculated as:

$$\Delta \delta_b = \frac{\Delta F}{k} = \frac{\Delta F_b}{k_b} \qquad (13.12)$$

and

$$\Delta F_b = \frac{k_b}{k} \Delta F \qquad (13.13)$$

The compressive force between the joined plates due to ΔF is:

$$F_c = F_0 - \Delta F_c \qquad (13.14)$$

where:

$$\Delta F_c = \frac{k_c}{k} \Delta F \qquad (13.15)$$

If F_c becomes negative, then the external load is so large compared with the bolt pretension that the contact pressure between the plates is lost. This means that with further loading, all additional load will be transferred directly to the bolt as tensile load. This shows the importance of keeping a long-term pretension of bolted connections in order prevent detachment of the joint and to achieve a high fatigue capacity. After external loading by ΔF, the remaining compression of the joined plates is δ_{FC} in Figure 13.11.

Accurate pretension of bolted connections requires that a construction procedure is developed and followed. The pretension procedure may need to be tested before it is used on an actual installation. Such testing may require installation of strain gauges in the bolts to be pretensioned during the testing.

13.7 Technical Specification for Supply of Heavy-Duty Bolts

There are a number of different factors that should be considered when procuring heavy-duty bolts for marine structures. Therefore it is recommended that a technical specification be developed for procurement where delivery of high-quality bolts is important for structural integrity. The following list of points should be included in a technical description:

- Introduction
- General requirements
- Reference standards and applicable documents
- Material grade for different elements
- Chemical requirements
- Manufacture and heat treatment process(es)
- Requirements for mechanical properties
- Dimensions, workmanship, and tolerances
- Corrosion protection
- Quality assurance.

13.8 Pretensioning of Bolts

Bolt fatigue is a failure mode that is often seen in connections with insufficient pretension of the bolts, resulting in the connection becoming too flexible. To eliminate this failure mode, the designer should specify as high a pretension load as practical. Higher clamping forces make the joint more rigid and reduce the cycling stresses transferred to the bolts (see also Section 13.6.2). It is important that the pretension is maintained without significant loss during the lifetime of the structure; therefore, it is also important that certain geometrical conditions in the bolt design are fulfilled, as explained in Section 13.4.1. Pretensioning of bolts may be achieved by torque or by a more direct axial tensioning of the bolt. The torque method is normally used for pretensioning of bolts for use in land-based industries. However, in some more special design solutions, where large bolt sizes are used for structural application, direct tensioning of the bolts has also been used to ensure a controlled pretension level after installation. Optimum fastener torques have been determined for many high-strength fasteners. However, because torque resistance is dependent on the friction between the contacting threads and between the head and nut and the bearing surfaces, such information should be used with caution. Thus, use of lubricants may result in overloading of the bolt during torqueing, while use of less effective lubricants may result in a loose connection. For connections that are not subjected to significant temperature variations, by using proper material, proper design of the bolt and nut bearing, and potential locking devices, it should be possible to achieve initial clamping forces that will remain during the in-service life of the fastener joint. This is considered to be more difficult to achieve for a bolted connection that is subjected to larger differences in temperature, where the bolt pretension may decrease due to creep or local exceedance of the material yield strength in the thread roots. After a group of bolts have been pretensioned, the torque on all the bolts should be checked, as some pretension may be lost in one bolt due to interaction with the pretension of the other bolts.

The elongation capacity of steel material decreases as bolt strength increases. However, the strength grades considered here have sufficient elongation so that a high pretension can be made. For example, 8.8 bolts are considered robust in this respect, as they can normally be pretensioned to the yield strength without risk of failure. This high pretension level has been shown to be beneficial with respect to fatigue capacity.

When assessing a pretension level, it should be remembered that the bolt stud and nut capacities are reduced during torqueing, compared with direct tension capacities without any torque (Viner et al. 1961). Typical values may lie between 5% and 20%, depending on the coefficient of friction between contacting bodies during makeup.

As an example of a design of riser flanges, a pretension of 10.9 bolts was used, equaling 80% of the minimum specified yield strength for the bolts. This pretension was achieved using a hydraulic axial tensioner device; to achieve this pretension in all the bolts around the circumference, it was necessary to perform the tensioning in two sequential steps. The maximum long-term loading was limited to 85% of the minimum specified yield strength for the bolts in order to limit loss of pretension force.

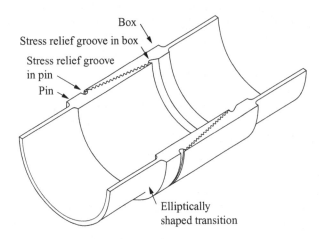

Box
Stress relief groove in box
Stress relief groove in pin
Pin
Elliptically shaped transition

Figure 13.13. Sketch of a connector used in the tethers of a tension leg platform.

For satisfactory performance of the joints, it is essential that the bolts have been tightened to the preloads specified by the designer, and that these high bolt preloads are maintained during the service life of the structure. Different types of locking devices are available to lock the nut relative to the bolt to avoid nut loosening during vibrations. Documentation of the locking capacity can be performed by vibration testing according to the DIN 25201-4 (2010) Standard. According to this standard, the securing effect of locking devices is considered "adequate" if the residual bolt preload after 2,000 load cycles is not less than 80% of the applied bolt preload.

13.9 Connectors for Tubular Structures

By using machined components, it is possible to design and fabricate connectors that can be used in tubular structures that have better fatigue capacity than the butt welds required in the tubular girth welds, even if these welds are ground flush on both the inside and the outside. A sketch of a typical connector used in tethers to connect tension leg platforms to the seafloor is shown in Figure 13.13. The connector consists of a pin and a box, which is designed to have a relatively even distribution of stress along the threads. This is achieved by the optimized geometry of the pin and box, by a varying stress area, elliptical stress relief grooves in the pin and the box, and also elliptically shaped transitions from the coupling body to the tubular sections. In addition, the pin and box may be machined with differences in pitch, such that a perfect fit is achieved after makeup of the connections and tether pretension. Thus, the calculated stress concentrations should refer to this condition also, as the stress cycling is around this pretension level in the tethers. An example of stress distribution in such a connector is shown in Gundersen et al. (1989). These connections are typically machined with very accurate tolerances, in the order of $\pm 1/100$ mm with respect to pitch (see definition in Figure 13.6). The connections can be made from forging using high-strength steel, with yield strength higher than 500 MPa, such that the high-strength S-N curve in Section 4.1.5 can be used for fatigue assessment. It is important to design the connection with appropriate corrosion protection in order to use this S-N curve. This is achieved by using a sealing compound in the threads and with an additional sealing at the pin groove. This sealing is also required for inside corrosion protection and to achieve the possibility of "leak-before-break" detection with empty tethers and for buoyancy. In addition, anodes are connected to the coupling bodies to provide corrosion protection on the outside. A large DFF is normally used because inspection and maintenance of the threaded sections can be very difficult.

14 Fatigue Analysis of Jacket Structures

14.1 General

This chapter is written with mainly jacket structures in mind. However, the same principles can also be used for analysis of other types of bottom-fixed structures such as jack-ups and support structures for wind turbines. Fatigue analysis of jacket structures should be performed for:

- transportation of the jacket from the construction yard to the installation site;
- driving of piles;
- the installed condition.

As fatigue damage is accumulated over different phases, the calculated fatigue damage at each hot spot from each of the operations can be added together.

Transportation of the jacket on a barge to the installation site requires special considerations with respect to fatigue. The most significant hot spots during transportation may differ from those in the installed condition. However, transportation can involve significant fatigue damage, and in some cases, repairs have been needed before the jacket could be installed. Therefore, it is important to plan transportation and assess relevant environmental criteria for the transportation period and route; this includes assessment of available sheltered harbors on route should transportation take longer timer than predicted. In fatigue analysis of the transportation phase, the question arises regarding which Design Fatigue Factor (DFF) should be used. This should be considered in conjunction with the criteria used for transportation. If fatigue damage is expected to accumulate at the same hot spot during transportation as during the installed condition, it is reasonable to use a DFF that is sufficiently large so that the allowable calculated fatigue damage during transportation is low. This should be described in a design specification before the detailed design commences.

It is also important to remember that slender members in a jacket lying on a barge during transportation may be subjected to vortex-induced vibrations in wind. Vortex-induced vibrations in water need to be considered in design for the installed condition. However, the driving forces and damping are different in wind than in conditions with wave and current loading.

Fatigue analysis of the piles in jacket structures shows that a significant part of the fatigue damage can be accumulated at thickness transitions during pile driving. The accumulated fatigue damage governs the probability of fatigue failure, as explained in Chapter 12. As the uncertainty in the load range from pile driving is similar to that from the installed condition without any information from the driving process as described in Section 12.7.2, it is reasonable to recommend the same DFF for both the pile driving and the installed condition, as recommended in NORSOK N-001 (2012) and in ISO 19902 (2007). However, if the uncertainty in the load effect from pile driving is reduced, due to information on blow counts and driving energy used during installation, it is acceptable to use a lower DFF for reassessment of calculated fatigue damage, as explained in Section 12.7.2.

Two different methods are used for fatigue analysis of installed jacket structures:

- deterministic discrete wave fatigue analysis (also denoted deterministic fatigue analysis);
- frequency response fatigue analysis (also denoted frequency domain fatigue analysis, spectral fatigue analysis and stochastic fatigue analysis).

These two analysis methods differ, and thus the fatigue lives calculated by each method also differ somewhat from each other. This indicates that uncertainties are associated with the fatigue analysis for the installed condition. The differences are related to the environment, the load and response calculation, and how the stress concentration factors (SCFs) at the tubular joints are calculated as a function of loading. The deterministic method has traditionally been preferred for fatigue analysis of jacket structures without significant dynamics in the North Sea. Structures in deeper water become more flexible, and the hot spot stresses are amplified due to dynamic structural behavior, as also described in more detail in the following sections (see also Naess and Moan 2013). By a stochastic fatigue analysis one may also understand a time-domain analysis that is being performed for structures with significant dynamic response such as deepwater structures and wind turbine support structures.

When designing jackets, horizontal framing is not placed at the water line in order to avoid variable buoyancy loads and slamming loads on horizontal members. There are various assessment procedures for slamming loads on such elements; see, for example, appendix A of DNVGL-RP-0001 (2015). This problem has been encountered for subsiding platforms at the Ekofisk field where the horizontal frames, which were placed well above the splash zone in the original design, have been gradually sinking through the mean water level during their in-service life.

Vortex-induced oscillations may occur due to wave and current loading, depending on member geometry and boundary conditions. The design should be such that vortex-induced vibrations are unlikely to occur. Guidance on analysis of vortex-induced vibrations can be found in DNV-RP-C205 (2014).

A finite element model of the jacket should be prepared for the fatigue analysis and should include a representation of the main frame structure, the foundation piles, and the stiffness of the deck structure. The stiffness of the foundation piles relative to the soil may be simulated by springs, as shown in appendix A of DNVGL-RP-0001 (2015), or by a matrix representing the coupled stiffness corresponding to each degree of freedom at the joints at the bottom of the pile-to-sleeve connection (the top of the pile at the mud line). New designs of piles commonly use piles without any, or possibly one, change in cross section over the pile length. In such cases it can normally be easily proven that the highest stresses and fatigue damage occur at the pile top. Here, the jacket-pile interaction can be represented by using a properly linearized stiffness matrix as a boundary at each pile connection. The matrix should be linearized to match resulting displacements and rotations for a typical shear force and corresponding pile top moment, and for a typical vertical force. This can be obtained from two pile analyses, where the typical shear force is applied with and without the corresponding pile top moment.

In cases where it is appropriate to perform specific fatigue damage evaluations not only in the pile top but also along the pile embedded in soil, an integrated complete pile model should be provided. This may be required when piles with multiple changes in cross section are used, as is often encountered in older jacket designs.

14.2 Deterministic Fatigue Analysis

By a deterministic fatigue analysis is understood stepping of a set of regular waves through the structure for calculation of stress ranges at relevant hot spots. For the following reasons, a deterministic fatigue analysis approach is recommended for jacket structures:

- It is more accurate than a frequency response analysis for the upper region of the jacket, as linearization of the drag term has to be performed in a frequency domain response analysis.
- Variable buoyancy may induce fatigue damage in the upper part of the jacket, and its representation is more physically accurate in a deterministic analysis than in a frequency domain analysis.
- The global dynamic effects are small if the first natural period is less than 2.5 sec for foundation springs corresponding to a sea state governing fatigue damage accumulation.
- The hot spot stress at any location varies with geometry, and the instantaneous force flow in all members entering the tubular joint as described in Section 8.2.1 can easily be included in a deterministic fatigue analysis approach. This is considered more complex in a frequency domain response analysis (see Section 14.3 for detailed explanation).

If the fundamental natural period (T) is larger than 2.5 sec, a deterministic discrete wave fatigue analysis may still be recommended. The dynamic effects can then be accounted for by weighted dynamic amplification factors that can be derived from frequency domain response analyses in Section 14.3. The dynamic amplification factors can be calculated for each member being considered as the ratio of the response with and without dynamics included in the analysis. The response is calculated for each wave direction by integration of the response over the sea scatter diagram. Thus, a weighted dynamic amplification factor for a considered wave direction can be derived as:

$$DAF_w = \left(\frac{D_{Dynamic\ included}}{D_{Quasistatic\ analysis}} \right)^{1/4.0} \tag{14.1}$$

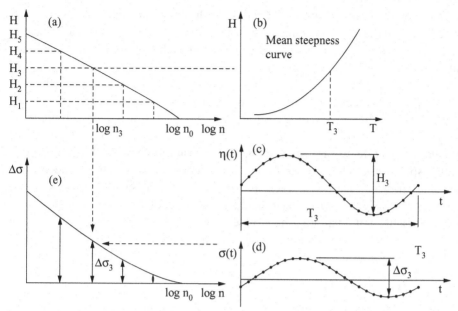

Figure 14.1. Schematic procedure for establishing the long-term stress range distribution in a deterministic fatigue analysis.

where:

$D_{\text{Dynamics included}}$ = Fatigue damage calculated (Palmgren-Miner summation) using an S-N curve with a constant slope, m = 4.0, where fatigue damage is calculated by integrating the damage over all sea states and where the dynamic response has been included in the frequency response function, which is also denoted transfer function for the response

$D_{\text{Quasistatic analysis}}$ = Fatigue damage calculated using an S-N curve with a constant slope, m = 4.0, where the fatigue damage is calculated by integration of damage over all sea states, without including the dynamic response in the frequency response function

m = 4.0 is used as a mean slope of the S-N curves considered to be the most representative of the contribution to fatigue damage (see also Section 1.2). For jacket structures in deeper waters, the dynamic effects may be significant. A frequency domain analysis is the preferred method, as the dynamic effects, as well as the geometric effects, are properly accounted for through identification of the "peaks and valleys" in the frequency response function.

The procedure schematically shown in Figure 14.1 is often followed when performing a deterministic fatigue analysis of a fixed structure without significant dynamic response (see NORSOK N-004 2013). The wave distribution and directionality is typically accounted for by considering various sectors – for example, eight. A wave height exceedance diagram is established within each sector, as indicated in Figure 14.1a. In each sector, several discrete wave heights are selected for analysis. For each wave height, H_i, a corresponding wave period, T_i, is determined, based on a mean wave steepness curve or on actual data for the area being considered (Figure 14.1b). Stokes's fifth-order theory is recommended for analysis, together with drag and mass coefficients from NORSOK N-003 (2007) for load calculation.

Forces on the structure are calculated using the Morison equation. Each wave is stepped through the structure at increments in the wave, at a phase angle as indicated in Figure 14.1c, for calculation of internal forces in each structural element at each joint (axial force, in-plane bending moments, and out-of-plane bending moments). Here the wave height H_3 is used as an example, and this wave is stepped through the structure in 24 steps that correspond to an increment in phase angle of 15°. The member forces at the tubular joint are used for estimation of the type of tubular joint: X-joint, Y-joint, or K-joint, or a fraction of these joints as described in Chapter 8 for calculation of an effective hot spot stress. This information is needed for calculating each hot spot stress to be used for fatigue life calculation on the chord side and the brace side on the tubular joint. Each step in the wave analysis results in a stress at each hot spot in Figure 14.1d, which includes the effect of the stress concentrations for the relevant joint, including the stresses from the axial force, in-plane bending moments, and out-of-plane bending moments. The stress range at the considered hot spot is then derived as the difference between the maximum and the minimum stress. This calculated stress range, $\Delta\sigma_i$, which corresponds to cycle n_i in the wave height diagram, is plotted into the graph for long-term stress ranges in Figure 14.1e; here $\Delta\sigma_3$ corresponding to the wave H_3 is shown for illustration.

The analysis procedure is repeated for all selected wave heights to establish a long-term stress range distribution as shown in Figure 14.1e. At least 10 wave heights should be selected for analysis, but the required number also depends on the geometry of the structure, especially the layout geometry in the waterline area. The analysis procedure is repeated for the other sectors, so that the long-term stress range distributions for all sectors are determined. The fatigue damage within each long-term stress distribution is calculated using the Palmgren-Miner rule. This is performed by numerical integration in which the long-term stress range distribution in Figure 14.1e is divided into a number of blocks – for example, 100–200. Different methods can then be used for integration, using either a trapezoidal integration or a higher-order method. Finally, the fatigue damage for the hot spot being considered is derived by summation of fatigue damage from the long-term stress range distribution within each sector.

14.3 Frequency Response Fatigue Analysis

Frequency response fatigue analysis of jacket structures was introduced by Vughts and Kinra (1976), and has been the recommended methodology for fatigue assessment of deepwater platforms with a high natural period, such as shown by Lotsberg et al. (1988). The methodology has also been used for fatigue assessment of jackets in the North Sea (Scherf and Thuestad 1987). However, most jackets in the moderate water depths are analyzed by deterministic fatigue analysis. Typical wave scatter diagrams used for stochastic fatigue analysis show wave periods in the 3–25 sec range. The energy in a wave increases with wave height and the wave period increases with increasing wave height. Then as the first natural period of a fixed platform is increased above 2.5–3 sec, the response increases due to dynamic amplification, which in turn increases the calculated fatigue damage. The dynamic amplification is increased by further increase in the natural period. A high natural period for a jacket may be in the 4–6 sec range. A high natural period for a jack-up is within the 6–8 sec range.

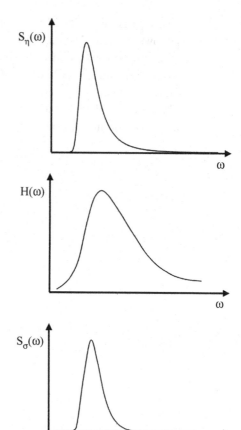

Figure 14.2. Schematic procedure for a stochastic fatigue analysis.

The dynamic amplification depends on the amount of damping. The damping is typically larger in jack-ups than jackets due to differences in foundation design and hydrodynamic behavior.

The methodology for stochastic fatigue analysis is described by Naess and Moan (2013). The Airy wave theory (linear wave theory) is used in the analysis, and this requires use of different hydrodynamic coefficients as compared with a deterministic fatigue analysis (see, e.g., NORSOK N-003 2007). The definition of the type of joints dependent on force flow may be more cumbersome in stochastic analyses than in deterministic fatigue analysis. One possible method for this is to establish a SCF function that is frequency dependent for each considered load direction, and to include this factor, together with the frequency response function, when the response in one sea state is calculated. In order to determine this SCF (ω) function, some separate analysis must first be performed for each load direction and then the response for each hot spot can be calculated as:

$$S_\sigma(\omega) = \int |H(\omega)SCF(\omega)|^2 S_\eta(\omega)d\omega \qquad (14.2)$$

From this response the fatigue damage within each sea state can be calculated based on equation (10.27) and a relevant S-N curve.

There is an effect of the width of the response on the calculated fatigue life as shown by Wirsching and Light (1980). The significance of not having a narrow-banded response depends on the slope of the S-N curve used for calculation of fatigue damage, and the effect becomes larger as the S-N curve becomes more horizontal. However, the error of neglecting this is to the safe side and is normally considered to be less than that due to other uncertainties related to fatigue assessment, and is thus not accounted for in most fatigue design standards for marine structures.

15 Fatigue Analysis of Floating Platforms

15.1 General

DNVGL has issued a number of guidelines for fatigue assessment of floating platforms. These documents include information that is related to:

- modeling environmental loading and response;
- modeling principles for analysis models of FPSO (floating production, storage, and offloading vessels);
- hydrodynamic analysis methodology;
- fatigue analysis methodology;
- documentation and verification of analysis methodology.

15.2 Semi-Submersibles

Guidelines for fatigue design of semi-submersibles are presented in DNV-OS-C103 (2014).

Guidelines on fatigue analysis for planning in-service inspection for fatigue cracks of semi-submersibles can be found in appendix B of DNVGL-RP-0001 (2015).

15.3 Floating Production Vessels (FPSOs)

Guidelines on how to perform a global and local fatigue analysis of a typical FPSO, with both turret-moored and fixed spreader-moored arrangements, can be found in DNV-RP-C206 (2012).

Guidelines on fatigue analysis of FPSOs for planning in-service inspection for fatigue cracks can be found in appendix C of DNVGL-RP-0001 (2015). The referred documents can be downloaded from the DNV GL web page for free. This is one reason for making this section short.

16 Fracture Mechanics for Crack Growth Analysis and Assessment of Fracture

16.1 Brittle and Ductile Failures

16.1.1 Introduction

The notation used to describe the development of fracture using adjectives as ductile or brittle depends on whether one refers to the micro or the macro scale. Thus, the wording brittleness and ductility may have different meanings depending on whether it is seen from a metallurgist or a designer. A metallurgist is considering how to best achieve ductile material that shows sufficient strength and elongation at fracture and that is robust with respect to fracture when defects are present in the material. Designers most often presume that the material they are using is ductile and that their main task is to avoid a brittle structural behavior through a less favorable design. However, the post-failure behavior of some structures is such that a global brittle development cannot be avoided and for these structures a larger safety factor is normally recommended than in design with development of a ductile failure mechanism. An example of this is shell structures and structural elements where local buckling may occur; see, for example, Eurocode 3 (EN 1993-1-1 2009). The terminology used by metallurgists and designers may also be combined in a matrix as shown in Table 16.1 for illustration of development of different failure mechanisms. However, there is a gradual development from brittle to ductile along both the horizontal and the vertical axes in Table 16.1.

16.1.2 Design of Ductile Structures

Design standards for marine structures recommend design and fabrication of ductile structures. This requires that ductile materials are used and that sound design principles are followed in order to avoid weak sections in structural elements due to locally reduced sectional area such as at bolt-holes. A requirement for material ductility is expressed through requirements for Charpy V values tested at different temperatures for different thicknesses. Crack Tip Opening Displacement (CTOD) testing is also recommended for documentation of sufficient material toughness for larger thicknesses and more complex components. Furthermore, use of overmatch material is required in butt welds. This means that the weld deposit should have a larger yield strength than the base material. This is normally specified in Welding

Table 16.1. *Illustration of material and design notation with respect to fracture*

		Design	
		Brittle behavior	Ductile behavior
Material	Brittle	Use of brittle material in design of marine structures is in general not recommended as fracture is associated with small deformation and energy absorption.	
	Ductile	Structures that show small deformation capacity at failure. Examples: Shell structures due to their post critical behavior during buckling and structures with joints showing less capacity than the adjoining members	This combination of material and design provides the best basis to achieve robust structures with respect to the Ultimate Limit States and the Accidental Limit States

Procedure Specifications. It is further documented during Welding Procedure Qualification Testing, when a welded butt joint is tested by bending to a specified radius without failure in the weld.

A global ductile fracture depends on the local deformation process during external loading being ductile. However, a global fracture may show brittle development if, for example, a butt weld has been fabricated using undermatch weld material or if a load-carrying cruciform joint has been designed with fillet welds that are too small; see, for example, the capacity of the fillet welds around the hydrophone support in the brace of the *Alexander L. Kielland* platform in Section I.2.1. Undermatch material is understood to mean that the yield strength of the weld material is lower than that of the base material.

Another example of brittle structural behavior is explosion walls that are fillet welded to the supporting structure where small fillet welds are made with less strength than the membrane yield strength of the plate. Thus this structure cannot absorb energy during an explosion as the fillet welds will be strained until failure while the plate still behaves elastic. Thus, the fillet welds need to be stronger than the plate to get a ductile structural behavior for such loads. Similarly it is problem to use high strength aluminum alloys that develops soft zones after fabrication in such structures.

16.1.3 Structural Strength of Connections with Defects

Components with planar defects like cracks or lack of fusion may fail for a load that gives stress levels significantly less than the material yield strength for less good material fracture toughness. The actual defect may be present from fabrication, see Section 11.6, or the crack has developed during fatigue loading. Marine structures are subjected to variable amplitude loading and at some stage during crack propagation there is a significant risk that the connection will fracture due to a combination of crack size, fracture toughness and external static and dynamic load. Also residual stress may influence the risk for unstable fracture, especially for connections with less good fracture toughness; it may even crack during fabrication due to residual stresses only for a low fracture toughness. The material fracture toughness and the load will govern the crack size at fracture. The time for crack growth until unstable fracture is important when assessing needs for in-service inspection of fatigue cracks as described in Chapter 18.

Brittle material is characterized by cleavage of grains (transcrystalline fracture) and/or separation along grain boundaries (intercrystalline fracture). A locally ductile fracture normally refers to growth and coalescence of voids formed around inclusions and is characterized by fibrous faces (Hellan 1984).

Thus, it is important to control the material fracture toughness and the size of defects in combination with loading from internal and external stresses. As the material fracture toughness is increased, the external loading can also be increased up to that of yielding of the net section. There is a gradual transition from a brittle fracture mechanism to a more ductile behavior and this is accounted for by using the Failure Assessment Diagram (FAD) in Section 16.8. Thus, both brittle and ductile fractures are included in this analysis procedure to avoid unstable fracture which can be used to describe fracture and rupture.

16.2 Stress Intensity Factors and Fatigue Crack Growth Equations

Fatigue of welded connections consists of a fatigue initiation phase and a crack growth phase. Fatigue test results from laboratory testing of welded connections include both the cycles due to fatigue crack initiation and the following crack growth until failure. Most fatigue life in welded structures is associated with fatigue crack growth. It may be practical to assume crack growth to occur from small (even fictitiously small) initial crack sizes, such that a similar fatigue life is calculated by fracture mechanics as that of S-N test data. If the initial defect size is known, then that should be used as the basis for the fatigue crack growth analysis. The stress condition at a cracked region can be described by the concept of stress intensity factors. The general expression for the stress intensity factor describing the stress condition at a crack tip region in a plane body with a through-thickness crack and a far-field stress normal to the crack 2a is:

$$K_g = \sigma Y \sqrt{\pi a} \tag{16.1}$$

where:

σ = remote stress as indicated in Figure 16.1
a = half crack length for an internal crack; a = crack depth for edge cracks
Y = geometry function.

The geometry function is 1.0 for a through-thickness internal crack in an infinite body. Otherwise, this is a function of geometry that is normally larger than 1.0 under tensile load; this function may be lower for a bending load. The elastic stress field at the crack tip region can be expressed by the following two-dimensional solution using polar coordinates:

$$
\begin{bmatrix} \sigma_x \\ \sigma_y \\ \tau_{xy} \end{bmatrix} = \frac{K_I}{\sqrt{2\pi r}} \begin{bmatrix} \cos\dfrac{\varphi}{2}\left(1 - \sin\dfrac{\varphi}{2}\sin\dfrac{3\varphi}{2}\right) \\ \cos\dfrac{\varphi}{2}\left(1 + \sin\dfrac{\varphi}{2}\sin\dfrac{3\varphi}{2}\right) \\ \cos\dfrac{\varphi}{2}\sin\dfrac{\varphi}{2}\cos\dfrac{3\varphi}{2} \end{bmatrix} \tag{16.2}
$$

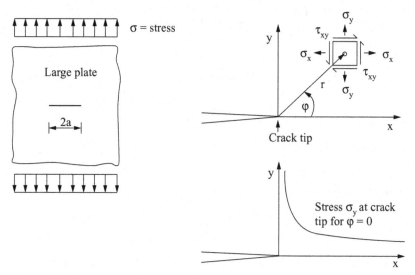

Figure 16.1. Illustration of stress field in front of a crack tip in a large plate.

where:

K_I is the stress intensity factor in the opening mode
Other symbols are shown in Figure 16.1.

The opening mode is understood to mean that the loading is such that the crack surfaces tend to separate in a direction normal to the crack surfaces. From equation (16.2) it can be seen that the stress field at a position (r, θ) at the crack tip is known if the stress intensity factor, K_I, is also known. And in principle this equation can be used to calculate K_I if the stress at a crack is known for example from a finite element analysis where the crack is included in the fine mesh analysis model. Normally Y is a function of the crack size and is written as $Y(a)$. In addition, Y is also a function of crack shape, boundary conditions, and type of loading (different in moment loading compared with tensile loading). A number of geometry functions can be found in handbooks on stress intensity factors. The more general equation for the stress intensity factor can be presented as:

$$K_I = \sigma Y(a) \sqrt{\pi a} \qquad (16.3)$$

From this equation it can be seen that the remote stress becomes a significant parameter governing the value of the stress intensity factor, in addition to the geometry function and the crack size. During cyclic loading from σ_{min} to σ_{max} the detail being considered is subjected to a stress range $\Delta\sigma = \sigma_{max} - \sigma_{min}$. From equation (16.3) it can be seen that this stress range also corresponds to a range in the stress intensity factor, $\Delta K = K_{max} - K_{min}$, where the index, I, for the opening mode is removed for simplicity. Modes II and III are also referred to in fracture mechanics, but these are normally less important in crack growth analysis of welded structures, as they are characterized by shear deformation along the crack surfaces in-plane and out-of-plane, respectively. During crack growth a crack tends to grow in a direction normal to the principal stress direction and this reduces the significance of modes II and III

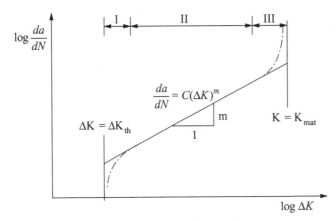

Figure 16.2 Illustration of the validy range for the crack growth equation.

with respect to fatigue. The following equation for range of stress intensity factor in mode I is then derived:

$$\Delta K = \Delta \sigma Y(a)\sqrt{\pi a} \tag{16.4}$$

An alternative to expressing fatigue load in terms of stress range is to express the same load by the range of the stress intensity factor when considering crack growth based on fracture mechanics. An expression for this was first formulated by Paris in the early 1960s (Paris et al. 1961) and is often called the Paris equation:

$$\frac{da}{dN} = C(\Delta K)^m \quad for \quad \Delta K_{th} \le \Delta K \le K_{mat} \tag{16.5}$$

where:

   da = increment in crack growth for dN stress cycles
   C and m are material parameters
   ΔK_{th} = threshold value for the stress intensity range
   K_{mat} = material fracture toughness.

Below the threshold range, crack growth does not occur, as indicated in region I in Figure 16.2. Unstable fracture of the connection may occur as the maximum stress intensity factor approaches the fracture toughness, as indicated in region III in Figure 16.2.

The following expression for increment in the number of cycles is derived from equations (16.4) and (16.5):

$$dN = \frac{da}{C\Delta \sigma^m \left(\sqrt{\pi a}\right)^m Y(a)^m} \tag{16.6}$$

By integrating the left and right sides in this equation for constant amplitude loading, from an initial crack size, a_i, to a final crack size, a_f, the following equation is derived:

$$N = \frac{1}{C\Delta \sigma^m} \int_{a_i}^{a_f} \frac{da}{(\pi a)^{m/2} Y(a)^m} \tag{16.7}$$

Here the following integral is introduced:

$$I = \int_{a_i}^{a_f} \frac{da}{(\pi a)^{m/2} Y(a)^m} \tag{16.8}$$

The equation for the number of cycles during fatigue crack growth is then derived as:

$$N = \frac{I}{C \Delta \sigma^m} \tag{16.9}$$

When taking the logarithm on both sides of this equation, the following equation is derived:

$$\log N = \log(I/C) - m \log \Delta \sigma \tag{16.10}$$

The similarity between an S-N formulation and fracture mechanics can be observed from this equation, where $\log(I/C)$ is similar to loga in an S-N formulation. This means that, in principle, S-N curves can be constructed for different details based on fracture mechanics, provided that proper geometry functions are available for the detail being considered. This also requires selection of an appropriate initial crack size. The selection of final crack size is normally of lesser importance for calculated the number of cycles to failure. This depends on type of detail to be analyzed. A crack propagation through the plate thickness is comparable with the failure criterion inherent in the S-N data; see also Section 4.2; however, in connections like girth welds in tether pipes supporting Tension Leg Platforms a significantly smaller defect size at failure will be estimated due to high maximum tensile stress. Furthermore, it is difficult to account for similar residual stresses in a fracture mechanics analysis as is inherent in fatigue tests and S-N data. The size of the crack at final fracture is important for in-service inspection planning as explained in Chapter 18.

The crack growth analysis procedure presented in this section can also be used for variable amplitude loading if the long-term stress range distribution first is transferred to an equivalent constant stress range as presented in Section 10.2.

16.3 Examples of Crack Growth Analysis

16.3.1 Assessment of Internal Defects in a Cruciform Joint

The following example of fatigue crack growth from an internal defect in a cruciform joint is used to present a relatively simple fracture mechanics analysis before considering more complex fracture mechanics analysis. Cruciform joints are often welded from both sides, and, in some cases, partial penetration welds are used. This leaves some lack of penetration in the root area as shown in Figure 16.3. Defects in the root may also occur in full penetration welds made from both sides if a back gouging/grinding of the root is not performed properly, such that impurities remain in the root before welding from the second side. For short defects in the root (in the width direction) of the connection in Figure 16.3, a geometry function, $Y = 1.0$, can be assumed and one dimensional crack growth analysis can be performed.

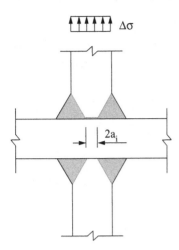

$\Delta\sigma$

$2a_i$

Figure 16.3. Cruciform joint with internal defect in the weld root.

Then, by integration of equation (16.7), the following expression is derived (as $m \neq 2.0$):

$$N = \frac{1}{C\pi^{m/2}\Delta\sigma^m} \left| \frac{a^{1-\frac{m}{2}}}{1-\frac{m}{2}} \right|_{a_i}^{a_f} \tag{16.11}$$

and

$$N = \frac{a_i^{1-\frac{m}{2}} - a_f^{1-\frac{m}{2}}}{C\pi^{m/2}\Delta\sigma^m \left(\frac{m}{2}-1\right)} \tag{16.12}$$

The following values for the material parameters $m = 3.0$ and $C = 5.21 \cdot 10^{-13}$ (N, mm) in the air environment are inserted into this equation to derive the number of cycles to failure calculated by crack growth analysis:

$$N = 6.894 \cdot 10^{11} \frac{a_i^{-0.5} - a_f^{-0.5}}{\Delta\sigma^{3.0}} \tag{16.13}$$

The C parameter used here represents the mean plus two standard deviations in crack growth test data for welds in air (BS 7910 2013). Thus, using this equation to calculate the number of cycles implies a safety level similar to that obtained from using mean minus two standard deviation S-N curves, as presented in design standards for fatigue.

Some uncertainty is associated with sizing of internal defects using ultrasonic testing, as shown in Section 11.6.6. This uncertainty should be kept in mind when acceptance criteria are being assessed. Also, regarding the probability of detecting fatigue cracks during service life, detecting cracks growing from internal defects before they become large is not easy. Therefore, assessment of such defects with a larger safety factor than for defects initiating from the outside weld toes may be preferable. A plate thickness, t = 25 mm, is considered. The yield strength of the weld material is usually higher than that of the base material. With some additional fillet welds as shown in Figure 16.3, the effective area at the defect region is bigger than in a section through the base plate. Thus, a fatigue crack can probably grow

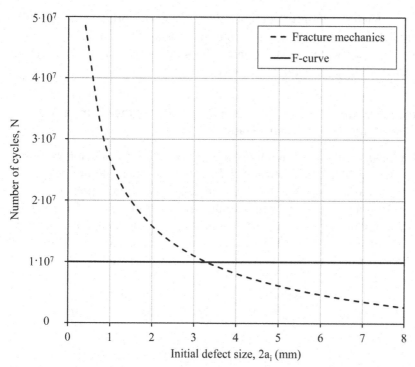

Figure 16.4. Calculated number of cycles as function of initial defect size in the weld root.

rather large before an unstable fracture of a cruciform joint is expected. However, this also depends on the fracture toughness of the material, service temperature, and loading. In the following, it is assumed that the maximum crack size, $2a_f$, is equal to half the plate thickness (12.5 mm). This may be acceptable if the initial defect is not very long, but this depends also on the maximum stress and the material fracture toughness. However, if a long defect is being considered, such as a continuous lack of penetration, the final crack size may be smaller. The crack length into the plane in Figure 16.3 is considered to be so long that this simplified two-dimensional model can be used, but the crack is not so long or large that the assumption that $Y = 1.0$ becomes significantly non-conservative.

It is assumed that the cruciform joint is fully utilized with respect to fatigue and that it is classified as F for fatigue cracking from the weld toe. From Table 4.1 a stress range of 41.52 is derived for 10^7 load cycles. The acceptable size of the initial defect without increasing the possibility of crack growth from the root, rather than from the weld toe, can be questioned. This can be solved using equation (16.13), and the results are shown in Figure 16.4. The same Design Fatigue Factor (DFF) ($= 1.0$) is here used for the weld root as for the weld toe. In this calculation it is assumed that the width of the weld at the defect is 35 mm such that the nominal stress in the weld at the defect is $\Delta\sigma = 41.52 \cdot 25/35 = 29.66$ MPa. It can be observed that the calculated crack growth from internal defects is longer than for the outer weld toe, until the initial crack size, $2a_i$, is approximately 3.3 mm. This is considered to be a conservative estimate as threshold in the crack growth has not been included. Furthermore, a possible effect of crack curvature as shown in Appendix B has not been accounted for as, for illustrative purposes, a simplified analysis has been performed. This shows that fracture mechanics analysis of internal defects can be performed relatively easily. For external defects, the geometry function depends more on crack size, and for defects at weld toes, the geometry function for the weld toe must also be accounted for. This is described in more detail in the Section 16.4.

16.3.2 Example of Crack Growth from the Crack around the Hydrophone Support in the *Alexander L. Kielland* Platform

After crack growth around the circumference of the hydrophone support in the *Alexander L. Kielland* platform, fatigue cracks began to propagate into the brace in a direction normal to the main stress direction. This crack initiation started on the corners in the braces at the hydrophone support shown in Figure I.1. This crack growth is considered to be governed by the local stress field due to the hole and the small initial crack (see, e.g., Bowie 1956). After some growth, the crack was through the brace thickness, and further crack growth is estimated in the following where the crack growth is mainly based on the total crack length, including that of the hole. The following stress intensity factor is used for a through-thickness center crack subjected to remote uniform stress for all a/W (relative crack sizes) under symmetry:

$$K_I = \sigma_\infty \sqrt{\pi a} \, \frac{1 - \dfrac{a}{2W} + 0.326 \dfrac{a^2}{W^2}}{\sqrt{1 - \dfrac{a}{W}}} \tag{16.14}$$

where:

$\Delta\sigma_\infty$ is the equivalent nominal axial stress range in the brace and $\Delta\sigma_\infty = 30.0$ MPa (NOU 1981). The equivalent stress range is based on the number of cycles and a long-term stress range distribution representative for one year.

W = half circumference of the brace (Circumference of brace with outer diameter 2600 mm: 2W = 8168 mm)

a = half crack width.

The hole for the hydrophone holder can be considered as part of the crack, such that the crack length now equals the hole diameter, and this part of the crack growth analysis is started, for illustration, at: $2a_i = d_h + 2t_{brace}$, see Figure 16.5 and Figure I.2. The diameter of hydrophone support is 325 mm and the thickness of brace is 26 mm. By using equation (16.14), the stress intensity factor for a crack length of $2 \cdot 190$ mm is K = 733.68 Nmm$^{-3/2}$. Then from equation (16.5), with a mean crack growth parameter, $C = 1.83 \cdot 10^{-13}$ (unit N, mm), da/dN = $7.22 \cdot 10^{-5}$ mm/cycle.

A mean cycling period of 6.3 sec is assumed, resulting in 13,714 cycles each day (= $60 \cdot 60 \cdot 24/6.3$). The number of cycles to grow the crack increments of 1.0 mm at each end is: $\Delta N = 1.0/7.22 \cdot 10^{-5} = 13836$ cycles. This corresponds to $13836/13714 = 1.008$ day, or that the mean expected crack growth was 1.0 mm per day in the initial phase of the crack growth into the brace. This increment in crack growth is then added to each crack tip end, and a new stress intensity factor is calculated and further crack growth is estimated in the same way. The resulting calculated crack growth is shown in Figure 16.6. It can be seen that the crack will grow to failure in less than 250 days. This crack growth life should be added to that of developing a full crack around

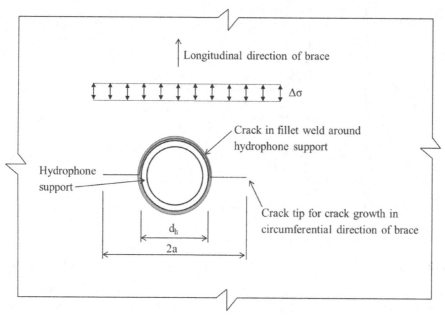

Figure 16.5. Crack growth from the hydrophone support into the brace of *Alexander L. Kielland*.

the hydrophone support and to initiate the further crack growth into the brace. This should also be considered in relation to seasonal weather, uncertainties in crack growth parameters, and other factors. Thus there are a number of uncertainties related to such analysis. However, the lesson is clear: it is dangerous to insert members into main carrying members that have not been properly designed and assessed with respect to fatigue.

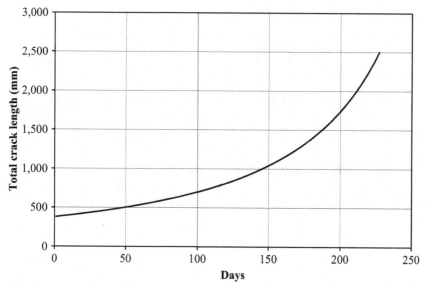

Figure 16.6. Fatigue crack growth from the hydrophone support around the brace in the *Alexander L. Kielland* platform.

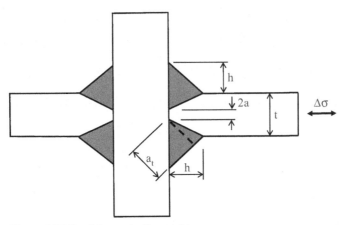

Figure 16.7. Partial penetration weld.

16.3.3 Example of Crack Growth from the Root of a Partial Penetration Weld

Fatigue assessment of a partial penetration weld shown in Figure 16.7 is performed. It is assumed that the size of the fillet weld, h, should be limited to t = 25 mm. The question then arises of the size of partial penetration that is considered acceptable, without fatigue failure from the weld root. The first assessment is based on nominal stress S-N curves. A stress range of $\Delta\sigma = 41.52$ MPa at 10^7 cycles, corresponding to the F-curve in Table 4.1 for fatigue crack growth from the weld toe, is used for crack growth analysis. 10^7 cycles is selected as reference; another value could have been selected without affecting the results of the analysis (when threshold is not included in the crack growth equation). It is assumed that a W3-curve is used for assessment of the weld root, and from Table 4.1 the stress range at 10^7 cycles for this curve is $\Delta\sigma_w = 21.05$ MPa. The same fatigue capacity for the fillet weld root as the weld toe is derived when

$$\Delta\sigma_w a_t 2 = \Delta\sigma t \tag{16.15}$$

This gives the following requirement regarding the thickness of the weld throat:

$$a_t = \frac{\Delta\sigma t}{2\Delta\sigma_w} = \frac{41.52 \cdot 25}{2 \cdot 21.05} = 24.65 \, mm \tag{16.16}$$

Based on geometric consideration in Figure 16.7, the throat thickness is derived as:

$$a_t = ((t - 2a)/2 + h)\sin 45° \tag{16.17}$$

The throat thickness of 24.65 mm is achieved for a lack of penetration of 2a = 5.26 mm. The capacity of this geometry is now assessed using fracture mechanics. The stress intensity range for the weld root for membrane loading is derived from BS7910 (2013) as:

$$\Delta K = \Delta\sigma\sqrt{\pi a} f_{wm} M_{km} \tag{16.18}$$

where:

$$f_{wm} = \left(\sec\left[\frac{\pi}{2}\left(\frac{2a}{W}\right)\right]\right)^{0.5} \qquad where \; W = 2h + t \tag{16.19}$$

Table 16.2. *Calculated number of cycles to failure*

a	f_{wm}	M_{km}	ΔK (N, mm)	$da/dN \cdot 10^7$ (mm)	$\Delta N \cdot 10^{-7}$
2.63	1.0030	0.594	71.05	2.28	0.162
3.00	1.0040	0.594	75.96	2.84	0.176
3.50	1.0054	0.591	81.76	3.45	0.145
4.00	1.0071	0.588	87.14	4.08	0.123
4.50	1.0090	0.585	92.18	4.75	0.105
5.00	1.0111	0.583	96.94	5.44	0.092
5.50	1.0135	0.580	101.48	6.18	0.081
6.00	1.0161	0.578	105.83	6.94	0.072
6.50	1.0189	0.576	110.03	7.74	0.065
7.00	1.0220	0.573	114.10	8.57	0.058
7.50	1.0254	0.571	118.06	9.45	0.053
8.00	1.0290	0.569	121.94	10.36	0.048
8.50	1.0329	0.567	125.76	11.32	0.044
9.00	1.0371	0.566	129.52	12.33	0.041
9.50	1.0415	0.564	133.25	13.38	0.037
10.00	1.0462	0.562	136.95	14.49	0.034
10.50	1.0513	0.561	140.64	15.67	0.032
11.00	1.0566	0.560	144.34	16.90	0.030
11.50	1.0623	0.558	148.04	18.21	0.027
12.00	1.0682	0.557	151.76	19.60	0.026
12.50	1.0746	0.556	155.51		
Sum					1.451

and

$$M_{km} = \lambda_0 + \lambda_1 (2a/W) + \lambda_2(2a/W)^2 \qquad (16.20)$$

where

$\lambda_0 = 0.956 - 0.343(h/t)$
$\lambda_1 = -1.219 + 6.210(h/t) - 12.220(h/t)^2 + 9.704(h/t)^3 - 2.741(h/t)^4$
$\lambda_2 = 1.954 - 7.938(h/t) + 13.299(h/t)^2 - 9.541(h/t)^3 + 2.513(h/t)^4$
$0.1 \leq 2a/W \leq 0.7$
$0.2 \leq h/t \leq 1.2$

From equations (16.4) and (16.18) it can be seen that the geometry function $Y(a) = f_{wm}(a) \cdot M_{km}(a)$.

For $h/t = 1.0, \lambda_0 = 0.613, \lambda_1 = -0.266$, and $\lambda_2 = 0.287$.
$W = 2 \cdot 25 + 25 = 75$ mm

Then crack growth is calculated as shown in Table 16.2. The range of stress intensity factor is calculated first for a crack length a = 2.63 mm; however, the stress intensity factor for a = 3.0 mm is used to calculate the number of cycles from a = 2.63 mm to a = 3.0 mm such that the calculated number of cycles is on the safe side (sec x in equation (16.19) is equal $1/\cos x$). Based on equation (16.5), with $C = 5.21 \cdot 10^{-13}$, da/dN is then calculated. The first considered increment in crack size is from $\Delta a = 3.0 - 2.63 = 0.37$ mm. The number of cycles required for the crack to grow by this increment is derived as $\Delta N = \Delta a/(da/dn)$ in the last column in Table 16.1. Thus, with

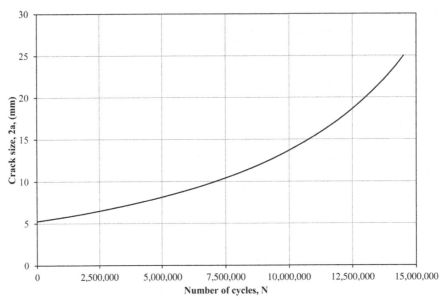

Figure 16.8. Calculated crack growth for partial penetration weld.

5.26 mm lack of penetration, a somewhat larger calculated fatigue life is estimated for the weld root as for the weld toe, by a factor of 1.451. However, this is not a significant difference keeping in mind that different analysis approaches are used: nominal stress S-N fatigue analysis and crack growth analysis based on fracture mechanics. Quite often, larger differences in calculated fatigue lives are expected. Calculated fatigue crack growth from the weld root is shown in Figure 16.8. For a fillet weld with $h/t = 1.0$ the weld toe is classified as FAT 63, according to Hobbacher (2009). This corresponds to S-N curve F1 and stress range 36.84 MPa at 10^7 cycles and thus less than that of the F-curve used in this example to represent the fatigue strength for the weld toe. It is therefore possible that the fatigue strength of the weld root is larger than the fatigue strength with fatigue crack growth from the weld toe. The residual stresses are also unlikely to be as severe in the root as at the weld toe for consideration of fatigue crack initiation.

16.3.4 Example of Crack Growth from the Root in a Single-Sided Girth Weld in a Pile Supporting a Jacket Structure

During reassessment of fatigue analysis based on S-N data it was found that piles supporting an older jacket structure did not meet the assessment criterion for fatigue damage in NORSOK N-006 (2015) with a Design Fatigue Factor (DFF) equal 3. The most critical hot spot is at a girth weld at a thickness transition on the inside of the pile, as shown in Figure 16.9. The weld is performed from the outside without any back welding from the inside. This results in use of the F3-curve and an additional stress concentration factor (SCF) for the root based on DNVGL-RP-C203 (2016). The largest thickness of the pile is $T = 50.8$ mm and the smallest thickness of the pile is $t = 38.1$ mm. The outer pile diameter is $D = 1374$ mm, and the mean diameter $D_m = d - t = 1335.9$ mm. This gives $D_m/t = 1335.9/38.1 = 35.06$ (with large

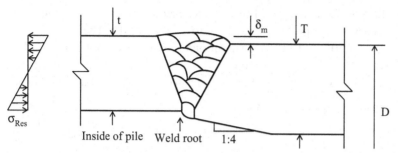

Figure 16.9. Single-sided girth weld in pile supporting an older jacket structure.

diameter ratios, normally no distinction is made between mean and outer diameters; see also equations used for calculation of SCFs in Section 6.3). The specified fabrication tolerance $\delta_m = 3.0$ mm. The thickness transition is made over a length $L = 4 \cdot (T - t) = 4 \cdot (50.8 - 38.1) = 50.8$ mm. The shift in neutral axis due to thickness difference is $\delta_t = 0.5(50.8 - 38.1) = 6.35$ mm. This results in the following SCF for the weld root from equation (6.23):

$$\alpha = \frac{1.82 \cdot 50.8}{\sqrt{1335.9 \cdot 38.1}} \frac{1}{1 + \left(\dfrac{50.8}{38.1}\right)^{2.11}} = 0.1446$$

$$\beta = 1.5 - \frac{1}{\log(1335.9/38.1)} + \frac{3}{(\log(1335.9/38.1))^2} = 2.11 \tag{16.21}$$

$$SCF = 1 + \frac{6\,(6.35 + 3.0)}{38.1} \frac{1}{1 + \left(\dfrac{50.8}{38.1}\right)^{2.11}} e^{-0.1446} = 1.45$$

The main contribution to calculated fatigue damage is due to pile driving during the installation phase. The fatigue damage accumulated during in-service life due to wave action is not large and, in this example, can be neglected. It is assumed that the pile-driving history is available and that the stress cycling during a hammer blow is as shown in Figure 16.10. It is mainly the first stress cycle that contributes to the fatigue damage and is used in the following analysis. The effect of more cycles can be included if these are assessed to contribute significantly to the fatigue damage.

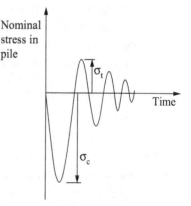

Figure 16.10. Nominal stress cycles in the pile during a single hammer blow.

During pile installation, 3,000 hammer blows are reported. The majority of the nominal largest stress cycle is compressive and equal to $\sigma_c = 200$ MPa. The first nominal tensile cycle is $\sigma_t = 80$ MPa. During assessment of a life extension of the platform it is reasonable to start with a fatigue analysis based on S-N data for air environment (see Section 4.10.1). This gives the number of cycles to failure (see equation 4.1):

$$N = 10^{(\log a - mk \log(t/t_{ref}) - m \log S)}$$
$$= 10^{11.546 - 3.0 \cdot 0.25 \log(38.1/25) - 3.0 \log((200 + 80) \cdot 1.45)} = 10^{3.583} = 3829 \tag{16.22}$$

This results in a calculated DFF of $3829/3000 = 1.27$. This does not meet the requirement in NORSOK N-006 (2015) with a DFF $= 3.0$. Thus, further assessment is made and, based on experience from similar connections, it is noted that residual stresses at the root region are compressive. This means that the stress condition at the root is considered to be similar to that after Post-Weld Heat Treatment (PWHT). Thus, with reference to Section 3.3.5, the compressive part of the stress cycle can be reduced by a factor of 0.80. This gives the number of cycles to failure as:

$$N = 10^{(\log a - mk \log(t/t_{ref}) - m \log S)}$$
$$= 10^{11.546 - 3.0 \cdot 0.25 \log(38.1/25) - 3.0 \log((200 \cdot 0.80 + 80) \cdot 1.45)} = 10^{3.784} = 6081 \tag{16.23}$$

The calculated DFF is then increased to $6081/3000 = 2.03$. However, the requirement in NORSOK N-006 (2015) is still not fulfilled.

The residual stresses at the root are expected to be more compressive than after PWHT. However, fatigue test data that are representative for this condition are not available. Therefore, crack growth analysis based on fracture mechanics is performed to investigate how much residual compressive stress is required to meet the acceptance criterion with respect to DFF in NORSOK N-006 (2015).

Based on potential defects in the weld root, the F3-curve in DNVGL-RP-C203 (2016) is recommended. With relatively large defects, the M_k factor can be approximated as equal to 1.0 for the weld root. The fatigue design curve F3 is derived for stress ranges that are tensile. Thus, an initial long defect size that corresponds to this loading is calculated. This initial defect is then used for calculation of crack growth, taking into account the actual residual stress, crack closure, and effective stress intensity factor that is assumed to be crack driving during the actual cycling loading. Often the best way to calibrate the methodology is to base it on the mean values of the parameters involved in the problem under consideration. Thus, the mean values of S-N data and crack growth parameters are used in the following (see also Section 4.1.3 for standard deviations in the S-N curves needed for derivation of mean S-N curves from the characteristic S-N curves). It is assumed that the initial crack size is calculated for a root crack in a single-sided weld subjected to tensile cyclic loading corresponding to the F3-curve for a reference thickness $t_{ref} = 25$ mm. If 10,000 load cycles are selected for the analysis, this gives from equation (4.1):

$$\Delta\sigma = 10^{(\log a_d + 0.2 \cdot 2 - \log N)/m}$$
$$= 10^{((11.546 + 0.2 \cdot 2 - \log 10000)/3.0)} = 445 \, MPa \tag{16.24}$$

The mean value of C is equal $1.83 \cdot 10^{-13}$ and $m = 3.0$. The crack size at fracture $a_f = 10$ mm is used for calculation of crack growth. It is further assumed that $Y = 1.0$ can be used as the crack is small compared to the total weld area. From equation (16.12) this gives:

$$a_i = \left(NC\pi^{m/2}\Delta\sigma^m \left(\frac{m}{2} - 1 \right) + a_f^{1-m/2} \right)^{1/(1-m/2)}$$

$$= \left(10^4 \cdot 1.83 \cdot 10^{-13} \cdot \pi^{3.0/2} 445^{3.0} \left(\frac{3.0}{2} - 1 \right) + 10^{1-3.0/2} \right)^{1/(1-3.0/2)} \quad (16.25)$$

$$= 1.7\,mm$$

Then the crack growth with the actual nominal mean stress level will be limited to a region without significant tensile residual stresses; therefore, $a_f = 6$ mm is selected and used for calculation of crack growth from the weld root (it can be assessed that the largest part of the fatigue life is associated with growth of small cracks). The following equation is derived from equation (16.12) for calculation of allowable stress range:

$$\Delta\sigma_{allowable} = \left(\frac{a_i^{1-m/2} - a_f^{1-m/2}}{C\pi^{m/2} \left(\frac{m}{2} - 1 \right) n} \right)^{1/m} \quad (16.26)$$

where:

n = number of cycles.

In order to document fatigue capacity, a design value for C is now used. This means that two standard deviations are added to the mean value of C to obtain a characteristic value, and for the weld this gives $C = 5.21 \cdot 10^{-13}$. With a DFF = 3.0 in the analysis, so that n in equation (16.26) becomes equal $3000 \cdot 3 = 9000$ cycles, this gives $\Delta\sigma_{allowable} = 301.76$ MPa. With SCF = 1.45, the allowable nominal stress range calculated by fracture mechanics is $301.76/1.45 = 208.11$ MPa without considering any mean stress effect. The yield strength of the material is equal 325 MPa. Following Section 3.3.2, the compressive stresses at the weld root may disappear during pile driving, as the compressive part of the stress cycle may exceed yield stress if a weld notch factor is also included. However, because of a rather deep crack, a notch factor for the weld root is not included. The resulting compressive stress, including SCF, becomes: $\sigma_c \cdot SCF = 200 \cdot 1.45 = 290$ MPa. This is less than the material yield strength and shakedown of the compressive stresses at the weld root is not expected. This means that only the tensile part of the stress cycle needs to be considered as part of the fatigue driving stress range and $\Delta\sigma = \sigma_t \cdot SCF = 80 \cdot 1.45 = 116$ MPa, which is less than the allowable stress range. Thus, because of crack closure during a significant part of the load cycle, it is possible to document that a DFF larger than 3.0 can be used for life extension of the pile being considered for a long defect in the root that is 1.7 mm deep. This is due to crack closure, because much of the stress cycle is compressive during pile driving. A more detailed analysis of the residual stresses during pile driving at the root may be performed by a nonlinear analysis in a real project for documentation of capacity.

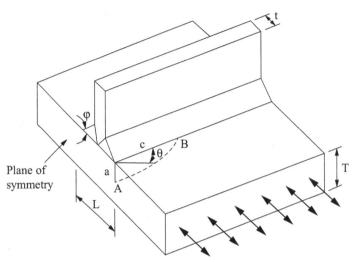

Figure 16.11. Semi-elliptic crack at a weld toe.

16.4 Fracture Mechanics Models for Surface Cracks at Weld Toes

Geometry functions for details with cracks can be found in handbooks on stress intensity functions, for example, Rook and Cartwright (1976) and Tada et al. (1973), and in some standards, such as BS 7910 (2013) and API 579-1/ASME FFS-1 (2007). The total stress intensity at a weld toe consists of two parts: a geometry part for the considered detail and a part describing the local stress at the weld. The first part corresponds to the geometry accounted for by the hot spot stress or geometrical stress and the second part corresponds to the notch stress increase at the weld that is normally included in a hot spot S-N curve. Thus, the total stress intensity factor at a weld can be presented as:

$$K = K_g M_k \qquad (16.27)$$

where:

K_g = function describing the stress intensity at the considered crack due to the geometry of the detail and type of loading (without the effect of the weld toe geometry).

M_k = function describing the stress field due to the weld notch and type of loading. M_k is a stress magnification factor that includes the effect of local stress concentration from the notch at the weld toe as presented by Bowness and Lee (2002); see example in Figure 4.12. M_k functions are presented in BS 7910 (2013) for fillet weld angles of 45°. More general equations are presented in Bowness and Lee (2002) (OTO 2000/077) published by the Health and Safety Executive (HSE) in London.

A semi-elliptic surface crack at a weld toe in Figure 16.11 is considered. The fatigue crack growth per stress cycle of this surface crack is assumed to follow the Paris equation at any point along the crack front, and an increment, dr (θ), in increased crack size during dN load cycles can be calculated as:

$$\frac{dr(\theta)}{dN} = C_r(\theta)(\Delta K_r(\theta))^m \qquad (16.28)$$

where:

θ is the location angle along the crack front as shown in Figure 16.11

$C_r(\theta)$ and m are material parameters for that specific point on the crack front

$\Delta K_r(\theta)$ is the range of the stress intensity factor for the considered load cycle.

It is assumed that the fatigue crack shape remains semi-elliptical as the crack propagates; thus, the crack depth, a, and the crack length, 2c, are sufficient to describe the crack front. The general differential equation (16.28) can then be replaced by two coupled differential equations:

$$\frac{da}{dN} = C_a(\Delta K_a)^m; \quad \Delta K_a > \Delta K_{th}; \quad a(t_i) = a_i \tag{16.29}$$

$$\frac{dc}{dN} = C_c(\Delta K_c)^m; \quad \Delta K_c > \Delta K_{th}; \quad c(t_i) = c_i \tag{16.30}$$

where:

t_i is the crack initiation time

ΔK_{th} is the threshold level for the stress intensity below which the crack does not propagate

a_i is initial crack depth

$2c_i$ is initial crack length.

The subscripts a and c refer to the deepest point, A, and the end point of the crack at the surface of the semi-elliptic crack at B, respectively.

The general expression for the stress intensity factor is:

$$K_g = \sigma_{tot} Y(a, c)\sqrt{\pi a} \tag{16.31}$$

where:

σ_{tot} is the total applied stress

$Y(a,c)$ is the geometry function accounting for the global geometry and the loading condition.

The stress intensity factors for a surface crack in a finite plate subjected to membrane and bending loads, as proposed by Newman and Raju (1981), are applied to represent the semi-elliptic crack:

$$Y(a, c) = Y_{ma}(a, c)\alpha + Y_{ba}(a, c)(1 - \alpha)$$
$$Y(a, c) = Y_{mc}(a, c)\alpha + Y_{bc}(a, c)(1 - \alpha) \tag{16.32}$$

where:

Y_m and Y_b are the geometry functions for pure membrane and pure bending loading, respectively

Suffix m refers to membrane loading and b refers to bending loading

Suffix a refers to the deepest point A and suffix c refers to point B at the surface

Factor α is the membrane to total stress ratio.

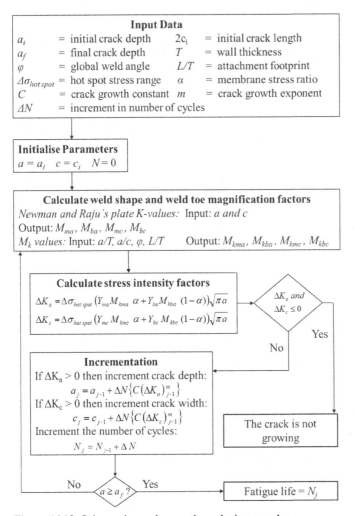

Figure 16.12. Schematic crack growth analysis procedure.

The stress intensity factors that also include the effect of the weld notch may then be calculated as:

$$K_a = \sigma_{tot} \left[Y_{ma}(a, c) M_{kma}(a, c)\alpha + Y_{ba}(a, c) M_{kba}(a, c)(1 - \alpha) \right] \sqrt{\pi a}$$
$$K_c = \sigma_{tot} \left[Y_{mc}(a, c) M_{kmc}(a, c)\alpha + Y_{bc}(a, c) M_{kbc}(a, c)(1 - \alpha) \right] \sqrt{\pi a} \quad (16.33)$$

These equations can be inserted into equations (16.29) and (16.30) and a numerical procedure is applied to solve these coupled ordinary first-order differential equations.

A schematic crack growth analysis procedure is shown in Figure 16.12. The threshold stress intensity factor has been set at zero. It is challenging to perform fracture mechanics analysis that corresponds with observed fatigue cracking of different details. One reason for this is the lack of reliable stress intensity factors for typical details used in marine structures.

The stress intensity factor as a function of crack size depends significantly on the local stress field, whether the stress is due to membrane loading or bending loading.

This stress field can be derived from a finite element analysis of a hot spot region without any cracks. The complexity increases when a change in stress distribution during crack growth must be accounted for. This can only be properly quantified by finite element analysis including a crack, so that appropriate boundary conditions are used in the analysis. This is important for assessment of larger cracks in different types of tubular joints.

It should be noted that the Newman-Raju equations for stress intensity factors have been derived from finite element analyses with cracks present, but subjected to a static determinate loading. This means that redistribution of stresses due to restraint at boundaries during crack growth is not included. Thus, use of these equations will normally lead to calculated fatigue lives that are on the safe side.

S-N data are generally more correct (or closer to the laboratory test data) than calculations based on fracture mechanics. Results from fracture mechanics analysis depend on more parameters than do S-N data: crack growth parameters, initial crack size, and stress intensity factors during crack growth. Therefore, the fracture mechanics model should be calibrated so that it provides fatigue lives that are in agreement with the test data and a similar calculated probability of failure is obtained as that derived from S-N data (test data that includes initiation and crack growth). This calibration approach was used in the assessment of the size effect, as explained in Section 4.7.4. It is important that the calculated crack growth period is realistic, because the crack growth curve provides the time estimate for detecting a growing crack during in-service inspection before it reaches a critical size. Due to the significance of the crack growth shape, performing a deterministic crack growth analysis for critical details is recommended for planning of in-service inspection for a better understanding of the criticality of the considered hot spot.

The fatigue initiation time for details that have been weld-improved may become a significant part of the fatigue life. Large hot spot stress ranges are a common reason for weld improvement. This can also imply that cracks may grow relatively fast at such regions immediately after crack initiation. Therefore, proper crack growth curves are import for planning inspection of such areas.

16.5 Numerical Methods for Derivation of Stress Intensity Factors

The crack-driving force at a crack tip can be represented by the energy available for a crack extension. This energy can be calculated by the J-integral presented by Rice (1968), which is path-independent for piece vice linear elastic stress strain material behavior:

$$J = \int_{\Gamma} \left(W \, dy - T \frac{\partial u}{\partial x} ds \right) \tag{16.34}$$

where:

Γ is an arbitrary contour around the crack tip, beginning at one crack face and ending on the opposite crack face

W = strain energy

T = traction vector along the integration path.

This can be included in finite element programs by integration of calculated stresses and displacement along a path from one crack surface around the crack tip to the other surface.

This J-integral is equal to the energy release rate:

$$G = \frac{\partial U}{\partial A} = J \tag{16.35}$$

where:

> dU is energy released through an increase in crack area dA; see, for example, Hellan (1984) or another fracture mechanics book

The relationship between J and the stress intensity factor for linear elasticity is expressed as:

$$J = \frac{\beta}{E} K_I^2 \tag{16.36}$$

where:

> $\beta = 1$ for plane stress
> $\beta = 1 - \nu^2$ for plane strain.

Examples of analysis using energy release rates by the finite element method have been presented by many authors and was used in, for example, Aamodt et al. (1973) for analysis of semi-elliptic cracks for assessment of leak-before-failure in tanks in ships used for transport of liquefied natural gas, see also Section I.1. The same methodology was also used by Lotsberg and Bergan (1979) for crack growth analysis of internal defects. Today, it is normal practice to define a J-integral path and calculate J using a standard finite element program and calculate the stress intensity factor from equation (16.36).

16.6 Crack Tip Opening Displacement

The Crack Tip Opening Displacement (CTOD) method has been used as a measure of fracture toughness since the first offshore structures were installed in the North Sea in the 1970s. The CTOD method is well described in textbooks on fracture mechanics. The test specimen is pre-cracked during fatigue loading, such that a sharp crack is introduced. The static capacity of the cracked specimen is then tested and the crack tip opening is measured and related to the position of the fatigue crack tip; reference is also made to assessment standards such as BS 7910 (2013) and API 579-1/ASME FFS-1 (2007). The critical values of the CTOD were previously usually determined by Single Edge Notched Bending (SENB) tests; however, in structures like girth welds in pipelines the loading is more axial and Single Edge Notched Tensile (SENT) specimens may be more representative of the actual behavior. SENT test specimens normally provide larger tested capacities than SENB tests; use of the most relevant test specimen for the actual structure under consideration is recommended.

The CTOD is related to energy release rates through the following equation:

$$J = \lambda \sigma_y CTOD \tag{16.37}$$

where:

λ has been shown to be between 1 and 2, equalling 1 for plane stress and 2 for plane strain

σ_y = yield strength.

It might be added that only the singular stress field is presented in equation (16.2). An additional term in the power series solution of the potential equation governing the stress field at a crack tip may be needed for inclusion of plastic behavior at the crack tip, and this may be needed for assessment of ductile fracture behavior with different constraints (see, e.g., Thaulow et al. 2004 and Macdonald et al. 2011). See Østby et al. (2005) for CTOD in assessment of large-scale plasticity in pipelines.

16.7 Fracture Toughness Based on Charpy V Values

In many projects, fracture toughness from CTOD testing is not available, as requirements for such testing depend on material thickness, service temperature, and consequences of failure. Charpy V tests are normally performed for documentation of the welding procedures used. However, Charpy V specimens do not have a sharp notch like that in CTOD specimens, and therefore it is difficult to derive fracture toughness from these specimens. Some relationships for deriving fracture toughness from Charpy V values have been presented in the literature (see, e.g., BS 7910 2013). These relationships often provide values on the conservative side, but nevertheless these values may be useful, as the calculated fatigue life is often less sensitive to fracture toughness. However, high fracture toughness may be useful for safe documentation of long inspection intervals during in-service life.

16.8 Failure Assessment Diagram for Assessment of Fracture

As explained in Section 16.1, a fracture can be brittle if the material fracture toughness is low, or it can be ductile where a nominal strength capacity larger than the yield strength can be achieved, even with a crack present in the structure. The brittle behavior of the material can be expressed along the vertical axis in a Failure Assessment Diagram (FAD), as is typically used to assess final fracture according to standards such as BS 7910 (2013) and API 579-1/ASME FFS-1 (2007). Ductile behavior is expressed along the horizontal axis in the FAD.

An example of a FAD is shown in Figure 16.13, where the limiting curve from BS 7910 (2013) is presented as:

$$
\begin{aligned}
f(L_r) &= \left(1 + 0.5 L_r^2\right)^{-1/2} \left(0.3 + 0.7 \exp\left(-\mu L_r^6\right)\right) && \text{for } L_r \leq 1 \\
f(L_r) &= f(1.0) L_r^{(N-1)/(2N)} && \text{for } 1 < L_r < L_{r,\max} \qquad (16.38) \\
f(L_r) &= 0 && \text{for } L_r \geq L_{r,\max}
\end{aligned}
$$

where:

$\mu = \min(0.001 E/\sigma_y, 0.6)$
$N = 0.3 \left(1 - \sigma_y/\sigma_u\right)$

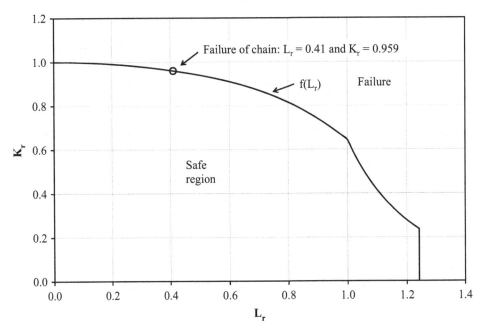

Figure 16.13. Failure assessment diagram (FAD) for failure of chain from example in Appendix A.

L_r = ratio of applied load to yield load and the limiting capacity can be expressed by:

$$L_{r\,max} = \frac{\sigma_f}{\sigma_y} = \frac{\sigma_y + \sigma_u}{2\sigma_y} \qquad (16.39)$$

where:

σ_y = material yield strength
σ_u = material ultimate strength
E = Young's modulus.

For material with a Lüder plateau (which is a horizontal part of the stress-strain curve after first yielding before strain hardening of the material occurs), L_r should be limited to 1.0.

The fracture ratio of applied elastic K_I value to K_{mat} is defined as:

$$K_r = \frac{K_I}{K_{mat}} \le f(L_r) \qquad (16.40)$$

and is plotted along the vertical axis in the FAD. An additional term may be added to K_I in equation (16.40), representing secondary stresses such as residual stresses. The material fracture toughness measured by stress intensity factor, K_{mat}, can be derived from the material toughness measured by J-methods, J_{mat}, as:

$$K_{mat} = \sqrt{\frac{J_{mat} E}{1 - \nu^2}} \qquad (16.41)$$

or by CTOD method, δ_{mat}, as:

$$K_{mat} = \sqrt{\frac{m\sigma_y \delta_{mat} E}{1 - v^2}} \tag{16.42}$$

where

$$m = 1.517\left(\frac{\sigma_y}{\sigma_u}\right)^{-0.3188} \quad for \quad 0.3 < \sigma_y/\sigma_u < 0.98 \tag{16.43}$$

$m = 1.5$ may be used where σ_y/σ_u is not known (BS 7910, 2013).

Combinations of K_r and L_r inside the limiting curve of the FAD indicate that the connection is safe and a point outside the curve indicates that the connection is unsafe. When more detailed stress strain curves are known, a more refined FAD can be established based on, for example, BS 7910 (2013). A number of different FAD diagrams have been published; the largest differences between these are for inter-action between L_r and K_r in the upper right part of the FADs. An example of the analysis of unstable fracture of a chain using the FAD in Figure 16.13 is shown in Appendix A.

16.9 Effect of Post-Weld Heat Treatment and Effect of Crack Closure

PWHT can have a metallurgical effect on the material, which is not considered fur-ther here, and can also remove residual stresses at welded connections. Removal of residual stresses may be important for assessment of a final fracture if the material is brittle or less ductile, while for ductile material it has less influence on fracture capacity. If the residual stresses at the welded connection are removed, crack closure can be considered to occur for the compressive parts of the stress cycles, and this can significantly improve fatigue life in some situations. This implies that only the tensile part of the stress cycle is entered into equation (16.4) for calculation of an effective stress intensity factor after PWHT. This requires that PWHT is performed by putting the component into an oven and that the treatment is performed according to a well-defined and approved procedure. In some cases, heating mats are also placed on the structural detail for PWHT. However, this is more difficult to control, and the effect with respect to residual stresses, fatigue, and fracture should be assessed on a case-by-case basis.

16.10 Alternative Methods for Derivation of Geometry Functions

Accurate analyses to determine geometry functions or stress intensity factors for different details may require considerable work. Therefore, simplifications are often used, such that practical solutions are achieved. The following levels of accuracy may be described:

Level 1 Geometry of the considered detail is known without any further infor-mation about the geometry function of the stress intensity factor.
Level 2 Finite element analysis of the hot spot region is performed without including any crack in the analysis model.

Level 3 Finite element analysis of the hot spot region is performed with different crack sizes included in the analysis model.

For less experienced analysts it may be easier to perform detailed analyses than make an engineering assessment, and the level 3 method would be expected to provide the most accurate results. However, this is also the most time-consuming method. Inclusion of the geometry of the welds, accurate boundary conditions, and relevant loading is also recommended for the level 3 method, so that further assessment of the geometry function, Y, in equation (16.32) is not required. Level 2 is considered to be acceptable for smaller cracks that have not grown through the plate thickness and where redistribution of stresses during crack growth is not as important for the analysis results. Here, the hot spot stress can be divided into a membrane part and a bending part, and superposition of stress intensities from these parts (Newman and Raju 1981) can be assumed to govern crack growth. This solution will represent a two-dimensional stress field through the plate thickness. Thus, this may be a conservative approach if the hot spot stress is local in nature and the stress decreases away from the hot spot, such as at bracket toes ended on plates (see Figure 1.3 such that the half axis, c, in Figure 16.11 has grown away from the main hot spot area into a region with less stress).

Level 1 is the simplest way to derive a geometry function. It is assumed that an S-N classification can be based on the actual geometry. When the S-N curve is known, the hot spot stress can also be derived from Table 4.1, from the column for structural stress concentration embedded in the detail. Deciding how to split the stress into a membrane part and a bending part may be difficult, and here it should be remembered that the membrane part implies a larger crack growth than that of bending. Therefore, in cases of uncertainty, more weight should be placed on the membrane part than the bending part.

Example of a Level 1 method

As an example of a level 1 method, assume that a hot spot at the end of a long attachment is being considered. The following steps are then followed for assessment:

- Reference is made to table A.7 and end of detail 1 in DNVGL-RP-C203 (2016). This gives S-N class F3.
- From Table 4.1 a SCF of 1.61 is derived for the F3 class.
- Fracture mechanics are calibrated against an F-curve to account for the weld notch geometry. The F-curve includes a SCF of 1.27 according to the same Table 4.1. It is now assumed that the fracture mechanics part is performed with the same attachment length and thickness ratio, $L/T = 40/25 = 1.6$, as used for calibration to the F-curve, where L is attachment length and T is plate thickness. The geometry function in equation (16.32) can then be increased, with the resulting stress increase for detail 1 by a factor of $1.61/1.27 = 1.27$.
- The fracture mechanics analysis can then be continued as outlined in Section 16.4.

Table 16.3. *Crack growth parameters (DNVGL-RP-0001 2015)*

Environment	Position of potential crack	Median value of C (Unit N and mm)	Standard deviation in log C	Median value + 2 st. dev. of C	m
Air	Base material	$1.83 \cdot 10^{-13}$	0.11	$3.00 \cdot 10^{-13}$	3.0
	Weld metal	$1.83 \cdot 10^{-13}$	0.22	$5.21 \cdot 10^{-13}$	3.0
Seawater with cathodic protection	Base material	Air value multiplied by factor from equation (16.44)	0.11	Air value multiplied by factor from equation (16.44)	3.0
	Weld metal	Air value multiplied by factor from equation (16.44)	0.22	Air value multiplied by factor from equation (16.44)	3.0
Free corrosion	Base material and weld	$8.35 \cdot 10^{-13}$	0.22	$2.30 \cdot 10^{-12}$	3.0

16.11 Crack Growth Constants

The crack growth parameters in Table 16.3 can be used for crack growth analysis. For the purpose of probabilistic analysis, log C is assumed to be normally distributed. The material parameter, m, is denoted as a crack growth exponent and assumed to be constant.

It should be noted that most fatigue cracks are initiated at weld toes and grow into the base material. In some cases, cracks may also grow from internal imperfections in highly dynamically loaded butt welds. The possibility of crack growth from internal cracks should be considered for the higher S-N curves, above curve D, when flush grinding of the butt welds is needed for documentation of acceptable fatigue life.

The ratio between crack growth in seawater with cathodic protection and air has been derived from S-N data and can be calculated from:

$$f(F, h) = (0.07(\log F)^3 - 0.275(\log F)^2 - 0.245 \log F + 2.38)^{2.0429 - 1.1523h} \quad (16.44)$$

where:

F = calculated fatigue life in years
h = Weibull shape parameter (Section 10.1).

This function is valid from 1 to 1,000 years. For calculated fatigue lives longer than 1,000 years, this ratio should be put as being equal to 1.0.

Alternatively, the crack growth parameters in BS 7910 (2013) and API 579-1/ASME FFS-1 (2007) can be used. A threshold value, ΔK_{th}, may be used in accordance with BS 7910 (2013) as a fixed value: $\Delta K_{th} = 63$ Nmm$^{-3/2}$. A higher threshold value can be used if the stress range is not purely tensile. However, at welds with residual stresses, it is difficult to account for this fully. Use of a threshold value also complicates the analysis under variable long-term loading. One possibility is to neglect the threshold value when calculating crack growth. Note also that the threshold level in BS 7910 (2013) is rather low and does not normally have much influence on the calculated results. It is important to use the same threshold value for calibration of initial crack growth parameters in fracture mechanics as is used later for crack growth analysis. For assessment of cracks in a structure, it is conservative to set the threshold value at zero for crack growth analysis.

See King et al. (1996) and HSE OTH 511 (1998) for background on crack growth parameters used in BS 7910 (2013). The scatter in crack growth data in the base material is considered to be less than in the weld material. Thus, one may use a lower fatigue design crack growth parameter for the base material than for the weld (it is noted that a smaller standard deviation for crack growth in PD 6493 1991 and in BS 7608 (1993) than in BS 7910 (2013)). Most often fatigue cracks are initiated at weld toes but propagate further into the base material.

16.12 Link between Fracture Mechanics and S-N Data

The similarity between the equation for crack growth based on fracture mechanics and the S-N data is shown in Section 16.2. By calibrating the parameters needed for fracture mechanics analysis, correspondence with S-N data can be achieved. Such calibration was used for assessment of the size effect in Section 4.7 and is recommended for planning in-service inspection, as explained in Chapter 18.

17.1 Jacket Structures

17.1.1 Background for Design Standards for Grouted Connections

Cylindrical grouted connections between piles and jacket legs have a positive track record, stretching back over many years, in jacket structures in the oil and gas industry. In older jacket structures, the piles were driven through the jacket legs and then welded to the top of the legs. Grouting was performed by filling the space between the outer surface of the pile and the inner surface of the leg. The grout was made as mix between water and cement. This methodology was used for the first platforms installed in the North Sea in the 1970s. In these structures shear keys were not required in order to achieve sufficient structural capacity. Shear keys are understood to refer to circumferential weld beads on the outer surface of the pile and on the inside of the sleeve in the jacket structure, which increases the shear capacity of a grouted connection when the pile is subjected to axial loading. In older design recommendations from the American Petroleum Institute (API) (1977) and UK Department of Energy (1977), no explicit guidance was provided on how to determine the capacity when mechanical shear keys were used. However, these recommendations did allow the use of shear keys, and the following section is quoted from the UK Department of Energy document (1977):

> The following recommendations for design can be made:
>
> (1) Mechanical shear keys should be used on both the sleeve and the pile surfaces and should be spaced uniformly along the length of the connection.
> (2) Full-scale static tests (or reduced scale tests in which all geometries and material properties are accurately modelled) should be used as a basis for design.
> (3) To determine long-term design bond strength from static ultimate load tests, a safety factor of at least 6 should be used to allow for unknown effects such as cycling loading, both during the grouting operation and in the long term, and offshore construction conditions.

In the same document it was stated that the API specifications at the time were based on tests conducted on relatively small-diameter and radially stiff piles. Furthermore, it was noted that recent laboratory testing had indicated that there was a severe reduction in joint strength with large-diameter piles, such as those used for North Sea structures. It was found that the design clauses in API RP2A (1977) were, in certain cases, unsafe and were generally inadequate for describing bond strength, even for relatively simple cases of bond strength between plain tubular members. At this time API RP2A (1977) included only allowable bond strength for plain tubular members of 0.138 MPa for the operating condition and 0.184 MPa for the extreme condition. However, the design of mechanical shear keys was provided in a separate clause where equivalent or higher bond strength could be achieved.

At the end of 1970s, several tests on the capacity of grouted connections were performed at Wimpey Laboratories in London, and a design equation accounting for the capacity with shear keys was developed (Billington and Lewis 1978; Billington and Tebbett 1980). This methodology was used in the design of the pile-to-sleeve connections in the Magnus platform around 1980, and it was concluded that a significant amount of steel was saved (Lewis et al. 1980). This design methodology became more formalized with the issue of Amendment No. 4 to the UK Department of Energy's Guideline in 1982, in which the design equation from Billington and Tebbett (1980) was included. An assessment of the capacity of grouted connections was also performed within API (Karsan and Krahl 1984), and this resulted in new design recommendations in API RP2A (1984). Some joint industry projects were also performed by DNV at the end of the 1970s and the beginning of the 1980s, in which the capacity of grouted connections was investigated (Løseth 1979; Ingebrigtsen et al. 1990). The second reference is a summary of work performed in these joint industry projects. In clause 2.8.5 in API RP2A (1984) it is stated: "Grouted pile to sleeve connections will be subjected to loading conditions other than axial load, such as transverse shear, bending moment and torque. The effect of such loadings, if significant, should be considered in design of connections by appropriate analytical or testing procedures." The design guidance in API RP2A (1984) is, however, only given with respect to axial load, due to the lack of relevant test data for other loads according to Karsan and Krahl (1984). Three ultimate capacity tests were reported from test specimens subjected to combined axial load and bending moment (Lamport et al. 1987). It was assessed that in these static tests, the bending moment did not reduce the axial capacity. During the 1990s the content of ISO 19902 was developed (Harwood et al. 1996). This standard was issued in 2007 and provides similar, or somewhat larger, interface shear capacity for axial load, depending on grout strength and radial stiffness, as compared with the UK Department of Energy (1982) and HSE UK (2002) documents. The permitted geometry of the shear keys, in terms of shear key height divided by the distance between the shear keys, is increased from 0.04 in HSE (2002) to 0.10 in ISO 19902 (2007). The latest version of API RP2A-WSD (2014) includes the same formulation as in API RP2A (1984) with respect to the capacity of cylindrical connections with shear keys.

17.1.2 Grouted Connections in Newer Jackets

The development of ISO 19902 (2007) was based on traditional jacket design, as reflected in the Lloyds's database on existing structures at the time this standard was developed in the 1990s. This implied design of jacket structures with a significant topside weight, and it was assessed that a design criterion for fatigue due to wave

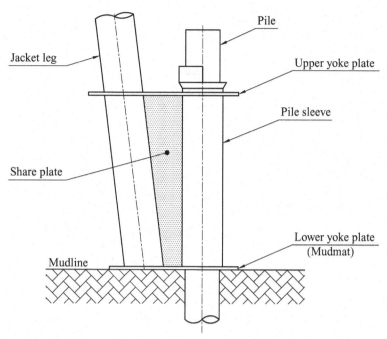

Figure 17.1. Sketch of jacket leg with pile and pile sleeve as typically used in 2015.

action was unnecessary, provided that the grouted connections were designed for the Ultimate Limit State. However, it has since been found that for some jackets designed later, a design criterion for fatigue should also have been included in the design basis. This is related to connections subjected to alternating dynamic axial loading and bending moment. By alternating loading is understood that the net load direction changes from pure compression loading to tension loading during a load cycle. With vertical piles as shown in Figure 17.1, the horizontal shear force on the jacket structure has to be transferred as moments in the piles. This leads to larger moments in these piles than in the inclined piles that were used in older jacket designs, as indicated in Figure 17.2. The moments in the piles in newer jackets are of significant magnitude and can result in stresses in the piles that are more than twice as large as that due to the axial force. Furthermore, factors such as optimization of weights, development of pile clusters, and use of underwater hammers for installation of the piles has resulted in decreased sleeve lengths compared with jacket design in the 1980s.

17.1.3 Assessment of Load Effects and Failure Modes

Due to negative experiences in the wind industry on the use of grouted connections, in around 2009, the design criteria for grouted connections in jacket structures were revisited. Critical reviews questioned whether the design standards included the necessary design requirements to avoid all relevant failure modes for grouted connections in pile-to-sleeve connections in jacket structures. Although NORSOK N-004 was originally similar to ISO 19902 (2007) with respect to design of grouted connections, this assessment resulted in a revision of NORSOK N-004 Annex K in 2013. The basis for this work is explained in this section.

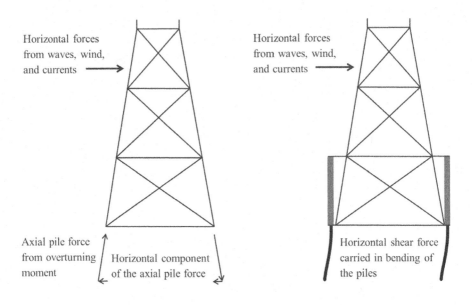

a) Older jacket with battered piles along the legs   b) Newer jacket with vertical piles

Figure 17.2. Difference in pile forces in traditional older jackets as compared with newer jackets.

Finite element analysis of grouted pile-to-sleeve connections shows significant compressive stresses in the lower region of the pile sleeve when there is a significant bending moment and shear force in the pile. This bending moment is gradually transferred from the pile to the sleeve structure through stresses in the grout. This may be illustrated by considering the pile as a beam on elastic springs, as indicated in Figure 17.3. The supporting springs are due to the radial stiffness of the connection resulting from the stiffness in the circumferential direction of the pile and the sleeve, and from the transverse stiffness of the grout. However, actual connections in jacket structures are somewhat more complex when the yoke and shear plates are also considered, as shown in Figure 17.1. In this example, the lower part of the grout is assumed to end well above the lower yoke plate. The moment and shear force from an actual jacket are used in an example calculation in Figure 17.4. The contact stress between the steel and grout also creates a shear stress in the lower region of the grout when it is combined with a friction coefficient between the steel and grout, as shown in Figure 17.5. This shear stress, together with the contact stress, gives a relatively large tensile stress in the grout, and the tensile strength of the grout may be exceeded. This is particularly concerning if the connection is subjected to alternating dynamic loading with cracking, as indicated in the right part of Figure 17.5. The principal tensile stress in the grout can be calculated as:

$$\sigma_{II} = \frac{p_{Local}}{2} \left(1 - \sqrt{1 + 4\mu^2} \right) \tag{17.1}$$

where:

p_{Local} = maximum local contact stress between steel and grout
μ = friction coefficient for steel against grout.

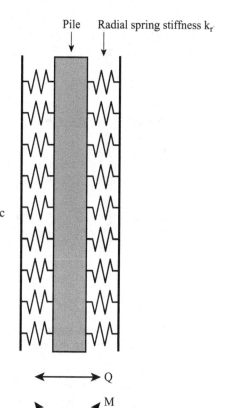

Figure 17.3. Pile in a sleeve simulated as a beam on elastic supports.

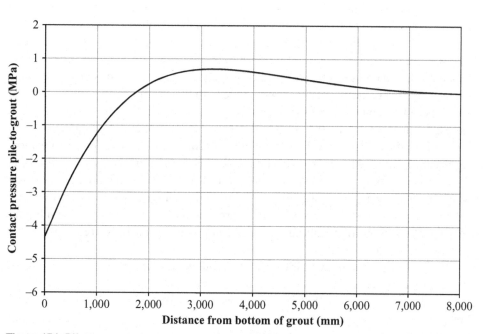

Figure 17.4. Pile-to-grout contact pressure in a typical jacket.

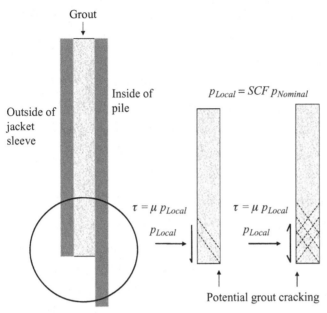

Figure 17.5. Stress condition at the lower region of grout (Lotsberg et al. 2012) (reprinted with kind permission from Ernst & Sohn).

Equation (17.1) can be derived from equation for smallest principal stress in equation (9.5) by putting the normal stress equal the pressure p, the stress in the vertical direction equal zero and the shear stress equal $p \cdot \mu$. The amount of sliding between the steel and grout depends on the diameter of the connection, as well as the other geometry of the connection and the loading. It is therefore difficult to extrapolate from experiences with small-scale laboratory tests to real, large-diameter connections. For large-diameter connections, significant shear stress cannot be transferred between the pile and the sleeve structure before the bond between the steel and the grout is broken. When the bond is broken, sliding will start, and wear in the connection may occur. This has been observed in jack-up type platforms with large-diameter grouted legs, where the connections were subjected to significant alternating moments. The estimated sliding between the grout and the steel for maximum bending moment on an 84-inch pile in a jacket is shown in Figure 17.6. Sliding due to the moment is most significant within the lowest region of the grouted connection. Therefore, positioning of shear keys in this region is questionable, as they will start to take up force from the moment. Based on results of laboratory simulations of full-scale tests, only limited relative displacement can be absorbed before cracking of grout occurs. From Figure 17.6 it is also observed that sliding in an actual pile is significantly greater than in a test specimen scaled at 1:6.

The axial capacity of cylindrical grouted connections without shear keys should depend on the inverse of the diameter of the connection. The capacity of plain surfaces is not very large for the ultimate capacity of grouted connections with shear keys, as the capacity due to shear keys in such connections is significantly larger than that due to the plain surface. However, plain surface capacity is important when assessing fatigue capacity for axial loads, as fatigue damage is not accumulated in grouted connections with shear keys, provided that the reversed axial load is lower than the resistance due to surface irregularity, neglecting the capacity from the shear

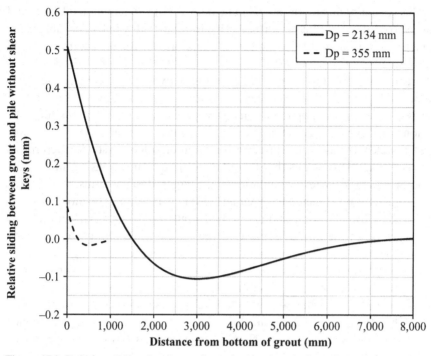

Figure 17.6. Relative sliding between pile and grout in a jacket pile, compared with a scaled test specimen.

keys (ISO 19902 2007). Surface irregularity is understood to refer to the surface roughness of the plates and production tolerances introduced during fabrication of tubular sections. This implies that the reversed cyclic load is not so large that sliding would occur without shear keys in the connection. From experiences from settlement of transition pieces on monopiles detected in around 2010, it can be seen that there is a size effect for axial capacity in connections without shear keys. Most of the fatigue test data were derived for specimens with diameters less than 400 mm (Ingebrigtsen et al. 1990; Harwood et al. 1996). There is a significant axial resistance for such connections due to surface irregularity, and the amount of dynamic axial load that was transferred to the shear keys in these test specimens has been questioned. Reliable test data for grouted connections with shear keys subjected to alternating dynamic loading are not extensive. However, some data were derived 17–30 years ago from testing connections with small-diameter pipes under constant amplitude loading in the air environment. The capacity depends on the fabrication tolerances of the pipes, which have probably changed over time. The fatigue capacity of grouted connections is also considered to be a function of load history and environment (lower capacity when submerged in water), in addition to geometry (DNV-OS-C502 2012; Waagaard 1981; Sørensen 2011).

17.1.4 Recommended Design Practice in NORSOK N-004 and DNV-OS-J101

The fatigue capacity of grouted connections is significantly lower for reversed loading that involves some sliding between the steel and the grout than in connections subjected to net loading in only one direction with the same magnitude of dynamic loading. This means that fatigue assessment due to both dynamic axial load and to bending moment should be performed. This is important for slender jackets for oil and gas exploitation, and for wind turbine support structures with preinstalled and post-installed piles (see NORSOK N-004 2013 and DNV-OS-J101 2014, respectively). Furthermore, limiting the contact pressure between the steel and the grout is recommended to avoid wear and crushing of the grout during reversed loading. Based on assessment of the fatigue capacity curves for grouted connections, the main contribution to fatigue damage is found to be in the range from a few cycles to fewer than 100,000 cycles. This is rather different from that of fatigue in steel structures, as shown in Section 1.2. This means that failure of grouted connections might be most likely to occur during damage accumulation in the largest design storms. Thus, design of grouted connections subjected to dynamic loads may be seen to be more as a part of control of the Ultimate Limit State than the Fatigue Limit State. This also means that the left region of the fatigue capacity curve, with a small number of cycles to failure, should be the target region for planning more fatigue tests on grouted connections; see also Lotsberg and Solland (2013) for recommended further research related to capacity of grouted connections. The laboratory tests on which recommendations on early age cycling were based were small-scale tests with rather small shear keys. The absolute height of the shear keys can have a significant influence on the disturbance of the grout at the shear keys during an early curing process. To avoid excessive sliding and lack of long-term fatigue capacity, it is necessary to have close contact between the shear keys and the grout in connections that are subjected to alternating loading. Furthermore, the allowable early-age cycling as a function of diameter, as stated in ISO 19902 (2007), can be questioned, and this has been revised in NORSOK N-004 (2013). A rather limited early-age cycling criterion of 1.0 mm relative sliding is included in DNV-OS-J101 (2014) and some early-age cycling is allowed during installation of wind turbine structures. However, use of pile grippers is a natural choice to avoid this failure mode for installation of jacket structures for oil and gas exploitation. The Ultimate Limit State is controlled by use of modified capacity equations from ISO 19902 (2007). The interface shear capacity in a grouted connection with shear keys is derived as:

$$f_{bk} = \left[\frac{800}{D_p} + 140 \left(\frac{h}{s} \right)^{0.8} \right] k^{0.6} f_{ck}^{0.3} \qquad (17.2)$$

where:

h = height of shear key
s = distance between shear keys
f_{ck} = characteristic cube strength of the grout in MPa
k = radial stiffness parameter defined as:

$$k = \left[(2R_p/t_p) + (2R_s/t_s) \right]^{-1} + (E_g/E)\left((2R_s - 2t_s)/t_g \right)^{-1}$$

$D_p = 2R_p$ = outer diameter of pile
t_p = thickness of pile
R_s = radius of sleeve
t_s = thickness of sleeve
t_g = thickness of the grout

E_g = Young's modulus of the grout
E = Young's modulus of the steel.

The interface shear capacity is limited by grout matrix shear failure by:

$$f_{bk} = \left[0.75 - 1.4\left(\frac{h}{s}\right) \right] f_{ck}^{0.5} \tag{17.3}$$

The characteristic capacity per unit length of one shear key is derived as:

$$F_{V1\ Shk\ cap} = f_{bk}S \tag{17.4}$$

The design capacity for one shear key is derived as:

$$F_{V1\ Shk\ cap,d} = \frac{F_{V1\ Shk\ cap}}{\gamma_m} \tag{17.5}$$

where:

material factor $\gamma_m = 2.0$ is used for the Ultimate Limit States.

The design requirement for the shear keys reads:

$$F_{V1\ Shk} \leq F_{V1\ Shk\ cap,d} \tag{17.6}$$

The requirement for the distance between shear keys is presented as:

$$s \geq \min \left\{ \begin{array}{l} 0.8\sqrt{R_p t_p} \\ 0.8\sqrt{R_s t_s} \end{array} \right\} \tag{17.7}$$

The distance between the shear keys can be smaller for the purpose of fulfilling the fatigue limit state. However, if smaller distances are used, the minimum distance between shear keys, as presented by equation (17.7), shall be used for control of capacity in the Ultimate Limit State.

The requirement for the geometry of shear keys is presented as:

$$1.5 \leq \frac{w}{h} \leq 3.0 \tag{17.8}$$

where h and w are height and width of shear keys respectively.

For assessment of capacity it is assumed that the height of the shear keys is at least 5 mm in DNV-OS-J101 (2014) and that:

$$\frac{h}{s} \leq 0.10 \tag{17.9}$$

where s is the distance from the center to the center of the shear keys.

Furthermore it is recommended that:

$$\frac{h}{D_p} \leq 0.012 \tag{17.10}$$

The following recommendation is given with respect to grout thickness:

$$10 \leq \frac{D_g}{t_g} \leq 45 \tag{17.11}$$

where D_g is the outer diameter of the grout section.

However, an alternative to this requirement may be to perform a fatigue assessment of the grout based on S-N curves for concrete, as described in Section 4.1.17. The following limits on length of grouted connection are given:

$$1 \leq \frac{L_e}{D_p} \leq 10 \tag{17.12}$$

This requirement is based more on testing with static loading than with dynamic loading.

The requirement for the geometry of the pile is presented as:

$$10 \leq \frac{R_p}{t_p} \leq 30 \tag{17.13}$$

Meeting this requirement for large-diameter connections such as monopiles is difficult, and an alternative to this criterion may be to perform a nonlinear buckling analysis of the connection. This can be used to document that local buckling due to the slenderness of the pile will not occur when it is subjected to ultimate loading. The geometry requirement for the transition piece is presented as:

$$15 \leq \frac{R_s}{t_s} \leq 70 \tag{17.14}$$

Shear keys placed within half an elastic length from the lower end of the grouted connection should not be relied upon with respect to fatigue from dynamic axial force.

Documentation of a detailed design should be based on calculation of fatigue damage using an S-N curve (or force to number of cycles relationship) presented as:

$$\begin{array}{ll}
\log N = 5.400 - 8y & \text{for} \quad y \geq 0.30 \\
\log N = 7.286 - 14.286y & \text{for} \quad 0.16 < y < 0.30 \\
\log N = 13.000 - 50y & \text{for} \quad y \leq 0.16
\end{array} \tag{17.15}$$

where, for each cycle (or cycles in a stress block):

$$y = \frac{F_{V1 \; Shk \; \gamma m}}{F_{V1 \; Shk \; cap}} \tag{17.16}$$

where:

$\gamma_m = 1.5$ is a recommended material factor or safety factor with respect to fatigue in DNV-OS-J101 (2014).

17.2 Monopiles

17.2.1 Experience with Plain Cylindrical Grouted Connections

Grouted connections are used to connect transition pieces to monopiles in wind turbine structures, as indicated in Figure 17.7. The transition piece is installed on top of the monopile resting on temporary supports. The transition piece is then jacked up to the correct verticality before grouting. After curing, the jacks are removed, leaving a gap between the supports and the monopile. The influence from dynamic

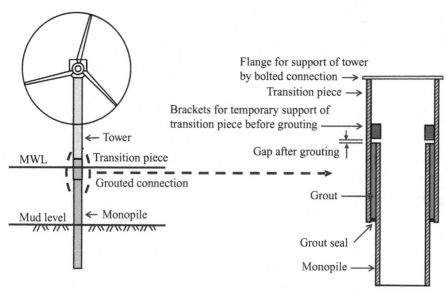

Figure 17.7. Principle of a grouted connection in a monopile structure.

moment seen in large-diameter monopile connections in the wind industry is significantly larger than from the axial force (see Schaumann et al. 2010; Lotsberg et al. 2011; Prakhya et al. 2012). Some results from tests on plain cylindrical grouted connections performed at the University of Aalborg, Denmark in around 2000 indicated that sufficient capacity could be achieved in these connections without the use of shear keys. In addition, due to significant dynamic moments in these structures, the use of shear keys in relation to grout capacity was questioned from a fatigue perspective. As settlement was observed in some support tower foundations during 2009, the axial capacity of large-diameter plain grouted connections was questioned. Due to the uncertainty related to the capacity of these connections, a joint industry project was conducted (November 2009 – January 2011) regarding the capacity of large-diameter grouted connections in offshore wind turbine structures (for plain connections without shear keys). A design methodology based on grouted conical connections was established in this project (Lotsberg et al. 2013a). During early 2010, inspection confirmed that a number of wind farms with plain grouted connections had settled, such that the temporary supports were again resting on top of the monopiles, as shown in Figure 17.7. The force flow through the structures therefore differed from that assumed in the design. In January 2011, another joint industry project was initiated with the objective of providing a sound basis of data and establishing a reliable design methodology for cylindrical grouted connections with shear keys subjected to alternating loading (Lotsberg et al. 2013b).

17.2.2 Moment Capacity of Plain Connections

In a wind turbine structure, the axial stresses in the transition piece and the monopile due to bending moment loading are usually much larger than that from the axial permanent load from the weight of the structure above the connection. The moment action leads to a tension load in the circumferential direction of the grouted

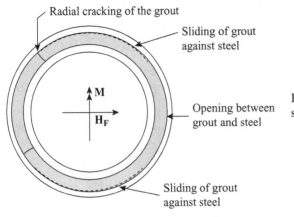

Figure 17.8. Behavior of grouted connection subjected to a large bending moment.

connection that may exceed the tensile capacity of the grout. This may result in cracking of the grout in radial and vertical sections, as indicated in Figure 17.8. Due to the relatively high local slenderness (diameter to thickness ratio) of the pile and the transition piece, ovalization of the cylinders will also occur and a gap will open up between the grout and the steel in the case of large moments. This will lead to a relative sliding between the steel and grout. Thus, it can be argued that the main purpose of the grout is to transfer pressure from the transition piece to the pile.

The moment is transferred from the transition piece to the monopile through horizontal contact forces, as indicated in Figure 17.9. There will also be vertical friction forces, due to the contact pressure, which contribute to the moment capacity of the grouted connection (vertical arrows in Figure 17.9). Based on test results, the

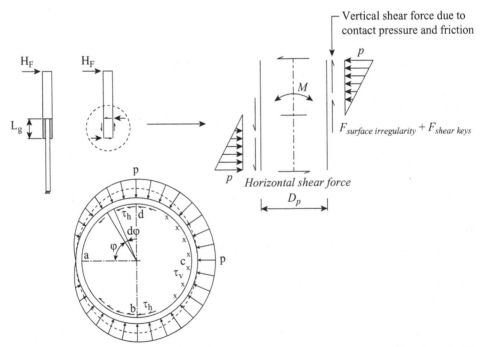

Figure 17.9. Pressure distribution used for calculation of moment resistance (Lotsberg 2013) (reprinted with permission from Elsevier).

mean friction coefficient between steel and grout is around 0.70 and a characteristic friction coefficient is in the 0.40–0.50 range. This can also be applied to grouted connections that are subjected to long-term sliding. As long as there is friction force between the steel and grout due to contact pressure, there will also be vertical friction forces due to the surface irregularity (or fabrication tolerances) in the connection, as indicated in Figure 17.9. Although this effect is not recommended for use in design, it must be kept in mind when assessing laboratory test data. If shear keys are installed around the circumference of the monopile and the transition piece, these will also transfer vertical shear forces, thereby contributing to the moment capacity of the grouted connection. Grouted connections without shear keys are considered in the following. The contact pressure shown in Figure 17.9 will act around most parts of the circumference and will provide some horizontal shear resistance due to the friction between the steel and grout. The horizontal shear forces shown in Figure 17.9 at positions b and d will also contribute to the moment capacity of the grouted connection. The actual behavior of the grouted connections subjected to a bending moment may be simulated by finite element analysis that accounts for compressive contact between the steel and the grout, but without transferring tensile stresses, and with a suitable friction coefficient where contact pressure is present. For design purposes, an analytical approach, as considered in the following part of this section, may also be efficient. An analytical expression of the relationship between the contact pressure and bending moment acting on the grouted connection can be derived based on certain assumptions concerning pressure distribution, as indicated in Figure 17.9. The moment resistance due to contact pressure is derived by integration around half the circumference b-c-d in Figure 17.9 as:

$$M_p = 2 \int_0^{\pi/2} \sin \varphi \, R_p \, d\varphi \, p \frac{L_g}{2} \frac{1}{2} \left(\frac{L_g}{2} \frac{2}{3} \right) 2 \tag{17.17}$$

where the term in brackets is the moment arm about the center of the grouted connection. The left side of this equation is force, and there is one reaction force at the top and one at the bottom of the grouted connection in Figure 17.9. The moment resistance due to contact pressure is then derived by integration as:

$$M_p = p \frac{R_p L_g^2}{3} \tag{17.18}$$

where:

p = maximum nominal pressure at the top and bottom of the grouted section, as shown in Figure 17.9
R_p = outer radius of pile
L_g = effective height of grouted section.

The pressure distribution within the dotted line around the circumference in Figure 17.9 can be expressed as:

$$p(\varphi) = p \frac{3}{2} \frac{\varphi}{\pi} \tag{17.19}$$

for $0 \leq \varphi \leq \pi/2$. This equation provides a pressure corresponding to 0.75p at positions b and d in Figure 17.9. The moment resistance due to the horizontal friction force in Figure 17.9 can be derived by integrating the contact pressure and area multiplied by the moment arm within the dotted line from a to c in Figure 17.9, with pressure 0.75p at positions b and d as:

$$M_{\mu h} = 4 \int_0^{\pi/2} \sin\varphi \, R_g \, d\varphi \, p(\varphi) \mu \frac{L_g}{2} \frac{1}{2} \left(\frac{L_g}{2} \frac{2}{3}\right) 2 \tag{17.20}$$

where the term in brackets is the moment arm to the center of the connection.

The following equation is added for assisting with integration of equation (17.20):

$$\int_0^{\pi/2} \varphi \sin\varphi \, d\phi = [\sin\varphi - \varphi \cos\varphi]_0^{\pi/2} = 1.0 \tag{17.21}$$

Then the moment due to horizontal friction force is derived as:

$$M_{\mu h} = \mu p \frac{R_p L_g^2}{\pi} \tag{17.22}$$

Similarly, the moment resistance due to the vertical friction force is derived as:

$$M_{\mu v} = 2 \int_0^{\pi/2} R_p \, d\varphi \, p\mu \frac{L_g}{2} \frac{1}{2} \left(R_p \sin\varphi\right) 2 \tag{17.23}$$

where the term in brackets is the moment arm.

The left side of this equation is force, and there is one reaction force at the top and one at the bottom of the grouted connection in Figure 17.9. The integration is performed around half the circumference b-c-d, with a constant contact pressure, p. This corresponds to integration around the full circumference, where the contact pressure is zero at position a, equal to p at c, and equal to 0.5p at b and d.

The moment due to vertical friction force is then derived as:

$$M_{\mu v} = \mu p R_p^2 L_g \tag{17.24}$$

The moment resistances calculated were compared with finite element analysis before a decision was reached on the pressure values to be used at positions b and d in Figure 17.9. The selected pressures at b and d for derivation of the two resistance friction moments were also based on the consideration that the force due to friction in any direction cannot be larger than pressure multiplied by friction multiplied by area. Thus, there cannot be a full friction force in both vertical and horizontal directions around the circumference, as the action from the two components cannot be larger than that of a vector representing full friction; see Lotsberg (2013) for further details.

Moment equilibrium with the external bending moment gives:

$$M_{tot} = M_p + M_{\mu h} + M_{\mu v} \tag{17.25}$$

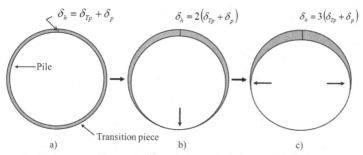

Figure 17.10. Illustration of derivation of equation for opening (Lotsberg 2013) (reprinted with permission from Elsevier).

In addition to these moments, there is a friction moment due to surface irregularity that is neglected for large diameter connections. From equations (17.18), (17.22), (17.24), and (17.25), the maximum nominal contact pressure can be estimated as:

$$p = \frac{3\pi M_{tot}}{R_p L_g^2 (\pi + 3\mu) + 3\pi \mu R_p^2 L_g} \tag{17.26}$$

The contribution from the global shear force at the grouted connection in a monopile to the nominal contact pressure is considered small, and is neglected in the further development of a design procedure.

17.2.3 Opening between the Steel and the Grout in the Connections due to Moment Loading

The maximum opening between the steel and the grout at the top and bottom of the grouted connection can be estimated from the radial flexibility of the pile and the transition piece, in addition to geometric considerations, as:

$$\delta_h = 3(\delta_{TP} + \delta_p) \tag{17.27}$$

where:

δ_p = radial decrease in pile diameter for contact pressure p
δ_{TP} = radial increase in diameter of the transition piece for contact pressure p

Here, compression of the grout is neglected as the contribution from high-strength grout, typically used in wind turbine connections, to the total deformation is small. The change in radius due to contact pressure for the pile and the transition piece is derived from Hooke's law as:

$$\delta_{TP} = \frac{pR_{TP}^2}{Et_{TP}}$$
$$\delta_p = \frac{pR_p^2}{Et_p} \tag{17.28}$$

The total radial displacement, δ_h, is derived as the sum of displacements from the transition piece and the pile, as illustrated in Figure 17.10a. For moment loading, the transition piece will move on the pile until contact around the circumference is achieved, as illustrated in the steps in Figures 17.10b and 17.10c. The last step, from

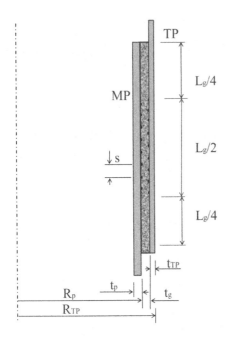

Figure 17.11. Example of a grouted connection with shear keys in a monopile.

position b to position c, implies some ovalization of the transition piece. Furthermore, based on geometrical considerations, the vertical relative displacement between the transition piece and the pile is derived as:

$$\delta_v = \delta_h \frac{2R_p}{L_g} \qquad (17.29)$$

The basis for this equation is that a small rotation of the transition piece about the center of the grouted connection gives a horizontal displacement at top of pile of δ_h, and a vertical displacement of δ_v at a distance, R_p, from the rotation center. The derived analytical approach has been compared with finite element analysis for a full-size model with pile diameter of 5,000 mm. Significant sliding between the steel and grout is expected to occur when there is a combination of a flexible transition piece (large diameter-to-thickness ratio) and large bending moments, even with a high friction coefficient. Thus, it is unrealistic to try to improve the structural behavior of large-diameter connections subjected to dynamic bending moments by increasing the roughness of the steel surfaces.

17.2.4 Load on Shear Keys in Grouted Connections with Shear Keys

In the following example, a cylindrical grouted connection with shear keys, as illustrated in Figure 17.11, is considered. A resistance moment from the shear keys adds to the resistance moments considered in Section 17.2.2, and the total resistance moment can be presented as:

$$M_{tot} = M_p + M_{\mu h} + M_{\mu v} + M_{shear\,keys} \qquad (17.30)$$

where the notations are as explained in Section 17.2.2, except for $M_{shear\,keys}$, which is the resistance moment due to the shear keys.

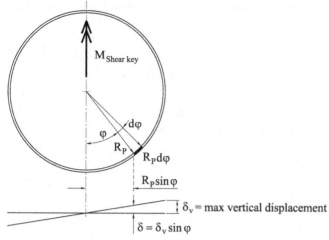

Figure 17.12. Model for analysis of moment resistance from shear keys (Lotsberg 2013) (reprinted with permission from Elsevier).

The opening between the pile and the transition piece at the top and bottom of the grouted connection can be expressed by the same equation as for a condition without shear keys; however, the contact pressure and ovalization are reduced due to force action from the shear keys. The relative vertical displacement between the pile and the monopile is derived as previously from equation (17.29). This relative vertical displacement can be further expressed by use of equations (17.27) and (17.28) as:

$$\delta_v = \frac{6p}{E} \frac{R_p}{L_g} \left(\frac{R_p^2}{t_p} + \frac{R_{TP}^2}{t_{TP}} \right) \tag{17.31}$$

where:

E = Young's modulus for steel

p = maximum nominal contact pressure at top and bottom of the grout section, as indicated in Figure 17.9.

The moment resistance due to the shear keys can be calculated based on Figure 17.12 as:

$$M_{shear\ key} = 4 \int_0^{\pi/2} k_v R_p \, d\phi \, R_p \, \sin\varphi \, (\delta_v \sin\varphi) \tag{17.32}$$

where:

k_v = spring stiffness of a shear key per unit length of that shear key around the circumference acting in the vertical direction

δ_v = maximum relative vertical displacement calculated at the positions of the shear keys.

The following capacity equation is derived by integration of equation (17.32):

$$M_{shear\ key} = \pi \delta_v k_v R_p^2 \tag{17.33}$$

It is assumed that the resistance from several shear keys can be added together, such that:

$$k_{vn} = k_v n \qquad (17.34)$$

It is assumed that the shear keys are placed in the center region of the grouted connection of the monopile (see Figure 17.11). This is because it is preferable to place the shear keys in a region without significant opening during a moment loading. The vertical stiffness of the transition piece, representing half the effective height in the grouted connection, reads:

$$k_{TP} = \frac{2t_{TP}E}{L_g} \qquad (17.35)$$

The inverse of stiffness is flexibility (in general terms):

$$f = \frac{1}{k} \qquad (17.36)$$

It is assumed that flexibilities in a series system can be summed, according to the following:

$$\frac{1}{k_{eff}} = \frac{1}{k_{vn}} + \frac{1}{k_{TP}} \qquad (17.37)$$

From this equation, k_{eff} can be derived. Equation (17.33) is then modified to include the stiffness of n effective shear keys and the axial stiffness of the transition piece. The resulting resistance moment due to shear keys can then be expressed as:

$$M_{shearkey} = \pi \delta_v k_{eff} R_p^2 \qquad (17.38)$$

The units of k_{eff} are: $\frac{N}{mm^2}$

From equations (17.18), (17.22), (17.24), (17.30), (17.31), and (17.38), the following expression for the nominal contact pressure is derived:

$$p = \frac{3\pi M_{tot}EL_g}{EL_g\left\{R_p L_g^2 (\pi + 3\mu) + 3\pi\mu R_p^2 L_g\right\} + 18\pi^2 k_{eff} R_p^3 \left(\dfrac{R_p^2}{t_p} + \dfrac{R_{TP}^2}{t_{TP}}\right)} \qquad (17.39)$$

where:

μ = friction coefficient between the grout and steel.

From finite element analysis of grouted connections it is observed that when the transition piece is moved vertically relative to the monopile, compressive struts develop between shear keys on the transition piece and shear keys on the monopile. Furthermore, tensile stresses develop in the grout normal to the compressive strut, and, as loading from the transition piece increases, these will lead to the grout cracking in the direction parallel with the compression strut. The spring stiffness, representing the stiffness of the grouted connection with shear keys, is derived from an analysis model, as shown in Figure 17.13. It is assumed that a compression strut develops between a shear key at the inner side of the transition piece and a shear key at the outer side of the monopile, as shown in Figure 17.13a.

Young's modulus of high-strength grout used in monopile structures is large, typically in the order of 50,000 MPa. Thus, the grout is stiff with respect to radial

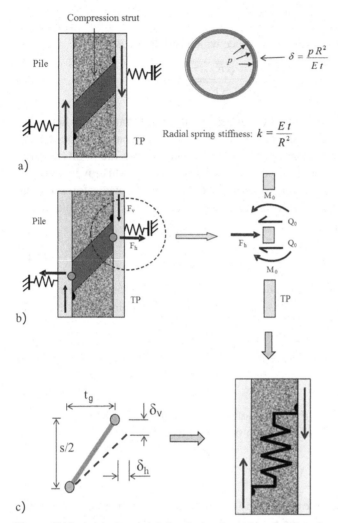

Figure 17.13. Analysis model for derivation of expression for stiffness for vertical springs between shear keys (Lotsberg 2013) (reprinted with permission from Elsevier).

compression and shear deformation. The main flexibility of the connection is due to radial deformation of the pile and the transition piece at the ends of the compression strut, as shown in Figure 17.13a. These flexibilities can be represented by springs, as is illustrated in this figure. The relative sliding between the transition piece and the monopile is in the vertical direction, and therefore it is most convenient to define a spring that also acts vertically (Figure 17.13c). See Figure 17.13b for assessment of the reaction force between the compression strut and transition piece. From this figure the following relationship applies (force per unit length of circumference):

$$Q_0 = \frac{F_h}{2} \qquad (17.40)$$

For the moment in the transition piece, the following relationship applies:

$$M_0 = -\frac{Q_0 l_{eTP}}{2} \qquad (17.41)$$

where:

l_{eTP} is elastic length of the transition piece which is defined as:

$$l_{eTP} = \frac{\sqrt{R_{RP} t_{TP}}}{\sqrt[4]{3(1 - v^2)}} \tag{17.42}$$

See Section 6.1 presenting basis for the classical shell theory that is used here with respect to equations and notations. Equation (17.41) is derived from equation (6.8) by setting dw/dx = 0, due to the symmetry of the deflection shape at the point of action for the force, F_h. Now $g_1(\xi) = 1.0$ and $g_3(\xi) = 1.0$ for $\xi = 0$. This gives equation (17.41). Equation (6.3) is then used to derive a relationship between force F_h and the displacement δ_{hTP}. $g_4(\xi) = 1.0$ for $\xi = 0$. (See equation 6.6). This results in the following equation:

$$\delta_{hTP} = \frac{l_{eTP}^2}{2D_{kTP}} (M_0 + Q_0 l_e) \tag{17.43}$$

By inserting for the moment from equation (17.41) and for the force from equation (17.40), the following relationship between the force and the radial displacement of the transition piece is derived:

$$F_h = \delta_{hTP} \frac{8D_{kTP}}{l_{eTP}^3} \tag{17.44}$$

where:

δ_{hTP} = radial deflection of the transition piece for horizontal force F_h
D_{kTP} = shell stiffness of the transition piece defined as:

$$D_{kTP} = \frac{E t_{TP}^3}{12(1 - v^2)} \tag{17.45}$$

The relationship between the force and radial displacement of the pile can be similarly developed as for the transition piece as:

$$F_h = \delta_{hp} \frac{8D_{kp}}{l_{ep}^3} \tag{17.46}$$

where:

δ_{hp} = radial deflection of the pile for horizontal force F_h
D_{kp} = shell stiffness of the pile defined as:

$$D_{kp} = \frac{E t_p^3}{12(1 - v^2)} \tag{17.47}$$

The resulting deflection from the transition piece and the pile is derived as:

$$\delta_h = \delta_{hTP} + \delta_{hp} \tag{17.48}$$

The following relationship is obtained from the moment equilibrium of the compression strut in Figure 17.13c:

$$\frac{s}{2} F_h = t_g F_v \tag{17.49}$$

where:

s is the distance between shear keys (or s/2, as shown in Figure 17.13c, as it is assumed that the shear keys on the monopile side are between the shear keys on the transition piece side).

From Figure 17.13c, the following relationship with respect to vertical and horizontal displacements is derived (assuming the displacements lead to a small rotation of the compression strut):

$$\frac{\delta_h}{s/2} = \frac{\delta_v}{t_g} \tag{17.50}$$

where:

δ_v is relative vertical displacement of the transition piece versus the pile.

From equations (17.44)–(17.50) the following vertical spring stiffness, representing a unit length of a shear key, is derived as:

$$k_v = \frac{s^2 E}{2\sqrt[4]{3(1-v^2)}t_g^2 \left\{ \left(\dfrac{R_p}{t_p}\right)^{3/2} + \left(\dfrac{R_{TP}}{t_{TP}}\right)^{3/2} \right\}} \tag{17.51}$$

where:

$$k_v = \frac{F_v}{\delta_v} \tag{17.52}$$

There is also some flexibility in the transition piece from the top of the pile to the center of the location of the shear keys, as represented by equation (17.35). By combining stiffness, as in equation (17.37), the following expression is now derived for the effective spring stiffness for n shear keys:

$$k_{eff} = \frac{2t_{TP}s_{eff}^2 nE}{4\sqrt[4]{3(1-v^2)}t_g^2 \left\{ \left(\dfrac{R_p}{t_p}\right)^{3/2} + \left(\dfrac{R_{TP}}{t_{TP}}\right)^{3/2} \right\} t_{TP} + ns_{eff}^2 L_g} \tag{17.53}$$

This spring stiffness can be entered into equation (17.39) for calculation of nominal contact pressure.

The relative vertical displacement between the transition piece and the pile can then be calculated from equation (17.31). The loading on the shear keys can then be calculated as:

$$F_{v\,Shk} = k_{eff}\delta_v \tag{17.54}$$

The action force (per unit length around the circumference) from bending moment and vertical force transferred to the shear keys is derived from equation (17.31) as:

$$F_{v\,Shk} = \frac{6p k_{eff}}{E} \frac{R_p}{L_g} \left(\frac{R_p^2}{t_p} + \frac{R_{TP}^2}{t_{TP}}\right) + \frac{P}{2\pi R_p} \tag{17.55}$$

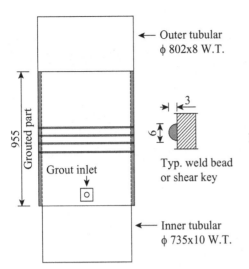

Figure 17.14. Geometry of cylindrical test specimen.

where:

P = self-weight of structure above the pile, including full weight of the transition piece. This load is assumed to be evenly distributed to the shear keys around the circumference of the connection. The loading on one shear key is obtained as:

$$F_{v1\,Shk} = \frac{F_{v\,Shk}}{n} \qquad (17.56)$$

The effective distances between the shear keys are somewhat shorter than the distances measured from center to center. The following expression for effective distance between shear keys is used:

$$s_{eff} = s - w \qquad (17.57)$$

where:

w = width of one shear key.

Laboratory testing should be performed to investigate the effective distance between shear keys when the distance between the shear keys is increased. For design purposes it is necessary to assess the above average action force against the capacity from ISO 19902 (2007) for the Ultimate Limit State. For the Fatigue Limit State, the capacity should be assessed against an S-N curve (or a force–N relationship) for grouted connections with shear keys (Lotsberg et al. 2013b).

A cylindrical test specimen with three effective shear keys is shown in Figure 17.14 (four shear keys on the inner tubular and three shear keys on the outer tubular). The grouting was performed in a vertical position with water-filled tubulars, and the grout was inserted through the grout inlet from below in a similar way as performed for grouting of structural connections in the sea. After curing, the specimen was placed in a test rig and instrumented with strain gauges and displacements transducers, as shown in Figures 17.15a and 17.15b. Then the test specimen was subjected to some constant axial force and an alternating dynamic bending moment until failure. After testing, the specimen was cut along the longitudinal direction and a

a) Specimen in test laboratory

b) Measurement of opening and
relative displacement between tubulars

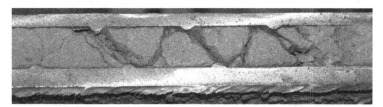

c) Section through cylindrical test specimen after fatigue testing
showing cracking along compression struts

Figure 17.15. Fatigue testing of cylindrical test specimen (Lotsberg 2013) (reprinted with permission from Elsevier).

section after cutting is shown Figure 17.15c, showing cracking of the grout along the compressive struts that were developed during the test.

Use of this design methodology presented in this section is shown in Figure 17.16 for different numbers of shear keys. This example represents the geometry of test specimen in Figure 17.14 and the analysis results are shown in Figure 17.16, with three effective shear keys. The geometry for this specimen is: $D_p = 735$ mm, $t_p = 10.0$ mm, $D_{TP} = 802$ mm, $t_{TP} = 8.0$ mm, $t_g = 25.5$ mm, $L_g = 955$ mm, $h = 3.0$ mm, $w = 6$ mm, $s = 60$ mm (this corresponds also to Specimen 7 (DNV test) in Figures 17.18 and 17.19). The analysis methodology described in this section is followed. This gives $k_{eff} = 1234.75$ N/mm/mm from equation (17.53). Moment loading $M_{tot} = 894.38$ kNm. This gives pressure $p = 1.90$ MPa from equation (17.39). This gives $F_{v\,Shk} = 867.68$ N/mm and $F_{v1\,Shk} = 289.23$ N/mm. The relative displacement (δ_v) between the sleeve and pile from equation (17.31) is 0.70 mm, and the opening is derived as $\delta_h = 0.91$ mm from equation (17.29). The contribution to moment resistance from pressure is 212 kNm,

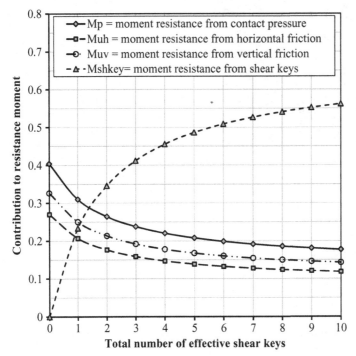

Figure 17.16. Contribution to resistance moment as a function of the number of effective shear keys.

moment resistance from pressure combined with friction coefficient in the tangential direction is 142 kNm, moment resistance from pressure combined with friction coefficient in the axial direction is 172 kNm, and the moment resistance from the shear keys is 368 kNm.

17.2.5 Design of Box Test Specimens

Most laboratory fatigue tests on grouted connections with shear keys reported in the literature were performed using specimens with diameters less than 400 mm (see, e.g., Ingebrigtsen et al. 1990). Whether these specimens are representative for the structural behavior of large-diameter grouted monopile connections that are subjected to a static axial load and a significant dynamic bending moment can be questioned. Therefore, efforts were made to design test specimens with representative radial stiffness similar to that of grouted connections of large-diameter piles in the order of 5 m or larger.

Laboratory testing of grouted connections demonstrated that compression struts develop in the grout between shear keys on the transition piece and the pile, as indicated in Figure 17.13a. As previously described, the struts are rather stiff compared with the radial stiffness of the pile and the transition piece. Therefore, the radial deflection is mainly governed by the thickness of the steel, the geometry of the connections in terms of radius, grout thickness, shear key geometry, and Young's modulus of the steel. From this information, box specimens could be designed that could simulate the structural behavior of large-diameter connections; see the photograph of a box specimen in Figure 17.17. This figure also shows a vertical and a horizontal section

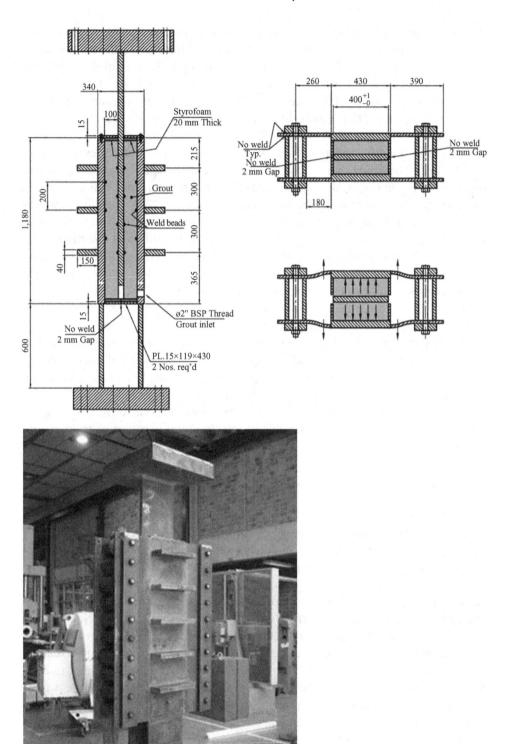

Figure 17.17. Box test specimen for simulation of a large-diameter connection.

through the grouted section. The box specimens were designed with a full-size grout thickness of 100 mm, full-size shear keys (height = 12 mm), and a distance between the shear keys similar to that used in full-size structures. Because of the symmetry, two volumes of grouted connections were tested, as shown in Figure 17.17. Four box specimens were designed, fabricated, and tested under reversed cyclic loading in a joint industry project. The stiffness of the box tests corresponds to that of cylindrical diameters of approximately 800 mm, 2,200 mm, and 5,000 mm.

The representative radial stiffness of the box specimens was achieved through the choice of governing dimensions (plate thickness and span length) of the plate segments between the grouted section and the bolt arrangements. This is shown in the horizontal section through the box specimen in Figure 17.17. The radial stiffness of the boxes was verified by measurements with hydraulic actuators inside the boxes before grouting. When the box specimens were loaded in the axial direction under reversed cyclic loading, the side plates of the boxes were subjected to "radial" contact pressure from the grout, and the behavior similar to that in large-diameter grouted connections was simulated, as illustrated in the lower part of the drawing in Figure 17.17. Note that there is no confinement in the circumferential direction in the box test specimens; this is of less significance due to a relatively low tensile capacity of the grout such that radial cracks in the meridian direction occur in cylindrical test specimens when they are subjected to large bending moments, as indicated in Figure 17.8. Laboratory testing of these box test specimens was considered successful. Using these test boxes it may also be possible to open up the sides (in the radial direction) after grouting such that degradation of the connections can be followed during testing to obtain a better understanding of the structural behavior of grouted connections with shear keys.

17.2.6 Comparison of Design Procedure with Test Data

The derived analytical procedure has been assessed against data obtained from testing the box test specimens in Section 17.2.5 and cylindrical test specimens subjected to bending moments. The calculated and measured relative displacements in the axial direction between the transition piece (outer sleeve) and the pile for the cylindrical test specimens are shown in Figure 17.18. Tests 1–6 were performed on cylindrical test specimens at the University of Leibniz, Hannover, Germany (Schaumann et al. 2010) and test 7 was performed at the DNV laboratory in Oslo, Norway. The cylindrical test specimens represent different grout capacities, different grout lengths, and different numbers of shear keys, as shown in Table 17.1. The measured relative axial displacement was derived from Linear Variable Transducers placed radially and horizontally at the end of the outer sleeve (or Transition Piece [TP]), as shown in Figure 17.15b. The amplitudes measured in test 7 were derived from one of the first load cycles. The plus and minus bending moments were approximately equal during the test. The measurements refer to the top and bottom of the specimen, which was positioned horizontally in the test laboratory, as these regions were subjected to the largest bending moments during testing. The correspondences between the calculated and measured displacements were better at the top than the bottom.

The calculated and measured openings between the transition piece (outer sleeve) and the pile for the cylindrical test specimens are shown in Figure 17.19. The

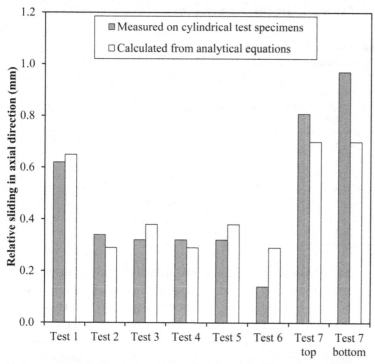

Figure 17.18. Relative sliding between sleeve (transition piece) and pile; comparison of measured results with those calculated based on the analytical approach.

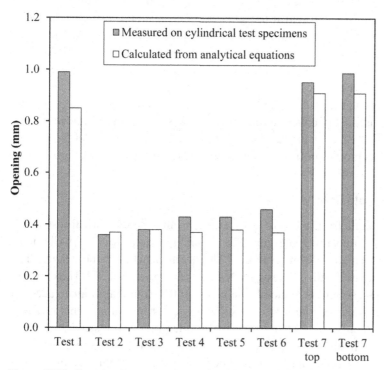

Figure 17.19. Opening between sleeve (transition piece) and pile; comparison of measured results with those calculated based on the analytical approach.

Table 17.1. *Parameters of the tested specimens*

Test no	Uniaxial compressive strength f_c (MPa)	Relative grout length $L/(2R_p)$	Number of shear keys
1	130	1.3	0
2	130	1.3	7
3	130	1.0	5
4	70	1.3	7
5	70	1.0	5
6	90	1.3	7
7	160	1.3	3

analytical approach presented in Section 17.2.4 provides displacements that generally agree closely with those measured. It should be realized that in laboratory testing on grouted connections some uncertainty in structural behavior and measurements should be expected due to other parameters than those accounted for in the analysis model, such as surface irregularity. Surface irregularity is defined as surface roughness and irregularity due to plate production deviations introduced during fabrication. In test 7 in Figure 17.18 the relative axial displacement obtained by the analytical approach is less than that measured. This may indicate that the spring stiffness derived in Section 17.2.4 may be on the high side, and this may provide forces on the shear keys that are on the high and safe side regarding design loads on the shear keys. However, this may also mean that if the shear keys represent a resistance capacity that is too large, the calculated contact pressure may be slightly too low. Therefore, in order to control the pressure, decreasing the spring stiffness by a factor of two was proposed for calculating the load effect related to the contact pressure in the design procedure for grouted connections in monopile structures. Test 7 was constructed with four shear keys on the pile side and three shear keys on the sleeve side, as shown in Figure 17.14. Thus, the effective number of shear keys for this specimen was three, as shown in Table 17.1. After fatigue testing this specimen, the derived number of cycles to failure was plotted into the fatigue capacity diagram shown in Figure 17.20.

17.2.7 Fatigue Tests Data

The fatigue test results from the box test specimens and the cylindrical test specimen tested at DNV are shown in Figure 17.20. Some test data from small-diameter fatigue testing performed by Wimpey in the 1980s are also included in the figure. The test results are compared with the design S-N curve from Section 17.1.4. The capacity curve for fatigue assessment is reduced by a factor of 0.8 from that of air to account for connections in water, as the fatigue tests were performed in air. This factor is based on the assessment of test data performed on concrete structures and the fatigue strength of concrete in air and water, as shown in Section 4.1.17. A somewhat similar S-N curve for grouted connections has been presented by Billington and Chetwood (2012), based on the assessment of older fatigue test data representative for air conditions. The presumptions for derivation of the curves differ somewhat with respect to environment and early-age cycling. Only very limited early-age cycling can be considered acceptable if the design S-N curve in Section 17.1.4 is considered

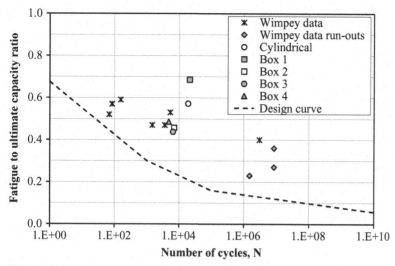

Figure 17.20. Fatigue test data for alternating loading.

valid. Early-age cycling is understood to refer to the relative movement in the grouted connection during early curing of the grout. The fatigue test data in Figure 17.20 were derived for a fully reversed load cycling; that is, the reversed load amplitude equaled the forward load amplitude. Sliding between the pile and the sleeve during testing was shown to degrade the connection by increasing cracking in the grout. Some deformation of the grout and the steel due to elasticity is needed to develop resistance against the external force by activating strut actions in the connections, as indicated in Figure 17.13. Thus, if the load changes direction then some sliding displacement cannot be avoided, as is also reflected by the relative displacements shown in Figure 17.18. The relative sliding during load cycling is insignificant provided that the net load does not change direction. Testing of load cycling with the net force in one direction only was reported by Ingebrigtsen et al. (1990), and for this loading the fatigue capacity was not significantly reduced below that of the ultimate capacity. This means that a large permanent axial load (vertical load in piles-to-sleeves in jackets and monopile connections) is beneficial with respect to fatigue capacity, as the probability of net load reversals becomes reduced. In DNV-OS-J101 (2014) for design of wind turbine structures this is now accounted for by a reduced load effect if load reversal, implying sliding in the connection, does not occur. If load reversals do not occur, the load amplitude can be replaced by half the load range before the S-N curve is entered for calculation of fatigue damage. This is exemplified further in the example that follows.

17.2.8 Illustration of Analysis for Long-Term Loads

The shear keys in wind turbine structures may be subjected to both a downward and a reversed upward load during dynamic loading. A Weibull long-term distribution for the force ranges from the dynamic behavior, as indicated in Figure 17.21a, may be considered for an initial fatigue assessment of a grouted connection. Adding the permanent load to the dynamic range for fatigue analysis, as shown in Figure 17.21a, is considered too conservative. However, if a reversed upward sliding occurs,

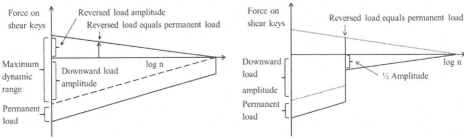

a). Long-term dynamic loading split into downward load and reversed load on shear keys

b). Long-term loading with the load amplitude half the range, where the reversed load is less than the permanent load

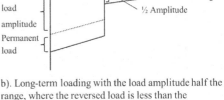

c). Resulting long-term downward loading on shear keys to be used in fatigue analysis

Figure 17.21. Derivation of long-term fatigue load distribution on shear keys from a Weibull distribution of long-term loading.

then the permanent load may also influence the fatigue capacity. Thus, including the permanent load where load reversals occur is recommended. The design S-N curve in Figure 17.20 is established for load reversals with force on shear keys for one environmental load direction (acting horizontal on the structure) at a time. Fatigue must be analyzed for force distribution both downward and upward. The fatigue capacity of grouted connections with shear keys is considered to be significantly larger than this S-N curve if load reversals do not occur as explained in Section 17.2.7. In a situation without load reversals, the S-N capacity curve is close to the ultimate load capacity, even with many load cycles. The transition in the long-term loading for load reversals that exceed the permanent load is indicated in Figure 17.21a. Due to the greater fatigue capacity without load reversals, the load amplitude for this condition is put as being equal to half the force range, as indicated in Figure 17.21b. This is performed instead of lifting the S-N curve for cyclic load in one direction only, resulting in the long-term force distribution in Figure 17.21c. As indicated in this figure, there is a step in the long-term distribution for force on shear keys. The left side represents situations where the load reversals are larger than the permanent load, while the right side represents cases where the load reversals are less than the permanent load. Figure 17.21 is mainly for illustrative purposes; the long-term force ranges in wind turbine structures are normally derived from time-domain simulation of the load response. Thus, using a more direct analysis approach with calculation of fatigue damage within each load cycle is recommended for the actual design of grouted connections in support structures for wind turbines.

18 Planning of In-Service Inspection for Fatigue Cracks

18.1 General

Degradation of offshore structures is mainly caused by corrosion and fatigue crack growth. The effects of corrosion are included in the design of these structures by incorporating a corrosion allowance or by use of a protection system by anodes and/or coating, enabling control of corrosion development also by possible replacement of anodes. In contrast, fatigue crack growth can be more critical, as cracks can result in a sudden rupture under conditions of large storm loads. Moreover, cracks are hard to detect because they are small for a significant part of crack growth time. Therefore the Fatigue Limit State is important for design of marine structures as well as during operation; see also Section I.4.

Defects much larger than those implicit in fatigue design curves are also of concern as some cracks found during inspection can be attributed to such defects. These defects are significantly larger than normal fabrication defects included in a probabilistic fatigue analysis, and are also sometimes denoted as gross errors, as explained in Section I.4. Therefore, the following safety principles should be implemented:

- Design for adequate fatigue life, including Design Fatigue Factors (DFFs) and a sound corrosion protection system.
- Design for robustness in relation to fatigue failure.
- Plan inspection of the structure during fabrication as well as during service life.

When inspection priorities are set, the potential for abnormal fabrication defects should also be considered. Since inspection after fabrication onshore can be performed much more cheaply and with higher reliability than during operation offshore, it is worthwhile emphasizing such inspections, particularly for components that are significant for the integrity of offshore structures.

As offshore structures possess different robustness with respect to fatigue cracking, and because inspection, repair, and failure costs vary significantly, different inspection strategies may be relevant for different types of structures.

Jackets with four or more legs show a larger reserve strength with X-type bracing than with K-type bracing which was frequently used in older jackets, but less so in new structures. The consequences of a fatigue crack will still depend on the position of the crack, type of loading such as amount of local bending stress over the

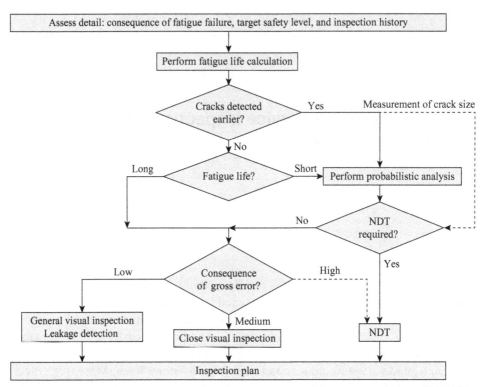

Figure 18.1. Schematic development of an inspection plan with respect to fatigue (DNVGL-RP-0001 2015).

plate thickness versus membrane stress at the hot spot, and the possibility of stress redistribution during crack growth. For most hot spots there is a significant period of crack growth before the integrity of the structure becomes a major concern. Flooded member detection can be used at these hot spots where potential fatigue cracks are expected to grow into air-filled members.

Crack control in semi-submersibles with slender braces is based on a basic fatigue design criterion and design for the accidental limit state (ALS), as well as leakage detection during operation. ALS is understood to mean that the structure shall be documented as damage tolerant in the ALS condition.

For floating production vessels (FPSOs) there is also significant residual strength with respect to fatigue cracks, and this normally means that it is possible to detect cracks using leak-before-break detection and by close visual inspection. However, it is difficult to document critical crack length based on existing assessment standards for similar crack lengths as sometimes have been observed in merchant ships. The reason for this is difficulty to estimate the critical crack length at welded connections when considering residual stresses, static loading and variable amplitude loading.

When planning inspection, it is important that the consequences of a potential fatigue crack at a considered hot spot are assessed. Different activities can be included to achieve an optimal inspection plan. Engineering assessment of the different methods, and the suitability of the methods for each hot spot should be evaluated. A schematic diagram illustrating the assessment and development of an inspection plan for a detail is shown in Figure 18.1. Cracks may have been detected during

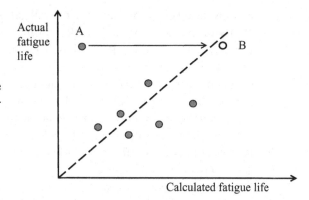

Figure 18.2. Example of calculated fatigue life versus actual fatigue life (DNVGL-RP-0001 2015).

former inspections, but have been assessed as not requiring a repair before another inspection is performed. If such cracks have been detected, this information should be used as the basis for planning another inspection.

All structural details should be evaluated in the development of an inspection plan. Consideration of each detail should be based on calculated fatigue lives and consequence of a potential fatigue failure. In addition, the probability of a gross error, related to load effect or capacity, should be kept in mind. The probability of such errors should also be considered when planning inspection for fatigue cracks. General visual inspection has traditionally been recommended for control of gross errors and may have a positive effect on reliability with respect to fatigue, even if the reliability of visual inspection is relatively low until the cracks have grown large. However, structural robustness is required in order to use this methodology.

Experience shows that gross errors are more likely to occur in new types of details than in details for which well-proven design and fabrication methodologies have been used. A new type of conductor support introduced in the 1990s is one example of this, where fatigue cracking occurred after a short period in service. For such errors, visual inspection and leakage detection is considered to be the most effective detection methodology. Probabilistic analyses are performed for selected details to take into account the best estimate of long-term loading and fatigue capacity and to account for uncertainties in design, fabrication and in-service inspection. The acceptance criteria are related to the consequences of failure. Based on this, it should be determined whether a detailed inspection by non-destructive testing (NDT) is required, or whether close visual inspection is sufficient for long calculated fatigue lives, as indicated in Figure 18.1. Due to the redundancy of the structures, it is assumed that NDT will only be required in special cases where the consequences of an error are large or catastrophic, as indicated by the dotted line in Figure 18.1 for details with long calculated fatigue lives.

If reliable fatigue analysis has been performed for design, this analysis may also be used for planning an in-service inspection. However, the focus during design often differs from that of inspection planning, where "consistent" calculated fatigue lives are preferred in order to achieve a reliable relative ranking of where fatigue cracks are most likely to occur. This is illustrated in Figure 18.2, where joint A could be selected for inspection as this joint show the shortest calculated fatigue life. However, if a more refined fatigue analysis was performed, the calculated fatigue life may be moved to B. This means that, in reality, this joint has a long fatigue life; the probability of a fatigue crack is small, and it would be expected that cracks would be detected earlier at other joints shown in Figure 18.2. Thus, in order to learn as much as possible from an in-service inspection of a hot spot in a structure, inspection points should be selected on the basis of fatigue analyses that are made for this purpose and that are consistent as far as possible (such that the ranking of the joint in terms of actual fatigue lives is correct). Realistic absolute values of calculated fatigue lives for different details are also required as a basis for probabilistic inspection planning. This is of most importance for new platforms for which the time until the first inspection is being planned.

Thus, in fatigue assessment of an existing structure, the best available data and information about its fatigue condition, as derived from the fatigue analysis described in appendices A, B, and C of DNVGL-RP-0001 (2015), which are made for purpose of inspection planning, should be used. However, it should also be realized that assessing all details by probabilistic analysis is impractical; for example, in a ship where longitudinals cross transverse frames with thousands of welded joints, it is not realistic to perform detailed inspections of all welds.

Due to the nature of fatigue and the number of uncertain parameters involved, when and where fatigue cracks will occur in a structure subjected to significant dynamic loading involves some uncertainty. The more information that is available, the more reliable are predictions of future behavior with respect to fatigue cracking. For example, if stress measurements have been performed over a sufficient period, these values, together with environmental data, may enable the uncertainty related to the long-term loading to be significantly decreased. When inspection of an as-built structure is performed and properly reported, it may be possible to include fabrication quality in the assessment. If the fabrication quality is low, the fatigue strength may be downgraded by, for example, one or more S-N curves; this requires skilled surveyors and engineers. Good documentation from as-is fabrication of local geometry, such as photographs, is recommended as this can contribute to improving engineering assessments. Better guidelines on how to relate the level of fabrication quality to S-N data would also be useful.

The methodology for planning an in-service inspection with respect to fatigue cracks in marine structures has gradually improved since the first estimated probability of detection (PoD) curve: a Bayesian updating approach was presented by Sletten et al. (1982) until a Recommended Practice on use of probabilistic methods for planning inspection for fatigue cracks was issued (DNVGL-RP-0001 2015). In this period, several research papers have been presented that have advanced the methodology, such as Itagaki et al. (1983), Madsen (1985), Madsen et al. (1987), Lotsberg and Kirkemo (1989), Lotsberg and Marley (1992), Moan et al. (1993), Cremona (1996), Vårdal and Moan (1997), Moan et al. (1997), Moan and Song (1998), Lotsberg et al. (1999), Moan (2005), Dalsgaard Sørensen and Ersdal (2008), Lotsberg and Sigurdsson (2014) and Lassen and Recho (2015).

18.2 Analysis Tools

When a crack is detected in a structure, the cause is investigated, and it is frequently concluded that the main reason is a poor designed fatigue detail or a proper fatigue analysis has not been performed. The reason for cracks can be inferior fabrication quality, or simply that it has not been analyzed properly. If cracks are detected, more information about the detail and the surrounding structure is obtained that can be used for calibration of analysis methodology and for further planning of inspection. Similar assessments can also be made for details that have been inspected but without finding any fatigue cracks, which may be explained by a long calculated fatigue life. However, hot spots with very long calculated fatigue lives would not normally be inspected. If the calculated fatigue life is short and fatigue cracks might be expected but are not detected, the reasons for not detecting a crack may include:

- a better detailed design, with a lower stress concentration for the considered detail than estimated;
- a higher S-N class than assumed;
- a lower dynamic loading than assumed at the design stage;
- need to wait for another inspection period in order to obtain more information about potential crack development.

Two different analysis tools can help the analyst assess the need for inspection:

- fracture mechanics, enabling establishment of crack growth curves;
- probabilistic analysis, including uncertainties in parameters used for calculation of fatigue crack growth and enabling the probability of detecting fatigue cracks by a specified inspection to be linked to that of estimating probability of a fatigue failure.

When planning an inspection, it is often useful to consider a deterministic crack growth development before starting with a probabilistic analysis approach. This is because a deterministic analysis approach is easier to understand and can be more easily related to actual physical behavior than probabilistic analysis.

Crack growth in a detail depends on the stress distribution through the thickness of the plate, as shown in Section 2.3.5. Furthermore, it depends on the possibility for redistribution of stress flow during crack growth. Typical crack growth developments are shown in Figure 18.3. The crack growth behavior can vary, depending on the structural detail and the type of loading. The crack growth curve in Figure 18.3a is representative for simple tubular joints with the possibility of redistribution of stress flow during crack growth. This means that there is a significant time interval from when the crack can be detected until failure of the joint. For joints with higher loading, the hot spot areas may need to be ground in order to achieve sufficient calculated fatigue life, as indicated in Figure 18.3b. This means that with a higher hot spot stress range, the crack will grow faster after initiation, and the period available for crack detection is reduced. This time interval is further decreased for elements that are mainly subjected to dynamic tensile stresses, like tethers of a tension-leg platform (TLP) as illustrated in Figure 18.3c. This development is also similar to that observed in the *Alexander L. Kielland* platform. Even higher dynamic stresses can occur in ground butt welds that also are used in tethers of TLPs, as indicated in Figure 18.3d. The development of the crack growth in Figure 18-d can be explained by a large membrane stress range as there after grinding is no longer a notch effect at the weld and the M_k factor as defined in Section 16.4 is equal 1.0 as compared with a M_k factor at a weld as shown in Figure 4.12 at a weld toe showing the stress increase at the

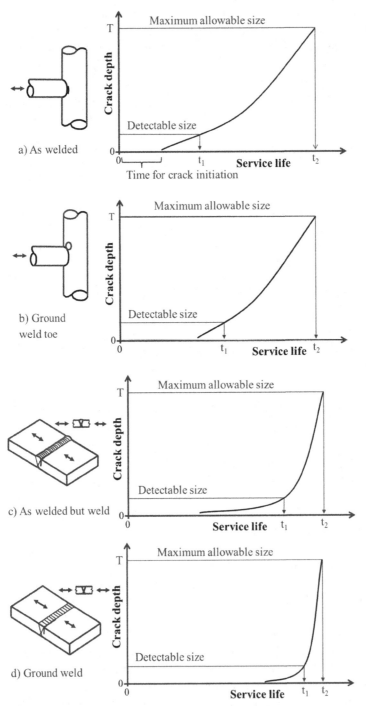

Figure 18.3. Schematic crack growth curves for various as-welded and ground details (Lotsberg 2008b) (reprinted with permission from Taylor & Francis Ltd., http://www.tandfonline.com).

notch where the stress at the crack tip is reduced with increasing crack depth. This is one reason for differences in crack growth curves. Other differences are due to stress distribution through the plate thickness at the hot spots, difference in possibility of redistribution of stress flow during crack growth and longer crack initiation time in ground details than in as-welded details. The different crack growth curves in Figure 18.3 show a wide variation regarding the growth period after which a crack becomes sufficiently large that it can be detected.

In order to include the effect of uncertainty of an inspection, a description of the probability of the presence of a crack at the hot spot is needed, together with a description of the uncertainty of the inspection method. Detection or not of a crack does not provide enough information for directly updating all parameters in a fatigue analysis, where several parameters are uncertain and contribute to the probability of a fatigue crack. However, the probability of not finding a crack can be formulated as a limit state function. This function can, in principle, be fulfilled by a number of combinations of the different parameters involved. The most likely of these combinations is the most relevant. The parameter values for this most likely combination can be determined on the basis of first-order reliability methods (FORM/SORM and may be verified by simulation; Madsen et al. 1986). The parameters for this combination can also be defined as "design values" of the parameters. These parameters need not necessarily be the physically correct values, but are considered to be the best estimates that an engineer can make on the basis of the available information.

18.3 Assessment of the Probability of Fatigue Failure

One of the most important objectives of a reliability evaluation of a structure is to identify potential failure modes and to define failure mechanisms in terms of mathematical models, that is, define the limit state function or safety margin. The failure criteria are modeled through limit state functions, defined so that the occurrence of failure is represented through a negative limit state function value. There are considerable uncertainties associated with fatigue of offshore structures, but these may be reduced over the service life through monitoring and in-service inspections. The additional information obtained through these actions will allow more confident estimates to be made, both with respect to the present state of the structure and on the expected future behavior. In this context, additional information refers to, for example, data such as the outcome from NDT indicating the status with respect to fatigue damage accumulation and potential fatigue cracks in the structure. Procedures to account for this additional information in determining the reliability of the structure are presented in Sections 18.5, 18.6, and 18.9.

An example of the development of a crack, from an initial defect distribution to a critical crack size, is indicated schematically in Figure 18.4. Crack growth is shown on the positive vertical axis and the time for crack growth is shown on the horizontal axis. The development of the calculated probability of fatigue failure is shown in the downward direction of the vertical axis. At time t_1 the initial defect distribution has probably grown wider due to uncertainties in the crack growth parameters and crack-driving stress range. It is assumed that an inspection is performed at time t_1, and in this example it is assumed that fatigue cracks are not found. The inspection is

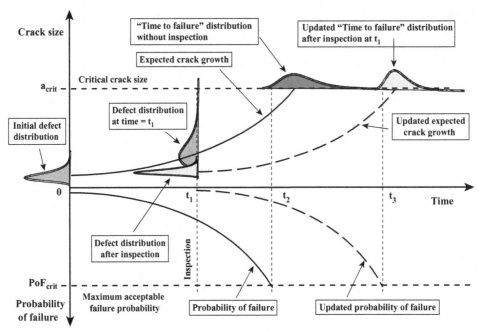

Figure 18.4. Schematic illustration of crack growth and probability of fatigue failure before and after inspection (DNVGL-RP-0001 2015).

assumed to be associated with a significant probability of detecting large cracks. Thus, provided that cracks are not detected, it is likely that large cracks are not present at the considered hot spot. Due to the actual PoD curve, a narrower defect distribution after inspection is indicated, as shown in the figure; if present, it is likely that the largest defects would have been detected. This also means that the probability of fatigue failure in the immediate future can be considered to be lower than was assumed before the inspection. This is also indicated by an updated probability of the failure curve in the figure. If inspection at t_1 was not performed, then the reliability of the detail would be unacceptable at time t_2. Now, after inspection at time t_1, the reliability of the detail with respect to fatigue is considered to be acceptable until time t_3. At this time a new inspection will be required to fulfill the acceptance criterion (maximum acceptable failure probability).

18.4 Implementation of Monitoring Data

Measurements of platform response, which might become available over the service life, will provide more knowledge on the structural behavior and thereby reduce the predicted uncertainties related to the loading. This will increase the confidence in the evaluation of the structural integrity. Monitoring results can be applied directly in the assessment of actual fatigue damage accumulation and in the prediction of degradation rate by improving estimates of the long-term loading, as well as by decreasing uncertainty related to the long-term loading. This additional information can be applied, together with data from inspections of the structure, for further assessment of the structural integrity.

18.5 Inspection Planning and Inspection Program

In inspection planning, the inspection has not yet been made, and therefore the outcome of the planned inspection is unknown and cannot be accounted for in the calculation of the probability of failure. Inspection planning does not influence the estimated future failure probabilities unless it is associated with a specific action, such as, for example, that all detected cracks will be repaired, or that all detected cracks above a critical predefined acceptance level will be repaired. This means that, in practice, only the first inspection can be planned initially. The inspection results must then be assessed before the next inspection interval is planned.

18.6 Reliability Updating

The difference between reliability or inspection updating and inspection planning should be emphasized. Inspection updating is using information that has been obtained through inspections, while inspection planning is preparing the future inspection program and how different possible outcomes from these inspections should be assessed, such as repair philosophy. Reliability updating is based on utilization of information that becomes available at discrete time intervals over the service life when the structure is inspected for fatigue cracks. This information is applied both in the assessment of the present condition and in the prediction of future behavior. Reliability updating determines the calculated updated future failure probability of the structure based on additional information after an inspection. When inspection for fatigue cracks has been carried out, three different levels of information may become available:

- *No detection*: potential fatigue cracks are smaller than the detection ability of the inspection equipment used.
- *Detection*: fatigue cracks are observed.
- *Detection with sizing measurement*: fatigue cracks are observed and their sizes measured.

In addition, there is the possibility of false identifications, but this is not considered further here. In order to illustrate this information, the following events are defined:

In a similar way to the limit state event, as presented by equation (12.1) where M below zero implies failure, a detection event can be formulated as:

$$H(t_i) = a_d - d(t_i) \tag{18.1}$$

where:

a_d is the detectable size of a crack

$d(t_i)$ defines the level of damage accumulation at the time of inspection, t_i.

The detectable size of a crack is obtained directly from the PoD curve for the inspection equipment used. If H is larger than zero, then the crack is smaller than the detection ability of the inspection tool, and a crack will not be detected. If H is less than zero, the crack is larger than the smallest detectable crack.

The sizing of a crack is formulated as:

$$D(t_i) = a_m - d(t_i) \tag{18.2}$$

where:

a_m is the measured crack size, which may be associated with uncertainty.

The sizing is zero as the crack size at time t_i is measured to be a_m.

The probability of failure occurring at time t prior to an inspection is then:

$$P_F(t) = P(M(t) \leq 0) = P(a_f - a(t) \leq 0) \tag{18.3}$$

where

a_f = crack size at fracture
$a(t)$ = crack size at time t.

The calculated probability of failure occurring at time t following an inspection at time t_1 in which cracks were not detected can be expressed as:

$$P_F(t) = P(M(t) \leq 0 | H(t_1) > 0) \tag{18.4}$$

For N multiple inspections, without any crack detection, the following formulation applies:

$$P_F(t) = P(M(t) \leq 0 | H(t_1) > 0 \cap H(t_2) > 0 \cap \ldots \cap H(t_N) > 0) \tag{18.5}$$

The calculated probability of failure at time t after an inspection at time t_1, with crack detection but without any sizing measurement is then:

$$P_F(t) = P(M(t) \leq 0 | H(t_1) \leq 0) \tag{18.6}$$

The additional information from the inspection is included in the probability formulation through conditioning, implying that the estimated failure probability is conditioned by the outcomes observed during inspections. The more information that is available to include in the modeling of the failure probability, the more accurately can the integrity of the structure be assessed.

18.7 Description of Probabilistic Fatigue Analysis Models

Probabilistic fatigue analysis can be based on standard S-N data and use of the Palmgren-Miner rule, together with a long-term stress range distribution as presented in Section 12.1.1. The limit state function applied in the probabilistic analysis is expressed as:

$$g(D(t), \Delta) = \Delta - D(t) \tag{18.7}$$

where:

Δ is a random variable describing the general uncertainty associated with the fatigue damage accumulation (this is the accumulated fatigue damage at failure which in deterministic fatigue design is equal 1.0).
$D(t)$ is the accumulated fatigue damage at time t as calculated by equation (3.40).

Defining v_0 as the mean number of stress cycles per unit time over the service life (up-crossing frequency), the total accumulated fatigue damage in a service period, t, can be expressed as:

$$D(t) = t \cdot v_0 \cdot D_{cycle} \tag{18.8}$$

where:

D_{cycle} is the expected fatigue damage per stress cycle, depending on the local stress range response process and the associated S-N curve.

Applying a bilinear S-N curve as presented in Section 4.1 and assuming the stress range distribution to have a Weibull distribution, the expected damage per stress cycle is calculated as:

$$D_{cycle} = \frac{1}{a_1} q^{m_1} \Gamma \left(1 + \frac{m_1}{h} ; \left(\frac{s_1}{q}\right)^h\right) + \frac{1}{a_2} q^{m_2} \gamma \left(1 + \frac{m_2}{h} ; \left(\frac{s_1}{q}\right)^h\right) \qquad (18.9)$$

where:

q and h are distribution parameters in the Weibull distribution
$\gamma(;)$ and $\Gamma(;)$ are the Incomplete and Complementary Incomplete Gamma functions
s_1 is the transition from one part of the S-N curve to the other part.

Reference is made to Section 10.2 for closed form equations, and also to Table 4.1 for constants in the two-sloped S-N curve for the air environment and to Table 4.4 for seawater with cathodic protection. Mean S-N curves are used here, that is, log a_1 and log a_2 represent the mean S-N curve cutoff with the log N axis.

The fatigue damage for floating offshore structures is derived from the weighted sum of the accumulated fatigue damage within each short-term stationary condition (sea state), for which the stress process is assumed to be stationary Gaussian and narrow-banded. This assumption implies that the stress ranges fit a Rayleigh distribution. The expected damage per stress cycle within each short-term condition, j, is then defined from the above expression with the shape parameter $h = 2$ and the scale parameter $q = 2\sqrt{2}\sigma_j$, where σ_j is the standard deviation of the stress response process in short-term condition j.

18.8 Description of Probabilistic Crack Growth Analysis

Probabilistic crack growth analysis can be based on the same equations that are used for deterministic crack growth analysis, as presented in Section 16.2. The variables in the differential equation (16.6) for non-threshold crack growth models can be separated and integrated to give:

$$\int_{a_i}^{a(t)} \frac{da}{Y^m \left(\sqrt{\pi a}\right)^m} = C \sum_{i=1}^{N(t-t_i)} (\Delta \sigma_i)^m \qquad (18.10)$$

where:

$a(t)$ is the crack depth at time t
$N(t - t_i)$ is the total number of stress cycles in the time period (t_i, t)

C and m are crack growth parameters, as defined in Section 16.11. The sum in equation (18.10) can be estimated by:

$$D(a_N) = C N(t - t_i) E[\Delta \sigma^m] \qquad (18.11)$$

where:

D(a_N) is an indicator of the damage accumulated by the growth of a crack from its initial value, a_i, to a crack size, a_N, after N stress cycles.

For fatigue crack growth models including thresholds, the fatigue damage indicator can be expressed as:

$$D(a_N) = \int_{a_i}^{a_N} \frac{da}{G(a)\, Y^m \left(\sqrt{\pi a}\right)^m} \qquad (18.12)$$

where:

G(a) is a reduction factor in the range 0–1, depending on the threshold level, ΔK_{th}, and the stress range process, $\Delta\sigma$ (Madsen et al. 1987).

18.9 Formulation of Reliability Updating

The effect of inspection on the fatigue reliability of structures depends on the detection ability of the particular NDT method used. Detection ability as a function of a defect size (crack depth, a, or crack length, 2c) is defined by a PoD curve (see Section 11.6). Regardless of whether cracks are detected or not, each inspection provides additional information to that which was available at the design stage, and this can be used to update the reliability. Reliability or inspection updating is based on the definition of conditional probability:

$$P(F|I) = \frac{P(F \cap I)}{P(I)} \qquad (18.13)$$

where:

P(F|I) is the probability that event F occurs given that event I occurs, for example an inspection result.

An inspection result is either "no detection" or "detection of a crack" as explained in Section 18.6:

$$\begin{aligned} a(t_i) &\leq A_{di} \\ a(t_j) &= A_j \end{aligned} \qquad (18.14)$$

In the first case, no cracks were found in the inspection after time t_i, implying that any cracks were smaller than the smallest detectable crack size, A_{di}. A_{di} is a random variable, since a detectable crack is only detected with a certain probability. The distribution function for A_{di} is equal to the PoD function. When more inspections are performed, the random variables A_{di} are mutually independent.

In the second case, a crack size, A_j, is observed after time t_j. A_j is also a random variable due to possible measurement errors and/or uncertainties in the interpretation of the crack measurements. For each inspection that does not result in crack detection, an event margin, M_i, can be defined that is similar to the event margin used

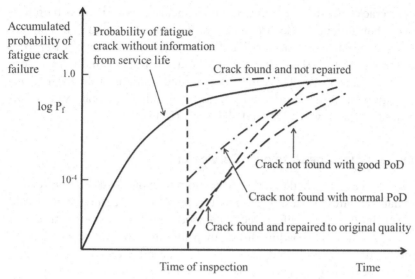

Figure 18.5. Schematic illustration of calculated accumulated probability of fatigue failure depending on inspection findings and repair (DNVGL-RP-0001 2015).

to describe fatigue failure. The event margins for a one-dimensional crack growth model may be formulated as:

$$M_i = \int_{a_0}^{A_{di}} \frac{1}{Y(x)^m \left(\sqrt{\pi x}\right)^m} dx - C N(t_i - t_{ini}) E[\Delta\sigma^m] \qquad (18.15)$$

These event margins are positive. An event margin for each measurement resulting in detection of a crack can similarly be defined as:

$$M_j = \int_{a_0}^{A_j} \frac{1}{Y(x)^m \left(\sqrt{\pi x}\right)^m} dx - C N(t_j - t_{ini}) E[\Delta\sigma^m] \qquad (18.16)$$

These event margins are zero.

The failure probability in the time period after an inspection is derived by applying probabilistic conditioning to the inspection outcome. This is illustrated schematically in Figure 18.5. More knowledge about the structural behavior and calculated accumulated fatigue damage is accumulated through inspection. If cracks are not detected, this indicates that the structure is acceptable for another time period, the length of which depends on the calculated fatigue life, the shape of the crack growth curve (amount of local bending at the hot spot and possibility for redistribution of stresses during crack growth), and the reliability of the inspection method as indicated in the figure. This is because the actual situation is better than what was predicted prior to inspection. The improved situation may be due to less long-term loading, lower hot spot stress, better fabrication than expected when compared with the S-N data used, less damage accumulation than predicted by the Palmgren-Miner rule, or a combination of these effects.

The calculated failure probability from a crack that is detected and not repaired is also indicated in Figure 18.5. The development depends on the calculated fatigue life and further crack growth development. An engineering assessment is therefore recommended when fatigue cracks are detected. If cracks are found and repaired, then the curve for the calculated accumulated probability of failure may differ from that indicated in Figure 18.5. If a repair is carried out, the repair will often be more reliable than the original detail. This can be due to modifications of the local geometry or weld improvement, for example by weld toe grinding. Several examples of inspection planning are shown in DNVGL-RP-0001 (2015).

18.10 Change in Damage Rate over Service Life

When fatigue damages are added together from more than one analysis, transferring the different damage rates into a single timeline, representing the same damage rate as of today over the full service life, is recommended. This is relevant for structures that are subjected to different loading during their lifetimes such as jackets that are subsiding and floating structures that are operating in different environments where a number of different analyses may be needed for calculation of fatigue damage, This is done to take the effect of the inspection properly into account. A detailed procedure for this is provided in DNVGL-RP-0001 (2015).

18.11 Effect of Correlation

18.11.1 General

Correlation is understood to mean similarity in parameters involved in a fatigue analysis for different hot spots. For example, a full correlation in hot spot stress for two similar details means that the probability of a specific stress range occurring in one detail is the same as for the same stress range occurring in the other detail. Another example of correlation is similarity in fabrication of one detail as compared with another similar detail when the same welding procedure is used by the same welder. Textbooks in statistics may be consulted for a mathematical description of correlation, see for example, Madsen et al. (1986) or Naess and Moan (2013). The effect of correlation is also presented by Moan and Song (1998).

The correlation coefficient between two variables, X and Y, is defined by:

$$\rho_{xy} = \frac{E\left[(x - \mu_x)(y - \mu_y)\right]}{\sigma_x \sigma_y} \tag{18.17}$$

where μ = mean value and σ = standard deviation.

If $\rho_{xy} = 0.0$, then the variables X and Y are uncorrelated; if $\rho_{xy} = 1.0$, the variables X and Y are fully correlated. Two different components, C_x and C_y, where X and Y are input variables in the limit state for components C_x and C_y, respectively, are considered. In cases where uncertainties in X and Y are important for predicting the failure probabilities, C_x and C_y, and the two components are correlated, then inspection of component C_x may enable an update in the prediction of failure probability of component C_y (and vice versa).

Based on today's knowledge, proper inclusion of correlation in a standard for probabilistic methods for planning inspection for fatigue cracks is considered difficult. The purpose of this section is to remind the reader that correlation may have a significant effect if there are similar hot spots (geometry, fabrication, and loading) in the structure, and that engineering judgment should be used for assessment if analysis indicates a requirement for extensive inspection of these hot spots. The effect of correlation should also be remembered if fatigue cracks are detected (see also Section 18.12).

Fatigue crack growth limit states for two different locations are now considered. If the correlation between failure events at the two different locations is significant, a no-find inspection result for one location may update the prediction of failure of the detail at the other location sufficiently that a planned inspection can be postponed for this detail. Conversely, if a crack is found at one location, it may be recommended that inspection of other locations is conducted earlier than initially planned. The extent of this effect depends significantly on the magnitude of the correlation, the predicted failure probabilities, the quality of the inspections and the findings, and, not least, the shape of the fatigue crack growth curve.

18.11.2 Example of Analysis Where Correlation Is Included in Assessment of an FPSO

The effect of correlation in parameters was analyzed with respect to inspection for fatigue cracks at doubling plates in a FPSO (Lotsberg et al. 1999). The following assumptions were made for the analyses:

- The stress range was assumed to be fully correlated between different components. This is a reasonable assumption for the actual areas considered in the deck and partly also in the bottom of an FPSO. A long-term stress range distribution was established based on the assumption that the FPSO is fully loaded 50% of the time, and is in ballast for the other 50%.
- The welded plates were assumed to be from the same batch and it was assumed that the same welding procedure was used for the considered components. Therefore, the crack growth parameter was assumed to be fully correlated. Although this might not be a conservative assumption, it was used here to study the effect of correlation.
- A conservative assumption was that the initial defect distributions were not correlated.

Analysis of Welds That Are Not Ground

This work demonstrated that the effect of correlation is significantly larger for welds that are not ground than for those that have been ground to achieve a minimum required fatigue life. Inspection of more components leads to longer time intervals between inspections. This result is an effect of correlation that is related to the load effect and to the crack growth parameters. The necessary time between inspections increases significantly when several components, which are equal in geometry and with the same calculated fatigue life, are inspected. The effect of correlation also gives some information on calculated fatigue life that does not involve any inspection requirements. If there are several hot spots of similar geometry and loading, and with a calculated fatigue life equal to L_{min}, inspection of hot spots with a calculated fatigue life of $3 \times L_{min}$ should not be necessary, provided that those components with a fatigue life of L_{min} are inspected regularly. This information may be used as a screening criterion for selection of hot spots to be analyzed in greater detail by probabilistic methods.

Analysis of Ground Welds at Doubling Plates and Transverse Stiffeners in Deck Structures

The effect of correlation was analyzed with respect to inspection for fatigue cracks at doubling plates that are ground during fabrication; the same considerations are also valid for transverse stiffeners below the deck plating. This work demonstrated that if cracks are not found during inspection of a hot spot, inspection of neighboring hot spots could be postponed by only approximately six months. This relatively small effect is due to the long initiation time for fatigue cracks in the model. Because initiation is not correlated, the effect of inspection of one hot spot will have only a marginal influence on the reliability of other ground hot spots.

18.12 Effect of Inspection Findings

If cracks are detected during an inspection, this may mean that it is likely that there are also fatigue cracks or crack indications in other areas of the structure that are subjected to a similar hot spot stress range. The probability of this may be assessed by probabilistic methods. However, in most cases, each fatigue crack is found to be unique when its actual basis is investigated. Therefore, it is important that the reasons for a crack are explored in detail before further actions are taken.

The effect of correlation with respect to long-term fatigue loading and fatigue capacity at different hot spots is discussed in DNVGL-RP-0001 (2015). The effect of correlation is considered to be most significant for details with a long period of crack growth. In order to consider correlation with respect to the possibility of fatigue cracking at two different hot spots, the details considered should be of similar or equal geometry, fabricated in the same way, and the long-term loading should be approximately the same for the two hot spots. If the load effects differ by around 10–20%, the correlation effect is substantially reduced even if the geometry is the same. The effect of correlation also depends on the probability of detection with the inspection method used. The amount of inspection may be reduced if there are many similar details that are subjected to a similar long-term stress range loading, in which case it may be sufficient to select a limited number of details for inspection. The effect of correlation can lead to reduced inspection requirements, provided that fatigue cracks are not detected. However, if cracks are detected, then the amount of inspection must be increased.

18.13 Residual Strength of the Structure or System Effects with a Fatigue Crack Present

Fatigue crack growth through the thickness of a joint or connection does not necessarily mean that the considered structure is close to collapse. This is because jacket structures, FPSOs, and semi-submersibles are rather redundant structures, partly due to the requirements imposed in design related to the Accidental Limit State in Standards, such as NORSOK N-001 (2012). Residual capacity with a crack present in the structures is considered in Section 12.6. Section 12.6.3 provides some simplified guidance with respect to system failures. This is important for assessment of a target safety level; simplified procedures will normally be sufficient for documentation of a safety level. It is likely that a limit on the number of fatigue cracks in the structures will govern the assessment of a safety level more than an advanced assessment of system reliability. In this respect it is also important to select connections for inspection such that a possible progressive failure path, with progressive failure of several joints such that a failure mechanism is established, is avoided.

18.14 Inspection for Fatigue Cracks during In-Service Life

18.14.1 General

NDT is commonly used to localize and size defects in structures. The detection ability of the NDT method is defined as a function of a defect size through PoD curves (see also Section 11.6). A number of references to relevant literature is provided by Visser (2002) and BS 7910 (2013). PoD curves are provided for the following inspection methods:

- Flooded Member Detection (FMD)
- Eddy Current (EC)
- Magnetic Particle Inspection (MPI)
- Alternating Current Field Measurement (ACFM).

Although General Visual Inspection (GVI) and Close Visual Inspection (CVI) are efficient for general assessment of the condition of structures, they are not suitable for detecting fatigue cracks before the crack size has grown long or through the plate thickness. Thorough removal of marine growth is crucial for detecting fatigue cracks, and close photographs of cleaned hot spot areas may provide useful information about potential fatigue cracks.

18.14.2 Magnetic Particle Inspection Underwater

In underwater applications the magnetic field is usually produced by using current-carrying coils. The alternative is a magnetic yoke, using either a permanent magnet or a coil; the fluorescent particles can be made visible underwater using ultraviolet light. For surface applications, use of ordinary non-fluorescent light is more common.

18.14.3 Eddy Current

See Section 11.6.9.

18.14.4 Flooded Member Detection

Flooded member detection (FMD) is a technique that is being rapidly introduced by many North Sea operators of offshore steel platforms. The method employs a yoke with a transmitter and a receiver on either side of a tubular member. The signal received is compared with the signal calculated for empty and water-filled tubular members of the same diameter and wall thickness. The FMD method (for partial) flooding can be based on Ultrasonic Testing (UT) or Radiographic Testing (RT) techniques. RT is used in combination with an ROV (Remote Operated Vehicle) because of the potential radiation hazard, whereas UT can be used manually. The main problem with FMD is that not all through-thickness cracks result in water filling a tubular section; it may be already filled with water or the pressure may be too low for the water to flow. Nevertheless, through-thickness defects have been found that could have been missed by other methods.

FMD is used for inspection of through-thickness cracks in, for example, braces in jacket structures, semi-submersibles, and FPSOs. FMD can be used for members that are not water-filled from installation, like braces (with potential fatigue cracks on the brace side and not on the leg side that is normally water-filled) or joints that have not been reinforced with grout. FMD is considered to be of high reliability and a probability of detection of 0.95 may be assumed.

When FMD is used, it should be established whether through-thickness cracks at hot spots can be accepted on the basis of the capacity required for ultimate load. Experience shows that FMD is efficient for conductor frames in jacket structures, where out-of-plane moments contribute significantly to the calculated fatigue damage. The capacity for ultimate load is of less concern here than for the main load-carrying braces.

18.14.5 Leakage Detection

Leakage detection is a reliable barrier with respect to fatigue crack detection in semi-submersibles and FPSOs. It is assumed that this method can only be relied on in redundant structures, where the plated structures are made of material with appropriate fracture toughness. The fracture toughness can be derived from SENT (Single Edge Notched Tensile Test) type CTOD (Crack Tip Opening Displacement) specimens for assessment of capacity of through-thickness cracks in plated structures. SENT data normally provide larger fracture toughness than SENB (Single Edge Notched Bend Test) specimens, which are CTOD specimens tested in bending (see also Section 16.6). A critical crack length can then be determined from BS 7910 (2013) by neglecting the presence of residual stress as cracks often propagate into the base material. If it is likely that a potential crack will follow the weld toe, a relevant residual stress should be included in the analysis. When relying on leakage detection, sufficient time from a significant probability of detecting a fatigue crack until failure should be verified so that repair can be performed for documentation of structural integrity.

18.14.6 Acoustic Emission

Acoustic emission (AE) is a well-known method by which crack growth can be measured. However, in the field of crack detection and sizing of cracks, methods based on AE are of an ad hoc nature only, as, in order to generate AE at a measurable level, crack growth rate has to exceed a minimum level. Thus, AE is primarily used for monitoring a known crack or defect, but is not suitable for defect detection after component fabrication.

18.14.7 Inspection Methods for Jackets

The following inspection methods may be proposed for jacket structures:

- GVI by ROV; however, this is used more for assessment of the general structural condition than for detection of fatigue cracks.
- FMD of members, where the use of this technique is considered efficient
- Cleaning and CVI
- EC may be used for connections that are considered significant for the integrity of the structure. However, operators prefer to use inspection methods where divers are not needed.

18.14.8 Inspection Methods for Floating Structures

The following NDT inspection methods may be proposed for an FPSO:

- Internal details that are accessible: EC or equivalent
- External details above Mean Water Line (MWL): EC or equivalent
- External details below MWL
- UT from inside, or, alternatively, Time of flight or Phased Array
- MPI by diver from outside if possible
- Inaccessible details: examine for through-thickness cracks on accessible side (leakage detection). It should be ensured that a through-thickness crack can be repaired before it becomes critical with respect to global structural integrity.

18.15 Effect of Measurements of Action Effects

Measurements of action effects can significantly reduce the inherent uncertainty. It is preferable that measurements of the environmental conditions are performed at the same time as measurement of the action effects. If not, a longer measurement interval will be required (several years) in order to obtain reliable data that can be used to update the long-term action effects at a hot spot.

APPENDIX A

Examples of Fatigue Analysis

A.1 Example of Fatigue Design of a Pin Support for a Bridge between a Flare Platform and a Larger Jacket Structure

A bridge is constructed between a large jacket and a flare platform. The bridge is fixed connected to the jacket and it is sliding at horizontal supports at the flare platform. The fixed connection is constructed as a vertical tubular column that is welded into the deck structure as shown in sketch in Figure A.1.

The weight of the bridge is approximately 400 tonnes, (= P). It is assumed that the weight of the bridge is uniform along its length. The design life of the structure is specified as 30 years. Furthermore a Design Fatigue Factor equal 3.0 is prescribed.

The fatigue capacity of the pin support on the jacket is questioned and the maximum allowable friction coefficient, μ, at the sliding support at the flare platform is asked for.

The zero up-crossing period due to the wind response on the flare tower has been estimated to 2.3 sec. For fatigue design of the jacket supporting the flare tower a period of approximately 6.3 sec has been estimated due to wave action. Based on this, a conservative estimate of the number of cycles to be used for fatigue assessment is derived as $60 \cdot 60 \cdot 24 \cdot 365 \cdot 30 \cdot 3/2.3 = 1.23 \cdot 10^9$ cycles.

For simplicity, it is assumed that the load to be transferred through the bridge from one end to the other is so large that sliding at the sliding support will occur for all load cycles (this is a conservative assumption). This horizontal force is

$$H = \frac{P}{2}\mu \qquad (A.1)$$

The force range is $2H = P\mu$.

The equations needed for calculation of nominal stresses in the pin are for moment of inertia and elastic section modulus given as

$$I = \frac{\pi}{64}\left(D^4 - (D - 2t)^4\right)$$

$$W_e = \frac{2I}{D} \qquad (A.2)$$

where D = diameter and t = thickness of the tubular pin. Inserting values for the diameter and thickness of the pin in Figure A.1 gives: $W_e = 6.54 \cdot 10^6$ mm^3.

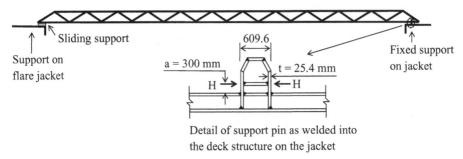

Detail of support pin as welded into
the deck structure on the jacket

Figure A.1. Bridge between two jacket structures.

From Section 6.7, a stress concentration SCF = 1.5 is derived at the weld between the pin and the upper deck plating (on the outside of the pin), which is the hot spot in this example; see equations (6.36), (6.35), and (6.40). Furthermore, it is assumed that there is a vertical weld along the pin between the two horizontal plates shown in the sketch. Thus, the welded connection is so long that it will be classified belonging to S-N curve F3 according to detail 3 in table A.7 in DNVGL-RP-C203.

Then the local stress at the hot spot can be calculated as

$$\Delta\sigma_{local} = \frac{\Delta M}{W_e} = \frac{P\mu a SCF}{W_e}$$
$$= \frac{400 \cdot 10^3 \cdot 9.81 \cdot \mu \cdot 300 \cdot 1.5}{6.54 \cdot 10^6} = 270\mu \qquad (A.3)$$

A way to assess this is to say that all stress cycles should be smaller than the fatigue limit at 10^7 cycles. The fatigue limit for the F3 curve is derived as 32.75 MPa from Table 4.1. This is reduced by the Design Fatigue Factor (DFF) as described in Section 12.7. See also Section 3.2.2 on how to use a DFF relative to a fatigue limit to derive an allowable stress range for an infinite number of cycles. Thus the fatigue limit is reduced to S = $32.75/DFF^{1/3.0}$ = 22.71 MPa. Then the maximum friction coefficient is calculated saying that the local stress range should not exceed the fatigue limit. This gives a friction coefficient $\mu = 22.71/270 = 0.084$. An alternative is also to use the second part of the F3 S-N curve with negative inverse slope equal m = 5.0 for a more conservative estimate of the friction coefficient. Then

$$\log(1.23 \cdot 10^9) = 14.576 - 5.0 \log(270\mu) \qquad (A.4)$$

From this equation the friction coefficient $\mu = 0.046$. This would require a sliding support with a low friction coefficient such as Teflon against stainless steel plates. Alternatively, the diameter of the support pin should be increased. It might be added that many bridges between platforms in the North Sea are designed with supports where steel is sliding against steel, and this implies a significantly larger friction coefficient.

A.2 Fatigue Design of Ship Side Plates

The required thickness t of side plates in an FPSO is asked for, see Figure A.2. The distance between side longitudinals s = 800 mm. The plates are subjected to a pressure range equal $\Delta p = 85.3 kN/m^2$ from Classification Note 30.7 (2010). This

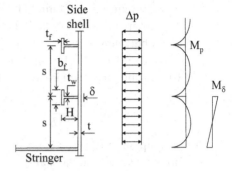

Figure A.2. Notations used for fatigue assessment of side shell in an FPSO.

pressure range is presented at 10^{-4} probability level. Then equation (10.35) is used to transfer the pressure range to 10^{-8} probability level as the allowable fatigue strength in Section 10.6 is related to this probability level.

$$\Delta\sigma_{10^{-4}} = \Delta\sigma_{10^{-8}} \left(\frac{\log 10^4}{\log 10^8}\right)^{1/h} = \Delta\sigma_{10^{-8}}\, 0.5^{1/h} \tag{A.5}$$

The bending moment in the plate at the longitudinal in Figure A.2 is

$$\Delta M_p = \frac{\Delta p s^2}{12} \tag{A.6}$$

And with section modulus $W_e = t^2/6$ along a unit plate length the stress range is derived as

$$\Delta\sigma = \frac{\Delta M_p}{W_e} = \frac{\Delta p s^2 6}{12 t^2} = \frac{\Delta p s^2}{2 t^2} \tag{A.7}$$

The weld between the side plate and the web in the longitudinal stiffener is classified as E due to its web thickness according to appendix A of DNVGL-RP-C203 (2016). The Weibull shape parameter is assumed equal 1.0. It might be questioned what the relevant environment here is as there is partly air and partly seawater in the ballast tanks. The fatigue strength for an E-detail in seawater with cathodic protection is used further and $\Delta\sigma_{allowable} = 215.3\,\text{MPa}$ is derived from table 5.3 in DNVGL-RP-C203 (2016). From equations (A.5) and (A.7), the required plate thickness is derived as

$$t = \left(\frac{(\Delta p/0.5^{1/h})s^2}{2\Delta\sigma_{allowable}}\right)^{0.5} = \left(\frac{(85.3/0.5^{1/1.0}) \cdot 800^2}{1000 \cdot 2 \cdot 215.3}\right)^{0.5} = 16.0\,mm \tag{A.8}$$

The stringer is considered to be rather stiff and the longitudinal stiffener above the stringer is somewhat flexible when subjected to transverse loading. Therefore, the effect of an additional moment in the plate at the stringer M_δ is considered as indicated in Figure A.2. First the deflection in the center of the longitudinal between transverse frames due to Δp is calculated. It is assumed that a region between typical transverse frames is considered such that a static model with fixed boundaries at the ends of the longitudinal stiffener can be assumed. The following geometry for the longitudinal stiffener is used:

$b_f = 100$ mm; $t_f = 19$ mm; $H = 400$ mm; $t_w = 11$ mm; the effective plate width is derived from DNV CN30.7 (2014) as $s_e = 0.67\, s = 536$ mm. The length of the stiffeners between transverse frames is $l = 3200$ mm. The effective section modulus is derived as: $I = 3.45 \cdot 10^8$ mm^4. Then the deflection range due to the pressure range at 10^{-8} probability level is calculated is

$$\Delta\delta = \frac{\Delta psl^4}{384EI} = \frac{(85.3/0.5^{1/1.0}) \cdot 800 \cdot 3200^4}{1000 \cdot 384 \cdot 2.1 \cdot 10^5 \cdot 3.45 \cdot 10^8} = 0.51\, mm \qquad (A.9)$$

The equation for the bending stress in the plate due to a displacement δ reads (see book on structural mechanics)

$$\Delta M_\delta = \frac{6EI_p}{s^2}\delta \qquad (A.10)$$

and with $I_p = t^3/12$ per unit length the additional stress range is derived as

$$\Delta\sigma_\delta = \frac{\Delta M_\delta}{W_e} = \frac{6Et^3 6}{s^2 12t^2}\delta = \frac{3Et}{s^2}\delta = \frac{3 \cdot 2.1 \cdot 10^5 \cdot 16}{800^2}0.51 = 8.03\, MPa \quad (A.11)$$

Thus if the maximum transverse pressure range is at a stringer, one may recommend to increase the plate thickness to 17 mm. This is a slightly larger thickness than in the shuttle tanker with fatigue crack shown in Section I.2.3.

Following the same principles, one may also perform a fatigue assessment of the longitudinal stiffeners. Here one should keep in mind that the transverse frames are more flexible than the bulkheads when calculating effective moments. Also many longitudinals are made as bulb sections or angles. Then it is important to consider the skew bending as explained in Section 8.12.2. More guidance on this can be found in, for example, Classification Note CN 30.7 (2014).

The considered longitudinal stiffener is symmetric with a section modulus equal $W = 1153428$ mm^3. The moment range at 10^{-4} probability level is derived as

$$\Delta M_\delta = \frac{\Delta pl^2}{12} = \frac{85.3 \cdot 3200^2 \cdot 800}{12 \cdot 1000} = 58231467\, Nmm \qquad (A.12)$$

And the stress range at 10^{-8} probability level is derived as

$$\Delta\sigma = \frac{\Delta M_\delta}{W_e} = \frac{58231457/0.5^{1/1.0}}{1153428} = 100.97\, N/mm^2 \qquad (A.13)$$

This means that detail no 4 from Section 2.4 can be used even with a bulb section with an additional stress concentration factor around 1.4 as the maximum allowable stress range is approximately 191 MPa for an F-detail in seawater with cathodic protection according to DNVGL-RP-C203 (2016) for 20 years of service life, and Palmgren-Miner damage equal 1.0.

A.3 Fatigue and Unstable Fracture of a Chain

A.3.1 Problem Definition

Reference is made to the fatigue and fracture of the mooring chain in Section I.2.2 where it was reported that a mooring chain used to anchor an FPSO failed in a large storm after approximately eight years of in-service life. The reason for the failure was considered to be a loose stud. There are several unknowns related to this chain failure:

- The load at failure is not well known.
- The stress concentration at a loose stud is not known and it may be difficult to estimate a stress concentration factor as it may be a nonlinear function of the loading and the stud opening.
- The material fracture toughness was not measured.

Thus, an exact reassessment of this failure cannot be made. However, it may be used as an exercise to illustrate how fracture can be calculated using Failure Assessment Diagrams (FADs) and fatigue analysis using S-N curves for chains and crack growth analysis based on fracture mechanics. To get some idea about the loading in the chain at failure, the 100-year load on a similar K4 chain used to anchor an FPSO in the North Sea is given as $F = 3038$ kN. The diameter $D = 84$ mm. This gives total area $A = \pi \cdot (D/2)^2 \cdot 2 = \pi \cdot (84/2)^2 \cdot 2 = 11083$ mm^2. Thus, $\sigma_{100} = F/A = 3038 \cdot 10^3/11083 = 274$ MPa. If one assume that a "large storm" corresponds to the largest expected storm within one year, the tensile stress in the chain at failure can be estimated using equation (10.35) together with an assumption of a Weibull shape parameter $h = 1.0$:

$$\Delta\sigma_1 = \Delta\sigma_{100}\left(\frac{\log n_1}{\log n_0}\right)^{1/h} = 274\left(\frac{\log(5 \cdot 10^6)}{\log(5 \cdot 10^8)}\right)^{1/1.0} = 274 \cdot 0.77 = 211 MPa \quad (A.14)$$

A.3.2 Assessment of Unstable Fracture Using Failure Assessment Diagram

Reference in the following is made to Section 16.8 where use of Failure Assessment Diagrams was explained. The failure assessment diagram in Figure 16.13 is made for the purpose of reassessment analysis of this chain with yield strength $\sigma_y = 580$ MPa and ultimate strength $\sigma_u = 860$ MPa. The amount of bending at the failed section is uncertain for calculation of resulting stress intensity factor. However, at plastic collapse the bending moment will disappear as it is not needed for equilibrium. The area of the semi-elliptical crack in Figure I.3 is calculated with $a = 19$ mm and $c = 22$ mm as $A_c = (\pi/2) \cdot ((a+c)/2)^2 = (\pi/2) \cdot ((19+22)/2)^2 = 660$ mm^2. Thus the effective area is reduced by a factor $R_c = (A/2 - A_c)/A/2) = (11083/2 - 660)/(11083/2) = 0.88$. Then it becomes rather simple to calculate the load ratio L_r:

$$L_r = \frac{\sigma_1}{\sigma_y R_c} = \frac{211}{580 \cdot 0.88} = 0.41 \quad (A.15)$$

As the chain failed, the design point is lying on the failure curve in Figure 16.13 and $K_r = 0.959 = f(0.41)$ from equation (16.38). The large K_r value (and low L_r ratio) indicates that the structural behavior is rather brittle. The fracture is considered to be due to external loading as the chain is made with heat treatment such that significant residual tensile stresses are not expected to be present at the cracked area. Thus, only the elastic stress intensity factor needs to be considered for derivation of K_{mat} from equation (16.40).

The stress intensity factor for an edge crack subjected to axial load can be calculated from equation (B.5) for an edge crack with $a/w = 19/84 = 0.226$:

$$K_{Ia} = \sigma_{1a}\sqrt{\pi a}\left(1.12 - 0.23\frac{a}{W} + 10.6\frac{a^2}{W^2} - 21.7\frac{a^3}{W^3} + 30.4\frac{a^4}{W^4}\right)$$

$$= \sigma_{1a}\sqrt{\pi \cdot 19}\left(1.12 - 0.23 \cdot 0.226 + 10.6 \cdot 0.226^2 - 21.7 \cdot 0.226^3 + 30.4 \cdot 0.226^4\right)$$

$$= 11.11\sigma_{1a} \tag{A.16}$$

The stress intensity factor for an edge crack subjected to bending load can be calculated from equation (B.6) for an edge crack with $a/w = 19/84 = 0.226$:

$$K_{Ib} = \sigma_{1b}\sqrt{\pi a}\left(1.12 - 1.39\frac{a}{W} + 7.3\frac{a^2}{W^2} - 13\frac{a^3}{W^3} + 14\frac{a^4}{W^4}\right)$$

$$= \sigma_{1b}\sqrt{\pi \cdot 19}\left(1.12 - 1.39 \cdot 0.226 + 7.3 \cdot 0.226^2 - 13 \cdot 0.226^3 + 14 \cdot 0.226^4\right)$$

$$= 8.23\sigma_{1b} \tag{A.17}$$

Based on the photo of the cracked area, it is seen that the crack front is not straight, and based on a semi-elliptical crack, the half axis ratio $a/c = 19/22 = 0.86$. Then from equation (B.2):

$$E_2 = \sqrt{1 + 1.464\left(\frac{a}{c}\right)^{1.65}} = \sqrt{1 + 1.464(0.86)^{1.65}} = 1.46 \tag{A.18}$$

Now it is a problem that one does not know how large the bending moment over the chain section has been at the loose stud. If the stud was completely loose, one might assess the stress condition to be similar to that of a studless chain. Vargas et al. (2004) calculated SCF at the straight section of a larger studless chain equal SCF = 3.39. If one uses this SCF for calculation of stress intensity factor, the bending moment $\sigma_{1b} = (SCF - 1)\sigma_1$. This equation can be deduced from definition of a stress concentration factor. Then one can use the superposition principle to calculate the elastic stress intensity factor:

$$\begin{aligned} K_I &= K_{Ia} + K_{Ib} \\ &= (11.11 + 8.23(SCF - 1))\sigma_1/E_2 \\ &= (11.11 + 8.23 \cdot (3.39 - 1)) \cdot 211/1.46 = 4448\,Nmm^{-3/2} \end{aligned} \tag{A.19}$$

From equation (16.40): $K_{mat} = K_I/K_r = 4448/0.959 = 4638\,\mathrm{Nmm}^{-3/2}$.

A material fracture toughness $\delta_{mat} = 0.10 - 0.20$ mm may be considered relevant for this type of chain. Therefore, equation (16.42) is used for calculation of δ_{mat}. First equation (16.43) is used to calculate the factor m:

$$m = 1.517\left(\frac{\sigma_y}{\sigma_u}\right)^{-0.3188} = 1.517\left(\frac{580}{860}\right)^{-0.3188} = 1.72 \tag{A.20}$$

Then

$$\delta_{mat} = \frac{K_{mat}^2(1 - \nu^2)}{m\sigma_y E} = \frac{4638^2(1 - 0.3^2)}{1.72 \cdot 580 \cdot 2.1 \cdot 10^5} = 0.093\,mm \tag{A.21}$$

This indicates a material fracture toughness for the failed chain that is to the low side, as typical values for more ductile material is values of δ_{mat} larger than 0.25 mm. It might also be added that the crack shape in Figure I.3 is close to half a circular, which is typical crack development for an axial stress range loading. If the bending over the thickness were less than assumed here, an even lower value of δ_{mat} would have been calculated.

A.3.3 Fatigue Assessment of the Chain Based on S-N Data

A fatigue assessment of the chain is performed based on S-N data listed in Table 4.7. The fatigue life calculations are based on equation (10.18). Now the mean values of the S-N curves are used for analysis. For studless chain, $\log a = \log a_d + 2$st. dev. $= 10.778 + 2 \cdot 0.20 = 11.178$. For stud chain, $\log a = 11.079 + 2 \cdot 0.20 = 11.479$. First a Weibull distribution of stress ranges with a shape parameter $h = 1.0$ is assumed. The 100-year stress range can now be used for calculation of fatigue damage together with stress cycles during 100 years: $\Delta\sigma_0 = 274$ MPa (from section A.3.1) and $n_0 = 5 \cdot 10^8$ cycles are conservatively assumed. This gives a calculated fatigue damage during 100 years for stud chain from equation (10.18) as:

$$D = \frac{n_0}{a} \frac{\Delta\sigma_0^m}{(\ln n_0)^{m/h}} \Gamma\left(1 + \frac{m}{h}\right)$$

$$= \frac{5 \cdot 10^8}{10^{11.479}} \frac{274^{3.0}}{(\ln 5 \cdot 10^8)^{3.0/1.0}} \Gamma\left(1 + \frac{3.0}{1.0}\right) \qquad \text{(A.22)}$$

$$= 25.49$$

Thus, the calculated fatigue life is equal $100/25.49 = 3.92$ years. Here it has been assumed that the chain has been subjected to all load cycles during 8 years. However, considered actual wave directionality around the FPSO it may be relevant to reduce the number of effective stress cycles to approximately half this value. This would result in a calculated fatigue life close to that observed.

Instead of using the 100-year loading as above, one may use the 1-year loading with corresponding number of cycles for calculation of fatigue damage (to get the same result). This gives calculated fatigue damage corresponding to 1-year service life:

$$D = \frac{n_0}{a} \frac{\Delta\sigma_0^m}{(\ln n_0)^{m/h}} \Gamma\left(1 + \frac{m}{h}\right)$$

$$= \frac{5 \cdot 10^6}{10^{11.479}} \frac{211^{3.0}}{(\ln 5 \cdot 10^6)^{3.0/1.0}} \Gamma\left(1 + \frac{3.0}{1.0}\right) \qquad \text{(A.23)}$$

$$= 0.255$$

Thus, the calculated fatigue life equals $1/0.255 = 3.92$ years.

In this calculation $\Gamma(1+3) = 6.0$ is used. This may also be derived from the methodology defined in Table 10.2 (Gamma $(1 + m/h)$). The calculated fatigue lives are sensitive to the Weibull shape parameter. Therefore, a calculation with $h = 1.1$ is also presented. $\Gamma(1 + 3/1.1) = 4.306$. Then a fatigue damage $D = 41.42$ for all cycles during 100 years is calculated for the stud chain, and this corresponds to a fatigue life of $100/41.42 = 2.4$ years and 4.8 years is derived for half the load cycles for this long-term response. Some uncertainty in calculated fatigue lives is expected, and therefore redundancy and use of Design Fatigue Factors larger than 1.0 is recommended in addition to use of design S-N curves as explained in different sections of this book.

A.3.4 Fatigue of the Chain Assessed by Fracture Mechanics

For purpose of fracture mechanics analysis, it is convenient to transfer the long-term stress range loading into constant stress cycles in equation (10.20). From equation (10.21) it is noted that $\Delta\sigma_{eq}$ is a function of the scale parameter q, the shape parameter h and m only. Thus it is independent of the number of cycles. For a shape parameter of $h = 1.0$ the following equivalent stress range is derived from equation (10.20):

$$\Delta\sigma_{eq} = \frac{\Delta\sigma_0}{(\ln n_0)^{1/h}} \sqrt[m]{\Gamma\left(1 + \frac{m}{h}\right)}$$

$$= \frac{274}{(\ln 5 \cdot 10^8)^{1/1.0}} \sqrt[3]{\Gamma\left(1 + \frac{3}{1.0}\right)}$$

$$= 24.86 \, MPa \tag{A.24}$$

For a Weibull shape parameter of $h = 1.1$ the equivalent stress range is increased to $\Delta\sigma_{eq} = 28.91$ MPa.

The geometry function for an edge crack subjected to axial stress

$$f\left(\frac{a}{w}\right) = 1.12 - 0.23\frac{a}{W} + 10.6\frac{a^2}{W^2} - 21.7\frac{a^3}{W^3} + 30.4\frac{a^4}{W^4} \tag{A.25}$$

is not significantly larger than 1.12 while the crack is small and it is a maximum of 1.44 at the end of crack growth. The geometry function for bending is decreasing with crack growth. The footprint of the stud has introduced a notch that indicates that one should have included an M_k function for a small crack growth. Thus the resulting geometry function can be derived as

$$Y\left(\frac{a}{w}\right) = f\left(\frac{a}{w}\right) M_k\left(a/w\right) / E_2\left(a/c\right) \tag{A.26}$$

One might use an M_k for a weld toe; however, without more knowledge, it is associated with significant uncertainty. For further illustration of crack growth it is assumed that the Y is a constant equal to 0.80 due to a rather large E_2 value from the shape of the crack. (This corresponds to values for $f(a/w) = 1.17$; $M_k(a/w) = 1.0$ and $E_2(a/c) = 1.46$.) Then from equation (16.12) by assuming crack growth from an initial defect depth equal to 0.1 mm to a crack depth at fracture equal to 19 mm, the following number of cycles are derived with the SCF at the crack region as an unknown:

$$N = \frac{a_i^{1-m/2} - a_f^{1-m/2}}{C\pi^{m/2}\Delta\sigma_{eq}^m SCF^m Y^m \left(\frac{m}{2} - 1\right)} \tag{A.27}$$

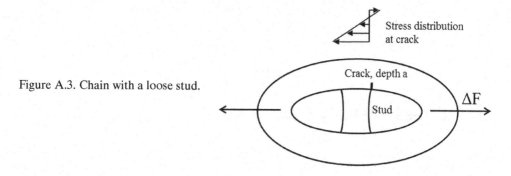

Figure A.3. Chain with a loose stud.

And by solving for the stress concentration factor and putting the number of cycles N to correspond to half the cycles in 8 years in service gives:

$$SCF = \sqrt[m]{\frac{a_i^{1-m/2} - a_f^{1-m/2}}{C\pi^{m/2}\Delta\sigma_{eq}Y^m\left(\frac{m}{2} - 1\right)N}}$$

$$= \sqrt[3.0]{\frac{0.10^{1-3.0/2} - 19^{1-3.0/2}}{8.35 \cdot 10^{-13}\pi^{3.0/2} \cdot 24.86^{3.0} \cdot 0.80^{3.0} \cdot \left(\frac{3.0}{2} - 1\right) \cdot 8 \cdot 5 \cdot 10^6 \cdot \frac{1}{2}}} \qquad (A.28)$$

$$= 2.0$$

Here it is assumed an initial crack depth equal to 0.1 mm and a mean value of C in corrosive environment as chains are subjected to free corrosion in seawater. These values are also uncertain parameters. However, the result is interesting as it may be considered to be in line with the observed crack shape as indicated above. The fatigue analysis based on both S-N and fracture mechanics seems to be in line with the observed crack growth. The low calculated SCF value, however, indicates that the bending over thickness has been less than that in a studless chain, and that the material fracture toughness has been lower than that estimated in Section A.3.2.

This example shows that there is a methodology available for assessment of unstable fracture at the end of a period of fatigue crack growth. This assessment could have been improved by a detailed finite element analysis of the chain. However, some uncertainty about the actual stress concentration factor might still remain due to uncertainty about the gap at the loose stud and possible contact with the remaining part of the chain at higher tensile forces. Also it would have been useful for a more accurate assessment to get more direct information about the fracture toughness δ_{mat} with a few CTOD tests of the chain material.

APPENDIX B

Stress Intensity Factors

Equations for stress intensity factors can be found in a number of handbooks such as Rooke and Cartwright (1976) and Tada et al. (1973) and other literature for a number of different geometries.

Elliptical cracks:

For elliptical and semi-elliptical edge crack under uniform remote normal stress, K_I refers to the point on the crack front where its value is greatest. This is the symmetry point on the minor axis. The formulas were derived for a « W, but may be used within some 10% error for all a ≤ 0.3W according to Hellan (1984).

Elliptical internal crack:

$$K_I = \frac{1}{E_2}\sigma_\infty\sqrt{\pi a} \tag{B.1}$$

where

E_2 = complete elliptic integral of the second kind with argument $\sqrt{1 - a^2/c^2}$
a = smallest half axis of the ellipse in the thickness direction as shown in Figure 16.11
c = longest half axis in ellipse
σ_∞ = far field stress normal to the crack.

The following approximation can be used for E_2:

$$E_2 = \left\{1 + 1.464(a/c)^{1.65}\right\}^{0.5} \tag{B.2}$$

This function is shown in Figure B.1.

Semi-elliptical surface crack:

$$K_I = \frac{1.12}{E_2}\sigma_\infty\sqrt{\pi a} \tag{B.3}$$

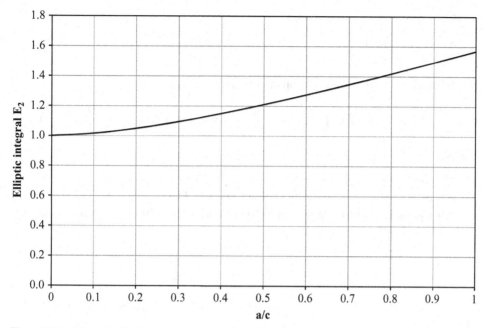

Figure B.1. Value of elliptic integral as function of crack curvature.

For through thickness center crack in a plate subjected to remote uniform stress for all a/W under symmetry:

$$K_I = \sigma_\infty \sqrt{\pi a} \; \frac{1 - \dfrac{a}{2} + 0.326\dfrac{a^2}{W^2}}{\sqrt{1 - \dfrac{a}{W}}} \qquad (B.4)$$

where

  a = half crack length
  W = half width of plate
  σ_∞ = far field stress normal to the crack.

For through thickness edge crack in a plate subjected to remote uniform axial stress for a/W < 0.7:

$$K_I = \sigma_\infty \sqrt{\pi a} \left(1.12 - 0.23\frac{a}{W} + 10.6\frac{a^2}{W^2} - 21.7\frac{a^3}{W^3} + 30.4\frac{a^4}{W^4} \right) \qquad (B.5)$$

where

  a = crack length
  W = width of plate.

For through thickness edge crack in a plate subjected to remote bending stress for a/W < 0.7:

$$K_I = \sigma_\infty \sqrt{\pi a} \left(1.12 - 1.39\frac{a}{W} + 7.3\frac{a^2}{W^2} - 13\frac{a^3}{W^3} + 14\frac{a^4}{W^4} \right) \qquad (B.6)$$

where

    a = crack length

    W = width of plate.

Radial crack around cylinders subjected to axial stress:

$$K_I = \sigma_\infty \sqrt{\pi a} \left(\frac{D}{d} + \frac{1}{2} + \frac{3d}{8D} - 0.36\frac{d^2}{D^2} + 0.73\frac{d^3}{D^3} \right) \frac{1}{2}\sqrt{\frac{D}{d}} \qquad (B.7)$$

where

    D = diameter of cylinder

    a = depth of crack around the cylinder and d = D−2a.

    For the same cylinder subjected to torsion moment T the stress intensity factor in mode III reads:

$$K_{III} = \frac{16T}{\pi D^3} \sqrt{\pi a} \left(\frac{D^2}{d^2} + \frac{1}{2}\frac{D}{d} + \frac{3}{8} + \frac{5}{16}\frac{d}{D} + \frac{35}{128}\frac{d^2}{D^2} + 0.21\frac{d^3}{D^3} \right) \frac{3}{8}\sqrt{\frac{D}{d}} \quad (B.8)$$

The following geometry function is proposed for calculation of stress intensity factor in general:

$$Y_m = M f_w M_m \qquad (B.9)$$

where

    M = bulging factor used for cracks in curved structures like shell structures; BS7910 (2013). For plated structures M = 1.0. The equations provided below apply to flat plates with semi-elliptical cracks. The finite-width correction function f_w is given by:

$$f_w = \left[\sec\left(\frac{\pi c}{W}\sqrt{\frac{a}{T}} \right) \right]^{1/2} \quad \text{for } (2c/W) \leq 0.8$$

where

    a = smallest half axis in ellipse

    c = longest half axis in ellipse

    W = width of plate.

Normally the cracks are small as compared with the global geometry in structures where the forces are transferred such that $f_w = 1.0$ in real plated structures. Thus, this factor is most relevant for assessment of smaller laboratory fatigue tests.

Geometry Function for Semi-Elliptical Surface Cracks, Y_m, for Membrane Loading

The stress intensity magnification factor for semi-elliptical cracks loaded by membrane stress is equal to (BS 7910 2013):

$$Y_m = \left[M_1 + M_2\left(\frac{a}{T}\right)^2 + M_3\left(\frac{a}{T}\right)^4 \right] g f_\theta / \Phi \qquad (B.10)$$

where:

$$M_1 = 1.13 - 0.09\,(a/c) \quad \text{for } 0 \le a/2c \le 0.5$$

$$M_1 = \sqrt{(c/a)}\,[1 + 0.04(c/a)] \quad \text{for } 0.5 < a/2c \le 1.0$$

$$M_2 = \frac{0.89}{0.2 + (a/c)} - 0.54 \quad \text{for } 0 \le a/2c \le 0.5$$

$$M_2 = 0.2(c/a)^4 \quad \text{for } 0.5 < a/2c \le 1.0$$

$$M_3 = 0.5 - \frac{1.0}{0.65 + (a/c)} + 14[1.0 - (a/c)]^{24} \quad \text{for } a/2c \le 0.5$$

$$M_3 = -0.11(c/a)^4 \quad \text{for } 0.5 < a/2c \le 1.0$$

$$g = 1 + \left[0.1 + 0.35(a/T)^2\right](1 - \sin\theta)^2 \quad \text{for } a/2c \le 0.5$$

$$g = 1 + \left[0.1 + 0.35\,(c/a)\,(a/T)^2\right](1 - \sin\theta)^2 \quad \text{for } 0.5 < a/2c \le 1.0$$

$$f_\theta = [(a/c)^2\cos^2\theta + \sin^2\theta]^{1/4} \quad \text{for } 0 \le a/2c \le 0.5$$

$$f_\theta = [(c/a)^2\sin^2\theta + \cos^2\theta]^{1/4} \quad \text{for } 0.5 < a/2c \le 1.0$$

The complete elliptic integral of the second kind Φ is given by:

$$\Phi = \sqrt{1 + 1.464\left(\frac{a}{c}\right)^{1.65}} \quad \text{for } 0 \le a/2c \le 0.5$$

$$\Phi = \sqrt{1 + 1.464\left(\frac{c}{a}\right)^{1.65}} \quad \text{for } 0.5 < a/2c \le 1.0$$

The definitions of a, c, and θ are shown in Figure 16.11. The following simplifications can be made:

At the deepest point on the crack front:

$$g = 1.0$$
$$f_\theta = 1.0 \qquad \text{for } 0 \le a/2c \le 0.5$$
$$f_\theta = (c/a)^{0.5} \quad \text{for } 0.5 < a/2c \le 1.0$$

At the ends of the crack, $\theta = 0$, so that:

$$g = 1.1 + 0.35(a/T)^2 \qquad \text{for } 0 \le a/2c \le 0.5$$
$$g = 1.1 + 0.35\,(c/a)\,(a/T)^2 \quad \text{for } 0.5 < a/2c \le 1.0$$
$$f_\theta = (a/c)^{0.5} \qquad \text{for } 0 \le a/2c \le 0.5$$
$$f_\theta = 1.0 \qquad \text{for } 0.5 < a/2c \le 1.0$$

If $a/2c > 1.0$ use solution for $a/2c = 1.0$.

Geometry Function for Semi-Elliptical Surface Cracks, Y_b, for Bending Loading

$$Y_b = HY_m \tag{B.11}$$

where

Y_m is calculated from equation (B.10).

$$H = H_1 + (H_2 - H_1)\sin^q\theta$$

where

$$q = 0.2 + (a/c) + 0.6\,(a/T)) \quad \text{for } 0 \le a/2c \le 0.5$$
$$q = 0.2 + (c/a) + 0.6\,(a/T)) \quad \text{for } 0.5 < a/2c \le 1.0$$
$$H_1 = 1 - 0.34\,(a/T) - 0.11\,(a/c)\,(a/T) \quad \text{for } 0 \le a/2c \le 0.5$$
$$H_1 = 1 - \{0.04 + 0.41(c/a)\}(a/T) + \{0.55 - 1.93(c/a)^{0.75} + 1.38(c/a)^{1.5}\}(a/T)^2$$
$$\text{for } 0.5 < a/2c \le 1.0$$

$$H_2 = 1 + G_1(a/T) + G_2(a/T)^2$$

where

$$G_1 = -1.22 - 0.12\,(a/c) \quad \text{for } 0 \le a/2c \le 0.5$$
$$G_1 = -2.11 + 0.77\,(c/a) \quad \text{for } 0.5 < a/2c \le 1.0$$
$$G_2 = 0.55 - 1.05(a/c)^{0.75} + 0.47(a/c)^{1.5} \quad \text{for } 0 \le a/2c \le 0.5$$
$$G_2 = 0.55 - 0.72(c/a)^{0.75} + 0.14(c/a)^{1.5} \quad \text{for } 0.5 < a/2c \le 1.0$$

The following simplifications can be made:

At the deepest point on the crack front, $\theta = \pi/2$ so that $H = H_2$ and:

$$g = 1.0$$
$$f_\theta = 1.0 \qquad \text{for } 0 \le a/2c \le 0.5$$
$$f_\theta = (c/a)^{0.5} \quad \text{for } 0.5 < a/2c \le 1.0$$

At the ends of the crack, $\theta = 0$, so that:

$$g = 1.1 + 0.35(a/T)^2 \qquad \text{for } 0 \le a/2c \le 0.5$$
$$g = 1.1 + 0.35\,(c/a)\,(a/T)^2 \quad \text{for } 0.5 < a/2c \le 1.0$$
$$f_\theta = (a/c)^{0.5} \qquad\qquad\quad \text{for } 0 \le a/2c \le 0.5$$
$$f_\theta = 1.0 \qquad\qquad\qquad\quad \text{for } 0.5 < a/2c \le 1.0$$

and

$$H = H_1$$

If $a/2c > 1.0$, use solution for $a/2c = 1.0$.

References

Aamodt, B., P. G. Bergan and H. F. Klem (1973). Calculation of Stress Intensity Factors and Fatigue Crack Propagation of Semi-Elliptical Part-Through Surface Cracks. 2nd Int. Conf. on Pressure Vessel Technology, part 2, 911–921, San Antonio, USA.

ABAQUS/Standard Users Manual Vol. 1 Ver. 6.4, 2003. Hibbit, Karlson & Sorensen Inc.

Abramowitz, M. and I. Stegun (1970). *Handbook of Mathematical Functions with Formulas, Graphs, and Mathematical Tables*. Dower Publications, Inc., New York.

Almar-Naess, A. (1985). *Fatigue Handbook: Offshore Steel Structures*. Tapir Publishers, Trondheim, Norway.

Anderson, T. L. (2005). *Fracture Mechanics: Fundamentals and Applications*. Third edition. Taylor & Francis Group, New York.

Andrews, R. M. (1996). The Effect of Misalignment on the Fatigue Strength of Welded Cruciform Joints. *Fatigue Fracture Engineering of Material Structures* 19 (6), 755–768.

Ang, A. H-S. and W. H. Tang (1975). *Probability Concepts in Engineering Planning and Design, Volume I – Basic Principles*. John Wiley and Sons, New York.

API RP2A (1977). Recommended practice for Planning, Designing and Constructing Fixed Offshore Platforms, Ninth Edition. Issued by the American Petroleum Institute, Dallas, TX.

API RP2A (1984). Recommended practice for Planning, Designing and Constructing Fixed Offshore Platforms, Fifteenth Edition. Issued by the American Petroleum Institute, Dallas, TX.

API RP2A-WSD (2014). Recommended Practice for planning, designing and Constructing Fixed Offshore Platforms – Working Stress Design, Twenty-second edition, November 2014. Issued by the American Petroleum Institute, Dallas, TX.

API 579-1/ASME FFS-1 (2007). Fitness-For-Service. ISBN 0-7918-3102-7.

ASTM E837-01e1 (2001). Standard Test Method for Determining Residual Stresses by the Hole-Drilling Stain-Gauge Method.

ASME Boiler and Pressure Vessel Code (BPVC), Section VIII, Division 1: Rules for Construction of Pressure Vessels, 2013.

ASME B31.3-2014 Process Piping. ASME Code for Pressure Piping, B31, ASME.

ASTM A153/A153M – 09 Standard Specification for Zinc Coating (Hot-Dip) on Iron and Steel Hardware. 2009.

ASTM A193/A193M-14a Standard Specification for Alloy-Steel and Stainless Steel Bolting for High Temperature or High Pressure Service and Other Special Purpose Applications. 2014.

ASTM A194/A194M-14a Standard Specification for Carbon and Alloy Steel Nuts for Bolts for High Pressure or High Temperature Service, or Both. 2014.

ASTM A320 / A320M – 2015 Standard Specification for Alloy-Steel and Stainless Steel Bolting for Low-Temperature Service. 2015.

ASTM A453 / A453M – 12 Standard Specification for High-Temperature Bolting, with Expansion Coefficients Comparable to Austenitic Stainless Steels.

ASTM E739-91 (1998). Standard Practice for Statistical Analysis of Linear or Linearized stress-Life (S-N) and Strain-Life (e-N) Fatigue Data. ASTM International, West Conshohocken, PA, 1998, www.astm.org.

Ayala-Uraga, W. and T. Moan (2007). Fatigue Reliability-Based Assessment of Welded Joints Applying Consistent Fracture Mechanics Formulations. *International Journal of Fatigue* 29, 444–456.

Aygül, M., M. Al-Emrani, and S. Urushadze (2012). Modeling and Fatigue Life Assessment of Orthotropic Bridge Deck Details Using FEM. *International Journal of Fatigue* 40, 129–142.

Aygül, M., M. Bokesjø, M. Heshmati, and M. Al-Emrani (2013). A Comparative Study on Different Fatigue Failure Assessments of Welded Bridge Details. *International Journal of Fatigue* 49, 62–72.

AWS D1.1/D1.1M (2010). Structural Welding Code – Steel, 22nd Edition. American National Standards Institute.

Baptista, C., L. Borges, S. Yadav, and A. Nussbaumer (2015). Fatigue Behaviour and Detailing of Slotted Tubular Connection. ISTS15 Rio. 15th International Symposium on Tubular Structures, Rio de Janeiro, Brazil.

Basquin, O. H. (1910). The Exponential Law of the Endurance Tests. *Annual Meeting, American Society for Testing Materials* 10, 625–630.

Bauschinger, J. (1881). Über die Veränderung der Elastizitätsgrenze und der Elastizitätsmodulus verschiedener Metalle. *Civilingenieur XXVII*, pp. 289–348, Felix, Leipzig, Germany.

Bell, R., O. Vosikovski, and S. A. Bain (1989). The Significance of Weld Toe Undercuts in the Fatigue of Steel Plate T-Joints. *International Journal of Fatigue* 11(1), 3–11.

Berge, S. (1984). Effect of Plate Thickness in Fatigue Design of Welded Structures. OTC Paper no. 4829. Houston, May.

Berge, S. (1985). On the Effect of Plate Thickness in Fatigue of Welds. *Engineering Fracture Mechanics* 21 (2), 423–435.

Berge, S., O. C. Astrup, T. Simonsen, and B. Lian (1987b). Effect of seawater and cathodic protection in fatigue of welded low carbon. *Steel in Marine Structures*, edited by C. Noorhook and J. deBack Elsevier Science Publishers B.V., Amsterdam, 729–735.

Berge, S., O. Eide, O. C. Astrup, S. Palm, S. Wästberg, Å. Gunleiksrud, and B. Lian (1987a). Effect of Plate Thickness in Fatigue of Welded Joints in Air and in Sea Water. *Steel in Marine Structures*, edited by C. Noorhook and J. deBack. Elsevier Science Publishers B.V., Amsterdam, 799–810.

Berge, S., O. I. Eide, and P. J. Tubby (1994). Fatigue Strength of Tubular Joints: Some Unresolved Problems. *OMAE – Volume III*, Materials Engineering, ASME, pp. 195–201.

Berge, S. and S. E. Webster (1987). The Size Effect on the Fatigue Behaviour of Welded Joints. *Steel in Marine Structures*, edited by C. Noorhook and J. deBack. Elsevier Science Publishers B.V., Amsterdam, 179–203.

Bickford, J. H. and S. Nassar (1998). *Handbook of Bolts and Bolted Joints*. Marcel Dekker, Inc., New York.

Billington, C. J. and J. Chetwood (2012). Lessons from previous research for the design of grouted connections for wind foundations. A new interpretation. IQPC Conference "Offshore Foundations for Wind Turbines", 2–5 July, Bremen, Germany.

Billington, C. J. and G. H. G. Lewis (1978). The Strength of Large Diameter Grouted Connections. Paper presented at 10th Offshore Technology Conference, Houston, Texas, 8–11 May, OTC Paper No. 3083, 291–301.

Billington, C. J. and I. E. Tebbett (1980). The basis for new design formulae for grouted jacket to pile connections. Paper presented at 12th Offshore Technology Conference, Houston, Texas, 5–8 May, OTC Paper No. 3788, 449–458.

Boardman, H. C. (1948). Stresses at Junction of two right Cone Frustums with a common Axis. The Water Tower.

Bokalrud, T. and A. Karlsen (1981). Probabilistic fracture mechanics evaluation of fatigue failure from weld defects in butt weld joints. Proceeding on Fitness for purpose validation of welded constructions. London, paper 28.

Bokalrud, T. and A. Karlsen (1982). Control of Fatigue Failure in Ship Hulls by Ultrasonic Inspection. *Norwegian Maritime Research* 10 (1), 9–15.

Boulton, C. F. (1976). Acceptance Levels of Weld Defects for Fatigue Service. Paper presented at the AWS 57th Annual Meeting in St. Louis, Missouri.

Bowie, O. L. (1956). Analysis of an Infinite Plate Containing Radial Cracks Originating at the Boundary of an Internal Hole. *Journal of Mathematics and Physics* 35, 60–71.

Bowness, D. and M. M. K. Lee (2002). Fracture mechanics assessment of fatigue cracks in offshore tubular structures. Offshore Technology Report 2000/077. HSE.

Brennan, F. P., W. D. Dower, R. F. Karé, and A. K. Hellier (1999). Parametric Equations for T-Butt Weld Toe Stress Intensity Factors. *International Journal of Fatigue* 21, 1051–1062.

Broek, D. (1986). *Elementary Engineering Fracture Mechanics*. Fourth revised edition. Kluwer Academic Publishers, Dordrecht.

Bruder, T., K. Störzel, J. Baumgartner, and H. Hanselka (2012). Evaluation of Nominal and Local Stress Based Approaches for the Assessment of Seamwelds. *International Journal of Fatigue* 34, 86–102.

BS 5400 Part 10 (1980). Code of Practise for Fatigue. Steel, Concrete and Composite Bridges. British Standard Institution.

BS EN ISO 21329 (2004). Petroleum and natural gas industries – Pipeline transportation systems – Test procedures for mechanical connectors.

BS 7608 (1993). Code of Practice for Fatigue Design and Assessment of Steel Structures, BSI, London.

BS 7608 (2014). Code of Practice for Fatigue Design and Assessment of Steel Structures, BSI, London.

BS 5500:2012+A2:2013. Specification for Unfired Fusion Welded Pressure Vessels, BSI Standards Publication, British Standard Institution 2013.

BSI (1972). Methods for Crack Opening Displacements (COD) Testing. DD 19:1972, London.

BS 7910 (2013). Guidance on Methods for Assessing the Acceptability of Flaws in Fusion Welded Structures. BSI.

Buitrago, J., K. Doynov and A. Fox (2006). Effect of Reeling on Welded Umbilical Tubing, OMAE2006-92579 Proceedings of OMAE2006 25th Int. Conf. on Offshore Mechanics and Arctic Engineering, June 2006, Hamburg, Germany.

Buitrago, J., K. Doynov and A. Fox (2008). Effect of Reeling on Small Umbilical Tubing Fatigue, OMAE2008-57544 Proceedings of OMAE2008 27th Int. Conf. on Offshore Mechanics and Arctic Engineering, June 2008, Estoril, Portgal.

Buitrago, J., M. S. Weir, and W. C. Kan (2003). Fatigue design and performance verification of deepwater risers. OMAE2003-37492. Proceedings of OMAE03 22nd International Conference on Offshore Mechanics and Arctic Engineering, Cancun, Mexico.

Buitrago, J., M. S. Weir, W. C. Kan, S. J. Hudak, and F. McMaster (2004). Effect of Loading Frequency on Fatigue Performance of Risers in Sour Environment. Proceedings of OMAE04 23rd International Conference on Offshore Mechanics and Arctic Engineering, Vancouver, British Columbia, Canada.

Buitrago, J. and P. C. Wong (2003). Fatigue Design of Driven Piles for Deepwater Applications. Int. Conf. ISOPE – Honolulu, Hawaii.

Buitrago, J. and N. Zettlemoyer (1997). Fatigue of welded joints peened underwater. Complex Joints. Proc. of 16th Int. Conf. on Offshore Mechanics and Arctic Engineering, Materials Engineering Symposium, Vol. 3, 187–196, Yokohama, Japan.

Buitrago, J. and N. Zettlemoyer (1999). Fatigue of Tendon Welds with Internal Defects, OMAE99/MAT-2001. Proceedings of OMAE99, 18th International Conference on Offshore Mechanics and Arctic Engineering, St. Johns, Newfoundland, Canada.

Buitrago, J., N. Zettlemoyer, and J. L. Kahlish (1984). Combined Hot-Spot Stress Procedures for Tubular Joints. OTC Paper No. 4775. Houston, TX.

Burdekin, F. M. (1981). The PD6493 approach to significance of defects, paper 37, Proceedings Conference Fitness for Purpose Validation of Welded Structures, November, London.

Burdekin, F. M. and M. G. Daves (1971). Practical application of fracture mechanics to pressure vessel technology/ a conference arranged by the Applied Mechanics Group of the London Institution of Mechanical Engineers. Paper C5/71.

Bærheim, M., A. Stacey, and N. Nichols (1996). Proposed Fatigue Provisions in the New ISO Code for Offshore Structures. *OMAE-Volume* III, Materials Engineering ASME. 513–519.

Callan, M. D., A. C. Wordsworth, I. G. Livett, R. H. Boudreaux, and F. J. Huebsch (1981). BP Magnus platform internally stiffened braceing node studies. OTC Paper No. 4109, 13th Annual OTC in Houston, TX.

Carlsson, J. (1976). Brottmekanikk. Ingenjørforlaget (In Swedish).

Chandawanisch, N. (1979). An Experimental Study of Fatigue Crack Initiation and Growth from Cold Worked Holes. *Engineering Fracture Mechanics* 11, 609–620.

Chattopadhyay, A., G. Glinka, M. El-Zein, J. Qian, and R. Formas (2011). Stress Analysis and Fatigue of Welded Structures. *International Journal Welding in the World* 55 (7/8), 2–21.

Chen, W. and E. Landet (2001). Stress Analysis of Cut-outs with and without Reinforcement. OMAE01/MAT-3013, Proceedings of OMAE'01, 20th International Conference on Offshore Mechanics and Arctic Engineering, Rio de Janeiro, Brazil.

Chen, W.-C. (1989). Fatigue-Life Predictions for Threaded TLP Tether Connector. OTC 5938. 21st Annual OTC in Houston, TX.

Cheng, X., J. W. Fischer, H. J. Prask, T. Gnäupel-Herold, B. T. Yen, and S. Roy (2003). Residual Stress Modification by Post-Weld Treatment and Its Effect on Fatigue Strength of Welded Structures. *International Journal of Fatigue* 25 (9–11), 1259–1269.

Coffin, L. F. (1954). Trans. ASME, 76, 931.

Coffin, L. F. (1984). Low cycle fatigue – a thirty year perspective. Presented at the 2nd International Conference on Fatigue and Fatigue Thresholds in Birmingham, UK. Engineering Materials Advisory Services Ltd.

Collins, J. A. (1993). *Failure of Materials in Mechanical Design*. John Wiley & Sons, New York.

Connelly, L. M. and N. Zettlemoyer (1993). Stress Concentration at Girth Welds of Tubulars with Axial Wall Misalignment. Proceedings Int. Conf. on "Tubular Structures", E & F N Spon, London.

Connolly, M. P., A. K. Hellier, W. D. Dover, and J. Sutomo (1990). A Parametric Study of the Ratio of Bending to Membrane Stress in Tubular Y- and T-Joints. *International Journal of Fatigue* 12, 3–11.

Cremona, C. (1996). Reliability Updating of Welded Joints Damaged by Fatigue. *International Journal of Fatigue* 18 (8), 567–575.

Dahle, T. (1993). Spectrum Fatigue of Welded Specimens in Relation to the Linear Damage Rule. *Fatigue under Spectrum Loading and in Corrosive Environments*, edited by A. F. Blom. EMAS Worley, West Midlands, UK, 133–147.

Dalsgaard Sørensen, J. and G. Ersdal (2008). Risk based Inspection Planning of Ageing Structures. OMAE08-57404, 27th Int. Conf. on Offshore Mechanics and Arctic Engineering, Estoril, Portugal.

Dalsgaard Sørensen, J. D., J. Tychsen, J. U. Andersen, and R. D. Brandstrup (2006). Fatigue Analysis of Load-Carrying Fillet Welds. *Journal of Offshore Mechanics and Arctic Engineering* 128, 65–74.

De Back, J. (1981). Strength of Tubular Joints. *International conference Steel in Marine Structures*, Paris, 53–95. Proceedings published by: Comptoir des Produits Sidèrurgiques, 1 rue P. Cèzanne, 75008 Paris.

Department of Energy (1977). Offshore installations: Guidance on Design and Construction. Second edition.

Department of Energy (1982). Amendment No. 4 to Offshore installations: Guidance on Design and Construction.

Department of Energy (1984). Offshore Installations: Guidance on Design, Construction and Certification. Third Edition.

Dexter, R. J., J. E. Tarquinio, and J. W. Fisher (1994). Application of Hot Spot Stress Fatigue Analysis to Attachments on Flexible Plate. *OMAE – Volume* III, Materials Engineering, ASME, pp. 85–92.

Dexter, R. J. and J. M. Ocal (2013). Manual for Repair and Retrofit of Fatigue Cracks in Steel Bridges. FHWA Publication No. FHWA-IF-13-020. US Depart of Transportation.

Dijkstra, O. D. and J. de Back (1980). Fatigue strength of welded tubular T- and X-joints, Proceedings Offshore Technology Conference (OTC 3696), Houston, TX.

DIN 2510 (1971). Schraubenverbindungen mit Dehnschaft. (Bolted Connections with Reduced Shank, Parts 1–8).

DIN 25201–4 (2010). Annex B Test specification for demonstrating the resistance to self-loosening of secured bolted joints, Deutsches Institut für Normung c.V., Berlin.

Ditlevsen, O. and H. O. Madsen (1996). *Structural Reliability Methods*, John Wiley and Sons, Chichester, West Sussex, England.

DNV (1977). *Rules for the Design Construction and Inspection of Offshore Structures*. DNV, Oslo, Norway.

DNV CN 30.2 (1984). *Classification Note No 30.2, Fatigue Strength Analysis for Mobile Offshore Units*, DNV, Oslo, Norway.

DNV CN 30.7 (2001). *Fatigue Assessment of Ship Structures*. DNV, Oslo, Norway.

DNV CN 30.7 (2014). *Fatigue Assessment of Ship Structures*. DNV, Oslo, Norway.

DNV-OS-C103 (2014). Structural Design of Column Stabilized Units. LRFD method.

DNV-OS-C502 (2012). *Offshore Concrete Structures*. DNV, Oslo, Norway.

DNV-OS-E301 (2010). *Position Mooring*. DNV, Oslo, Norway.

DNV-OS-F101 (2013). Submarine Pipeline Systems. Det Norske Veritas.

DNV-OS-J101 (2014). *Design of Offshore Wind Turbine Structures*. DNV, Oslo, Norway.

DNV-RP-B101 (2007). *Corrosion Protection of Floating Production and Storage Units*. DNV, Oslo, Norway.

DNV-RP-B401 (2011). *Cathodic Protection Design*. DNV, Oslo, Norway.

DNV-RP-C203 (2001). *Fatigue Design of Offshore Steel Structures*. DNVGL, Oslo, Norway.

DNV-RP-C205 (2014). *Environmental Conditions and Environmental Loads*. DNV, Oslo, Norway.

DNV-RP-C206 (2012). *Fatigue Methodology for Offshore Ships*. DNV, Oslo, Norway.

DNV-RP-C207 (2012). *Statistical Representation of Soil Data*. DNV, Oslo, Norway.

DNV-RP-F201 (2002). *Design of Titanium Risers*. DNV, Oslo, Norway.

DNV Rules for Classification of Ships (2012). Compressed Natural Gas Carriers. Part 5, Chapter 15. Special Service and Type – Additional Class.

DNVGL-OS-C401 (2015). *Fabrication and Testing of Offshore Structures*. DNV GL, Oslo, Norway.

DNVGL-RP-0001 (2015). *Probabilistic Methods for Planning of Inspection for Fatigue Cracks in Offshore Structures*. DNVGL, Oslo, Norway.

DNVGL-RP-C203 (2016). *Fatigue Design of Offshore Steel Structures*. DNVGL, Oslo, Norway.

Dobson, A. (2007). Effect of Strain History on Steel Tube Umbilical, Proceedings of OMAE2007 26th Int. Conf. on Offshore Mechanics and Arctic Engineering, June 2007, San Diego, CA.

Doerk, O., W. Fricke, and C. Weissenborn (2003). Comparison of Different Calculation Methods for Structural Stresses at Welded Joints. *International Journal of Fatigue* 25, 359–369.

Dong, P. (2001). A Structural Stress Definition and Numerical Implementation for Fatigue Analysis of Welded Joints. *International Journal of Fatigue* 23, 865–876.

Dover, W. D., R. F. Karé, and M. S. Hall (1991). The Reliability of SCF Predictions using Parametric Equations: A Statistical Analysis, In Proceedings of the Tenth International Conference on Offshore Mechanics and Arctic Engineering, Stavanger, Norway, pp. 453–459.

Dowling, N. E. (1998). *Mechanical Behavior of Materials: Engineering Methods for Deformation, Fracture, and Fatigue*, Second edition. Prentice Hall, Upper Saddle River, NJ.

Edwards, P. R. (1988). Full-Scale Fatigue Testing of Aircraft Structures. *Full-Scale Fatigue Testing of Components and Structures*, edited by K. J. Marsh. Butterworths & Co Ltd., London, 16–43.

Efthymiou, M. (1988). *Development of SCF Formulae and Generalised Influence Functions for Use in Fatigue Analysis*, OTJ'88 Recent Developments in Tubular Joints Technology, Surrey, UK.

Efthymiou, M. and S. Durkin (1985). *Stress Concentrations in T/Y and Gap/Overlap K Joints, in: Proceedings of the Conference on Behaviour of Offshore Steel Structures*, Delft, Elsevier Science Publishers, Amsterdam, The Netherlands, 429–440.

Eggen, T. G. and E. Bardal (1994). Corrosion Fatigue Crack Growth Rates in Offshore Structural Steels. *OMAE – Volume* III, Materials Engineering, ASME, pp. 207–213.

Eide, O. I. and S. Berge (1990). Fatigue Capacity of Stiffened Tubular Joints. Proc. Int. Conf. Offshore Mechanics and Arctic Engineering OMAE, Houston, TX.

Engesvik, K. (1981). Analysis of Uncertainties in the Fatigue Capacity of Welded Joints. Dissertation. Report No UR-82-17. Marine Technology Centre, Trondheim, Norway.

Engesvik, K. and Moan, T. (1983). Probabilistic Analysis of the Uncertainties of the Fatigue Capacity of Welded Joints. *Engineering & Fracture Mechanics* 18 (4), 743–762.

EN 1990 (2002). Eurocode – Basis for structural design. European Committee for Standardization, Brussels.

EN-1090-2 (2008). Execution of steel structures and aluminium structures. Part 2: Technical requirements for steel structures. European Committee for Standardization, Brussels.

EN 1993–1–9 (2009). Eurocode 3: Design of Steel Structures. Fatigue strength of steel structures, European Committee for Standardization, Brussels.

EN 1999–1–3 (2007). Eurocode 9: Design of Aluminium Structures. Part 1–3: Structures Susceptible to Fatigue. European Committee for Standardization, Brussels.

EN 1993–1–4 (2006). Eurocode 3: Design of Steel Structures – Part 1–4: General rules – Supplementary rules for stainless steels. European Committee for Standardization, Brussels.

EN 1993–1–8 (2009). Eurocode 3: Design of Steel Structures. Part 1–8: Design of joints. European Committee for Standardization, Brussels.

EN 1999-1-1 (2009). Eurocode 9: Design of aluminium structures – Part 1-1: General structural rules. European Committee for Standardization, Brussels.

Etterdal, B., H. Grigorian, D. Askheim, and R. Gladsø (2001). Strengthening of Offshore Steel Components Using High-Strength Grout: Component Testing and Analytical Methods, OTC, Paper 13192.

Fergestad, D., Løtveit S. A. and B. J. Leira (2014). Life-Cycle Assessment of Flexible Risers. 33rd International Conference on Ocean, Offshore and Arctic Engineering, San Francisco, USA. OMAE2014-24135.

Fessler, H., T. H. Hyde, and A. W. Buchan (1991). Assessment of Parametric Equations for Stress Concentration Factors in Welded Tubular Connections. *Proceedings of the Fourth International Symposium on Tubular Structures*, Delft University Press, Delft, The Netherlands, 219–228.

Fjeld, S. (1978). Reliability of Offshore Structures. *Journal of Petroleum Technology*, 30, 1486–1496. Also OTC Paper no 3027 (1977), Houston, TX.

Fisher, J. W. (1984). *Fatigue and Fracture in Steel Bridges*. John Wiley & Sons, New York, Chichester, Brisbane, Toronto, Singapore.

FKM-Guideline (2003). *Analytical strength assessment of components in mechanical engineering*, 5th Edition, Editor: Forschungslaboratorium Mashinenbau (FKM), VDMA Verlag, Frankfurt, Germany.

Flugge, W. (1973). *Stresses in Shells*. 2nd edition. Springer Verlag, Berlin Germany.

Forrest, P. G. (1962). *Fatigue of Metals*. Pergamon Press, Oxford.

Fricke, W. and H. Petershagen (1992). Detail Design of Welded Ship Structures based on Hot-Spot tresses. Practical design of ships and mobile units. Amsterdam: Elsevier Science, 2.1087–2.1099.

Fricke, W. (2001). Recommended Hot Spot Analysis Procedure for Structural Details of FPSO's and Ships Based on Round-Robin FE Analyses. *Proc. 11th ISOPE, Stavanger. Also Int. J. of Offshore and Polar Engineering* 12 (1), 40–48.

Fricke, W. (2003). Fatigue Analysis of Welded Joints: State of Development. *Marine Structures* 16, 185–200.

Fricke, W. (2006). Weld root fatigue assessment of fillet-welded structures based on structural stress. OMAE2006–92207. *Proceedings of OMAE2006, 25th International Conference on Offshore Mechanics and Arctic Engineering*, Hamburg, Germany.

Fricke, W. (2013). IIW Guideline for the Assessment of Weld Root Fatigue. *Welding in the World* 57, 753–791.

Fricke, W. and O. Feltz (2009). Fatigue Tests and Numerical Analysis of Partial-Load and Full-Load Carrying Fillet Welds at Cover Plates and Lap Joints. Doc. XIII-2278–09/XV-1320-09.

Fricke, W. and A. Kahl (2005). Comparison of Different Structural Approaches for Fatigue Assessment of Welded Ship Structures. *Marine Structures* 18, 473–488.

Fricke, W. and H. Paetzold (1995). Fatigue Strength Assessment of Scallops – An Example for the Application of Nominal and Local Stress Approaches. *Marine Structures* 8, 423–447.

Fujimoto, Y., S. C. Kim, and K. Hamada (1997). Inspection Planning of Fatigue Deteriorating Structures Using Genetic Algorithm. *Journal of the Society of Naval Architects of Japan* 1997 (182), 729–739.

Fujimoto, Y., S. C. Kim, E. Shintaku, and K. Ohtaka (1996). Study on Fatigue Reliability and Inspection of Ship Structures Based on Enquete Information. *Journal of the Society of Naval Architects of Japan* 181, 601–609. (In Japanese).

Gadelmawla, E. S., M. M. Koura, T. M. A. Maksoud, I. M. Elewa, and H. H. Soliman (2002). Roughness Parameters. *Journal of Materials Processing Technology* 123, 133–145.

Gibstein, M. B. (1978). Parametric Stress Analysis of T-joints. Paper 26 in European Offshore Steels Research Seminar, Cambridge, UK.

Gibstein, M. G. (1981). Stress Concentration in Tubular Joints. Its Definition, Determination and Applications. Paper 1.4, International Conference on Steel in Marine Structures, Paris. Proceedings published by: Comptoir des Produits Sidèrurgiques, 1 rue P. Cèzanne, 75008 Paris.

Glinka, G. (1985). Energy Density Approach to Calculation of Inelastic Stress-Strain Near Notches and Cracks. *Engineering & Fracture Mechanics* 22(3), 485–508.

Gordon, J. R., W. C. Mohr, S. D. Dimitrakis and F. V. Lawrence (1997). Fatigue Design of Offshore Structures. Int. Conf. on Performance of Dynamically Loaded Welded Structures, San Francisco, CA. Edited by S. J. Maddox and M. Prager.

Gran, S. (1992). A Course in Ocean Engineering. *Development in Marine Technology*, Vol. 8. Elsevier Science Publishers B.V, Amsterdam.

Griffith, A. A. (1920). The Phenomena of Rupture and Flow in Solids. *Philosophical Transactions* (Series A) 221, 163–198.

Grøvlen, M., E. Bardal, S. Berge, O. Eide, K. Engesvik, P. J. Haagensen, and O. Ørjasæther (1989). Localised corrosion on offshore tubular structures: Inspection and repair criteria. OTC 5987. Houston, TX.

Guedes Soares, C. and T. Moan (1991). Model Uncertainty in the Long-term Distribution of Wave-Induced Bending Moment for Fatigue Design of Ship Structure. *Marine Structures* 4, 295–315.

Guennec, B., A. Ueno, T. Sakai, M. Takanashi, and Y. Itabashi (2013). Effect of Loading Frequency in Fatigue Properties and Micro-Plasticity Behaviour of JIS S15C Low Carbon Steel. 13th Int. Conference on Fracture, Beijing, China.

Gulati, K. C., W. J. Wang, and K. Y. Kan (1982). An Analytical study of Stress Concentration Effects in Multibrace Joints under Combined Loading. OTC Paper no. 4407, Houston, TX.

Gundersen, R. H., J. Buitrago, S. A. Fox, and J. Q. Burns (1989). A Large-Diameter Threaded Connector for Tension Leg Platform Tendons. OTC Paper no. 5937, presented at 21st Annual OTC in Houston, TX.

Gurney, T. R. (1976). Fatigue Design Rules for Welded Steel Joints. *The Welding Institute Research Bulletin* 17 (5), 115–124.

Gurney, T. R. (1979). *Fatigue of Welded Structures*. Second Edition. Cambridge University Press, Cambridge.

Gurney, T. R. (1979). The Influence of Thickness on the Fatigue Strength of Welded Joints. Boss'79. Second Int. Conf. on Behaviour of Offshore Structures, London, UK.

Gurney T. R. (1989). "The Influence of Thickness on Fatigue of Welded Joints – 10 years on." 8th OMAE Conf., The Hague.

Gurney T. R. (1991). *The Fatigue Strength of Transverse Fillet Welded Joints: A Study on the Influence of Joint Geometry*. Abington Publishing, Cambridge.

Gurney, T. R. (2006). *Cumulative Damage of Welded Joints*. Woodhead Publishing Limited, Cambridge.

Haagensen, P. J. (1994). Methods for Fatigue Strength Improvement and Repair of Welded Structures, OMAE – Volume III, Materials Engineering, ASME, pp. 419–427.

Haagensen, P. (1997). IIIWs Round Robin and Design Recommendations for Improvement Methods. In Proceeding from international conference on performance of dynamically loaded welded structures held in San Fransisco, CA, Edited by S. J. Maddox and M. Prager.

Haagensen, P. and S. J. Maddox (2013). *IIW Recommendations on Post Weld Fatigue Life Improvement of Steel and Aluminium Structures*, Woodhead Publishing Ltd., Cambridge.

Haghani, R., M. Al-Emrani and M. Heshmati (2012). Fatigue-Prone Details in Steel Bridges. Buildings 2, 456-476; ISSN 2075-5309.

Haibach, E. (1970). Modifizierte Lineare Schadenakkumulations – Hypotese zur Berücksichtigung des Dauerfestigheitsabfalls mit Fortschreitender Schädung. LBF, Technische Mitteilung TM 50/70. Darmstadt, Germany.

Haibach, E. (2006). Betriebsfestigkeit. Verfahren und Daten zur Bauteilberechnung. 3. korrigierte und ergänzte Auflage. Springer-Verlag.

Hannus, H. (1985). On the Fatigue Design Procedure of Marine Structural Details. Report No TRITA-SKP 1056. The Naval Architecture at the Royal Institute of Technology, Stockholm.

Harrison, J. D. (1981). The effect of residual stresses on fatigue behaviour. *Proc. Welding Inst. (Abington)*, 47, 9–16.

Harrison, J. D., G. L. Archer, and C. F. Boulton (1978). Significance of Weld Defects in Pressure Vessels under Fatigue Loading. *Conference on Tolerance of Flaws in Pressurized Components*; London. IMechE Conference Publications; no. 10; pp. 255–270.

Harwood, R. G, C. J. Billington, J. Buitrago, A. Sele, and J. V. Sharp (1996). Grouted Pile to Sleeve connections: Design Provisions for the New ISO Standard for Offshore Structures. OMAE 1996, ASME.

Haslum, K., K. Kristoffersen, and L. A. Andersen (1976). Stress Analysis of Longitudinal/Girder Connections. *Norwegian Maritime Research* 4 (3), 13–28.

Hellan, K. (1984). *Introduction to Fracture Mechanics*. McGraw-Hill Book Company, New York.

Hellier, A. K., M. P. Conolly, and W. D. Dower (1990a). Stress Concentration Factors for Tubular T and Y Joints. *International Journal of Fatigue* 12 (1), 13–23.

Hellier, A. K., M. P. Connolly, R. F. Karé, and W. D. Dover (1990b). Prediction of the Stress Distribution in Tubular Y- and T-Joints. *International Journal of Fatigue* 12, 25–33.

Heo, J.H., J.-K. Kang, Y. Kim, I.-S. Yoo, K.-S. Kim, and H.-S. Urm (2004). A Study on the Design Guidance for Low Cycle Fatigue in Ship Structures. 9th Symposium on Practical Design of Ships and Offshore Floating Structures, Luebeck-Travemuende, Germany.

Ho, N. J. and F. V. Lawrence, Jr. (1984). Fatigue Test Results and Predictions for Cruciform and Lap Welds. *Theoretical and Applied Fracture Mechanics* 1 (1), 3–21.

Hobbacher, A. (1996). Fatigue Design of Welded Joints and Components. *Recommendations of IIW Joint Working Group* XIII-1539–96/XV-845–96. Abington Publishing, The International Institute of Welding, Oxford.

Hobbacher, A. and Kassner, M. (2012). On Relation between Fatigue Properties of Welded Joints, Quality Criteria and Groups in ISO 5817. *Welding in the World* 56 (11–12), 153–169.

Hobbacher, A. (2009). *Recommendations for Fatigue Design of Welded Joints and Components*. The Welding Research Council, New York.

Holtsmark, G. (1993). The Bending Response of Laterally Loaded Panels with Unsymmetrical Stiffeners, DNVC Report No. 93–0152. Høvik, Norway.

HSE (1990). *Offshore Installations: Guidance on Design, Construction and Certification*. Fourth edition. London.

HSE (1995). *Offshore Installations: Guidance on Design, Construction and Certification*. Third Amendment to Fourth edition. London.

HSE OTH 354 (1997). Stress Concentration Factors for Simple Tubular Joints: Assessment of Existing and Development of New Parametric Formulae.

HSE OTH 511 (1998). Review of Fatigue Crack Growth Rates in Air and Seawater.

HSE OTH 92390 (1999). Background to New Fatigue Guidance for Steel Joints and Connections in Offshore Structures. Offshore technology report, HSE.

HSE OTO 1999 022 (1999). Fatigue Life Implications for Design and Inspection for Single Sided Welds at Tubular Joints. Offshore Technology Report, HSE.

HSE OTO 98 044 (1998). Stress Concentrations in Single-Sided Welds in Offshore Tubular Joints. Offshore Technology Report, HSE.

HSE (2002). Offshore Technology Report 2001/016 Pile/Sleeve Connections. ISBN 0 7176 2390 4.

Huo, L., D. Wang, and Y. Zhang (2005). Investigation of the Fatigue Behaviour of the Welded Joints Treated by TIG Dressing and Ultrasonic Peening under Variable-Amplitude Load. *International Journal of Fatigue* 27, 95–101.

IACS (2013). Common Structural Rules for Bulk Carriers and Oil Tankers.

Iida, K. (1987). State of the Art in Japan. *Steel in Marine Structures*, edited by C. Noorhook and J. deBack. Elsevier Science Publishers B.V., Amsterdam, 71–98.

Iida, K. and T. Uemura (1996). Stress Concentration Factor Formulae Widely used in Japan. *International Journal of Fatigue & Fracture of Engineering Materials & Structures* 19 (6), 779–786.

Ingebrigtsen, T., Ø. Løset, and S. G. Nielsen (1990). Fatigue Design and Overall Safety of Grouted Pile Sleeve Connections. Paper presented at 22th Offshore Technology Conference, Houston, Texas, 7–10 May, OTC Paper No, 6344, 615–628.

Irwin, G. R. (1957). Analysis of Stresses and Strains near the End of a Crack Traversing a Plate. *Journal of Applied Mechanics* 24, 361–364.

ISO 8501 (2012). Preparation of steel substrates before application of paints and related products – visual assessment of surface cleanliness – Part 1: Rust grades and preparation grades of uncoated steel substrates and of steel substrates after overall removal of previous coatings. Informative supplements to part 1: Representative photographic examples of the change of the appearance imparted to steel when blast-cleaned with different abrasives.

ISO 2394 (1998). International Organization for Standardization (ISO), General principles on reliability for structures.

ISO 12107 (2003). International Organization for Standardization (ISO), Metallic materials – Fatigue testing – Statistical planning and analysis of data.

ISO 14347 (2008). Fatigue – Design Procedures for welded hollow section joints – recommendations.

ISO 19902 (2007). Fixed Steel Structures.

ISO 5817 (2014). Welding – Fusion-welded joints in steel, nickel, titanium and their alloys (beam welding excluded): Quality levels for imperfections.

ISO 68–1 (1998). ISO general purpose screw threads – Basic profile – Part 1 Metric screw threads.

ISO 261 (1998). ISO general purpose metric screw threads – General plan.

ISO 898–1 (2013). Mechanical properties of fasteners – Part 1: Bolts, screws and studs with specified property classes – Coarse thread and fine pitch thread.

ISO 898–1 (2012). Mechanical properties of fasteners – Part 2: Nuts with specified proof load values – Coarse thread.

ISO 965–1 (2013). ISO general purpose metric screw threads – Tolerances – Part 1: Principles and basic data.

ISO 1891 (2009). Fasteners – Terminology. Second edition.

ISO 15156–2 (2009). Petroleum and natural gas industries – Materials for use in H2S-containing environments in oil and gas production – Part 2: Cracking-resistant carbon and low alloy steels, and the use of cast irons.

ISO 15156–3 (2009). Petroleum and natural gas industries – Materials for use in H2S-containing environments in oil and gas production – Part 3: Cracking-resistant CRAs (corrosion-resistant alloys) and other alloys.

ISO 21457 (2010). Petroleum, petrochemical and natural gas industries – Materials selection and corrosion control for oil and gas production systems.

Itagaki, H, Akita, Y, and Nitta, A. (1983). Application of Subjective Reliability Analysis to the Evaluation of Inspection Procedures on Ship Structures, Role of Design, Inspection, and Redundancy in Marine Structural Reliability, Proc. Int. Symp., ISBN: 0-309-03488-4, National Academy Press.

Jaccard, R. and Ogle, M. H. (1997). Aluminium Eurocode Design for Fatigue. In *Proceeding from international conference on performance of dynamically loaded welded structures held in San Francisco, CA*, Edited by S. J. Maddox and M. Prager.

Johnson, W. (1961). *Stress Concentration around Holes*. Trans. G. N. Savin. Pergamon Press, New York.

Jones, R., D. Karunakaran, J. Mair, and H. Wang (2011). Reeled Clad SCR Weld Fatigue Qualification. OTC Paper No. 21655. OTC, Houston, TX.

Jonsson, B., J. Samuelsson, and G. Marquis (2011). Development of Weld Quality Criteria based on Fatigue Performance. *Welding in the World*, 55, 79–89.

Kang, S. W., W. S. Kim, and Y. M. Paik (2002). Fatigue Strength of Fillet Welded Steel Structure under Out-of-Plane Bending, International Welding / Joining Conference – Korea.

Kang, S. W. and W. S. Kim (2003). A Proposed S-N Curve for Welded Ship Structure, *Welding Journal* 82, (7), 161–169.

Karsan, D. I. and N. W. Krahl (1984). New API Equation for Grouted Pile-to-Structure. Paper presented at 16th Offshore Technology Conference, Houston, TX, 7–9 May, OTC Paper No. 4715, 49–54.

Katsumoto, H., N. Konda, K. Arimochi, K. Hirota, A. Isoda, H. Kitada, M. Sakano, and H. Yajima (2005). Development of Structural Steel with High Resistance to Fatigue Crack Initiation and Growth, Part 3. Proceedings of OMAE'05 24th Int. Conf. on Offshore Mechanics and Arctic Engineering, Halkidiki, Greece.

Kim, D. S. and J. D. Smith (1994). Residual Stress Measurements of Tension Leg Platform Tendon Welds. *OMAE – Volume* III, Materials Engineering, ASME, pp. 31–36.

Kim, I. T. and K. Yamada (2004). Fatigue behaviour of fillet welded joints inclined to a uniaxial load. Document No. XIII-2021–04. The International Institute of Welding.

Kim, W. S., Y. Tomita, K. Hashimoto and N. Osawa (1997). Effects of Static Load on Fatigue Strength of Ship Structures. Proceedings of the Seventh International Offshore and Polar Engineering Conference, Honolulu, USA. ISBN 1-880653-32, 565–571.

Kim, W. S., D. H. Kim, S. G. Lee, and Y. K. Lee (2001). Fatigue Strength of Load-Carrying Box Fillet Weldment in Ship Structure. Proc. of 8th PRADS, Shanghai.

Kim, W. S. and I. Lotsberg (2004). Fatigue Test Data for Welded Connections in Ship Shaped Structures". OMAE-FPSO'04–0018, Int. Conf. Houston, 30 August – 2 September. Also in *Journal of Offshore and Arctic Engineering* 127 (4) (November 2005), 359–365.

King, R. N., A. Stacey and J. V. Sharp (1996). A Review of Fatigue Crack Growth Rates for Offshore Steels in Air and Seawater Environments. 1996 OMAE – Volume III, Material Engineering ASME 1996, pp. 341–348.

Kirch, G. (1898). Die Theorie der Elastizitat u. d. Bedurinisse d. Festigkeitslehre, V.D.J., m. 42, No. 29, s. 799.

Kirkemo, F. (2014). Stud and Nut Structural Tension Capacity Matching. WRC Bulletin 542.

Kitagawa, H., K. Tsuji, T. Hisada, and Y. Hashimoto (1983). An analysis of Random Pits in Corrosion Fatigue: A Statistical Three-Dimensional Evaluation of an Irregularly Corroded Surface. In Corrosion Fatigue: Mechanics, Metallurgy, Electrochemistry and Engineering, ASTM STP 801, T. W. Croocker and B. N. Leis, Eds., American Society for Testing and Materials, 1983, pp. 147–158.

Knott, J. F. (1973). *Fundamentals of Fracture Mechanics*. Butterworth & Co Ltd., London.

Konda, N., A. Inami, K. ArimochiT. Takaosha, T. Yoshida, and I. Lotsberg (2010). A Proposed Design S-N Curve for Steels with Improved Fatigue Resistance (FCA Steels), PRADS, Rio de Janeiro, September. In Proceedings, pp. 1233–1242.

Kosteas, D. and B. Gietl (1996). A Comparative Analysis of Existing Test Data on Welded Aluminium Tubular Joints. *Fatigue & Fracture of Engineering Materials & Structures* 19(6), 731–738.

Kristofferesen, S. and P. J. Haagensen (2004). Fatigue Design Criteria for Small Super Duplex Steel Pipes. Proceedings OMAE, Vancouver.

Kuang, J. G., A. B. Potvin, and R. D. Leick (1975). Stress Concentration in Tubular Joints. Proceedings of the Seventh Annual Offshore Technology Conference, Houston, TX, Paper OTC 2205, 593–612.

Kudryavtsev, Y., J. Kleiman, L. Lobanov, V. Knysh, and G. Prokopenko (2004). Fatigue Life Improvement of Welded Elements by Ultrasonic Peening, IIW Document XIII-2010-04.

Kulak, G., J. W. Fisher, and J. H. A. Struik (1987). *Guide to Design Criteria for Bolted and Riveted Joints*. Second edition. John Wiley & Sons, New York.

Kvamsdal, R. S. and J. L. Howard (1972). Moss Rosenberg's Spherical Tank LNG – Carrier. First International Liquefied Gas Transportation Conference, London, England.

Kvitrud, A., G. Ersdal, and R. L. Leonhardsen (2001). On the Risk of Structural Failure on Norwegian Offshore Installations. Int. Conf. ISOPE, Stavanger.

Lachman, E., W. Schütz, and C. M. Sonsino (1987). Corrosion Fatigue of V-Shaped Welded and Cast Specimens in Seawater under Variable Amplitude Loading. *Steel in Marine Structures*, edited by C. Noorhook and J. deBack Elsevier Science Publishers B.V., Amsterdam, 551–564.

Lamport, W. B., J. O. Jirsa, and J. A. Yura (1987). Grouted Pile-to-Sleeve Connection Tests. Paper presented at 19th Offshore Technology Conference, Houston, TX, 27–30 April, OTC Paper No 5485, pp. 19–26.

Langer, B. F. (1937). Fatigue Failure from Stress Cycles of Varying Amplitude. *ASME Journal of Applied Mechanics* 59, A160–A162.

Lassen, T. and N. Recho (2006). *Fatigue Life Analysis of Welded Structures*. ISTE Ltd., London.

Lassen, T. and N. Recho (2015). Risk Based Inspection Planning for Fatigue Damage in Offshore Steel Structures. Proceedings of the 34nd International Conference on Ocean Offshore and Arctic Engineering, OMAE 2015, Paper OMAE2015–41165, Saint John, Canada.

Lee, M. K. K. (1999). Estimation of Stress Concentrations in Single-Sided Welds in Offshore Tubular Joints. *International Journal of Fatigue* 21, 895–908.

Lewis, G. H. G., J. G. Livett, R. T. P. McLaughlin, and K. C. Mead (1980). A Cost Saving Design for Pile to Structure Connections as Applied to BP Magnus. Paper presented at 12th Offshore Technology Conference, Houston, TX, OTC Paper No. 3789, 459–466.

Liebowitz, H. (1968). *Fracture: An Advanced Treatise, Volume II: Mathematical Fundamentals*. Academic Press, New York and London.

Lindemark, T., I. Lotsberg, J-K. Kang, K-S. Kim, and N. Oma (2009). Fatigue Capacity of Stiffener to Web Frame Connections. OMAE2009–79061.

Lopez Martinez, L. (2010). Fatigue Life Extension Procedure for Offshore Structures by Ultrasonic Peening. OTC Paper no. 20368, Offshore Technology Conference, Houston, TX.

Løseth, Ø. (1979). Grouted Connections in Steel Platforms – Testing and Design, Institute of Structural Engineers Informal Study Group – Model Analysis as a Design Tool, Joint I. Paper no. 8 in seminar on the use of physical models in the design of offshore structures.

Lotsberg, I. (1998). Stress Concentration Factors at Circumferential Welds in Tubulars. *Journal of Marine Structures* 11, 203–230.

Lotsberg, I. (2004a). Fatigue Design of Welded Pipe Penetrations in Plated Structures. *Marine Structures* 17 (1), 29–51.

Lotsberg, I. (2004b). Recommended Methodology for Analysis of Structural Stress for Fatigue Assessment of Plated Structures. OMAE-FPSO'04–0013, Int. Conf. Houston, TX, 30 August – 2 September.

Lotsberg, I. (2005). Background for Revision of DNV-RP-C203 Fatigue Analysis of Offshore Steel Structures. Paper no OMAE-67549, Int. Conf. Halkidiki, Greece.

Lotsberg, I. (2006a). Assessment of Fatigue Capacity in the New Bulk Carrier and Tanker Rules, *Marine Structures* 19 (1), 83–96.

Lotsberg, I. (2006b). Fatigue Capacity of Fillet Welded Connections subjected to Axial and Shear Loading. Presented at OMAE in Hamburg, 4–9 June. OMAE paper no 2006–92086. Also in *Journal of Offshore and Arctic Engineering* 131 (4) (2009), 041302-1-9.

Lotsberg, I. (2008a). Stress Concentration Factors at Welds in Pipelines and Tanks subjected to Internal Pressure. *Journal of Marine Structures* 21, 138–159.

Lotsberg, I. (2008b). *Current Practices in Condition Assessment of Aged Fixed-Type Offshore Structures*. Woodhead, London.

Lotsberg, I. (2009a). Assessment of Design Criteria for Fatigue Cracking from Weld Toes subjected to Proportional Loading. *Journal of Ships and Offshore Structures* 4 (2), 175–187.

Lotsberg, I. (2009b). Stress Concentration due to Misalignment at Butt Welds in Plated Structures and at Girth Welds in Tubulars. *International Journal of Fatigue* 31, 1337–1345.

Lotsberg, I. (2011). On Stress Concentration Factors for Tubular Y- and T-joints. *Journal of Marine Structures* 20, 60–69.

Lotsberg, I. (2013). Structural Mechanics for Design of Grouted Monopile Connections. *Marine Structures* 32, 113–135.

Lotsberg, I. (2014). Assessment of the Size Effect for Use in Design Standards for Fatigue Analysis. *Journal of Fatigue* 66, 86–100.

Lotsberg, I. and P. G Bergan (1979). Calculation of Fatigue Growth of Internal Cracks by the Finite Element Method. *Engineering Fracture Mechanics* 92, 34–47.

Lotsberg, I., E. Cramer, G. Holtsmark, R. Løseth, K. Olaisen, and S. Valsgård (1997). Fatigue Assessment of Floating Production Vessels. BOSS 97, July.

Lotsberg, I., A. Fjeldstad, M. Ro Helsem, and N. Oma (2014). Fatigue Life Improvement of Welded Doubling Plates by Grinding and Ultrasonic Peening. *Welding in the World* 58 (6), 819–830.

Lotsberg, I. and S. Fredheim (2009). Assessment of Design S-N curve for Umbilical Tubes. OMAE2009–79201. Int. Conf. Honolulu, HI.

Lotsberg, I. and P. A. Holth (2007). Stress Concentration Factors at Welds in Tubular Sections and Pipelines. OMAE paper no 2007–29571. 26th International Conference on Offshore Mechanics and Arctic Engineering, San Diego, California, June.

Lotsberg, I. and F. Kirkemo (1989). A Systematic Method for Planning In-Service Inspection of Steel Offshore Structures. Int. Conf. Offshore Mechanics and Artic Engineering. The Hague, March.

Lotsberg, I and E. Landet (2005). Fatigue Capacity of Side Longitudinals in Floating Structures. *Marine Structures* 18, 25–42.

Lotsberg, I. and P. K. Larsen (2001). Developments in Fatigue Design Standards for Offshore Structures. ISOPE, Stavanger, Norway.

Lotsberg I., and M. Marley (1992). In-Service Inspection – Planning for Steel Offshore Structures Using Reliability Methods. BOSS 92, London.

Lotsberg, I., M. Nygård, and T. Thompson (1998). Fatigue of Ship Shaped Production and Storage Units. OTC Paper No. 8775. Houston, TX.

Lotsberg, I., O. Olufsen, G. Solland, I .J. Dalane, and S. Haver (2005a). Risk Assessment of Loss of Structural Integrity of a Floating Production Platform due to Gross Errors. *Marine Structures* 17, 551–573.

Lotsberg, I. and K. O. Ronold (2011). On the Derivation of Design S-N Curves based on Limited Fatigue Test Data. OMAE2011–49175. Int. Conf. Rotterdam, The Netherlands.

Lotsberg, I. and H. Rove (2014). Stress concentration factors for butt welds in plated structures. Paper no. OMAE2014–23316. Int. Conf. OMAE, San Francisco, CA.

Lotsberg, I., T. A. Rundhaug, H. Thorkildsen, Å. Bøe, and T. Lindemark (2008c). A Procedure Fatigue Design of Web Stiffened Cruciform Connections. *Journal of Ships and Offshore Structures* 3 (2), 113–126.

Lotsberg, I., A. Serednicki, H. Bertnes, and A. Lervik (2012). Design of Grouted Connections for Monopile Offshore Structures. *Results from Two Joint Industry Projects. Stahlbau* 81 (9), 695–704.

Lotsberg, I., A. Serednicki, E. Cramer, and H. Bertnes (2013a). Behaviour of Grouted Connections of Monopile Structures at Ultimate and Cyclic Limit States. The Structural Engineer (February), 51–57.

Lotsberg, I., A. Serednicki, E. Cramer, H. Bertnes, and P. Enggaard Haahr (2011). On Structural Capacity of Grouted Connections in Offshore Structures. OMAE 2011–46169. Int. Conf. Rotterdam, The Netherlands.

Lotsberg, I., A. Serednicki, R. Oerleans, H. Bertnes, and A. Lervik (2013b). Capacity of Cylindrical Shaped Grouted Connections with Shear Keys in Offshore Structures reported from a Joint Industry Project. *The Structural Engineer* 91 (1), 42–48.

Lotsberg, I. and G. Sigurdsson (2004). Hot Spot S-N Curve for Fatigue Analysis of Plated Structures. OMAE-FPSO'04–0014, Int. Conf. Houston, TX. Also in *Journal of Offshore and Arctic Engineering* 128 (November 2006), 330–336.

Lotsberg, I. and G. Sigurdsson (2014). A New Recommended Practice for Inspection Planning of Fatigue Cracks in Offshore Structures based on Probabilistic Methods. OMAE2014–23187. Int. Conf. San Francisco, USA.

Lotsberg, I., G. Sigurdsson, K. Arnesen, and M. E. Hall (2008b). Recommended Design Fatigue Factors for Reassessment of Piles subjected to Dynamic Actions from Pile Driving. OMAE 2008–57251. Also in *Journal of Offshore and Arctic Engineering* 132 (4), 041603-1-8.

Lotsberg, I., G. Sigurdsson, and A. Crocker (2005b). Fatigue Analysis of Gusset Plate Joints in a Floating Platform, Int. Conf. on Computational Methods in Marine Engineering MARINE 2005. Edited by P. Bergan, J. García, E. Oñate, and T. Kvamsdal.

Lotsberg, I., G. Sigurdsson, and P. T. Wold (1999). Probabilistic Inspection Planning of the Åsgard A FPSO Hull Structure with respect to Fatigue. OMAE'99 St. John's, Newfoundland, July. Also in *Journal of Offshore and Arctic Engineering* 122 (May 2000), 134–140.

Lotsberg, I. and G. Solland (2013). Assessment of Capacity of Grouted Connections in Piled Jacket Structures. Int. Conf. OMAE, Nantes, France. Paper no. OMAE2013–10850.

Lotsberg, I., Å. Tårnes, and P. A. Simensen (1988). Design of the Karin Deepwater Platform with Respect to Fatigue. *Journal of Structural Engineering* 114 (6), 1211–1229.

Lotsberg, I., S. Wästberg, H. Ulle, P. Haagensen, and M. E. Hall (2008a). Fatigue Testing and S-N data for Fatigue Analysis of Piles. OMAE 2008–57250. Also in *Journal of Offshore and Arctic Engineering* 132 (4), 041602-1-7.

Lu, J. (1996). *Handbook of Measurement of Residual Stresses*. Liburn, GA: The Fairmont Press, Inc.

Macdonald, K. A. (2011). *Fracture and Fatigue of Welded Joints ans Structures*. Woodhouse Publishing Limited, Cambridge.

Macdonald, K.A., E. Østby, and B. Nyhus (2011). Constraint fracture mechanics: Test methods. Fracture and Fatigue of Welded Joints and Structures, 31–59. DOI: 10.1016/B978-1-84569-513-2.50002-2.

Maddox, J. (1991). *Fatigue Strength of Welded Structures*, Second edition. Woodhead Publishing Limited, Cambridge.

Maddox, S. J. (1985). *Fitness for Purpose Assessment of Misalignment in Transverse Butt Welds Subjected to Fatigue Loading*. IIW Document XIII-1180–1985. International Institute of Welding, London.

Maddox, S. J., J. G. Wylde, and N. Yamamoto (1995). Significance of Weld Profile on the Fatigue Lives of Tubular Joints. 1995 OMAE – Volume III, Materials Engineering, ASME.

Maddox, S. J. (2001). Recommended Hot-Spot Stress Design S-N Curves for Fatigue Assessment of FPSOs. *Proceedings of the Eleventh International Offshore and Polar Engineering Conference*, Stavanger, Norway.

Maddox, S. J. (2003). Key developments in the fatigue design of welded constructions. IIW Int. Conference, Bucharest Romania.

Maddox, S. J. (2006). Status Review on Fatigue Performance of Fillet Welds. Int. Conf. Offshore and Arctic Engineering, Hamburg, Germany. Paper no. OMAE2006–92314.

Maddox, S. J. and C. Johnston (2011). Factors Affecting the Fatigue Strength of Girth Welds: An Evaluation of TWI's Resonance Fatigue Test Data Base. OMAE2011–49192. Proceedings of the ASME 2011 30th Int. Conf. on Ocean, Offshore and Arctic Engineering OMAE 2011, Rotterdam, The Netherlands.

Madsen, H. O. (1985). Model Updating in First-Order Reliability Theory with Application to Fatigue Crack Growth. *Proc. of 2nd Int. Workshop on Stochastic Methods in Structural Mechanics*, University of Paris, Italy.

Madsen, H. O., S. Krenk, and N. C. Lind (1986). *Methods of Structural Safety*, Prentice-Hall, Upper Saddle River, NJ.

Madsen, H. O., R. Skjong, and F. Kirkemo (1987). Probabilistic Fatigue Analysis of Offshore Structures – Reliability Updating through Inspection Results, IOS'87, Glasgow, UK.

Majzoobi, G. H., G. H. Farrahi, and N. Habibi (2005). Experimental Evaluation of the Effect of Thread Pitch on Fatigue Life of Bolts. *International Journal of Fatigue* 27, 189–196.

Manson, S. S. (1954). Behaviour of Materials under Condition of Thermal Stress. *NACA Report 1170*, Lewis Flight Propulsion Laboratory, Cleveland, OH.

Marines, I., X. Bin, and C. Bathias (2003). An understanding of very high cycle fatigue of metals. *International Journal of Fatigue* 25, 1101–1107.

Marquis, G. B. (1996). Long Life Spectrum Fatigue of Crarbon and Stainless Steel Welds. *Fatigue & Fracture of Engineering Material & Structures* 19 (6), 739–753.

Marquis, G. B. and T. P. J. Mikkola (2001). Effect of Mean Stress Changes on the Fatigue Strength of Spectrum Loaded Welds. *Proceedings PRADS* 2001, 1113–1120.

Marquis, G. B., E. Mikkola, H. C. Yildirim, and Z. Barsoum (2013). Fatigue Strength Improvement of Steel Structures by High-Frequency Mechanical Impact: Proposed Fatigue Assessment Guidelines. *Welding in the World* 57 (6), 803–822.

Marsh, K. J. (1981). The Fatigue Strength of Tubular Welded Joints. Fatigue in Offshore Structural Steels. Proceedings of a Conference organized by the Institution of Civil Engineers, London.

Marshall, P. W. (1992). *Design of Welded Tubular Connections: Basis and Use of AWS Code Provisions*. Elsevier Science Publishers B.V., Amsterdam.

Marshall, P. W. (1993). API Provisions for SCF, S-N, and Size Effects. OTC Paper No. 7155. Houston, TX.

Matsuischi, M. and T. Endo (1968). Fatigue of Metals subjected to Varying Stress. Paper presented at the Kyushu District Meeting of the Japan Society of Mechanical Engineering. Japan Society of Mechanical Engineers, 37–40.

McDonald A. and J. G. Wylde (1981). Experimental Results of Fatigue Tests on Tubular Welded Joints. Fatigue in Offshore Structural Steels. Proceedings of a Conference organized by the Institution of Civil Engineers, London.

Melchers, R. E. and X. Jiang (2006). Estimation of Models for Durability of Epoxy Coatings in Water Ballast Tanks. *Ships and Offshore Structures* 1 (1), 61–70.

Miki, C. and K. Tateishi (1997). Fatigue Strength of Cope Hole in Details in Steel Bridges. *International Journal of Fatigue* 19 (6), 445–455.

Miner, M. A. (1945). Cumulative Damage in Fatigue. *Trans. ASME Journal of Applied Mechanics* 12, A159–A164.

Moan, T. (1999). Target Levels for Reliability-based Assessment of Offshore Structures during Design and Operation. OTH Report 99060, HSE, London.

Moan, T. (2005). Reliability-Based Management of Inspection, Maintenance and Repair of Offshore Structures. *Journal of Structure and Infrastructure Engineering* 1 (1), 33–62.

Moan, T., Z. Gao, and W. Ayala-Uraga (2005). Uncertainty of Wave-Induced Response of Marine Structures Due to Long-term Variation of Extratropical Wave Conditions. *Marine Structures* 18, 359–382.

Moan, T., G. O. Hovde, and A. M. Blanker (1993). Reliability-Based Fatigue Design Criteria for Offshore structures Considering the Effect of Inspection and Repair. OTC paper no. 7189, OTC in Houston, TX.

Moan, T. and R. Song (1998). Implication of Inspection Updating on System Fatigue Reliability of Offshore Structures. OMAE, Lisbon, Portugal. ASME.

Moan, T., O. T. Vårdal, N. C. Hellevig, and K. Skjoldli (1997). Inservice observations of cracks in North Sea Jackets – a study on initial crack depth and pod values. Proc. of 16th Int. Conf. on Offshore Mechanics and Arctic Engineering, Yokohama, Japan. ASME.

Mohaupt, U. H., O. Vosikovski, D. J. Burns, J. G. Kalbfleich, and R. Bell (1987). Fatigue Crack Development, Thickness and Corrosion Effects in Welded Plate to Plate Joints. *Steel in Marine Structures*, edited by C. Noorhook and J. deBack. Elsevier Science Publishers B.V., Amsterdam, 269–280.

Mori, T. and K. Arakawa (2013). Effective suppressing method for fatigue crack extension by threaded stud. IIW-XIII-2492-13.

Mudge, P. J., N. W. Nichols, and J. Sharp (1996). Internal Ultrasonic Examination of Ring Stiffened Tubular Joints, OMAE – Volume III, Materials Engineering, ASME, 473–481.

Naess, A. and T. Moan (2013). *Stochastic Dynamics of Marine Structures*. Cambridge University Press, New York.

Nakai, T., H. Matsushita, N. Yamamoto, and H. Arai (2004). Effect of Pitting Corrosion on Local Strength of Hold Frames of Bulk Carriers. *Marine Structures* 17, 403–432.

Nasution, F. P., S. Sævik, and S. Berge (2014). Experimental and Finite Element Analysis of Fatigue Strength for 300 Mm^2 Copper Power Conductor. *Marine Structures* 39, 225–254.

Neuber, J. (1961). Theory of Stress Concentrations for Shear-Strained Prismatical Bodies with Arbitrary Nonlinear Stress-Strain Law. *Journal of Applied Mechanics (ASME)* 28, 544–550.

Newman, J. C. and I. S. Raju (1981). An Empirical Stress Intensity Factor Equation for the Surface Crack. *Engineering Fracture Mechanics* 15 (1–2), 185–192.

Newman, J. C. and I. S. Raju (1983). Stress Intensity Factor Equation for Cracks in Three-Dimensional Finite Bodies. *Fracture Mechanics: 14th Symposium*. Vol. I: Theory and Analysis. Philadelphia: ASTM STP 791, 238–265.

Nibbering, J. J. W., B. C. Buisman, H. Wildshut, and E. Van Rietbergen (1987). Corrosion Fatigue Strength of T-Type Welded Connections of Thick Plates for High Numbers of Load Cycles. *Steel in Marine Structures*, edited by C. Noorhook and J. deBack. Elsevier Science Publishers B.V., Amsterdam, 773–786.

Nichols, N. W. and G. Slater (1996). An Assessment of the Fatigue and the Remaining Static Strength Performance of Ring Stiffened Joints. 1996 OMAE – Volume III, Material Engineering ASME 1996, 493–504.

Niemi, E. (1997). Random Loading Behaviour of Welded Components. In *Proceeding from International Conference on Performance of Dynamically Loaded Welded Structures held in San Francisco, CA*. Edited by S. J. Maddox and M. Prager.

Niemi, E., W. Fricke, and S. J. Maddox (2006). Fatigue Analysis of Welded Components. *Designer's Guide to the Structural Hot-spot Stress Approach*. Woodhead Publishing Limited, Cambridge.

Nishida, S. and C. Urashima (1984). Corrosion Fatigue Properties of High Tensile Steel for Offshore Structures. Fatigue 84. 2nd Int. Conf. on Fatigue and Fatigue Thresholds, Birmingham, UK.

Nishida, S., C. Urashima, and H. Tamasaki (1997). A New Method for Fatigue Life Improvement of Screws. In *Fatigue Design of Components*. Edited by G. Marquis and J. Solin. ESIS Publication 22, Elsevier, Amsterdam.

NORSOK M-101 (2011). Structural Steel Fabrication. Rev. 5.

NORSOK M-121 (1997). Aluminium Structural Material. Rev. 1.

NORSOK M-501 (2012). Surface preparation and protective coating, Edition 6.

NORSOK N-001 (2012). Structural Design. Rev. 8.

NORSOK N-003 (2007). Action and action effects. 2 Edition.

NORSOK N-004 (2013). Design of Steel Structures. Rev. 3.

NORSOK N-006 (2015). Assessment of structural integrity for existing offshore load-bearing structures. Rev. 2.

NOU (1981). "Alexander L. Kielland" ulykken. *Norges offentlige utredninger* 1981, 11. (In Norwegian).

NPD (1977). Regulations for the structural design of fixed structures on the Norwegian continental Shelf. The Norwegian Petroleum Directorate.

NS 3472 (1984). Design of Steel Structures. Rev. 02.

Nussbaumer, A., L. Borges, and L. Davaine (2011). Fatigue Design of Steel and Composite Structures. Eurocode 3: Design of Steel Structures Part 1–9-Fatigue. Eurocode 4: Design of Composite Steel and Concrete Structures. ECCS – European Convention for Constructional Steelwork.

Ogle, M. H. (1991). Weld Quality Specifications for Steel and Aluminium Structures. *Welding in the World* 29 (11/12), 341–362.

Ohta, A., N. Suzuki and Y. Maeda (1997). Effect of Residual Stresses on Fatigue of Weldment. In *Proceeding from International Conference on Performance of Dynamically Loaded Welded Structures held in San Francisco, CA*. Edited by S. J. Maddox and M. Prager.

Oliver, R., M. Greif, W. Oberparleiter, and W. Schütz (1981a). Corrosion behaviour of offshore steel structures under constant amplitude loading. Paper 2.4, International Conference Steel in Marine Structures, Paris, France.

Oliver, R., M. Greif., W. Oberparleiter, and W. Schütz (1981b). Corrosion behaviour of offshore steel structures under variable amplitude loading. Paper 7.1, International conference Steel in Marine Structures, Paris, France.

Ørjasæter, O., P. Haagensen, and H. O. Knagenhjelm (2007). Relative Importance of Defects in Girth Welds of Pipelines. Proceedings of the 26th Int. Conf. on Offshore Mechanics and Arctic Engineering OMAE2007, San Diego, CA.

Østby, E., K. R. Jayadevan and C. Thaulow (2005). Fracture response of pipelines subject to large plastic deformation under bending. *International Journal of Pressure Vessels and Piping*, 82 (3), 201–215.

Palm, S., S. Wästberg, and H. Ulle (1984). Veritec Report No 84–3106. *Fatigue of Heavy Welded Joints: Experimental Results*. Det Norske Veritas, Oslo, Norway.

Palmgren, A. (1924). Die Lebensdauer von Kugellagern. *VDI-Zeitschrift* 68, 339–341.

Paris, P. C., M. P. Gomez, and W. P. Anderson (1961). A Rational Analytic Theory of Fatigue. *The Trend in Engineering* 13, 9–14.

PD 6493 (1991). Guidance on methods for assessing the acceptability of flaws in fusion welded structures. BSI 1991.

Petershagen, H. (1991). A comparison of two different approaches to the fatigue strength assessment of cruciform joints, IIW doc. XIII-1410–91.

Peterson, R. E. (1974). *Stress Concentration Factors*. John Wiley & Sons, Inc, New York.

Pilkey, W. D. (1997). *Peterson's Stress Concentration Factors*. John Wiley & Sons, Inc, New York.

Polezhayeva, H., D. Howarth, M. Kumar, Bilal Ahmad, and M. E. Fitzpatrick (2014). The Effect of Compressive Fatigue Loads on Fatigue Strength of Non Load Carrying Specimens subjected to Ultrasonic Impact Treatment. IIW Document XIII-2530–14.

Potvin, A. B., J. G. Kuang, R. D. Leick, and J. L. Kahlich (1977). Stress Concentration in Tubular Joints. *Society of Petroleum Engineers Journal* 17, 287–299.

Poutiainen, I., P. Tanskanen, and G. Marquis (2004). Finite Element Methods for Structural Hot Spot Stress Determination – A Comparison of Procedures. *International Journal of Fatigue* 26, 1147–1157.

Pozzolini, P. F. (1981). Test on Tubular Joints. *International Conference Steel in Marine Structures*, Paris, 98–145. Proceedings published by: Comptoir des Produits Sidèrurgiques, Paris, France.

Prakhya, G., C. Zhang, and N. Harding (2012). Grouted Connections for Monopiles – Limits for Large Wind Turbines. *The Structural Engineer* (March), 30–45.

PROBAN (1996). *Theory Manual, General Purpose Probabilistic Analysis Program*. DNV Software Report 96–7017, Det Norske Veritas, Oslo, Norway.

Qian, X., Y. Petchdemaneengam, S. Swaddiwudhipong, P. Marshall, Z. Ou, and C. T. Nguyen (2013). Fatigue Performance of Tubular X-joints with PJP + Welds: I – Experimental Study. *Journal of Constructional Steel Research* 90, 49–59.

Radaj, D. (1996). Review of Fatigue Strength Assessment of Nonwelded and Welded Structures based on Local Parameters. *International Journal of Fatigue* 18 (3), 153–170.

Radaj, D., C. M. Sonsino, and W. Fricke (2006). *Fatigue Assessment of Welded Joints by Local Approaches*. Second edition. Woodhead Publishing Limited Abington Hall, Abington, Cambridge.

Radaj, D. and M. Vormwald (2013). *Advanced Methods of Fatigue Assessment*. Springer-Verlag Berlin Heidelberg.

Radenkovic, D. (1981). Stress Analysis in Tubular Joints. *International Conference Steel in Marine Structures*, Paris, 53–95. Proceedings published by: Comptoir des Produits Sidèrurgiques, Paris, France.

Razmjoo, G. R. (1995). Design Guidance on Fatigue of Welded Stainless Steel Joints. 1995 OMAE – Volume III, Materials Engineering ASME, 163–171.

Razmjoo, G. R. and P. J. Tubby (1997). Fatigue of Welded Joints under Complex Loading. *Fatigue Design of Components*. Edited by G. Marquis and J. Solin. ESIS Publication 22, Elsevier, Amsterdam.

Rice, J. R. (1968). A Path Independent Integral and the Approximate Analysis of Strain Concentrations by Notches and Cracks. *Journal of Applied Mechanics* 35, 379–386.

Ringsberg, J. W. and A. Y. J. Ulfvarson (1998). On Mechanical Interaction between Steel and Coating in Stressed and Strained Exposed Locations. *Marine Structures* 11, 231–250.

Roark, R. J. and W. C. Young (1975). *Formulas for Stress and Strain*. International Student Edition. McGraw-Hill, New York.

Ronold, K. O. and A. Echtermeyer (1996). Estimation of Fatigue Curves for Design of Composite Laminates, *Composites, Part A: Applied Science and Manufacturing, Elsevier* 27A (6), 485–491.

Ronold, K. O. and I. Lotsberg (2012). On the Estimation of Characteristic S-N curves with Confidence. *Marine Structures* 27, 29–44.

Ronold, K. O. and S. Wästberg (2002). Characteristic S-N Curves for Fatigue Design of Titanium Risers. OMAE2002–28475. *Proceedings of OMAE2002: 21st International Conference on Offshore Mechanics and Arctic Engineering*, Oslo, Norway.

Rooke, D. P. and D. J. Cartwright (1976). *Compendium of Stress Intensity Factors*. HMSO, London.

Rucho P., S. Maherault, W. Chen, A. Berstad, and G. E. Samnøy (2001). *Comparison of Measurements and Finite Element Analysis of Side Longitudinals*. ISOPE Stavanger, Cupertino, CA.

Schaumann, P., S. Lochte-Holtgreven, and F. Wilke (2010). *Bending Tests on Grouted Joints for Monopile Support Structures*. DEWEK2010–10th German Wind Energy Conference, Bremen, Germany.

Scherf, I. and T. Thuestad (1987). Fatigue Design of the Oseberg Jacket Structure. *Proceedings OMAE*, Houston, TX.

Schijve, J. (2003). Fatigue of structure and materials in the 20th century and the state of art. International journal of fatigue, 25, 679–702.

Schijve, J. (2009). *Fatigue of Structures and Materials*. Second edition. Springer Verlag, Berlin, Germany.

Shimokawa, H., K. Takena, F. Ito, and C. Miki (1984). Effects of Stress Ratios on the Fatigue Strengths of Cruciform Fillet Welded Joints. Proceedings of the Japan Society of Civil Engineers No. 344/I-1 (Structural Eng./Earthquake Eng.), 121–128.

Schneider, C. R. A. and S. J. Maddox (2006). Best practice guide on statistical analysis of fatigue data, International Institute of Welding, IIW-XIII-2138–06.

Scholte, H. G., J. L. Overbeeke, O. D. Dijkstra, H. Wildschut, and C. Noordhoek (1989). Fatigue and corrosion fatigue on flat specimens and tubular joints – Dutch results. OMAE The Hague, The Netherlands.

Schumacher, A., L. L. Costa Borges, and A. Nussbaumer (2009). A Critical Examination of the Size Effect for Welded Steel Tubular Joints. *International Journal of Fatigue* 31, 1422–1433.

Schütz, W. (1981). Procedures for the Prediction of Fatigue Life of Tubular Joints. International Conference Steel in Marine Structures, Paris, 98–145. Proceedings published by: Comptoir des Produits Sidèrurgiques, Paris, France.

Shütz, W. (1996). A History of Fatigue. *Engineering Fracture Mechanics* 54 (2), 263–300.

Schütz, W. (1987). Corrosion Fatigue. *Proceeding of the 2nd Int. Conf. on Structural Failure, Product Liability and Technical Insurance*. Edited by H. P. Rossmanith. Inderscience Publishers, London and Vienna.

Sendeckyj, G. P. (2001). Constant Life Diagrams – A Historical Review. *International Journal of Fatigue* 23, 347–353.

Sih, G. and H. Liebowitz (1968). Mathematical theories of brittle fracture. In Fracture: An Advanced Treatise, Volume II: Mathematical Fundamentals. Edited by Liebowitz, H. Academic Press, New York and London, 67–190.

Sines, G. and J. L. Waisman (1959). *Metal Fatigue*. McGraw-Hill, New York.

Skjeggestad, B., M. Ringard, and E. Bakke (1969). Fatigue Tests of Plates with Circular Cutouts. Skipsteknisk Forskningsinstitutt. Report no. 76.

Slater, G. and P. J. Tubby (1996). Fatigue behaviour of internally ring stiffened tubular joints. OMAE 1996. Volume III, Material Engineering ASME, 483–492.

Sletten, R., K. Mjelde, S. Fjeld, and I. Lotsberg (1982). Optimization of Criteria for Design Construction and Inservice Inspection of Offshore Structures Base on Resource Allocation Techniques. EW322. EUROPEC'82. European Petroleum Conference, London.

Slind, T. (1994). Long Life VA Testing of Welded Steel Specimens in Air and in Seawater with Cathodic Protection. *OMAE – Volume* III, Materials Engineering, ASME, 215–220.

Smedley, P. (2003). Advanced SCF Formulae for Simple and Multi-Planar Tubular Joints. *10th International Symposium on Tubular Structures*, Madrid, Spain.

Smedley, P. A. and P. J. Fisher (1991a). Stress Concentration Factors for Simple Tubular Joints. *Proceedings of the First International Offshore and Polar Engineering Conference, (ISOPE)*, Edinburgh, UK, 475–483. Publ. by Int. Soc. of Offshore and Polar Engineers (ISOPE), Golden, CO.

Smedley, S. and P. J. Fischer (1991b). Stress Concentration Factors for Ring-Stiffened Tubular Joints. *Proceedings of the First International Offshore and Polar Engineering Conference, Edinburgh*, 239–250. Publ. by Int. Soc. of Offshore and Polar Engineers (ISOPE), Golden, CO.

Sonsino, C. M. (1997). Multiaxial and Random Loading of Welded Structutes. *Proceeding from International Conference on Performance of Dynamically Loaded Welded Structures held in San Francisco, CA*. Edited by S. J. Maddox and M. Prager.

Sonsino, C. M. (2007). Course of SN-Curves Especially in the High-Cycle Fatigue Regime with Regard to Component Design and Safety. *International Journal of Fatigue* 29, 2246–2258.

Sonsino, C. M. (2009). A Consideration of Allowable Equivalent Stresses for Fatigue Design of Welded Joints According to the Notch Stress Concept with Reference Radii $r_{ref} = 1.00$ and 0.05 mm. *Welding in the World* 53 (4), R64–R75.

Sonsino, C. M., F. Müller, J. de Back, and A. M. Gresnigt (1996). Influence of Stress Relieving by Vibration on the Fatigue Behaviuor of Welded Joints in Comparison to Post-Weld Heat Treatment. *Fatigue & Fracture of Engineering Materials & Structures* 19 (6), 703–708.

Sørensen, E. V. (2011). Fatigue Life of High Performance Grout in Wet or Dry Environment for Wind Turbine Grouted Connection. EWEA Offshore November 2011, Amsterdam, The Netherlands. Also NCR No 44.

Sors, L. (1971). *Fatigue Design of Machine Components*. Pergamon Press, Oxford, New York, Toronto, Sydney, Braunscweig.

Spence, J. and A. S. Tooth (1994). *Pressure Vessel Design: Concepts and Principles*. E & FN Spon, an imprint of Chapman & Hall, London.

Stacey, A. and J. V. Sharp (1995). The Revised HSE Guidance. 1995 OMAE-Volume III, Materials Engineering, ASME, 1–16.

Stacey, A., J. V. Sharp, and N. W. Nichols (1996). The Influence of Cracks on the Static Strength of Tubular Joints. 1996 OMAE – Volume III, Materials Engineering ASME 1996, pp. 435–450.

Stacey, A., J. V. Sharp, and N. W. Nichols (1997). Fatigue Performance of Single-sided Circumferential and Closure Welds in Offshore Jacket Structures. 1997 OMAE, ASME 1997.

Stahl, B. (1986). Reliability Engineering and Risk Analysis. Planning and Designing of Fixed Offshore Structures. McClelland, B. and M. D. Reifel, (Eds), Van Nostrand Reinhold, New York, pp. 59–98.

Statnikov, E. (2004). Physics and Mechanism of Ultrasonic Impact Treatment. IIW Document XIII-2004-04.

Stephens, R. I., A. Fatemi, R. R. Spephens, and H. O. Fuchs (2001). *Metal Fatigue in Engineering*. Second edition. John Wiley & Sons, New York.

Storsul, R., E. Landet, and I. Lotsberg (2004a). Convergence Analysis for Welded Details in Ship Shaped Structures. OMAE-FPSO'04-0016, Int. Conf. Houston, TX.

Storsul, R., E. Landet, and I. Lotsberg (2004b). Calculated and Measured Stress at Welded Connections between Side Longitudinals and Transverse Frames in Ship Shaped Structures. OMAE-FPSO'04-0017, Int. Conf. Houston, TX.

Tada, H., P. C. Paris, and G. R. Irwin (1973). *The Stress Analysis of Cracks Handbook*. Del Research Corporation, Hellertown, P A.

Thaulow, C., E. Østby, B. Nyhus, Z. L. Zhang and B. Skallerud (2004). Constraint correction of high strength steel. Selection of test specimens and application of direct calculations. *Engineering Fracture Mechanics*, 71 (16-17), 2417–2433.

Taylor, D. (2007). *The Theory of Critical Distances: A New Perspective on Fracture Mechanics*. Elsevier Ltd., Amsterdam.

Tenge, P. and O. Solli (1973). Fracture Mechanics in the Design of Large Spherical Tanks for Ship Transport of LNG. Conference on Welding Low Temperature Containment Plant, Proceedings The Welding Institute, pp. 149–158, London, UK and Norwegian Maritime Research 1973, vol. 1 No 2, 1–18.

Tenge, P., O. Solli and O. Førli (1974). Significance of Defects in LNG-Tanks in Ships, ASTM Symposium on Properties of Materials for Liquid Natural Gas Tankage, Boston, USA. In ASTM Special Technical Publication 579.

Timoshenko, S. P. and J. N. Goodier (1970). *Theory of Elasticity*. Third edition. McGraw-Hill, New York.

Timoshenko, S. P. and S. Woinowsky-Krieger (1959). *Theory of Plates and Shells*. Second edition. McGraw-Hill, New York.

Tovo, R. and P. Lazzarin (1999). Relationships between Local and Structural Stress in the Evaluation of the Weld Toe Distribution. *International Journal of Fatigue* 21 1063–1078.

UEG (1985). Underwater Engineering Group, Design Guidance on Tubular Joints in Steel Structures, UR33.

Ulleland T., M. Svensson, and E. Landet (2001). *Stress Concentration Factors in Side Shell Longitudinals Connected to Transverse Webframes*. ISOPE Stavanger.

Urm, H. S., I. S. Yoo, J. H. Heo, S. C. Kim, and I. Lotsberg (2004). Low Cycle Fatigue Strength Assessment for Ship Structures. PRADS 12–17 September, Lüebeck – Travemunde, Germany. 9th International Symposium on Practical Design of Ships and Other Floating Structures. Edited by H. Keil and E. Lehman, Seehafen Verlag. Vol. 2.

Valsgård, S., I. Lotsberg, G. Sigurdsson, and K. Mørk (2010). Fatigue Design of Steel Containment Cylinders for CNG Application. *Marine Structures* 23, 209–225.

Vander Voort, G. F. (1990). Embrittlement of Steels, Properties and Selection. *Irons, Steel and High-Performance Alloys*, Vol. 1, ASM Handbook, ASM International, 689–736.

Van Wingerde, A. M., J. A. Packer, and J. Wardenier (1996). New Guidelines for Fatigue Design of HSS Connections. *Journal of Structural Engineering* 122 (2), 125–132.

Van Wingerde, A. M., J. A. Packer, and J. Wardenier (1997a). IIW Fatigue Rules for Tubular Joints. In Proceeding from International Conference on Performance of Dynamically Loaded Welded Structures held in San Francisco, CA. Edited by S. J. Maddox and M. Prager.

Van Wingerde, A. M., D. R. V. van Delft, J. Wardenier, and J. A. Packer (1997b). Scale Effects on the Fatigue Behaviour of Tubular Structures. In Proceeding from International Conference on Performance of Dynamically Loaded Welded Structures held in San Francisco, CA. Edited by S. J. Maddox and M. Prager.

Vårdal, O. T. and T. Moan (1997). Predicted versus Observed Fatigue Crack Growth. Validation of Fracture Mechanics Analysis in North Sea Jackets. Paper no. 1334, Proc. 16th OMAE Conference, Yokohama, Japan. Volume II, Safety and Reliability, 209–218.

Vargas, P.-M., T.-M. Hsu, and W. K. Lee (2004). Stress Concentration Factors for Stud-less Mooring Chain Links in Fairleads, OMAE2004-51376. Proceedings 23rd Int. Conf. on Offshore Mechanics and Arctic Engineering, Vancouver, British Columbia, Canada.

VDI 2230 (2003). *Systematic Calculation of High Duty Bolted Joints: Joints with One Cylindrical Bolt. Part 1*. Beuth Verlag GmbH, Berlin.

Viner, J. G., R. L. Dineen, and R. L. Chesson, Jr. (1961). A Study on the Behavour of Nuts for Use with High Strength Bolts. University of Illinois, Structural Research Series No. 212.

Visser, W. (2002). HSE Offshore Technology Report 2000/018: POD/POS curves for non-destructive examination.

Vosikovski, O., R. Bell, D. J. Burnsand, and U. H. Mohaupt (1987). Effects of Cathodic Protection on Corrosion Fatigue Life of Welded Plate T-Joints. *Steel in Marine Structures*. Edited by C. Noorhook and J. deBack. Elsevier Science Publishers B.V., Amsterdam, 787–798.

Vughts, J. H. and R. K. Kinra (1976). Probabilistic Fatigue Analysis of Fixed Offshore Structures. OTC Paper no. 2608, Offshore Technology Conference, Houston, TX.

Waagaard, K. (1981). Fatigue Strength of Offshore Concrete Structures. COSMAR Report, PP2-1 and PP2-2.

Wathne Tveiten, B. and T. Moan (2000). Determination of Structural Stress for Fatigue Assessment of Welded Aluminum Details. *Marine Structures* 13 (3), 189–212.

Wardenier, J. (1982). *Hollow Section Joints*. Delft University Press, Deflt, The Netherlands.

Wästberg, S. and M. Salama (2007). Fatigue Testing and Analysis of Full Scale Girth Weld Tubulars. OMAE Paper 2007–29399, Proceedings of the 26th International Conference on Offshore Mechanics and Arctic Engineering OMAE2007, San Diego, CA.

Weich, I., T. Ummenhofer, T. Nitschke-Pagel, K. Dilger, and H. Eslami (2007). Fatigue Behaviour of Welded High Strength Steels after High Frequency Mechanical Postweld Treatments. IIW Document no XIII-2154-07.

Wells, A. A. (1969). Crack Opening Displacement from Elastic-Plastic Analysis of Externally Notched Tension Bars. *Engineering Fracture Mechanics* 1, 399–410.

Westergaard, H. M. (1939). Bearing Pressures and Cracks. *Journal of Applied Mechanics* 6, 49–53.

Wiebesiek, J. and C. M. Sonsino (2010). New Results in Multiaxial Fatigue of Welded Aluminium Joints. IIW-Doc. No. XIII-2314-10/XV-1349-10.

Wiegand, H. and K.-H. Illgner (1962). *Berechnung und Gestaltung von Schraubverbindungen, 3. Auflage*, Springer-Verlag, Berlin. (In German).

Williams, M. L. (1957). On the Streess Distribution at the Base of a Stationary Crack. *Journal of Applied Mechanics* 24, 109–114.

Wirsching, P. H. (1984). Fatigue Reliability of Offshore Structures. *ASCE Journal of Structural Engineering* 110, 2340–2356.

Wirsching, P. H. and Y.-N. Chen (1988). Considerations on Probability-Based Fatigue Design for Marine Structures. *Journal of Marine Structures* 1, 23–45.

Wirsching, P. H. and M. C. Light (1980). Fatigue under Wide Band Random Stresses. *Journal of the Structural Division, ASCM* 106 (ST7), 1593–1607.

Woghiren, C. O. and F. P. Brennan (2009). Weld Toe Stress Concentrations in Multi-Planar Stiffened Tubular KK Joints. *International Journal of Fatigue* 31 (1), 164–172.

Woollin, P., S. Maddox, and D. J. Baxter (2005). Corrosion Fatigue of Welded Stainless Steels for Deepwater Riser Applications. Paper OMAE2005-67499. Proceedings of OMAE 2005: 24th International Conference on Mechanics and Arctic Engineering, Halidiki, Greece.

Wordsworth, A. C. (1981). Stress Concentration Factors at K – and KT-Tubular Joints. Proceedings Fatigue in Offshore Structural Steels, London.

Wordsworth, A. C. and G. P. Smedley (1978). Stress Concentrations at Unstiffened Tubular Joints, Paper no. 31 in European Offshore Steels Research Seminar, Cambridge, UK.

Wormsen A., M. Avice, A. Fjeldstad, L. Reinås, K. A. Macdonald, E. Berg, and A. D. Muff (2015). Fatigue Testing and Analysis of Notched Specimens with Typical Subsea Design Features. Submitted for publication in the *International Journal of Fatigu* 81, 275–298.

Wylde, J. G. (1983). Fatigue Tests on Welded Tubular T-joints with Equal Brace and Chord Diameters. OTC Paper no. 4527, 15th annual OTC, Houston, TX.

Wylde, J. G. and A. McDonald (1979). The Influence of Joint Dimensions on the Fatigue Strength of Welded Tubular Joints. Boss '79. Second Int. Conf. on Behaviour of Offshore Structures, London.

Wylde, J. G. and A. McDonald (1981). Modes of Fatigue Crack Development and Stiffness Measurements in Welded Tubular Joints. Fatigue in Offshore Structural Steels. Proceedings of a Conference organized by the Institution of Civil Engineers, London.

Xiao, Z. G. and K. A. Yamada (2004). A Method of Determining Geometric Stress for Fatigue Strength Evaluation of Steel Welded Joints. *International Journal of Fatigue* 26, 1277–1293.

Xu, T. (1997). Fatigue of Ship Structural Details – Technical Development and Problems. *Journal of Ship Research* 41 (4), 318–331.

Yamamoto, N., M. Mouri, T. Okada, and T. Mori (2012). Analytical and Experimental Study on the Thickness Effect to Fatigue Strength. IIW document XIII-2434–12.

Ye, N. and T. Moan (2008). Improving Fatigue Life for Aluminium Cruciform Joints by Weld Toe Grinding. *International Journal Fatigue & Fracture of Engineering Materials & Structures* 31 (2), 152–163.

Ye, X. W., Y. H. Su and J. P. Han (2014). A State-of-the-Art Review on Fatigue Life Assessment of Steel Bridges. Mathematical Problems in Engineering, Volume 2014, Article ID 956473, 13 pages.

Yildirim, H. C. and G. B. Marquis (2012a). Overview of Fatigue Data for High Frequency Mechanical Impact Treated Welded Joints. *Welding in the World* 56 (7/8), 82–96.

Yildirim, H. C. and G. B. Marquis (2012b). Fatigue Strength Improvement Factors for High Strength Steel Welded Joints Treated by High Frequency Mechanical Impact. *International Journal of Fatigue* 44, 168–176.

Yokobori, T., and K. Sato (1976). The Effect of Frequency on Fatigue Crack Propagation Rate and Striation Spacing in 2024-T3 Aluminium Alloy and SM-50 Steel. *Engineering Fracture Mechanics* 8, 81–88.

Yoneya, T., A. Kumano, N. Yamamoto, and T. Shigemi (1993). Hull Cracking of Very Large Ship Structures, IOS '93.

Zhang, Y.-H., S. J. Maddox, and R. Razmjoo (2003). Re-evaluation of fatigue curves for flush-ground girth welds. Research report 090, HSE, UK.

Zhang, Y.-H. and S. J. Maddox (2009a). Investigation of Fatigue Damage to Welded Joints under Variable Amplitude Loading Spectra. *International Journal of Fatigue* 31, 138–152.

Zhang, Y.-H. and S. J. Maddox, (2009b). Fatigue Life Prediction for Ground Welded Joints. *International Journal of Fatigue* 31, 1124–1134.

Zhang, Y.-H. and S. J. Maddox (2012). Fatigue testing of full scale girth welded pipes under variable amplitude loading. Paper no. OMAE2012-83054. Proceedings of the ASME 2012 31st International Conference on Ocean, Offshore and Arctic Engineering, OMAE2012, Rio de Janeiro, Brazil.

Zhao, X.-L. and J. A. Packer (2000). IIW. *Fatigue Design Procedure for Welded Hollow Section Joints*. Abington Publishing, Cambridge.

Index

hydraulic actuator, 31, 460
hydraulic axial tensioner device, 397
hydrodynamic analysis, 407
hydrodynamic behavior, 405
hydrodynamic coefficients, 405
hydrogen, 130
  cracking, 128, 329, 332, 384
  damage, 383
  embrittlement, 390
  induced stress cracking, 383
  resistant material, 384
  sources, 383
hydrophone support, 6, 9, 203, 274, 409, 416
hysteresis loops, 96

IACS, 185. *See also* International Association of
  Classification Societies
IIW, 49, 83, 114, 126, 306, 339, 350. *See also*
  International Institute of Welding
image on film, 336
impact strength, 386
imperfections, 19, 80, 238, 328, 330, 433
impressed current, 383
improvement, 339, 363
improvement by grinding, 351
improvement factor, 267, 339, 345, 350
impurity elements, 386
in way of, 181, 183
inclined piles, 437
inclusions, 20, 336, 410
Incoloy, 131
incomplete gamma function, 312
incomplete penetration, 330
Inconel, 131
increment in crack growth, 412
independent analysis, 16
independent testing, 335
independent variables, 371
inelastic strain, 3
inflection point, 209, 239
influence function, 259, 265
initial crack
  depth, 425
  length, 425
  size, 412
initial defect depth, 148, 492
initiation of crack, 19
inner bottom of bulk carriers, 107
in-plane bending, 254, 264, 267, 268, 404
in-plane loading, 380
insert gusset plates, 299
insert tubular, 202
in-service inspection, 14, 116, 267, 268, 324, 337,
  359, 364, 366, 367, 407, 427, 434, 467, 468
in-service life, 14, 33, 43, 319, 331, 343, 348, 350,
  368, 390, 397, 401, 421, 429, 489
inspection
  intervals, 429
  planning, 473
  updating, 473

installation of jacket structures, 442
installed condition, 400
instrumented test data, 185
integrated complete pile model, 402
interaction equation, 74, 89
intercept of log N axis, 124
intercrystalline fracture. *See* separation along
  grain boundaries
internal
  crack, 410
  defect, 339, 413
  imperfections, 332
  voids, 20
internal pressure, 237, 245
internal ring stiffeners, 252, 270, 305
International Association of Classification
  Societies. *See also* IACS
International Institute of Welding, 126, 280. *See
  also* IIW
intersection line, 90, 93, 194, 254, 285, 286, 297
intrusions, 19, 20
inverse negative slope, 27
iron particles, 333
iron-zinc alloys, 390
irregular slag, 139
ISO – International Standard Organization, 5
isotropic hardening, 119

jacket, 13, 205–227, 400, 435, 483, 485
jack-up, 13, 128, 400, 440
Japanese Ship Classification Society, 121
J-integral, 427
joint classification, 109, 256
joint flexibility, 263

K-joint, 252, 253, 257, 263, 270, 404

lack of fusion, 138, 139, 141, 328, 330, 332
lack of fusion between welding passes, 332
lack of parallelism, 387
lack of penetration, 138, 139, 210, 301, 328, 332,
  413, 415, 420
lamellar tearing, 140
land-based structures, 392
large diameter connections, 440
large-scale test specimen, 75, 137
large-scale tubular joints, 270
lateral pressure, 276, 279, 288
LCF. *See* low cycle fatigue
leak-before-break, 399, 466
leg length, 68, 75, 151, 283, 294, 297, 342, 344
legs with insert piles, 272
length effect, 374
length of bolt, 385
length of grouted connection, 444
life extension, 83
lighting, 333
limit state, 14
limit state function, 355, 376, 471
linear cumulative damage, 114

Printed in the United States
By Bookmasters